Soft and Biological Matter

More information about this series at http://www.springer.com/series/10783

"Soft and Biological Matter" is a series of authoritative books covering established and emergent areas in the realm of soft matter science, including biological systems spanning all relevant length scales from the molecular to the mesoscale. It aims to serve a broad interdisciplinary community of students and researchers in physics, chemistry, biophysics and materials science.

Pure research monographs in the series, as well as those of more pedagogical nature, will emphasize topics in fundamental physics, synthesis and design, characterization and new prospective applications of soft and biological matter systems. The series will encompass experimental, theoretical and computational approaches. Topics in the scope of this series include but are not limited to: polymers, biopolymers, polyelectrolytes, liquids, glasses, water, solutions, emulsions, foams, gels, ionic liquids, liquid crystals, colloids, granular matter, complex fluids, microfluidics, nanofluidics, membranes and interfaces, active matter, cell mechanics and biophysics.

Both authored and edited volumes will be considered.

Ariel Fernández

Physics at the Biomolecular Interface

Fundamentals for Molecular Targeted Therapy

Springer

Ariel Fernández
National Research Council (CONICET)
Buenos Aires, Argentina

Former Hasselmann Professor of Engineering
Rice University, USA

ISSN 2213-1736 ISSN 2213-1744 (electronic)
Soft and Biological Matter
ISBN 978-3-319-30851-7 ISBN 978-3-319-30852-4 (eBook)
DOI 10.1007/978-3-319-30852-4

Library of Congress Control Number: 2016936663

Printed on acid-free paper

This Springer imprint is published by Springer Nature
The registered company is Springer International Publishing AG Switzerland

Preface

Whoever cannot seek
the unforeseen sees nothing,
for the known way
is an impasse.

Heraclitus, sixth century B.C.

The biological functionality of a soluble protein can only be fully grasped when its aqueous interface becomes an integral part of the structural analysis. Furthermore, the acknowledgment of how exquisitely the structure and dynamics of proteins and their aqueous environment are related attests to the overdue recognition that biomolecular phenomena cannot be grasped without dealing with interfacial behavior at multiple scales. This is essentially the *dictum* that guided the writing of this book.

The book focuses primarily on the biological and pharmacological role of interfacial forces determined by the embedding of protein structure in a physiological aqueous environment. By providing a suitable statistical mechanical apparatus to handle epistructural ("around the structure") interfacial phenomenology, the book becomes uniquely positioned to address core problems in molecular biophysics. It highlights the importance of interrelated concepts like water hydrogen-bond frustration, interfacial tension and non-Debye dielectrics in delineating a solution to the protein folding problem, in unraveling the physicochemical basis of enzyme catalysis and protein associations, in delineating the molecular etiology of aberrant protein aggregation, and in rationally designing molecular-targeted therapies.

The book incrementally builds upon a statistical mechanical apparatus to deal with epistructural interfaces in a biological context. As it extends previous work in interfacial physics to the biological context, the book strives to maintain the level of rigor expected from a researcher devoted to interfacial physics, notwithstanding the daunting complexities of biomolecular systems. In the biological/biomedical context, the book introduces the necessary controls and experimental corroborations to validate the physical treatment. Furthermore, molecular dynamics and quantum mechanics computations are used to validate the theoretical advances. Such computations fulfill the tenets of statistical physics.

Grounded in recent advances in the statistical mechanics of epistructural aqueous interfaces, the book develops a technological platform for drug design termed *epistructure-based design*, above and beyond the prevailing paradigm of structure-based drug design. The book is set to inspire scientists at any level in their careers determined to address some of the major challenges in molecular biophysics and pharmacological engineering. Thus, the book is essentially interdisciplinary and covers vast conceptual territory, from statistical physics to molecular-targeted therapy. As a measure of its intellectual latitude, Chap. 1 introduces a statistical thermodynamics framework to handle the aqueous interface of a protein, while Chap. 17 describes the epistructure-based design of kinase inhibitors with controlled multi-target activity to treat cancer metastasis and overcome drug resistance. In spite of this diversity, the conceptual progression remains smooth throughout the presentation. For this reason, the book can serve as a textbook, as originally intended, and also as an advanced monograph for practitioners in drug design or molecular-targeted therapy interested in the translational aspects of their art. Fruitful reading requires a background in physical chemistry and some notions of biophysics. The selected problems at the end of the chapters and the progression in conceptual difficulty make it a suitable textbook for a graduate level course or an elective course for seniors majoring in chemistry, physics, bioengineering, or related disciplines.

Buenos Aires, Argentina
Rice University, USA
Ariel Fernández

Contents

About the Author

Ariel Fernández (born Ariel Fernández Stigliano) is an Argentine-American physical chemist and mathematician. He obtained his Ph.D. degree in chemical physics from Yale University in record time. He held the Karl F. Hasselmann endowed chair professorship in engineering at Rice University and was a professor of bioengineering until his retirement in 2012. To date, he has published over 350 scientific papers in professional journals including *Physical Review Letters*, *PNAS*, *Nature*, *Genome Research*, and *Genome Biology*. Ariel Fernández has also published two books *Transformative Concepts for Drug Design* (2010) and *Biomolecular Interfaces* (2015), both with Springer, and holds two patents (US 8,466,154 and 9,051,387) on biotechnological innovations. He is currently involved in research and entrepreneurial activities at various consultancy firms.

Chapter 1
Interfacial Physics for Water in Biology

Biological behavior at molecular scales remains opaque in spite of the enticing promises of biophysicists in the post-genomic era. The protein folding problem has not been solved from first principles, the physico-chemical basis of enzyme catalysis and protein interactions remains obscure, the molecular etiology of aggregation-related disease is still quite enigmatic, and we are still unable to design therapeutics based on our current understanding of biomolecular phenomena. This state of affairs brings us to the cross-roads of several disciplines and, more importantly, leaves us with a daunting question: Are these problems intractable in their complexity, or they require conceptual advances that have not yet been brought to fruition? This introductory chapter argues for the second alternative, and makes it a *leitmotiv* for the entire book.

The dynamic interaction between protein structure and its aqueous environment attests to the overdue appreciation that biomolecular phenomena cannot be properly grasped without a firm understanding of interfacial behavior. It becomes imperative that we provide a statistical physics framework to deal with the "epistructural" biological behavior mediated by the structuring of biomolecular interfaces. This chapter addresses this imperative and serves as an introduction to a new discipline that we have aptly named *epistructural physics*.

1.1 Physics of the Biological Interface

This book focuses on the delineation of the interfacial forces that steer biological phenomena, a knowledge that gets translated in the last ten chapters into design concepts for molecular medicine. By providing a suitable statistical mechanical apparatus to handle the biomolecular interface, the book becomes uniquely positioned to address core problems in molecular biophysics and biomedical molecular design. This chapter begins by introducing key statistical mechanical concepts such as frustration in water hydrogen-bonding, interfacial tension, non-Debye

A. Fernández, *Physics at the Biomolecular Interface*, Soft and Biological Matter,
DOI 10.1007/978-3-319-30852-4_1

dielectrics, and variational interfacial thermodynamics. These concepts span a physical framework and a novel intellectual discourse tailored for biological phenomena that will be adopted throughout the book. Epistructural physics, the discipline thus emerging, will enable us to delineate a solution to the protein folding problem, to unravel the physico-chemical basis of enzyme catalysis and protein associations, to identify the molecular etiology of aggregation disease, and to rationally design molecular targeted therapies.

The acknowledgment of the exquisite dynamic coupling between structural protein motions and their aqueous environment attests to the overdue recognition that biological phenomena cannot be effectively understood without dealing with interfacial behavior. There is an urge to grasp how biologically relevant behavior is shaped by the structuring of biomolecular interfaces and how interfacial tension affects the molecular events that take place in the cell. This chapter introduces these matters from a physicist perspective adopting a variational thermodynamics perspective to address the complexities of the protein-water interface.

Physicists have long realized that where different phases meet, unusual things are likely to happen. For interfaces representing solution of continuity between bulk phases, interfacial tension arises. Yet, when water interfaces with a more complex solute such as a biomolecule, endowed with subnanoscale detail and chemical heterogeneity, the concept of interfacial tension becomes much more blurred to the point of being physically elusive. The free energy cost of spanning the interface obviously makes it a locus of unexpected phenomena, and the singular interfacial behavior of water becomes all the more apparent when the interface is structured by the nanoscale and sub-nanoscale topographic and chemical complexities of biological macromolecules. *Episteric* ("around the solid") water is indeed quite different from its bulk counterpart, and even fails to align with the solute electrostatic field due to tight geometric confinement coupled with short-range intermolecular forces. These deviations from bulk properties can enhance the chemical inhomogeneity of protein surfaces by altering the dielectric properties of interfaces in ways that cannot be understood invoking conventional mean field approaches enshrined in a single permittivity coefficient. The goal of this chapter is bringing this conceptual leap to sharp focus as it is paramount to understand biology at the molecular scale.

This introductory chapter shows that the conceptual development of a statistical-mechanical framework for *epistructural physics* is essential to address the challenges in molecular biophysics and drug design that have so far proven unyielding to other research efforts. The chapter draws basic relationships between interfacial water and the structure of soluble proteins, providing a physical framework rooted in variational thermodynamics, to understand and make sense of interfacial tension in a biological context. In this way, the chapter sets the tone for the entire book.

Properly harvesting and ultimately exploiting the output of structural biology to make more efficacious and safer drugs has proven to be much more difficult than originally thought. We believe this is so because structure-based design requires nothing less than a revolution in molecular biophysics. In this chapter we put forth the view that structural biology *per se* cannot possibly provide a fully operational framework for the drug designer. It is not by chance that rational design efforts have not panned out to meet the current expectations. Translating biophysical concepts like interfacial tension or variational interface thermodynamics into the discovery platforms of the pharmaceutical industry, as this book attempts to do, will be paramount to truly expand its technological basis. As we argue for higher intellectual standards in molecular biophysics, we are constantly reminded that drug discovery and development remains essentially a serendipitous endeavor where high-throughput screening and toxicological studies take the upper hand while rational molecular design is often relegated to an academic curiosity. Thus, more than ever, the lead in the pharmaceutical industry will depend pivotally on our ability to harness innovative high-risk research. This chapter and ultimately this book may have a place in this scenario, as we shall introduce fundamental concepts that hold potential to become transformative for the pharmaceutical industry.

1.2 Water in Biology: A Molecular Stare

Water is known to provide two proton donor and two proton acceptor hydrogen bonding coordinations in a nearly tetrahedral arrangement dictated by the hybridized sp^3 atomic orbitals of the oxygen atom. As water interfaces with a complex solute endowed with nanoscale or subnanoscale surface detail, such as a biomolecule in the cytosolic cellular environment, the tetrahedral lattice may be perturbed, and this perturbation carries a thermodynamic cost. More specifically, certain hydrogen bond opportunities for interfacial water may not be fulfilled, leading to "*frustration*", a phenomenon of immense ramifications in interfacial physics, as we shall see throughout this book. Such frustration may occur, for example when a solvent-exposed backbone hydrogen bond in a soluble protein accommodates a water molecule that hydrogen bonds with the carbonyl oxygen, which has an unused lone electron pair. The confinement of the water molecule in the sub-nanocavity often hinders the complete fulfillment of the water coordination possibilities (Fig. 1.1). On the other hand, the cavity may be sealed by a ligand that has a good chance to stick if the water molecule in the cavity is too frustrated in the free nonbonded state, while the protein-ligand complex causes less of a distortion in the water lattice than the free subunits. This picture, while intuitively appealing and certainly correct, needs rigorous support and has profound consequences for the understanding of biological events at the molecular scale. The book will revisit this picture from different perspectives, scaffolding it further each time around.

Water is known to be the matrix of life, and soluble proteins, like other solutes in the cytosol or intercellular space, perturb this matrix. A way of quantifying the magnitude of this perturbation is to compute the free energy change ΔG_{if} associated with spanning the aqueous interface that envelops the protein. If we normalize this extensive thermodynamic quantity by the surface area of the protein, we obtain the interfacial tension. Since the structure of the interface must be in principle determined by the structure of the protein, we may adopt the term "epistructural tension".

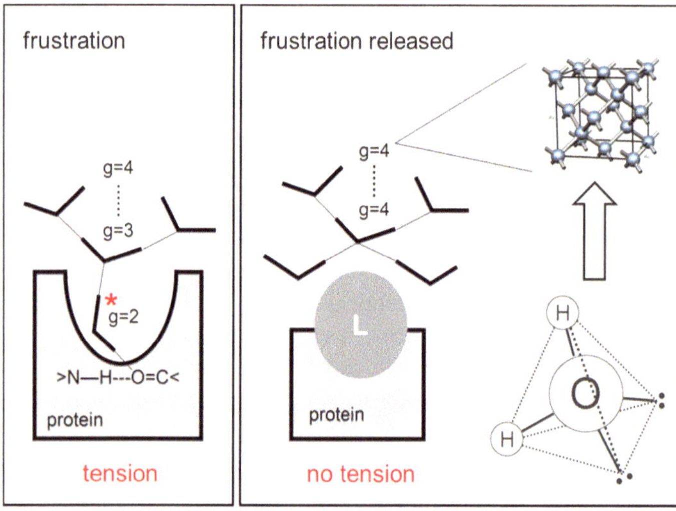

Fig. 1.1 Disruption of water structure at the interface with a soluble protein. A structural defect exposing the protein backbone produces "frustration" in vicinal water molecules (thick angular segments) as its hydrogen-bonding coordination (g) is reduced when solvating the backbone. The frustrated molecule is marked by an asterisk. Thus, interfacial tension arises due to the disruption of the dynamic tetrahedral hydrogen-bond lattice characteristic of bulk water (g = 4). The electronic sp^3-structure of the water molecule with two lone electron pairs and two proton donors in tetrahedral arrangement dictates the arrangement of the supramolecular hydrogen-bond lattice. When ligand (L) seals the sub-nanoscale cavity, the disruption of water structure is reduced and frustration is released with concurrent decrease in interfacial tension

It is evident that protein associations would tend to reduce ΔG_{if}, as such associations reduce the aqueous interface, so it is imperative that we find a way to compute epistructural tension and identify chemical and structural features in the protein that generate such tension. Thus, the first question we need to address in dealing with biomolecular interfaces is how are the structure and ultimately the thermodynamic properties of the interface determined by the structure of the soluble protein.

The multi-scale structure of water in the condensed phases of biological relevance remains a source of controversy [1, 2]. In bulk water, the controversy has been partly fuelled by the discovery of structural distortions in the fluctuating tetrahedral hydrogen-bond lattice due to asymmetric electron density [1]. On the other hand, the long-range tetrahedral hydrogen-bond network signatures are clearly present in the X-ray scattering intensity and structure factor data [2]. Thus, at nano-meter scales we may assume that the structure of water, with its resilient tetrahedral lattice of intermolecular hydrogen bonds, is largely responsible for its significant dielectric properties. As it is known, this resilience to the polarization induced by an external electrostatic field generated by a charge separation becomes apparent in the large permittivity coefficient of water.

Provided tetrahedral coordination possibilities are preserved as in bulk-like environments, polarization is believed to align with electrostatic fields determined by fixed charges, as first proposed by Peter Debye [3–7]. There are indications that this "Debye ansatz" may break down under confinement of water molecules in regions of nanoscale dimensions [4]. The breakdown is attributed to the fact that the bulk-like floppy tetrahedral lattice has been distorted to a point where water is effectively deprived of hydrogen-bond partnerships and tends to preserve its interfacial hydrogen-bond pattern in spite of the field-aligning torque imposed by the electrostatic field [4]. For instance, the complexities of dielectric properties at biological heterogeneous interfaces [8–10] bespeak of a spatial scale where the behavior of water dipoles cannot be accounted for through linear polarization relations [5, 11, 12]. The nanoscale structure of water must be inevitably incorporated into the electrostatic description of "*episteric*" fields at interfaces that confine the solvent to discrete levels [4, 5, 12]. By *episteric* we mean surrounding a solute particle with a defined interface and charge distribution, such as a soluble protein.

The laws of dielectric response are inherently continuous [7], at odds with the discrete structural distortions of water that promote anomalous polarization under nanoscale confinement. To exploit the mathematical apparatus of electrostatics, we face the conundrum of having to adopt continuous—actually differentiable—descriptors of water structure that are inadequate at scales where the discrete patterns of hydrogen-bond coordination becomes significant. To reconcile these conflictive aspects, we introduce a local indicator of time-averaged hydrogen-bonding patterns for water molecules at spatial locations. Specifically, we introduce a scalar field $g = g(\vec{r})$ indicating the expected number of hydrogen bonds of a water molecule computed while its oxygen atom is contained within a sphere of radius $r = 4$ Å (~van der Waals radius of carbon atom + thickness of one hydration layer) centered at position $\vec{r}$ and during a timespan $\tau = 10$ ps [4]. The (r,τ)-parametrization of this scalar field has been calibrated empirically, so that the choices guarantee second-order differentiability for g. This level of smoothness is required to introduce a nanoscale dielectric relation. The g-function indicates in a coarse grained way the number of interactions of a water molecule within a neighborhood of each point in space, remaining informative at ~1 nm levels. By definition, $g = g(\vec{r})$ represents also an average over all water molecules visiting the sphere within the 10 ps-timespan. The hydrogen bonds in the counting involve neighboring water molecules and polar groups from the protein.

To examine the aqueous interface of a soluble protein we may embed and equilibrate the protein chain conformation in an explicit water bath. The chain conformation is determined by structural coordinates from the corresponding protein data bank (PDB) entry. Details of such computations will be provided subsequently and the robustness to different solvent models will be discussed in this and the four chapters that follow. The zest of this section is to simply delineate a relationship between protein structure and interfacial tension that will be operative to address core problems in molecular biophysics.

To obtain a description of the aqueous interface of a protein after equilibrated embedding of the protein structure, we have computed g-values at every atom on the surface of the protein, so that water molecules are present in the solvation shells of 4 Å-radius. It is entirely expected that g-values in the range $g \geq 3$ will occur on the interface, as $g = 3$ is the lower bound at the interface for an aqueous bulk phase. We are interested in "unusually low" g-values, in the range $g < 3$, as those would entail the most significant perturbation of the water matrix. As an example, the surface g-values for *ubiquitin* are displayed in Fig. 1.2a, b. Figure 1.2a displays van der Waals radii of all surface atoms and in lighter color the 4 Å-radius solvation shells with $g < 3$. Figure 1.2b shows the location of the hot $g < 3$-spots in relation to a ribbon rendering of the ubiquitin backbone in the native fold. Examination of the hot spots prompts us to ask: How does this interfacial pattern relate to the 3D native structure of ubiquitin? What are the structural features causative of such hot spots? The $g < 3$ values result from severe confinement of water molecules at the interface as it would be feasible at sub-nanoscale cavities. Such cavities would involve a special kind of packing defect involving unusual arrangements of the side chains of the protein. This is as far as intuition takes us, so we computed a strictly structure-based feature: The backbone hydrogen bonds that are incompletely shielded from the solvent, also known as *dehydrons*. The partial exposure of the backbone to water as represented by a dehydron clearly constitutes a packing defect since intramolecular backbone hydrogen bonds exposed to water are inherently weak and prone to cause a disruption of the protein structure as the amides and carbonyls become maximally hydrated. Thus, we contrasted the dehydron pattern for ubiquitin (Fig. 1.2c, details on dehydron computation will be revealed subsequently) with the $g < 3$ hot spots displayed in Fig. 1.2a, b. A clear correlation emerged: *The $g < 3$ hot spots invariably corresponded to the solvation shells of the atoms closest to dehydrons.* This result can be carried over to all soluble proteins reported in PDB and clearly implies that protein *dehydrons play a decisive role in shaping the interface as they promote the most significant distortions in the water matrix.*

In fact, the solvation shells with the highest vacating probability $P(N = 0)$ (i.e. the probability of the solvation shell to be spontaneously deprived of water molecules, with $N =$ number of water molecules in solvation shell) are precisely those with the lowest g-values, in accord with the empirical relation $P(N = 0) = 1 - g/4$ hereby established and found across all soluble protein interfaces with reported structure (problems section, Chap. 2). Since $P(N = 0) = \exp\left[- <N>^2/2\sigma^2\right]$ with $\sigma^2 = <(N - <N>)^2>$, we may assert that *dehydrons generate the largest fluctuations in interfacial water density and promote the most significant local dehydration of the aqueous interface.* This result sets the agenda for the rest of the chapter, as it strongly suggest that dehydrons are crucial determinants of protein associations and therefore deserve to be studied closely from a physico-chemical perspective.

1.3 Interfacial Tension of Biological Water

At the time when Debye's "molecular theory" of dielectric response emerged [7], the nanoscale structure of water, with its resilient tetrahedral lattice of intermolecular hydrogen bonds [1, 2, 4], was largely unknown and the importance of hydrogen bonds as determinants of water structure was not yet recognized. With

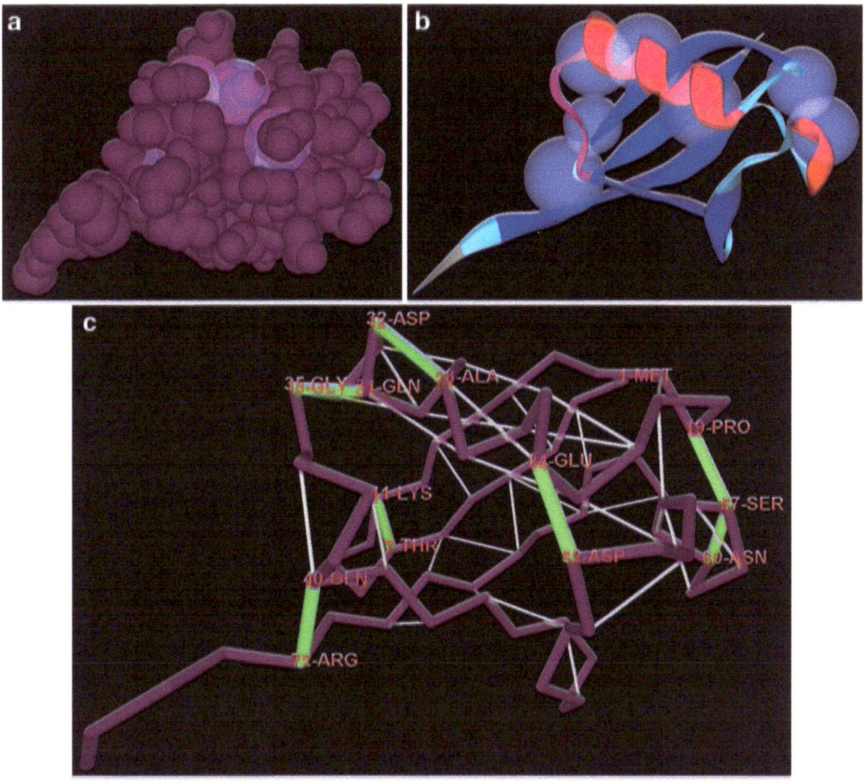

Fig. 1.2 (**a**) The structure of *ubiquitin* (PDB entry 1UBI) with atoms represented by their van der Waals spheres (*magenta*) with semi-transparent *blue spheres* indicating 4 Å-solvation shells with g < 3 (hot spots). (**b**) Location of the solvation shells with g < 3 within a ribbon rendering of the ubiquitin backbone in the native structure. (**c**) Dehydron pattern for ubiquitin. The backbone is represented by virtual bonds (*magenta*) joining α-carbons of residues adjacent in the chain, dehydrons are represented by *green sticks* joining the α-carbons of hydrogen-bonded paired residues, while well-wrapped backbone hydrogen bonds are displayed as *grey sticks* joining paired residues

virtually no information on the nanoscale structure of water and merely assuming the existence of a polarizable "sea of dipoles", Peter Debye derived a theory applicable in contexts where the dielectric response could be treated as a bulk attribute. The theory postulates an alignment of water polarization $\vec{P}$ with the internal electrostatic field $\vec{E}$. This ansatz enables the computation of the effective permittivity ε of the dielectric medium by assuming that the proportionality between the vector fields $\vec{P}$ and $\vec{E}$ can be expressed without loss of generality as $\vec{P} = (\varepsilon - \varepsilon_0)\vec{E}$, where ε_0 is the vacuum permittivity. The ansatz yields a simplified Poisson electrostatic equation $\vec{\nabla}.(\varepsilon_0\vec{E} + \vec{P}) = \vec{\nabla}.(\varepsilon\vec{E}) = \rho$ (ρ = fixed charge distribution) that holds for an electrostatic condenser with planar featureless parallel plates and for virtually all aqueous dielectric media provided it lacks nanoscale detail.

The success of Debye's simplification encouraged researchers to extrapolate his linear polarization assumption to biophysical contexts such as protein-water interfaces [13, 14]. This extrapolation proves to be unwarranted and ultimately incorrect, as we shall presently demonstrate. The theoretical results obtained thereof proved to be nearly impossible to reconcile with experiment, as recent work suggests [8–11]. In the case of biological water, the supramolecular structure of the solvent and the nanoscale distortions relative to a bulk-like organization cannot be ignored. The crux of the problem is that confinement of water at the biological interface often creates dipole organizations that result from deprivation of hydrogen bonding opportunities, thereby creating uncompensated effective charges that are not a resultant of polarization along the direction of the internal field $\vec{E}$ of the protein [4].

In critically assessing the applicability of Debye's model to protein-water interfaces we first note that a single parameter, the effective permittivity coefficient ε, cannot possibly quantify the dielectric response of water. Recent computations reveal that there is a component of polarization generated by the nanoscale topography and chemical composition of the protein surface that offers resistance to the torque imposed by the field $\vec{E}$ on the water dipoles [15].

The biological interfaces confine water into patterns that cannot simultaneously minimize the disruption of the hydrogen-bond lattice and align the water dipoles with the electrostatic field. There are polarization components that do not align with the field and arise from tight local nanoscale organizations of water molecules deprived of hydrogen bonding opportunities [4]. This picture brings about a complex dielectric structure that cannot be subsumed into a single permittivity parameter. The nanoscale water structure must be incorporated into the electrostatic description of the resultant fields at interfaces that confine solvent to discrete levels.

We expect and hereby postulate that nanoscale local organization will introduce a significant component of polarization, $\vec{P^{\#}}$, orthogonal to the protein internal field $\vec{E}$. This "anomalous" polarization is a predictable resultant of the partial hindrance in the alignment of water dipoles with the field $\vec{E}$ due to nanoscale confinement. The departure from the "linear dielectrics" picture is described by writing the Poisson equation in its utmost generality as follows [4]:

$$\vec{\nabla}.\left(\varepsilon_0\vec{E} + \vec{P^{\parallel}} + \vec{P^{\#}}\right) = \rho \tag{1.1}$$

where, without loss of generality, $\vec{P}$ has been decomposed into a field-aligned component $\vec{P^{\parallel}} = \left(\vec{P}.\vec{e}\right)\vec{e}\ ;\ \vec{e} = \vec{E}/\left|\left|\vec{E}\right|\right|$, and a component $\vec{P^{\#}}$, orthogonal to $\vec{E}$. Accordingly we obtain:

$$\vec{\nabla}.\left(\vec{P^{\#}}\right) = \rho - \vec{\nabla}.\left(\varepsilon_0 \vec{E} + \vec{P^{\parallel}}\right) \tag{1.2}$$

In (1.2) the term $\rho - \vec{\nabla}.\left(\varepsilon_0 \vec{E} + \vec{P^{\parallel}}\right)$ measures the departure from the Debye dielectrics, with $\rho - \vec{\nabla}.\left(\varepsilon_0 \vec{E} + \vec{P^{\parallel}}\right) = 0$ when the Debye ansatz $\vec{P} = \vec{P^{\parallel}}$ is fulfilled, i.e. in the Debye limit conventionally adopted in molecular biophysics.

Defining the electrostatic potential φ through the relation: $-\vec{\nabla}\varphi = \vec{E} + \varepsilon_0{}^{-1}\vec{P}$, the total electrostatic energy becomes

$$U = -\frac{1}{2}\varepsilon_0 \int \varphi \Delta \varphi d\vec{r} = \frac{1}{2}\varepsilon_0 \int \left|\left| \vec{\nabla}\varphi \right|\right|^2 d\vec{r} = U_D + U^{\#} \tag{1.3}$$

where $\Delta = \nabla^2$ is the Laplace operator, and

$$U_D = \frac{1}{2}\varepsilon_0 \int \left|\left| \vec{E} + \varepsilon_0{}^{-1}\vec{P^{\parallel}} \right|\right|^2 d\vec{r} \tag{1.4}$$

is the Debye contribution, and

$$U^{\#} = \frac{1}{2}\varepsilon_0{}^{-1} \int \left|\left| \vec{P^{\#}} \right|\right|^2 d\vec{r} = \frac{1}{2}\varepsilon_0{}^{-1} \int \left|\left| \vec{P} - \left(\vec{P}.\vec{e}\right)\vec{e} \right|\right|^2 d\vec{r}. \tag{1.5}$$

is the "orthogonal" polarization contribution to the electrostatic energy.

Since departures from bulk water structure (spatially measured by $\nabla g \neq 0$) induce orthogonal polarization, we adopt the ansatz $\vec{P^{\#}} \propto \vec{\nabla} g$ or $\vec{P^{\#}} = \xi \vec{\nabla} g$ [4] where $\xi = (\lambda \varepsilon_o)^{1/2}$ the parameter λ is obtained by estimation of the interfacial free energy of a nonpolar sere with radius θ and contrasting in the macroscopic limit the result with the elastic integral $\int \frac{1}{2} \left|\left| \vec{\nabla} g \right|\right|^2 d\vec{r}$. The elastic integrand $\frac{1}{2} \left|\left| \vec{\nabla} g \right|\right|^2$ accounts for tension-generating reductions in water coordination and vanishes everywhere except at the solute-water interface. The comparison is valid since interfacial tension is related to the distortion of the tetrahedral coordination structure of bulk water. In the macroscopic limit, where $\theta \gg 1$ nm, we get $\lambda = \frac{9.0 \text{ mJ}}{\text{m}} = limit \left\{ \frac{\gamma\left(4\pi\theta^2\right)}{\int \frac{1}{2} \left|\left| \vec{\nabla} g \right|\right|^2 d\vec{r}} \right\}$ where $\gamma = 72$ mJ/m^2 is the macroscopic surface tension of water at 298 K. From the ansatz $\vec{P^{\#}} = \xi \vec{\nabla} g$, we obtain the relation:

$$U^{\#} = \frac{1}{2}\varepsilon_0{}^{-1} \int \left|\left| \vec{P^{\#}} \right|\right|^2 d\vec{r} = \int \frac{1}{2}\lambda \left|\left| \vec{\nabla} g \right|\right|^2 d\vec{r}. \tag{1.6}$$

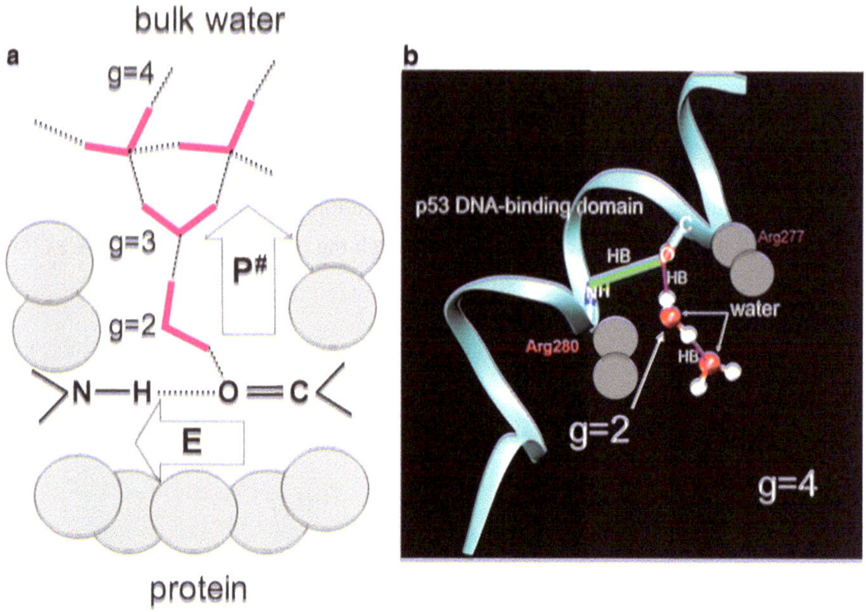

Fig. 1.3 (**a**) Schematic representation of the non-Debye polarization of interfacial water constrained by the sub-nanoscale cavity created by a dehydron. The orthogonal polarization results from a deprivation in hydrogen-bonding opportunities for water molecules in the dehydron cavity. (**b**) Hydration pattern for dehydron Arg277-Arg280 in the DNA-binding domain of p53, displaying the paired backbone amide and carbonyl

This equation asserts that the interfacial tension has an electrostatic origin and may be characterized as the energy per unit of protein surface stored in the anomalous non-Debye polarization resulting from sub-nanoscale confinement at the interface (Fig. 1.3a, b). Since dehydrons create the biggest drops in g-value, we may intuitively assert that dehydrons cause the breakdown of the Debye electrostatic ansatz and at theame time, represent the culprit of interfacial tension. This in turn implies that dehydrons are the dominant features promoting protein associations (Fig. 1.1), an assertion that will be proven in Chaps. 2 and 4.

1.4 Variational Thermodynamics for the Biomolecular Interface

As water interfaces with a solute endowed with sub-nanoscale detail, its hydrogen-bonding opportunities are often curtailed (Fig. 1.1). Interface confinement generates a frustration field. In this section we show that the frustration dynamics are governed by a variational principle that defines a microstate sampling strategy. The sampling enables a search for the least-frustrated state within relaxation times for coupled protein-water systems. The frustration field is shown to steer protein-protein (P-P) associations, a result validated vis-à-vis experimental affinity scanning of P-P interfaces that assemble to mitigate frustration.

As discussed in the previous section, frustrated interactions, like unmet opportunities for hydrogen-bonding involving water at a protein interface [15, 16], contribute to the interfacial tension that may be alternatively stored as electrostatic energy from anomalous polarization. This contribution is not included in molecular dynamics (MD) computations [17–20]. It would be logically inconsistent to energetically penalize missing opportunities for favorable interactions even when the solvent is incorporated explicitly [17–19]. Frustration is not incorporated in implicit-solvent computations either, since the latter do not incorporate information on coordination of individual water molecules [20]. It is also unclear—and unlikely—that the global free energy minimum would be a state of minimal interfacial frustration [21]. Yet, as shown in this section, a microstate sampling steering MD computations towards a least-frustrated interface proves essential to identify the molecular factors that trigger elementary events in biology. Accordingly, we implement an explicit-solvent MD algorithm with microstate sampling guided by a variational principle to enable a search for minimally frustrated interfaces. We demonstrate that the resulting frustration field drives protein-protein (PP) associations with high specificity. The validation of this assertion entails contrasting the computational results vis-à-vis affinity scanning of P-P interfaces of protein complexes [22] that assemble to mitigate the interfacial tension of free subunits.

To describe the local structural distortion of water, we introduce a "frustration scalar field" $\phi = \phi(\vec{r})$ that quantifies the expected number of unmet hydrogen bonds of a test water molecule placed at spatial location $(\vec{r})$ relative to the tetrahedral lattice representing bulk water structure. Thus we get $\phi(\vec{r}) = 4 - g(\vec{r})$, where $g(\vec{r}) \leq 4$ is the time-averaged number of hydrogen bonds (coordination number) that a water molecule sustains while it visits a sphere of radius $r = 4$ Å centered at position $\vec{r}$ for a minimum timespan 1 ps. The temporal averaging parameter is of the magnitude of the relaxation timescale for a decoupled water lattice [23], and the averaging required to operate with a smooth scalar field [21]. A hydrogen bond is operationally defined by geometric constraints [21]: O-O distance <3.5 Å and O-H-O angle α_{HB} satisfying $120° \leq \alpha_{HB} \leq 180°$. Compared with bulk water, interfacial water may have reduced hydrogen-bonding opportunities, and may partly mitigate such losses by hydrogen bonding with polar groups on the protein surface.

A frustration functional $\mathfrak{I}(\phi) = \int \mathcal{L}(\phi) d\vec{r}$ is introduced to compute the free energy cost $\Delta G_{if}(\phi)$ of generating the water frustration field ϕ when spanning an aqueous interface with a solute molecule

$$\Delta G_{if}(\phi) = \mathfrak{J}(\phi) = \int \mathcal{L}(\phi) d\vec{r}$$
$$= \int \left[\frac{1}{2}\omega || \vec{\nabla}\phi ||^2 - k_B T(\rho_0 - \rho)\ln\left(1 - \frac{\phi}{4}\right) \right] d\vec{r} \quad (1.7)$$

where the elastic energy functional $\int \frac{1}{2}\omega || \vec{\nabla}\phi ||^2 d\vec{r}$ is a Dirichlet-type functional [21, 24] penalizing distortions of the water structure, and the entropy differential $dS_{if}(\phi) = k_B ln\left[\frac{4-\phi}{4}\right]^{(\rho_0-\rho)d\vec{r}} = \left[k_B(\rho_0 - \rho)\ln\left(1 - \frac{\phi}{4}\right)\right] d\vec{r}$ corresponds to the transference of a water molecule from bulk solvent with number density $\rho_o = 0.033$ Å^{-3} to a differential volume region $d\vec{r}$ at position $\vec{r}$ with number density $\rho(\vec{r})$ and frustration $\phi(\vec{r})$. For small structural distortions with $\phi \ll 4$, the linear approximation becomes valid, yielding: $dS_{if}(\phi) \approx -\left[k_B(\rho_0 - \rho)\left(\frac{\phi}{4}\right)\right] d\vec{r}$.

The least-frustrated water interface is described by a frustration field $\phi = \phi(\vec{r})$ that satisfies the Euler-Lagrange equation [24]:

$$\nabla \partial \frac{\mathcal{L}(\phi)}{\partial \nabla \phi} = \partial \frac{\mathcal{L}(\phi)}{\partial \phi}, \quad (1.8)$$

written explicitly as:

$$\Delta\phi = \left(\frac{k_B T}{4\omega}\right)(\rho_0 - \rho)\left(1 - \frac{\phi}{4}\right)^{-1}, \quad (1.9)$$

where $\Delta = \nabla^2$ is the Laplace operator, and hence $\Delta\phi$ gives the curvature of the frustration field. For small distortions $\phi \ll 4$, we get

$$\Delta\phi = \left(\frac{k_B T}{4\omega}\right)(\rho_0 - \rho). \quad (1.10)$$

To obtain ω, we use (1.9) applied to an equilibrating computation of ϕ obtained as a function of the distance X to a rigid nonpolar interface [25] consisting of a concave region fixed curvature radius $\theta = 3.25$ Å, with X = 0 = center of curvature. As described in Chap. 2, an MD-equilibration of the NPT canonical ensemble is carried out for 1 ns (Fig. 1.4), a time incommensurably larger than the relaxation timescale for the decoupled water lattice dynamics (~1 ps) [23]. Thus, at T = 298 K (1.9) yields $\frac{1}{9}$ Å$^{-2} = \omega^{-1} \times 1.02 \times 10^{-20}$ mJÅ^{-3} or $\omega = 9.18 \times 10^{-20}$ mJÅ^{-1}. To validate this estimation, we compute the interfacial tension $\Delta G_{if}(\phi(\theta))/4\pi\theta^2$ for a vapor spherical bubble B equilibrated at radius $\theta = 2.48$ nm within a simulation timespan of 1 ns [26]. The bubble dimensions approximate a macroscopic limit at T = 298 K within an NVT canonical ensemble of 29,328 TIP3P water molecules [27] in a $10 \times 10 \times 10$ nm^3 box with periodic boundaries. The equilibrated bubble is slightly larger than the original bubble at starting conditions, generated by water

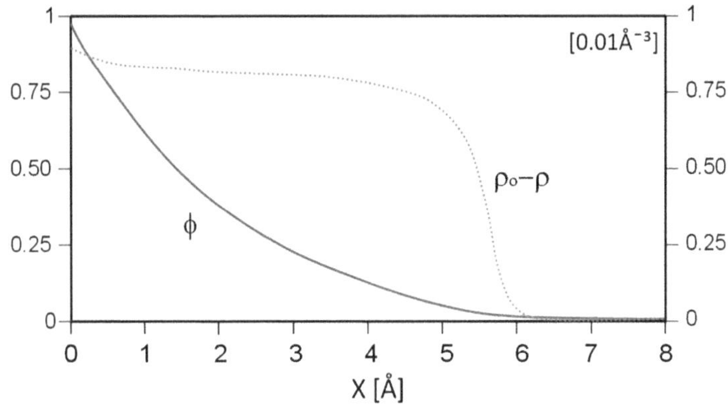

Fig. 1.4 Computation of water frustration ϕ (ordinates, *left axis*) and number density ρ (ordinates, *right axis*, dimensions in figure) relative to the distance X to a rigid nonpolar interface consisting of a concave region of fixed curvature radius θ = 3.75 Å, with X = 0 at center of curvature. We generated ten lattice-equilibrating MD 1 ns-trajectories for water dynamics within an isobaric/isothermal ensemble (1 atm, 298 K) consisting of 10,190 water packed against a featureless concave layer of hydrogenated graphene (6 × 6 supercell, 72C, 72H) representing a nonpolar surface [25]. The solvent bath provided at least four water layers of solvent envelope with water molecules extending 15 Å from the nonpolar surface. The solvent was modeled with TIP3P water molecules, employing periodic boundary conditions [26, 27]. An AMBER package [28] was adopted, and Ewald sums and an 8 Å-distance cutoff were adopted to treat Coulombic interactions [29]

molecule removal from a spherical region with radius $\theta_0 = 2.00$ nm from an initial ensemble with density 958.5 kg/m^3. Thus, we obtain

$$\frac{\Delta G_{if}(\phi(\theta))}{4\pi\theta^2} = \int_{\partial B(\theta)}^{\infty} \frac{\mathcal{L}(\phi)d\vec{r}}{4\pi\theta^2} = \gamma = 53.8 \ \ \mathrm{mJ/m^2}. \tag{1.11}$$

This value is in good agreement with the Young-Laplace estimation: $\gamma_{YL} = -\frac{p_l\theta}{2} = 55.2 \ \ \mathrm{mJ/m^2}$, resulting from the virial computation of the bulk liquid water pressure $p_l = -46 \times 10^6 \ \ \mathrm{N/m^2}$, given a negligible vapor pressure in the bubble: $p_v \approx 0$, as no water molecule was detected escaping to the vapor phase at T = 298 K during the 1 ns-computation.

The minimization of water frustration is now shown to translate into a major force steering protein associations, the elementary events in biology at the molecular scale. A major caveat holds as we incorporate water frustration into the canonical ensemble molecular dynamics of soluble proteins with explicit solvent: we cannot incorporate a frustration term directly into the potential energy since there is no explicit energetic penalty for missed interactions or unmet hydrogen-bonding possibilities. Yet, as shown in the microstate sampling scheme on Fig. 1.5, we can select at each stage of the simulation a representative from the set of *a-priori* equi-probable microstates to ensure a *probable* decrease in $\Delta G_{if}(\phi) = \mathfrak{I}(\phi)$ and, ultimately, an effective search for a state of minimal frustration. Let $[\mathbf{Q}_n, \mathbf{P}_n^p \bigoplus \mathbf{P}_n^w]$ denote the state of the system at stage n, specified by $\mathbf{Q}_n$ = atomic coordinates vector $(\mathbf{Q}_n = \mathbf{Q}_n^p \bigoplus \mathbf{Q}_n^w)$, and $\mathbf{P}_n = \mathbf{P}_n^p \bigoplus \mathbf{P}_n^w$ = momentum vector, both decomposable as direct sums of protein (p) and water (w) contributions. The next state $[\mathbf{Q}_{n+1}, \mathbf{P}_{n+1}^p \bigoplus \mathbf{P}_{n+1}^w]$ resulting after 5 ps of MD simulation is retained depending on a comparison of the scalar fields $\phi_n = \phi_n(\mathbf{Q}_n)$ and $\phi_{n+1} = \phi_{n+1}(\mathbf{Q}_{n+1})$. Retention of the destiny state $[\mathbf{Q}_{n+1}, \mathbf{P}_{n+1}^p \bigoplus \mathbf{P}_{n+1}^w]$ occurs provided that either of the following inequalities holds:

$\mathfrak{I}(\phi_{n+1}) \leq \mathfrak{I}(\phi_n)$ or $\mathfrak{I}(\phi_{n+1}) > \mathfrak{I}(\phi_n)$ and $r \leq \exp\{-\beta[\mathfrak{I}(\phi_{n+1}) - \mathfrak{I}(\phi_n)]\}$, with r = realization of random variable within the interval [0,1] and $\beta = [k_B T]^{-1}$. On the other hand, if $\mathfrak{I}(\phi_{n+1}) > \mathfrak{I}(\phi_n)$ and $r > exp\{-\beta[\mathfrak{I}(\phi_{n+1}) - \mathfrak{I}(\phi_n)]\}$, then we perform 5 ps MD runs on 1000 initial randomly selected *a-priori* equiprobable n-microstates $[\mathbf{Q}_n, \mathbf{P}_n^p \oplus \mathbf{P}_n^{w(J)}]$, $j = 1, 2, \ldots, 1000$, with randomly chosen water momenta $\mathbf{P}_n^{w(J)}$ $(\mathbf{P}_n^{w(1)} = \mathbf{P}_n^w)$ subject to the constraint of constant kinetic energy, and select the state $\left[\mathbf{Q}_{n+1}^*, \mathbf{P}_{n+1}^{p^*} \oplus \mathbf{P}_{n+1}^{w^*}\right]$ as the $(n+1)$-destiny state that realizes the lowest $\mathfrak{I}\left(\phi_{n+1}\left(\mathbf{Q}_{n+1}^{(J)}\right)\right)$ from within the set of 1000 destiny coordinate vectors.

To assess the steering power of the functional in (1.7) towards the minimally frustrated interface state, and illustrate the role of water frustration as enabler of protein associations, we generated 500 ps-trajectories for soluble proteins capable of associating with binding partners. The computations are restricted to proteins engaged in complexes with structures reported in the protein data bank (PDB). The AMBER package was adopted for MD simulation within the iterative scheme hybridized by the microstate sampling described in Fig. 1.5 for an isothermal/isobaric ensemble at T = 298 K [28]. Coulombic interactions were evaluated with the Ewald summation scheme [29], initial structural coordinates $\mathbf{Q}_1^p$ were obtained for the corresponding free subunit chain extracted from its respective PDB-reported complex and the water solvent was incorporated adopting the TIP3P model. To first illustrate our findings we report in Fig. 1.6a the time evolution of $\Delta G_{if}(\phi) = \mathfrak{I}(\phi)$ for the free human growth hormone (hGH) receptor subunit hGHbp that forms the receptor/hGH complex, adopting the structural coordinates provided in PDB file 3HHR. The thin solid line reports a single trajectory, the thick solid grey line, the average over 100 trajectories following the iterative scheme in Fig. 1.5, and the dashed line represents the control consisting of an average over 100 trajectories of 500 ps-MD simulation without microstate sampling. The minimal $\Delta G_{if}(\phi)$-value is $\Delta G_{if}^*(\phi) = 121$ kJ/mol, corresponding to a protein/water interfacial tension $\gamma = 2.13$ mJ/m^2. A significant relaxation time is observed in the 350–400 ps range (Fig. 1.6a), in consonance with experimental relaxation times associated with the coupling of protein motion and hydration pattern [30], significantly distinct from the decoupled water lattice relaxation time (~1 ps) [23].

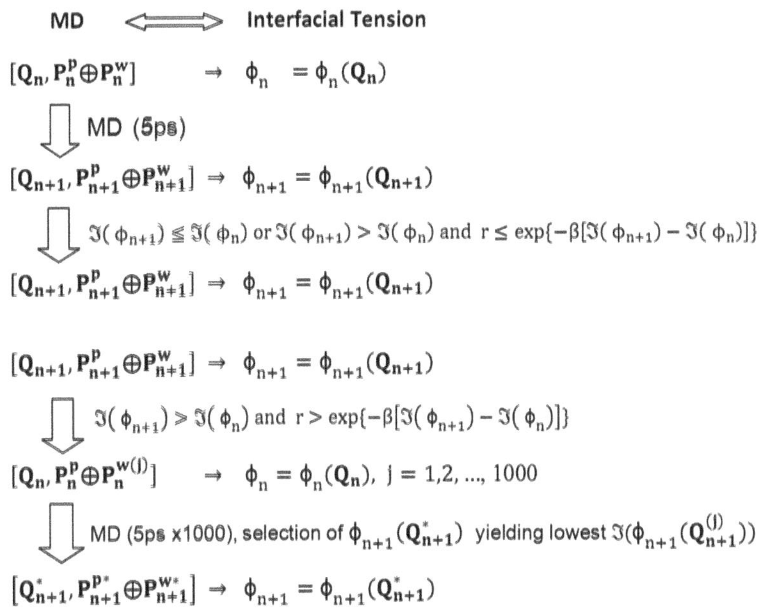

Fig. 1.5 Microstate sampling scheme for a 5 ps-MD iteration guided by interfacial frustration functional $\Im(\phi) = \int \mathcal{L}(\phi) d\vec{r}$. The *upper box* indicates flow under which the destiny microstate $[\mathbf{Q}_{n+1}, \mathbf{P}^{p}_{n+1} \oplus \mathbf{P}^{w}_{n+1}]$ is retained in the $n \to n+1$ iteration based on a comparison of scalar fields $\phi_n = \phi_n(\mathbf{Q}_n)$ and $\phi_{n+1} = \phi_{n+1}(\mathbf{Q}_{n+1})$. The *lower box* indicates the sampling procedure that takes place when the destiny state is not chosen to be retained

To assert the role of interfacial water frustration as promoter of PP association, we determine the effect of site-directed mutations on $\Delta G_{if}(\phi)$ and compare the $\Delta\Delta G_{if}(\phi)$ values with the change in affinity quantified by the calorimetrically measured $\Delta\Delta G_a = -RTln\left[\frac{K_a(m)}{K_a(0)}\right] = RTln\left[\frac{K_d(m)}{K_d(0)}\right]$, where $\Delta\Delta G_a$ is the change in association free energy arising from a site-specific mutation [22], and $K_d(m) = K_a(m)^{-1}$, $K_d(0) = K_a(0)^{-1}$ indicate the equilibrium dissociation constants for mutant (m) and wild-type (0), respectively. Hot spots in PP associations were experimentally identified by scanning the interface through site-specific alanine substitution of amino acids for alanine (effectively a side-chain truncation at the beta-carbon, except for glycine)The most extensively scanned interface corresponds to the 1:1 hGH-hGHbp receptor complex [22]. Thus, the hot-spot contributions $\Delta\Delta G_a$ obtained by alanine scanning were contrasted with an *in-silico* shaving procedure (truncation at beta-carbon, equivalent to alanine substitution) yielding the changes $\Delta\Delta G_{if} = \Delta G_{if}(m) - \Delta G_{if}(0)$. A tight correlation ($R^2 = 0.88$) exists between $-\Delta\Delta G_{if}$ and $\Delta\Delta G_a$ for substitution of interfacial residues in the receptor hGHbp (Figs. 1.6b and 1.7), supporting the approximate relation $-\Delta\Delta G_{if} \approx \Delta\Delta G_a$. *Thus, the results from Fig.* 1.6b, *based on the most comprehensively studied PP*

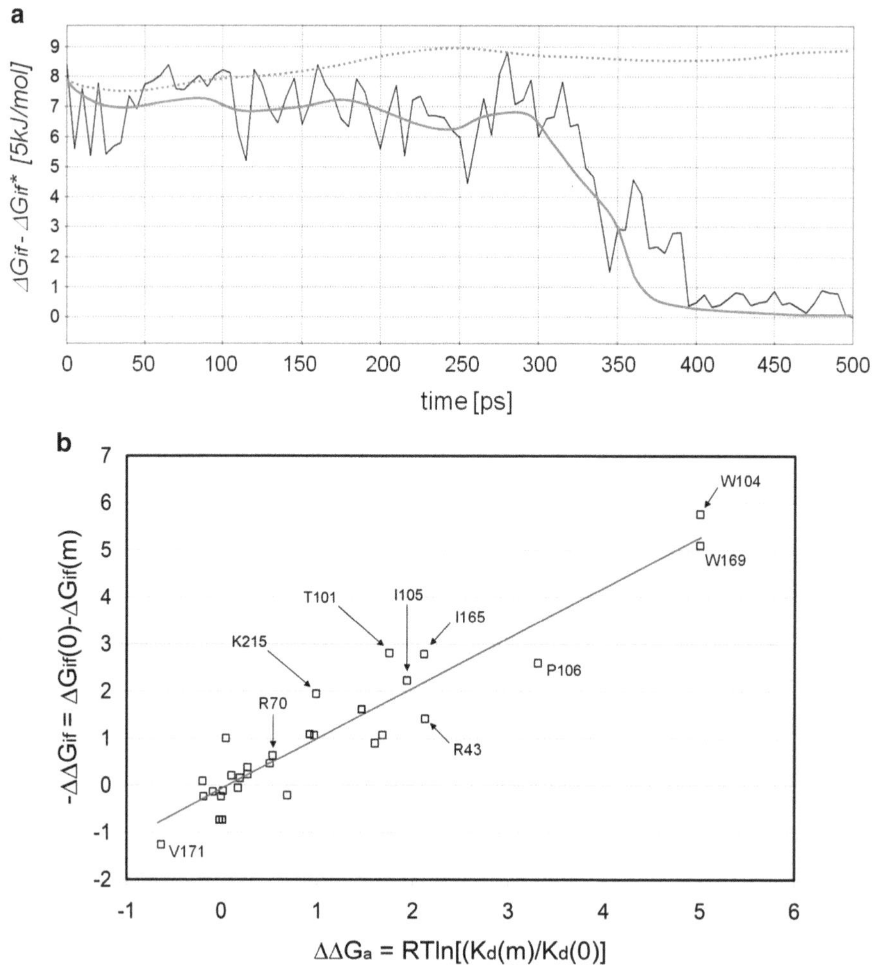

Fig. 1.6 Water frustration as enabler of protein associations. (**a**) Steering power of microstate sampling controlled by the frustration functional towards the minimally frustrated interface state, as evidenced in the time evolution of interfacial frustration free energy $\Delta G_{if}(\phi) = \Im(\phi)$ obtained for 500 ps-MD trajectories for a soluble protein with explicit solvent. The single trajectory is in *solid thin line*, the average over 100 trajectories in *thick grey line*, and the *dashed grey line* represents an average over 100 control 500 ps-trajectories with no microstate sampling. Initial structural coordinates $\mathbf{Q}_1^p$ were obtained for the corresponding free-subunit chain for the hGHbp receptor that forms a 1:1 complex with the human growth hormone (hGH) as reported in PDB file 3HHR. The microstate sampling described in Fig. 1.5 was enforced within an isothermal/isobaric ensemble at T = 298 K. Computational details are given in main text. The minimal $\Delta G_{if}(\phi)$-value is $\Delta G_{if}^*(\phi) = 121$ kJ/mol and is essentially reached within the 350–400 ps range. (**b**) Correlation ($R^2 = 0.88$) between $-\Delta\Delta G_{if}$ and $\Delta\Delta G_a$ for site-directed mutational substitution of interfacial residues in receptor hGHbp, yielding the approximate identification $-\Delta\Delta G_{if} \approx \Delta\Delta G_a$

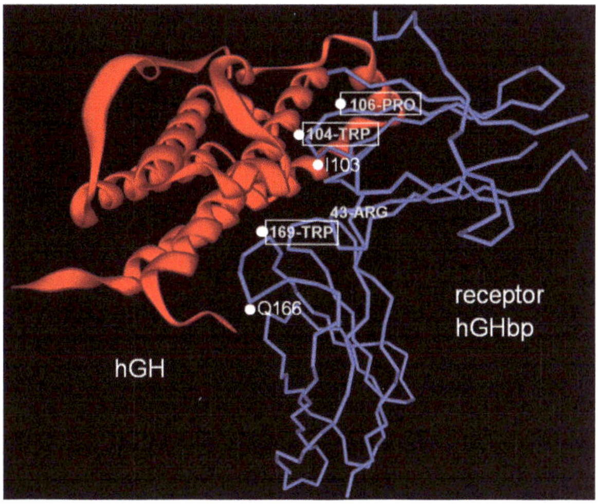

Fig. 1.7 Structure of hGH/receptor 1:1 complex, with hormone in ribbon representation and receptor in virtual-bond backbone representation. The alpha-carbons of certain residues at the PP interface are indicated as *white circles*, with the significant contributors to affinity ($\Delta\Delta G_a \geq 3$ kcal/mol) highlighted in *boxes*

interface, support the view that water frustration drives protein association and that the complex assembles to mitigate water frustration.

To further uphold the frustration-driven nature of PP association vis-à-vis alanine-scanning of PP interfaces, other well studied complexes are considered. We classify residues according to $\Delta\Delta G_a$ ranges and independently according to their respective $\Delta\Delta G_{if}$ ranges. Hot-spot residues are grouped (upper row for each PP interface, Fig. 1.8) according to the ranges $\Delta\Delta G_a \geq 3$ kcal/mol (red), 1 kcal/mol $\leq \Delta\Delta G_a < 3$ kcal/mol (white), and $\Delta\Delta G_a < 1$ kcal/mol (blue). Using the second classifier (lower row for each PP interface, Fig. 1.8), residues are labeled according to the ranges $-\Delta\Delta G_{if} \geq 3$ kcal/mol (red), $\frac{1\ \text{kcal}}{\text{mol}} \leq -\Delta\Delta G_{if} < 3$ kcal/mol (white), and $-\Delta\Delta G_{if} < 1$ kcal/mol (blue). The PP interfaces for the following complexes with available alanine scanning data were examined: hGH/hGHbp receptor (3HHR) [22], HIV-1-CD4/GP120 (1GC1) [31], trypsin inhibitor/beta-trypsin (2PTC) [32], colicin E9 immuno-protein/colicin E9 DNase domain (1BXI) [33] and p53/MDM2 (1YCR) [34]. The *correlation between $\Delta\Delta G_a$ and $-\Delta\Delta G_{if}$ classifiers* (*Fig.* 1.8) *enables us to assert* (*P-value* < *10*$^{-5}$) *that protein association is driven by the displacement of frustrated water.*

This section introduced interfacial frustration, a property of water under partial sub-nanoscale confinement. Frustration refers to the inability of water molecules to fulfill their full hydrogen-bonding coordination possibilities. A microstate sampling method determined by a variational principle is implemented to search for the state of minimal interfacial frustration within relaxation times for protein-water dynamic coupling. The identification of this state is essential to delineate the molecular factors that steer biomolecular events.

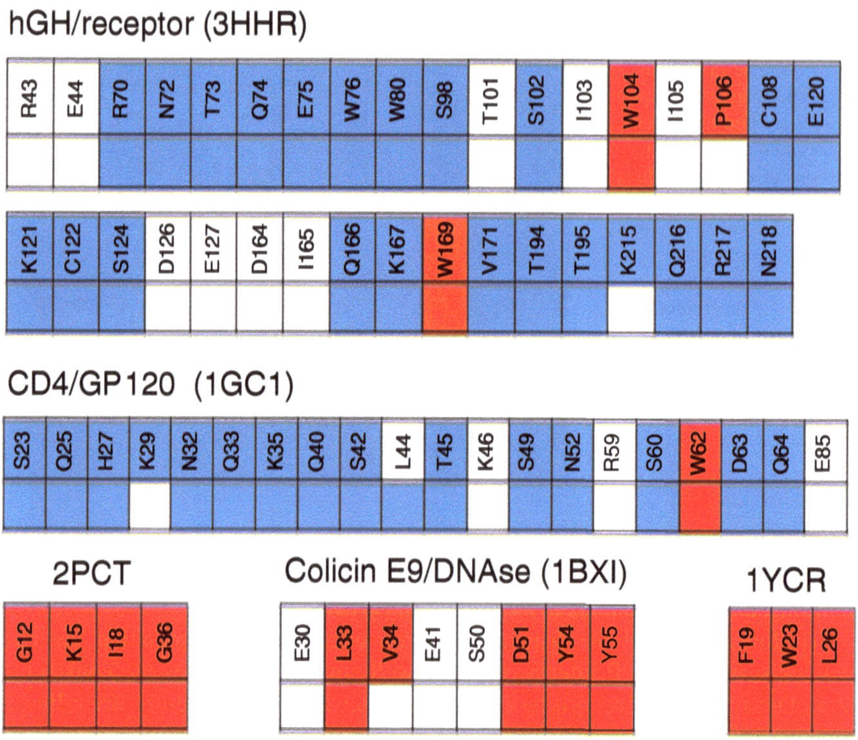

Fig. 1.8 Residue classification for PDB-reported protein complexes with available affinity scanning data according to $\Delta\Delta G_a$ range and, independently, according to $-\Delta\Delta G_{if}$ range. In the *upper row* for each PP interface, the classification is established according to the ranges $\Delta\Delta G_a \geq 3$ kcal/mol (*red*), 1 kcal/mol $\leq \Delta\Delta G_a < 3$ kcal/mol (*white*), and $\Delta\Delta G_a < 1$ kcal/mol (*blue*). The *lower row* informs on the grouping according to the second classifier ($\Delta\Delta G_{if}$), with residues labeled according to the ranges $-\Delta\Delta G_{if} \geq 3$ kcal/mol (*red*), $\frac{1\ \text{kcal}}{\text{mol}} \leq -\Delta\Delta G_{if} < 3$ kcal/mol (*white*), and $-\Delta\Delta G_{if} < 1$ kcal/mol (*blue*)

1.5 The Aqueous Interface from a Structure-Centric Perspective

To complement and enrich the solvent-centric perspective on the epistructural interface provided in the previous sections, we now examine the features in the protein structure that most decisively contribute to determine the interfacial properties, i.e. the dehydrons, from a structure-centric perspective. Since we have identified the dehydron as the structural feature that contributes most decisively to the interfacial free energy, we may at this stage briefly switch from the solvent-centric perspective on biomolecular interfaces to a structure-centric perspective and examine dehydrons within the context of protein structure.

This approach naturally leads to the concept of cooperativity in the assembly of the protein structure, or rather, leads us to regard dehydrons as local indicators of failed cooperativity. Thus, dehydrons are viewed as structural defects that may be "corrected" upon association with protein binding partners or with man-made ligands such as therapeutic drugs. This conceptual framework becomes most efficacious to help improve drug design, as we shall see shortly.

While the biophysical principles governing the affinity of a drug for a target biomolecule are believed to be understood, the control of specificity, the safety, and the idiosyncratic efficacy of the therapeutic agents remain opaque subjects. Clinical uncertainty and unpredictable adverse effects often hamper drug development and this situation is unlikely to change unless a higher level of conceptual innovation is effectively incorporated in the discovery pipeline.

We shall narrow down our treatment of these vast problems to small-molecule drugs purposely engineered to target human proteins and thereby inhibit their biological function. Aiming at a paradigm shift in the field, we advocate a translational top-down approach that takes us back to the very fundamentals of protein associations as we introduce a foundational platform for a next generation of safer and more effective drugs.

In molecular therapy, we often deal with water-soluble proteins that are targeted by man-made ligands, therefore the efficacy and target-specificity of a molecular design depends pivotally on our understanding of protein-ligand associations. In this regard, there is a crucial property that seems have been missed altogether in rational drug design: *cooperativity*. We somewhat narrowly define this property as the concurrent participation of different regions of the biomolecule to promote and sustain intramolecular or intermolecular interactions. In plain terms, "cooperativity is the nonadditive contribution to protein interactions", a peculiar property often illustrated by the phrase "the whole is more than the sum of the parts". In our context of interest, the nonadditive nature of protein associations implies that *the rational drug designer faces a many-body problem*: the interactions between the protein target and the drug/ligand involve more than groups matched up in a pairwise fashion at the target-ligand interface. Because protein-ligand interactions take place in an aqueous medium, this many-body problem is a very special one. As we shall advocate throughout this book, matching groups with complementary biochemical properties across the target/ligand interface is only one aspect of what rational design is about and by no means the decisive one, as the evidence attests. The next generation of molecular designs must take into account modes of association or binding above and beyond pair-wise intermolecular interactions involving groups in the ligand and their purported matched groups in the target.

Be as it may, the current design paradigm is unlikely to change anytime soon unless a clear case can be made for cooperativity, and the right computational tools are brought to fruition to operationally incorporate this concept in drug design. As we deal with cooperativity, perhaps the first core question that needs to be dealt with is: *what sort of many-body problem is the drug designer facing and how can this knowledge play advantageously to address the major therapeutic imperatives of today and tomorrow?*

Protein structure in solution is assumed to arise and be sustained by forces that are essentially electrostatic [35–39]. Even the hydrophobic attraction between two nonpolar groups, an entropic effect arising from the minimization of unfavorable interfaces with water, includes a major electrostatic contribution as it increases the extent of hydrogen bonding among surrounding water molecules [37]. The forces that drive protein folding and protein associations are actually modulated by an important factor often neglected: *the shaping of the solvent microenvironment wherein the forces become operational* [36]. Since the shaping of the microenvironment around a pair-wise intramolecular interaction requires the participation of other regions of the molecule, we may state that cooperativity is inherent to the folding of a protein chain [38–47].

To illustrate the importance of cooperativity, we may recall that an electrostatic interaction occurring in bulk water is 78 times weaker than the same interaction in an anhydrous medium [36, 44]. Thus, the stability and strength of pairwise interactions between different parts of the peptide chain is determined not only by the atomic groups directly engaged in the interaction, but also by the groups involved in shaping their microenvironment by promoting the expulsion of surrounding water [44]. The latter contributors are just as important, as they determine either the persistence or the ephemeral nature of the interactions and, ultimately, the integrity of the protein structure [36]. In fact, low-permittivity microenvironments around the backbone hydrogen bonds of a self-interacting polypeptide chain are essential to promote and sustain its structure and have been the focus of much attention as we attempt to underpin the physical basis of cooperativity [36, 38, 45].

The backbone of a protein or peptide chain is highly polar, comprising an amide and carbonyl group per residue. This chemical feature introduces constraints on the nature of the hydrophobic collapse [43] and on the chain composition of foldable proteins, i.e. those capable of sustaining such a collapse [46, 47]. Thus, the hydrophobic collapse entails the dehydration of backbone amides and carbonyls and such a process would be thermodynamically unfavorable unless amides and carbonyls engage in hydrogen bonding with each other [43]. Only a hydrophobic collapse that ensures the formation and protection of backbone hydrogen bonds is likely to be conducive to sustainable folding [45].

The hydration of amides and carbonyls competes with the formation of the intramolecular hydrogen bonds. Thus, the structural integrity of proteins is compromised by a "deficiently wrapped" backbone [45, 48]. Wrapping refers to a cluster of nonpolar groups around a pre-formed coulombic interaction [47]. The need for protection of intramolecular hydrogen bonds from water attack is an important factor in determining the chain composition of a *foldable* protein, that is, of a chain capable of sustaining a soluble structure and folding expeditiously and reproducibly [45].

As noted above, the strength and stability of backbone hydrogen bonds clearly depend on the microenvironment where they occur: The proximity of nonpolar groups to a hydrogen bond enhances the electrostatic interaction by de-screening the partial charges or lowering the local environment permittivity [44, 46]. These nonpolar groups also stabilize the hydrogen bond by destabilizing the nonbonded state, i.e. by hindering the hydration of the polar groups in the nonbonded state [46, 47]. Thus, to guarantee the integrity of soluble protein structure, most intramolecular hydrogen bonds must be surrounded or "*wrapped*" by nonpolar groups fairly thoroughly as to become significantly dehydrated [45–48].

To make the wrapping concept more precise, we need a definition that enables a direct assessment of the extent of hydrogen bond protection from structure coordinates. This parameter, denoted ρ, is given by the number of side-chain carbonaceous nonpolar groups (CH_n, n = 0, 1, 2, 3) contained within a desolvation domain that represents the hydrogen-bond microenvironment. This domain is defined as the reunion of two intersecting spheres of fixed radius (~thickness of three water layers) centered at the α-carbons of the residues paired by the hydrogen bond. In structures of PDB-reported soluble proteins, backbone hydrogen bonds are protected on average by $\rho = 26.6 \pm 7.5$ side-chain nonpolar groups for a desolvation sphere of radius r = 6 Å. The desolvation domain adopted for a hydrogen bond is a residue-based feature, incorporating a descriptor of the local environment of each of the paired residues [47]. It fully subsumes the local environment of the hydrogen bond itself since the heavy atoms N and O are invariably within 6 Å of the α-carbons of the paired residues and hence fully contained in the intersection of the desolvation spheres.

Accordingly, if we treat the solvent implicitly, we may identify a *wrapping interaction* as a three-body *ijk*-interaction, where nonpolar group *k* contributes to the removal of water from the microenvironment of an amide-carbonyl hydrogen bond that pairs groups *i* and *j*. In other words, the "wrapping" nonpolar group *k* lies within the desolvation domain of the *ij*-interaction. This type of three-body contribution is cooperative by definition and needed to maintain the structural integrity of the protein [36, 47]. An illustration of wrapping interactions in the native structure of human *ubiquitin* (PDB accession code PDB.1UBI) is given in Fig. 1.9. Thus, the wrapping of the Lys27-Gln31 backbone hydrogen bond by residues Ile36 and Lys29 is represented by thin blue lines. These residues contribute four and three nonpolar groups respectively to the desolvation domain (r = 6 Å) of the backbone hydrogen bond.

Taken together, the hydration propensity of amide and carbonyl and the dehydration-induced strengthening of their electrostatic association represent two conflictive tendencies, suggesting that there must be a crossover point in the dehydration propensity of a backbone hydrogen bond [49]. If the bond is poorly wrapped by a few nonpolar groups that cluster around it, then hydration of the paired amide and carbonyl is favored and prevails over a structured region, but as the hydrogen bond becomes better wrapped, the surrounding water loses too many hydrogen bonding partnerships and thus may be favorably removed [47]. This observation is essential to rationalize the cooperative two-state nature of the folding of single-domain proteins [50–52], as shown in Chap. 3: we may say that the state of hydration of a protein hydrogen bond is in a statistical sense a local reflection of the degree of progress of the folding process.

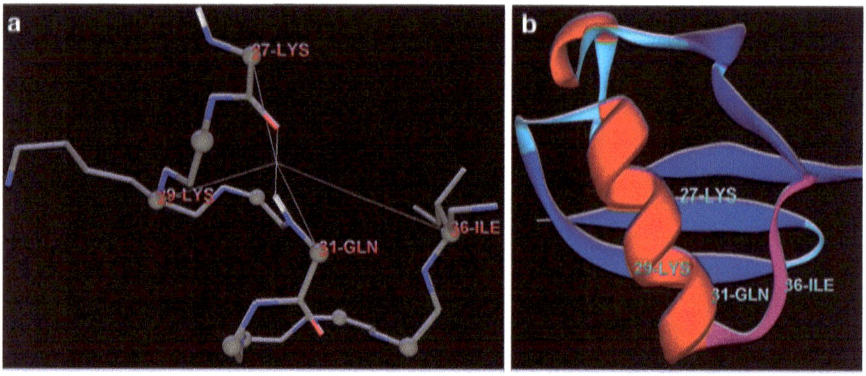

Fig. 1.9 (**a**) Cooperative "wrapping" interactions involving residues Lys29 and Ile36 that contribute to stabilize and enhance the backbone hydrogen bond Lys27-Gln31 (*thin white line*) in human ubiquitin (PDB.1UBI). Each interaction generates a three-body energy term and a residue contributing with nonpolar side-chain groups to the desolvation domain of the hydrogen bond is represented by a *thin blue line* joining the α-carbon (*grey sphere*) of the residue with the center of the amide-carbonyl hydrogen bond. Conventional colors are used for atom representation and the protein backbone is represented schematically, except for the two residues paired by the backbone hydrogen bond that are displayed in full backbone detail. Only the side chains of the wrapping residues 29Lys and 36Ile are shown. (**b**) Location of the residues in a ribbon rendering of the native structure of human ubiquitin

1.6 Dehydrons Promote Protein Associations: A Structural Perspective

The structural integrity of a soluble protein is contingent on its ability to exclude water from its amide-carbonyl hydrogen bonds [45, 46]. Thus, water-exposed intramolecular hydrogen bonds, the so-called *dehydrons*, constitute structural weaknesses taking the particular form of wrapping deficiencies [46, 53, 54]. On the other hand, these defects favor the removal of surrounding water as a means to strengthen and stabilize the underlying electrostatic interaction [47, 54, 55], and thus are predictably implicated in protein associations [55], aberrant aggregation [56] and macromolecular recognition [57, 58]. By exogenously contributing to the wrapping of pre-formed hydrogen bonds, these associations in effect remove the wrapping defects, thereby stabilizing the structure.

Intramolecular hydrogen bonds that are not "wrapped" by a sufficient number of nonpolar groups in the protein itself may become stabilized and strengthened by the attachment of a ligand, i.e. a potential drug, or a binding partner that further contributes to their dehydration (Fig. 1.10) [53, 55]. Ample bioinformatics evidence on the distribution of dehydrons at the interface of protein complexes support this physical picture [47, 55]. Thus, dehydrons are decisive factors driving association in 38 % of the PDB complexes and constitute significant factors (interfacial dehydron density larger than average on individual partners) in about 95 % of all complexes reported in the PDB [47].

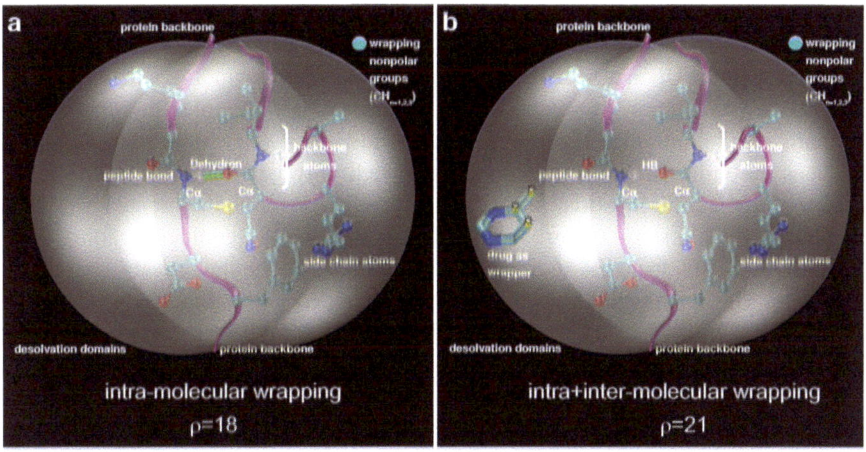

Fig. 1.10 (**a**) Dehydron in a soluble protein. The dehydron ($\rho = 18$), marked in *green*, pairs two backbone groups (amide and carbonyl, conventional colors for atoms). The microenvironment is indicated by *two intersecting grey spheres* centred at the α-carbons of the paired residues. Wrapping side-chain groups are shown in *light blue* and only side chains contributing (fully or partially) to the dehydration of the hydrogen bond are indicated. (**b**) The drug depicted in the figure acts as an exogenous wrapper of the hydrogen bond (*grey bond*, $\rho = 21$) turning the dehydron into a well-protected bond (the three atoms marked with *asterisk* complete the desolvation of the dehydron)

Dehydrons may be identified from atomic coordinates of proteins with reported structure, as illustrated in Fig. 1.10. Thus, we need to introduce an auxiliary quantity, the extent of hydrogen-bond wrapping, ρ, indicating the number of nonpolar groups contained within a "desolvation domain" around the bond. In structures of soluble proteins at least two thirds of the backbone hydrogen bonds are wrapped on average by $\rho = 26.6 \pm 7.5$ nonpolar groups for a desolvation ball radius 6 Å. Dehydrons lie in the tails of the distribution, i.e. their microenvironment contains 19 or fewer nonpolar groups, so their ρ-value is below the mean ($\rho = 26.6$) minus one standard deviation ($\sigma = 7.5$) [46, 47].

Thus, the overall under-wrapping of a protein may be assessed by determining the percentage of intramolecular hydrogen bonds satisfying the inequality $\rho \leq 19$, that is, the percentage of dehydrons in its structure. An example of the under-wrapping of a protein is given in Fig. 1.2c, where the dehydron pattern of human ubiquitin is displayed.

Dehydron-rich regions in soluble proteins are typical hot spots for protein associations because of their propensity towards further dehydration [47, 55]. A functional perspective reinforces this view, since dehydrons constitute vulnerabilities that need to be "corrected" to maintain the integrity of the protein structure and its functional competence. Thus, specific residues of the binding partner contribute to the desolvation of dehydrons, as they penetrate the desolvation domain of intramolecular hydrogen bonds upon association. This intermolecular wrapping is illustrated in Fig. 1.11a–c, displaying the functionally competent human HIV-1 protease dimer. The purported interfacial region of the homodimer contains seven dehydrons when the monomeric domains are considered in isolation and separated from each other. Upon association, specific residues of the binding partner contribute to the desolvation of some of the dehydrons, as they enter the desolvation domain of the intramolecular hydrogen bonds. This inter-molecular wrapping reduces the vulnerability of the protease, which only has three dehydrons at the interface upon complexation, instead of the original seven in the monomeric form. Similarly, the allosteric clam haemoglobin, a functionally competent homodimer utilizes five dehydrons of the monomeric state to accommodate its quaternary structure. These five dehydrons become well wrapped in the dimer (Fig. 1.11d).

Throughout the book, dehydrons will be referred to in different ways depending on the context. Thus, the terms *packing defect*, *wrapping defect*, *dehydron*,

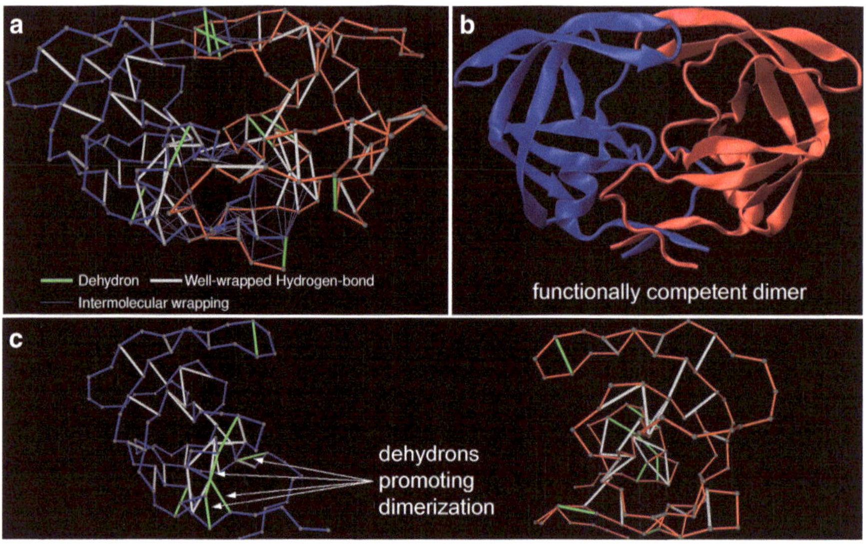

Fig. 1.11 (**a**) Inter-molecular wrapping in the human HIV-1 protease dimer (PDB.1A30) as a means of protecting the enzyme structure from water attack. Dehydrons are indicated as *green segments* joining the alpha-carbons of the paired units, well-wrapped hydrogen bonds are shown in *light grey*, and the protein backbone is conventionally shown as virtual bonds joining the alpha-carbons of consecutive amino acid units. Complexation reduces the structural vulnerability of the protease by reducing the number of dehydrons at the interface from 7 to 3. Intermolecular wrapping is depicted by *thin blue lines* from the alpha-carbon of the wrapping residue to the middle of the hydrogen bond that is intermolecularly protected. Thus, an intermolecularly wrapping residue contributes with nonpolar groups to the dehydration of the pre-formed hydrogen bond from the binding partner. (**b**) Ribbon display of the dimer. (**c**) Wrapping of separate domains of the human HIV-1 protease. (**d**) Allosteric clam haemoglobin, a functionally competent homodimer that utilizes five dehydrons (marked by *asterisks* in *lower panel*) of the monomeric state to create a favorable interface for its quaternary structure. The five dehydrons become well wrapped in the dimeric state (*upper panel*). The wrapping pattern of one monomer is displayed while the other monomer is shown in ribbon representation

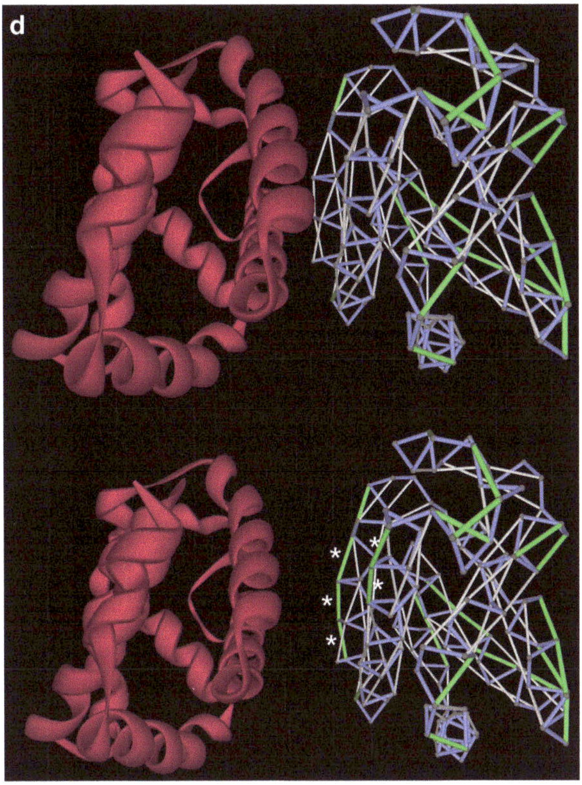

Fig. 1.11 (continued)

structural deficiency, *structural vulnerability*, will be used synonymously. Far from introducing a notational chaos, this name multiplicity bespeaks of the richness of the concept.

1.7 Epistructural Physics of the Protein Interface

As indicated above, dehydrons have unique physico-chemical properties: they represent structural vulnerabilities of the protein, but they also constitute sticky spots promoting the removal of surrounding water [46–48, 53, 55]. This latter property could only be established by addressing the following questions: How do we effectively demonstrate that a dehydron attracts nonpolar test groups? Can we measure the mechanical equivalent of its dehydration propensity?

Reported experimental work addressed these questions by measuring the adsorption of proteins with wrapping defects and equivalent surface hydrophobicity (area of solvent-exposed nonpolar surface) onto a "wrapping" layer [53]. This wrapping medium consists of a Langmuir-Blodgett phospholipid film coating a waveguide, as

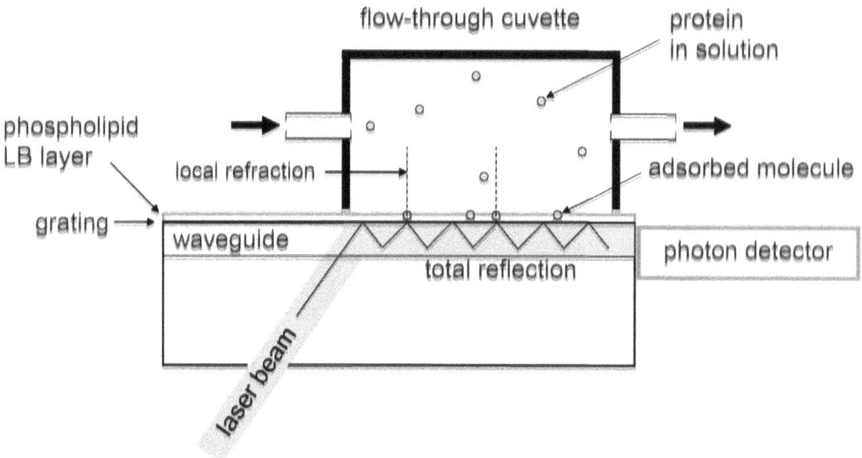

Fig. 1.12 High-precision total reflection set-up to measure the adsorption uptake of proteins onto a hydrophobic surface under controlled hydrodynamic conditions. The adsorption uptake is proportional to the photon loss due to local alterations in the refractive index of a Langmuir-Blodget (LB) layer that constitutes the wrapping medium for the protein

shown in Fig. 1.12. These high-precision experiments made use of evanescent-field spectroscopic interrogation of the wrapping medium enabling a direct measurement of the protein adsorption uptake. This observable is determined by detecting local changes in refractive index of the phase within which total reflection of the incident light occurred [53]. Thus, a beam from a He-Ne laser travels through a waveguide at an incidence angle suitable for total reflection within the medium. The adsorbed molecules alter the refractive properties of the hydrophobic layer and consequently alter the critical angle for total reflection. Thus, protein adsorption is commensurate with photon loss resulting from the extent of local refraction or "evanescent field". Hence adsorption uptake can be determined by the loss of photons due to refractive photon leakage from the total reflection pattern.

For proteins with comparable surface hydrophobicity, the adsorption uptake correlates strongly with the extent of protein under-wrapping [53]. As an adequate control, only proteins with the same extent of surface hydrophobicity or solvent-exposed nonpolar area were included in the comparative analysis. Hence, the attractive drag exerted by dehydrons on test hydrophobes became accessible. The net gain in Coulomb energy associated with wrapping a dehydron has been experimentally determined to be ~4 kJ/mol [53]. The adhesive force exerted by a dehydron on a hydrophobe at 6 Å distance is ~7.8 pN, a magnitude comparable to the hydrophobic attraction between two nonpolar moieties that frame unfavorable interfaces with water.

This study was motivated by the earlier observations that dehydrons play a pivotal role in driving protein associations, as such associations contribute intermolecularly to the wrapping of pre-formed structure [46, 48], as discussed in the previous section. In consistency with current terminology, the force stemming from the dehydration propensity of the partially wrapped hydrogen bond is hithertofore termed *dehydronic*. The dehydronic force arises as a nonpolar group approaches a dehydron with a net effect of immobilizing and ultimately removing surrounding water molecules. This displacement lowers the polarizability of the microenvironment which, in turn, de-shields the paired charges [46, 53]. Thus, a net attractive force is exerted by the dehydron on a nonpolar group and this force represents the mechanical equivalent of the dehydration propensity of an unburied preformed hydrogen bond. Since the water molecules solvating an amide and carbonyl paired by a dehydron are necessarily depleted of some hydrogen-bonding partners, the work required for their ultimate removal from the bond surroundings is minimal [46, 56]. The dehydronic field, denoted $\vec{\Phi}(\vec{R})$, is necessarily orthogonal to the Coulomb field generated by the polar (amide-carbonyl) pair, and may be described within a quasi-continuous treatment of the solvent by the equation:

$$\vec{\Phi}(\vec{R}) = -\vec{\nabla}_{\vec{R}}\left[4\pi\varepsilon(\vec{R})\right]^{-1} qq'/r \qquad (1.12)$$

where $\vec{R}$ represents the position vector of the hydrophobe or nonpolar group with respect to the center of mass of the hydrogen-bonded polar pair, $\vec{\nabla}_{\vec{R}}$ is the gradient taken with respect to this vector, r is the distance between the charges of magnitude q and q′ paired by the hydrogen bond (Fig. 1.13a), and the local permittivity coefficient $\varepsilon(\vec{R})$ subsumes the polarizability of the microenvironment, which is generically dependent on the position of the test hydrophobe [36, 48]. An appropriate expression for $\varepsilon(\vec{R})$, valid at nanoscales, is unavailable at present [59]. This is because of the discreteness of the dielectric medium and the need to include individual solvent dipole correlations [60]. Although a mean-field dielectric description is unsatisfactory, it is still possible to assert that $\vec{\Phi}(\vec{R})$ is an attractive force since a decrease in $||\vec{R}||$ entails a decrease in local polarization which, in turn, enhances the Coulomb attraction.

Building on this analysis, we may quantify the net hydrophobicity η of a hydrogen bond by taking into account the surface flux of the dehydronic field generated by the hydrogen bond. This field is given by $\omega^{-1}\vec{\Phi}(\vec{R})$ (ω = volume of test hydrophobe). Thus, in accord with Gauss theorem we obtain:

$$\eta = \oiint \vec{\Phi}(\vec{R}).d\vec{\sigma}(\vec{R}) \qquad (1.13)$$

where integration extends over the closed surface of the dehydration domain of the hydrogen bond (cf. Fig. 1.10) and $d\vec{\sigma}(\vec{R})$ is the differential surface area vector.

To fully understand dehydron stickiness, it is useful at this stage to switch back to the solvent-centric view of the protein-water interface. The object we need to focus on is the 4 Å-radius solvation shell of the atom closest to the center of a backbone hydrogen bond and containing at least one water oxygen in the shell. The

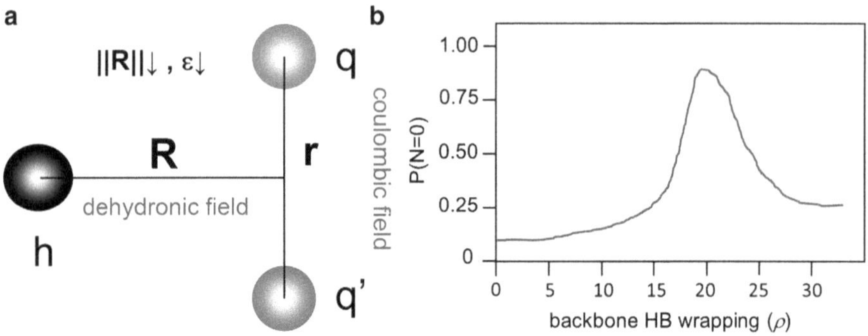

Fig. 1.13 (**a**) Orthogonality between *dehydronic* field exerted on the test hydrophobe (h) along coordinate **R** and the coulomb field exerted along coordinate **r** between two spherical charges q, q′. (**b**) Water-vacating probability of nearest water-occupied solvation shell (P(N = 0)) versus wrapping (ρ) of backbone hydrogen bond for an ensemble of soluble proteins (main text), signaling the dehydron stickiness and the role of dehydrons as promoters of protein associations

vacating probability P(N = 0) previously defined represents also a measure of the propensity for water exclusion from the protein surface, itself a local indicator of stickiness. The plot of P(N = 0) for the atom closest to a backbone hydrogen bond versus the wrapping of the backbone hydrogen bond (ρ) is displayed in Fig. 1.13b and obtained from 11,964 monomeric soluble proteins with PDB-reported structure (see Sect. 1.9). The plot reveals that dehydrons are indeed the most significant promoters of water exclusion of all structured regions (i.e. with hydrogen-bonded backbone) in the soluble protein.

1.8 Statistical Mechanics Apparatus for Drug-Target Associations

The analysis of biomolecular interfaces presented so far suggests that, like in any protein-ligand complex, the wrapping intermolecular interactions between a drug and its target protein are expected to play an important role in determining the affinity. As indicated in Sect. 1.7, such cooperative interactions between the drug and pre-formed hydrogen bonds in the target protein actually entail a net gain to the stability of the protein of ~4 kJ/mol per wrapping interaction. Thus, the drug contributes with nonpolar groups to the desolvation of pre-existing dehydrons in the target and each wrapping contribution translates into a ~4 kJ/mol decrease in the free energy change associated with drug binding. As an illustration, the wrapping contributions between the powerful anticancer drug imatinib (Gleevec®) and one of its primary therapeutic targets, the KIT kinase [61, 62], are depicted in Fig. 1.14. The wrapping interactions, extensively discussed in Chap. 9, are promoted by the dehydronic fields generated by the deficiently wrapped hydrogen bonds 673Cys-676Gly, 595Leu-603Val, 640Glu-644Leu and 811Phe-814Ala.

The kinase inhibitor imatinib was not *purposely* designed as a wrapper of its purported protein targets, yet, as shown in Fig. 1.14, five wrapping interactions contribute to its affinity for the primary target KIT kinase. A crude estimate would indicate that this contribution to the association free energy is actually significant

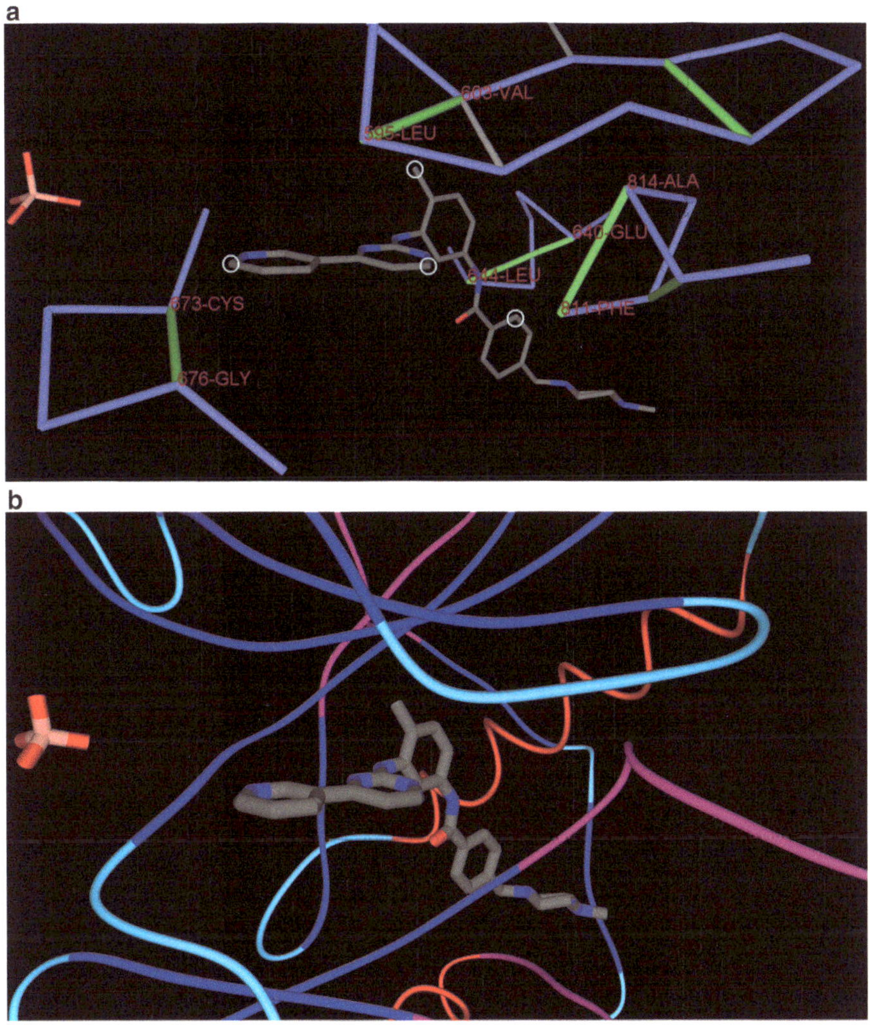

Fig. 1.14 (**a**) Dehydrons in the KIT kinase wrapped intermolecularly by the kinase inhibitor imatinib in the crystallized drug/target complex (PDB.1T46). The drug nonpolar groups contributing to the wrapping upon association are marked by circles. The dehydrons (*green*) wrapped by the drug involve residue pairs 673Cys-676Gly, 595Leu-603Val, 640Glu-644Leu and 811Phe-814Ala. (**b**) Simplified tube rendering of the protein backbone provided as visual aid. (**c**) Detail of intermolecular wrapping interaction between imatinib and the KIT kinase. The drug penetrates the desolvation domain (*intersecting pink spheres*) of KIT dehydron 640Glu-644Leu upon binding, contributing with two nonpolar groups to the desolvation of the pre-formed hydrogen bond

c

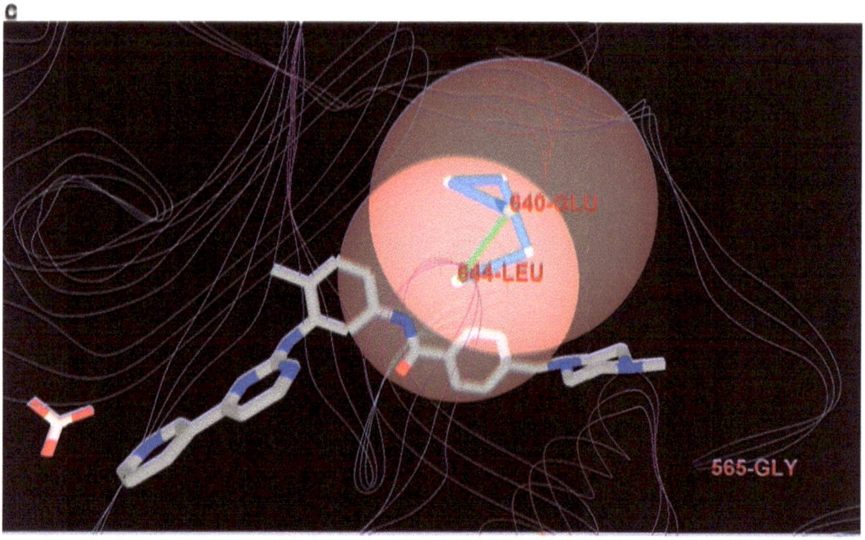

Fig. 1.14 (continued)

(~5 × 4 kJ/mol = 20 kJ/mol). Notice that while imatinib was selected for its pharmacological properties that conferred anticancer activity [61], it is unlikely to have emerged from a conventional structure-based design. Matching a nonpolar group in the drug against a carbonyl or amide in the target protein is hardly conceivable in conventional structure-based design as it would entail a hindrance to the hydration of the polar groups. Yet, as a three-body contribution, the proximity of a nonpolar group to an unburied polar pair is justified given the dehydronic field generated by the latter. Furthermore, given the extent to which wrapping interactions are likely to affect drug affinity, one wonders whether the wrapping concept can be fruitfully exploited to guide drug design.

This possibility naturally elicits some basic questions:

- Can we purposely design "cooperative drugs" that wrap dehydrons and thereby mitigate protein-water interfacial tension in the protein target?
- What is the potential impact of "epistructural design" on the technological base of drug discovery and what are the advantages it is likely to provide when compared with conventional structure-based design?
- Can this type of cooperativity-based design be integrated into an "epistructural bioinformatics platform" (cf. Chap. 10), leading to a better exploitation of genomic information to develop safer and more efficacious therapeutic agents?
- Can we harness idiosyncratic differences in the wrapping patterns of gene products across patients to develop personalized treatments?

These are all crucial problems in molecular targeted therapy that will be addressed in the subsequent chapters as we assess the transformative power of the wrapping concept through a careful examination of biomolecular interfaces.

1.9 The Principles of Epistructural Physics Translate into the Architecture of Soluble Proteins

This section introduces a basic design principle that can be rationalized through an analogy. Just like defiance of gravity in building engineering requires buttressing to preserve the integrity of the building, protein design allowing for backbone hydration (the force counteracting structural cohesion) requires disulfide bridges to maintain the structural integrity of the protein. In this regard, this section unravels two constants that define a fundamental architectural constraint in soluble proteins: after suitable normalization, a single disulfide bond stabilizes *five* structural deficiencies, and every soluble protein has a *20 %*-baseline level of structural deficiency. These constants define a design principle.

The baseline structural deficiency $Y = 20$ represents the maximum of a tight distribution (standard deviation $\sigma = 2.25$) of Y-values for the structural deficiency of soluble proteins with no disulfide bridges. This baseline Y-value implies that soluble proteins are not perfectly packed and maintain at least 20 % of unburied backbone hydrogen bonds. Since such structural deficiencies locally promote backbone hydration, they belong to an intermediate region between order and disorder and hence represent markers of structural flexibility. Thus, because of its universality, the $Y = 20$ constant may be interpreted as the baseline flexibility needed for protein function.

In this section, dehydrons are characterized purely as structural deficiencies [55]. These deficiencies are of a special kind: they are promoters of backbone hydration and hence, destabilizers of the native structure. On the other hand, disulfide bonds preformed in the denatured state reduce the structure-destabilizing conformational entropy cost associated with the folding process, hence stabilizing the native structure. Thus, it should be generally expected that the two major and opposite contributors to native-structure destabilization would be correlated, as revealed by the balance equation described in this section.

This relation is hence likely to assist the molecular engineering of soluble proteins. Furthermore, since disulfide bridges can be formed or dismantled in accord with redox environmental conditions, the relation presented is likely to enable the type of design fine tuning that may be required for an environmental modulation of the protein function.

As the dehydron/disulfide organizational principle is established and shown to hold for soluble proteins, we cannot fail to notice that it also introduces a new set of problems stemming from a basic question: *what is in physical terms the fate of a soluble protein whose structure significantly violates the architectural constraints defined by the balance equation?* This issue will be explored in Chap. 5.

Wrapping defects in soluble proteins represent local weaknesses of the native structures and have received little attention, especially by the drug design community. The protein structure may be inherently weak at sites where hydration of the backbone is locally hampered by formation of an intramolecular hydrogen bond which in turn is not stabilized through complete burial within a hydrophobic environment. This section explores the architectural implications stemming from the existence of these vulnerabilities. Thus, the unburied backbone hydrogen bonds or *dehydrons*, are shown to be compensated by disulfide bridges that are needed to maintain the structural integrity in extracellular environments. Examination of all reported soluble structures reveals that the number of disulfide bonds correlates tightly with the number of dehydrons in a 1:5 ratio. The results have implications for biomolecular design as they introduce universal constraints in the architecture of water-soluble proteins.

Backbone hydration, prevalent in the unfolded state of a polypeptide chain, is often hindered in a soluble folded state as backbone amides and carbonyls are paired through hydrogen bonds [63, 64]. Yet, the thermodynamic cost of dehydration is not always compensated, especially if the backbone hydrogen bond is not completely sequestered from solvent. In soluble proteins, such bonds may be readily identified from the structural coordinates by determining the number of nonpolar groups within the bond microenvironment [46, 47, 55]. These unburied backbone hydrogen bonds constitute structural deficiencies and represent markers for protein associations [55]. In turn, these associations are required to maintain the structural integrity of the protein through intermolecular protection of the pre-formed hydrogen bonds.

A thorough examination of the protein data bank (PDB) singles out toxin peptides with picomolar affinity for the Kv1.3 potassium channel [65], such as HsTX1 (PDB.1QUZ), as members of the protein family with the highest extent of structural deficiency. The unburied hydrogen bonds in such proteins can make up to 100 % of the backbone hydrogen bonds. This observation immediately suggests a reason for the extremely high target affinity of neurotoxins: we expect a huge *dehydronic field* for such biomolecules.

A separate analysis reveals that such proteins contain an inordinately large number of disulfide bonds, with an average of 11 when normalized to 100 amino acids. These observations prompt us to investigate the relation between structural deficiency and disulfide bonds in search for a balance equation that reflects a statistical correlation between structural strengths and vulnerabilities of soluble proteins and polypeptides. The balance equation unraveled in this chapter is likely to impact the design of soluble proteins and enable a better control of their functional modulation in relation to environmental redox conditions.

A comprehensive wrapping analysis of the PDB yields the structural deficiencies of an exhaustive nonredundant set of 2989 monomeric uncomplexed soluble proteins or peptides with disulfide links and 8975 proteins without disulfide links [66]. These data is compiled as illustrated in Table 1.1 for some selected PDB entries. Peptide chains were excluded from the analysis if their structural integrity required prosthetic groups or cation coordination. In order to compare protein or peptide chains of different lengths, two normalized parameters were used to characterize a protein structure: $Y =$ number of structural deficiencies per 100 backbone hydrogen bonds, and $X =$ number of disulfide bonds per 100 amino acids. Proteins were binned according to their X-value in integer groups with $n = 0, 1, \ldots,$ 18, where proteins with no disulfide bonds ($X = 0$) belong to group $n = 0$ and

Table 1.1 Structural parameters for few selected monomeric uncomplexed soluble proteins with disulfide bridges lacking prosthetic groups and scaffolding cation coordination

PDB entry	Disulfide bonds	Chain length	Structural deficiencies	Backbone HBs	X	Y	Non alpha/beta residues (%)
2PNE	2	81	37	37	2.47	100.00	100.00
1M4F	4	25	8	8	16.00	100.00	68.00
1EZG	8	81	33	48	9.87	68.75	77.78
1HOE	2	74	11	34	2.70	32.35	54.05
2Z9T	1	100	10	40	1.00	25.00	50.00
135L	4	129	35	109	3.10	32.11	63.57
153L	2	185	38	185	1.08	20.54	48.65
154L	2	185	39	194	1.08	20.10	48.65
172L	1	164	47	188	0.61	25.00	30.49
1A2J	1	189	57	196	0.53	29.08	39.15
1A39	9	402	69	250	2.24	27.60	59.70
1A3P	2	45	9	18	4.44	50.00	82.22
1A43	1	87	22	67	1.15	32.84	49.43
1A67	2	108	20	62	1.85	32.26	48.15
1A7M	3	180	37	155	1.67	23.87	41.67
1AC5	3	483	91	370	0.62	24.59	51.76
1ACJ	3	537	77	404	0.56	19.06	50.09
1ACW	3	29	18	18	10.34	100.00	41.38
1ACX	2	108	20	44	1.85	45.45	56.48
1ADX	3	40	5	7	7.50	71.43	100.00
1ADZ	3	71	11	35	4.23	31.43	76.06
1AE5	4	225	33	123	1.78	26.83	62.22
1AEC	3	218	33	155	1.38	21.29	55.96
1AFH	4	93	39	74	4.30	52.70	46.24
1AG2	1	103	15	80	0.97	18.75	43.69
1AGG	4	48	11	14	8.33	78.57	87.50
1AGI	3	125	18	82	2.40	21.95	48.80
1AGY	2	200	43	158	1.00	27.22	51.50
1AH1	2	129	15	51	1.55	29.41	60.47
1AHK	3	129	13	43	2.33	30.23	80.62
1AHL	3	49	6	14	6.12	42.86	87.76
3DHM	1	100	10	50	1.00	20.00	52.00
3DIH	7	122	35	97	5.74	36.08	52.46
3EGP	1	108	8	37	0.93	21.62	58.33
3EHS	3	476	93	359	0.63	25.91	40.34
3EMY	1	329	44	198	0.30	22.22	49.85
3ENG	7	213	34	119	3.29	28.57	66.67
3EO5	1	171	47	125	0.58	37.60	48.54
3EOW	2	221	21	75	0.90	28.00	70.14
3ETP	2	187	22	89	1.07	24.72	51.87
3EXD	4	129	35	113	3.10	30.97	63.57

(continued)

Table 1.1 (continued)

PDB entry	Disulfide bonds	Chain length	Structural deficiencies	Backbone HBs	X	Y	Non alpha/beta residues (%)
3EZM	2	101	31	57	1.98	54.39	42.57
3GF1	3	70	25	53	4.29	47.17	77.14
3LYM	4	129	36	109	3.10	33.03	63.57
3LZ2	4	129	32	91	3.10	35.16	64.34
3MAN	1	302	58	252	0.33	23.02	50.99
3PTE	1	349	70	297	0.29	23.57	50.43
3RAT	4	124	28	81	3.23	34.57	49.19
3RSD	4	124	25	78	3.23	32.05	49.19
3SEB	1	238	27	153	0.42	17.65	50.00
3SSI	2	113	26	69	1.77	37.68	59.29
3TGF	3	50	13	34	6.00	38.24	68.00
3TGL	3	269	42	221	1.12	19.00	49.81
4AIT	2	74	13	35	2.70	37.14	54.05
4APE	1	330	49	195	0.30	25.13	47.88
4CMS	3	323	53	203	0.93	26.11	42.72
4ENG	7	210	34	119	3.33	28.57	67.14
4RAT	4	124	29	82	3.23	35.37	49.19
4TGL	3	269	36	204	1.12	17.65	50.19
4TSV	1	150	15	70	0.67	21.43	50.67
5LYT	4	129	32	102	3.10	31.37	62.79
5LYZ	4	129	29	90	3.10	32.22	64.34
5PEP	3	326	51	200	0.92	25.50	46.01
5RAT	4	124	28	79	3.23	35.44	49.19
5RNT	2	104	21	62	1.92	33.87	64.42
6HIR	3	65	10	13	4.62	76.92	90.77
6LYT	4	129	32	103	3.10	31.07	62.79
6LYZ	4	129	27	95	3.10	28.42	62.79
7LYZ	4	129	24	88	3.10	27.27	62.79
7RAT	4	124	28	81	3.23	34.57	49.19
8PTI	3	58	10	29	5.17	34.48	65.52
8RAT	4	124	25	76	3.23	32.89	49.19
9RAT	4	124	28	76	3.23	36.84	49.19

proteins with X in the range $n < X \leq n + 1$ belonged to group n + 1. The mean Y-value and standard deviation were computed for each n-group and the results are shown in Fig. 1.15a. A tight X-Y linear correlation ($R^2 = 0.96$) results and is further corroborated by the raw X-Y-trendline generated by linear regression on all (X, Y) data points (Fig. 1.15b). Taken together, the results from Fig. 1.15 unambiguously reveal a simple balance relation $Y = 5X + 20$. *This statistical relation introduces a 1:5 ratio to buttress vulnerable proteins and allows for a 20 %-baseline in structural deficiency.*

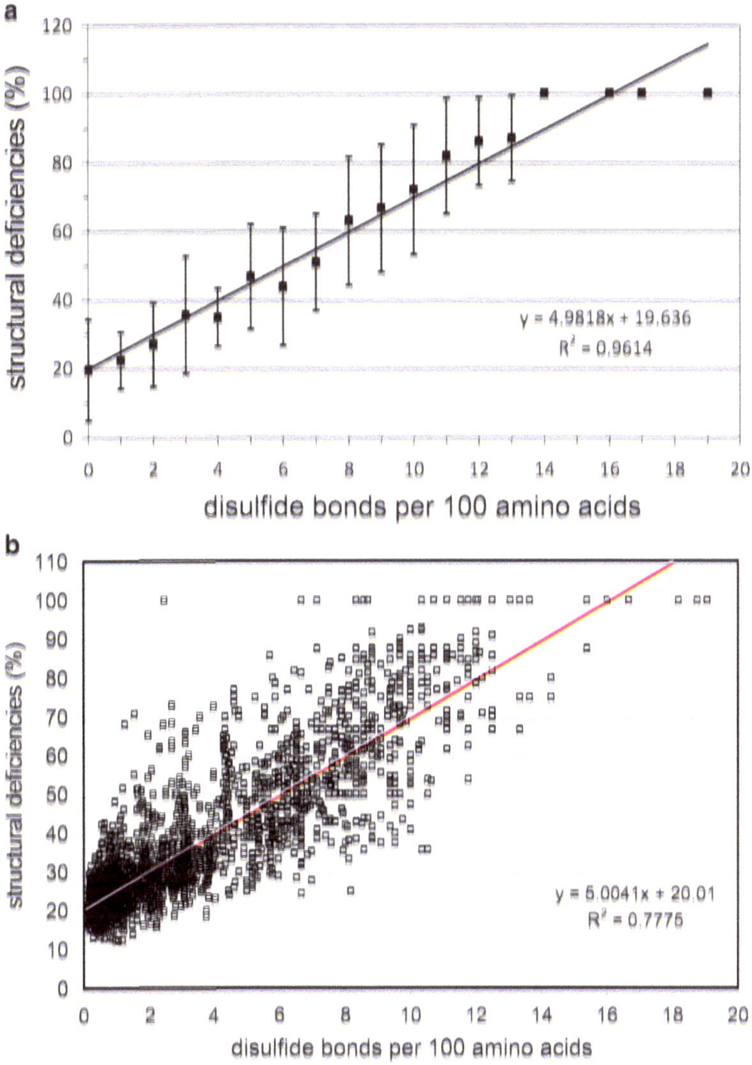

Fig. 1.15 Number of structural deficiencies (unburied backbone hydrogen bonds) normalized to 100 backbone hydrogen bonds (Y) plotted against number of disulfide bonds normalized to 100 amino acids (X) for PDB-reported soluble proteins. (**a**) Mean Y-value (*square*) and standard deviation (error bar) for proteins grouped according to their number of disulfide bonds. Proteins were binned according to their X-value in integer groups with $n = 0, 1, \ldots, 18$, where proteins with no disulfide bonds ($X = 0$) belong to group $n = 0$ and proteins with $n < X \leq n + 1$ belong to group $n + 1$. (**b**) All (X, Y)-data points from the nonredundant exhaustive set of PDB-entries for uncomplexed soluble proteins. Reprinted from [66], copyright 2009 with permission from the American Chemical Society

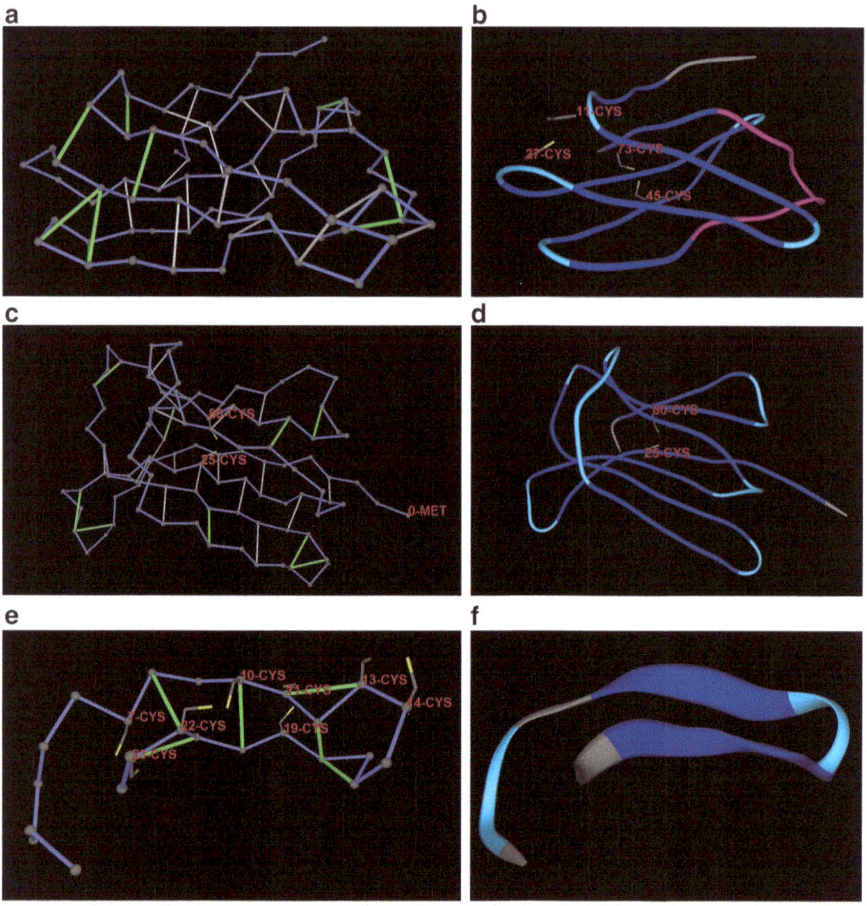

Fig. 1.16 Structural deficiencies in soluble proteins. The protein backbone is shown as virtual bonds (*blue*) joining consecutive α-carbons in the peptide chain. *Light-grey* segments joining α-carbons represent completely buried backbone hydrogen bonds, and *green* segments represent structural deficiencies (unburied backbone hydrogen bonds). A tube/ribbon representation is added for visual aid. Cysteines involved in disulfide bonds are identified by side-chain display. Structural deficiencies (**a**, **c**, **e**) and tube/ribbon representation (**b**, **d**, **f**) respectively of alpha-amylase inhibitor HOE-467A (PDB.1HOE) (**a**, **b**), beta-2 microglobulin (PDB.2Z9T) (**c**, **d**) and antimicrobial hormone hepcidin (PDB.1M4F) (**e**, **f**). Reprinted from [66], copyright 2009 with permission from the American Chemical Society

To illustrate the role of this relation in defining protein structure we focus on specific examples of proteins with widely diverse levels of structural deficiency. Figure 1.16a, b shows the structural deficiencies of the alpha-amylase inhibitor HOE-467A (PDB.1HOE; length N = 74; two disulfide bonds). This protein contains 23 fully buried backbone hydrogen bonds (5 are double bonds, with each paired residue contributing both proton donor and acceptor, and 13 are single

Table 1.2 Thermodynamic and structural parameters of soluble proteins

PDB entry	Y − (5X + 20)	ΔG (kcal/mol)	T (C)	pH	Reference
1BSQ	−9.83	11.10	40.00	7.00	INT J BIOL MACROMOL 38, 9–17 (2006)
1RTB	−5.00	10.10	25.00	8.40	BIOPHYS CHEM 127, 51–63 (2007)
4LYZ	−3.30	9.02	26.85	7.00	BIOPOLYMERS 85, 264–273 (2007)
1CX1	−3.26	5.38	24.85	7.09	BIOCHEMISTRY 37, 3529–3537 (1998)
1QG5	1.13	8.80	40.00	7.00	INT J BIOL MACROMOL 38, 9–17 (2006)
2AIT	2.39	6.70	25.00	5.00	J MOL BIOL 223, 769–779 (1992)
3SSI	8.82	4.07	20.00	7.00	J MOL BIOL 249, 625–635 (1995)
1HIC	25.58	5.02	25.00	7.00	EUR J BIOCHEM 202, 67–73 (1991)
1PMC	38.33	1.10	20.00	3.00	NAT STRUCT BIOL 3, 45–53 (1996)

Thermal denaturation free energy change, ΔG, under reducing conditions and comparable temperatures for an exhaustive set of monomeric uncomplexed proteins with disulfide bonds and without prosthetic groups or ion coordination [66]. Deviations from the balance relation are measured by Y − (5X + 20), and shown to anticorrelate tightly ($R^2 = 0.72$) with the denaturation free energies

bonds) and 11 structural deficiencies (2 double bonds, 7 single bonds). Hence, its parameters are $Y = 11 \times 100/(11 + 23) = 32.35$; $X = 2 \times 100/74 = 2.70$. Applying the balance relation, we get the estimate $Y = 5 \times 2.7 + 20 = 33.5$, differing in less than 1 % from the real Y-value. Beta2-microglobulin (PDB.2Z9T, N = 100, $X = 1$), has 30 fully buried hydrogen bonds (9 double bonds, 12 single bonds), and 10 structural deficiencies (1 double bond), as shown in Fig. 1.16c, d. Hence, following the balance equation, Y is estimated at $5 \times 1 + 20 = 25$, which agrees exactly with the actual value $Y = 10 \times 100/(10 + 30)$. Finally, the antimicrobial hormone hepcidin (PDB.1M4F, N = 25, 4 disulfide bonds or $X = 16.00$) has a fully defective structure with $Y = 100$ (Fig. 1.16e, f), fitting exactly the balance equation $100 = 5 \times 16 + 20$.

The dehydron/disulfide balance relation clearly identifies proteins with excess ($Y > 5X + 20$) or lack ($Y < 5X + 20$) of structural deficiencies, with the former likely to be more favorably denatured than the latter under equivalent redox and denaturation conditions. To test this prediction, thermodynamic data on thermal denaturation (Table 1.2) was obtained for an exhaustive set of proteins for which structural information was also available [66]. Thus, the thermal denaturation free energy change, ΔG, under reducing conditions and comparable temperatures [66, 67] was obtained for monomeric uncomplexed PDB-reported proteins with disulfide bonds and lacking prosthetic groups or ion coordination. A significant anticorrelation was found ($R^2 = 0.72$, Fig. 1.17) between the deviation from the balance equation, measured as Y-(5X + 20), and the thermal denaturation free energy (ΔG). This tight anticorrelation provides a thermodynamic validation of the balance equation.

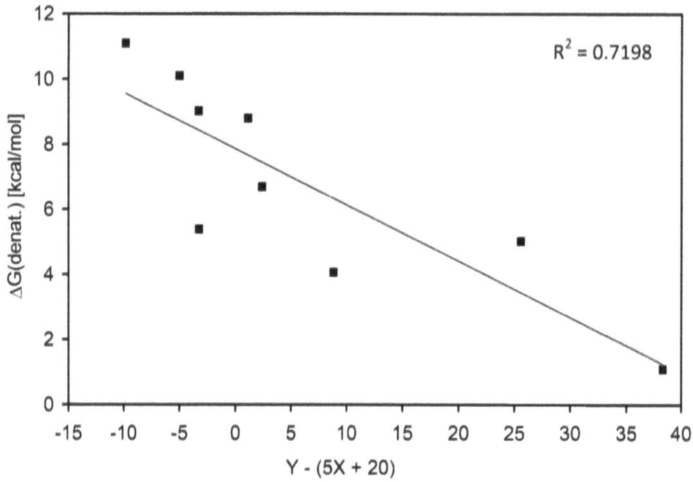

Fig. 1.17 Anticorrelation between denaturation free energy (ΔG) and excess structural defects with respect to the balance relation, measured by Y-(5X + 20). The exhaustive set of monomeric uncomplexed proteins with disulfide links and the respective denaturation conditions are given in Table 1.2. The coefficient $R^2 = 0.72$ for the linear fit was obtained by linear regression. Reprinted from [66], copyright 2009 with permission from the American Chemical Society

The 5:1 rule may be justified on thermodynamic grounds. Thus, Doig and Williams [68] addressed the inconsistencies in Flory's treatment of the entropic contribution to protein denaturation, calculating ΔΔG for denaturation for a cross-linked protein versus its non-cross-linked counterpart. At physiological temperature 300 K, they estimated $\Delta\Delta G \approx 4.4$ kcal/mol. This value is essentially independent of protein length and loop size and best represents the insensitivity of experimental values to loop size-dependent configurational entropies [69].

This 4.4 kcal/mol constant agrees reasonably well with the free energy contribution associated with the native-state destabilization brought about by five structural deficiencies. If we take into account that −0.93 kcal/mol is the free energy change associated with complete dehydration of an unburied backbone hydrogen bond [46, 53], we may estimate the net destabilization effect promoted by five structural deficiencies at $0.93\ \text{kcal/mol} \times 5 = 4.65$ kcal/mol. This value is in close proximity to the Doig-Williams constant. The thermodynamic agreement supports the 5:1 golden ratio for protein buttressing arising from structural analysis.

Since the evolutionary axis is germane to any biological analysis, the following question naturally arises: Is the architectural constraint defined by the dehydron/disulfide balance equation respected by evolution? Bioinformatics evidence on orthologous proteins (homologs across species) supports the tenet of evolutionary conservation. Thus, we may compare the structural deficiency and normalized number of disulfide bonds across 1105 homolog pairs that differ in at least one disulfide bond (one homolog may have no disulfide bond). The changes in X and Y (ΔX, ΔY) were obtained for homolog pairs identified by their respective PDB

Table 1.3 Evolutionary conservation of the balance relation Y = 5X + 20

SCOP family[a]	Sequence identity	ΔY − 5ΔX	ΔX	ΔY	Homolog 1	Homolog 2
Fibronectin type II module	63.125	−0.1	0.21	0.95	1QO6	1E8B
Beta-glycanases	40	−0.1	0.31	1.45	1FH8	1I1X
Beta-glycanases	42.13836478	−0.1	0.31	1.45	1E0V	1B30
C-type lysozyme	57.69230769	−0.09	0.79	3.86	1KXY	1LHM
C-type lysozyme	58.46153846	−0.09	0.79	3.86	1JIS	1LHM
Ribonuclease A-like	30.23255814	−0.09	0.79	3.86	1RUV	1B1E
Ribonuclease A-like	31.00775194	−0.09	0.79	3.86	1RUV	1H52
Acetylcholinesterase-like	30.82437276	−0.07	0.18	0.83	2CKM	1K4Y
C-type lysozyme	67.69230769	−0.07	0.79	3.88	1LMP	2BQM
C-type lysozyme	56.92307692	−0.07	0.79	3.88	1LSM	2BQK
Eukaryotic proteases	31.47410359	−0.07	0.29	1.38	1EX3	1DST
Snake venom toxins	32.89473684	−0.07	0.4	1.93	1TXA	1ERA
Pepsin-like	40	−0.06	0.31	1.49	1PSN	1FQ6
Ribonuclease A-like	31.74603175	−0.06	0.73	3.59	1RTB	1K5B
C-type lysozyme	59.23076923	−0.05	0.79	3.9	1IR7	1HNL
C-type lysozyme	59.23076923	−0.05	0.79	3.9	1XEK	1HNL
C-type lysozyme	59.23076923	−0.05	0.79	3.9	1UIH	1HNL
Ribonuclease A-like	31.00775194	−0.05	0.79	3.9	4RAT	1K5A
C-type lysozyme	58.46153846	−0.05	0.79	3.9	2HS7	2BQK
Plant proteinase inhibitors	47.74774775	−0.04	0.34	1.66	1TIH	1FYB
Papain-like	70.21943574	−0.03	0.84	4.17	1ITO	2PBH
Beta-glycanases	36.36363636	−0.03	0.31	1.52	2XYL	1B3X
C-type lysozyme	84.72222222	−0.03	1.02	5.07	3LZ2	1LSG
Ribonuclease A-like	31.00775194	−0.03	0.79	3.92	1RHB	1K5A

[a]*SCOP* structural classification of proteins (Murzin AG, Brenner SE, Hubbard T, Chothia C (1995) SCOP: a structural classification of proteins database for the investigation of sequences and structures. J Mol Biol 247:536-540)

accessions and the deviation from ideality was measured as $\Gamma = \Delta Y - 5\Delta X$. For all homolog pairs we obtained $\Gamma/Y < 11\ \%$, with Y associated with either homolog. Table 1.3 illustrates the tightest evolutionary conservation of the architectural constraint across homolog pairs with nontrivial buttressing differences.

1.10 Why Epistructural Physics?

The last decades have witnessed the meteoric success of structural biology, and yet this success has also helped establish an unrealistic picture of the protein in solution, as it is found in the cellular or intercellular environment. In Fig. 1.18 we

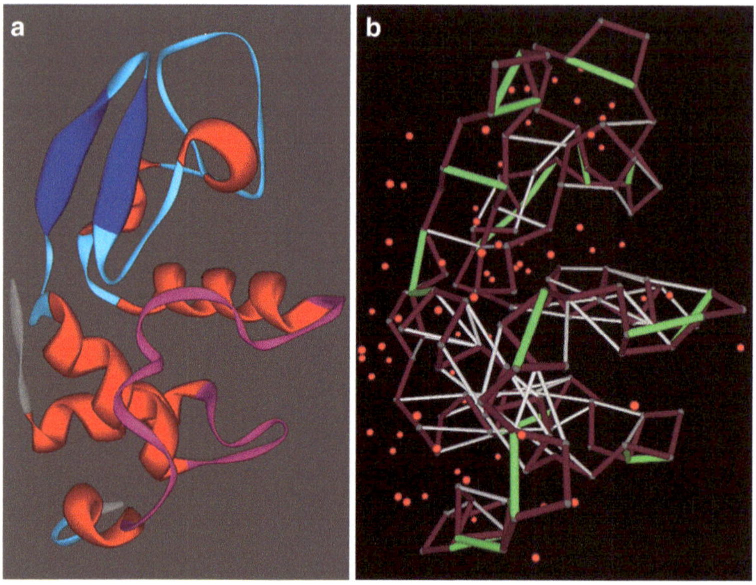

Fig. 1.18 Structural (**a**) and epistructural (**b**) representations of *lysozyme*, a soluble hydrolytic enzyme. (**a**) The schematic structural representation provides a ribbon rendering of the protein backbone. This representation is enabled because of the local planar character of the peptide bond, with planes articulated relative to one another at the α-carbon locations. The solvent is represented merely as *grey* background against which the structure is displayed. (**b**) Epistructural representation of lysozyme enshrined in its dehydron pattern. Well wrapped backbone hydrogen bonds are represented as *grey* segments joining the α-carbons of the paired residues, while dehydrons are shown in *green*. A simplified virtual-bond representation of the backbone chain is adopted and the oxygen atoms of interfacial water molecules are featured

see *lysozyme*, a widely studied hydrolytic soluble enzyme. The conventional structural-biology representation (Fig. 1.18a) reveals a 3D structural arrangement obtained from the X-ray diffraction pattern. Nothing in this static representation indicates how the protein relates to or interacts with its solvent environment since the solvent is simply represented as a background for the protein structure. With more or less atomic detail, this is the picture that prevails in fundamental or applied research labs that exploit the output of structural biology. In fact, structural coordinates provide much more information than indicating the relative position of the atoms in the chain, its folding topology, or the chemical composition of specific locations on the protein surface. Firstly, the tightness of the protein hydration shell is widely modulated by structural features of the protein, so that an interfacial tension driving protein associations is established between protein solute and water solvent that is essentially determined by protein structure but not captured by the standard structural renderings such as the one given in Fig. 1.18a. Certain structural defects in the protein structure named *dehydrons* are known to play a crucial role in promoting interfacial tension by causing water frustration at the interface, which in turn causes significant deviations from the standard dielectric picture introduced by Debye. As described in Chap. 2, dehydrons widely modify the dielectric properties of the aqueous interface above and beyond the classical Debye picture, amplifying electrostatic fields in the protein. Thirdly, as described in Chaps. 3 and 4, protein and aqueous interface are engaged in an exquisite dynamic entanglement reciprocally influencing and steering their respective motions. Finally, as explored in Chap. 7, dehydrons functionalize interfacial water promoting a catalytic role that arises from water frustration. These interrelated aspects of the aqueous interface shaped by the protein structure pertain to *epistructural biology* and build upon information obtained from structural biology, and yet they differentiate from the latter. Thus, *the dehydron pattern of the protein (Fig.* 1.18*b) enshrines the epistructural biology picture that complements the standard structural representation (Fig.* 1.18*a) as it subsumes vital information on the physical and catalytic properties of the protein-water interface determined by the protein structure.* The epistructural aspects have remained vastly unexploited by drug designers and molecular engineers, with hefty consequences for the pharmaceutical and biotechnological industries. Chapters 8–17 are intended to improve substantively this state of affairs.

1.11 Problems

1.1. *Interfacial Fugacity.* The fugacity of interfacial water is a key nanoscale parameter to define the entropic contribution to interfacial tension. To properly define thermodynamic fugacity in this context, we consider a water-confining osculating (first-order contact) sphere D with radius $\theta =$ curvature radius of the protein water-smeared envelope at the point of contact. A local fugacity *f*(D) is defined as f(D) = N(D)/N(D,b), where N(D) is the expected number of water molecules in D at equilibrium and N(D,b) is the number associated with the same volume of bulk solvent.

(a) Determine analytically the chemical potential μ(D) of interfacial water in D, that is, the free energy change associated with transferring a water molecule from D to the bulk, as a function of its fugacity.

(b) Determine the θ-dependence of f-values obtained at equilibrium within a particle number–pressure–temperature (NPT) ensemble for monomeric single-domain proteins with structural coordinates obtained from the PDB (the set of proteins may be found in [66]). The computations should start with PDB-reported protein structures embedded in a pre-equilibrated cell of water molecules [4, 15]. To determine the dependence of *f* on local curvature radius θ of the solvent-accessible envelope of the protein surface, the latter should be covered by a minimal set *W* of disjoint water-confining osculating (first-order contact) spheres *Dj s*, $j \in W$. The dynamics and equilibrium parameters for local hydration patterns within each osculating sphere will result from classical trajectories that should be generated by molecular dynamics (MD), as indicated in [4, 15]. The PDB structures for an exhaustive set of 2661 monomeric uncomplexed proteins [66] lacking prosthetic groups should be used in the computations. The MD trajectories should be generated by adopting an integration time step of 2 fs in a particle number–pressure–temperature ensemble with box size 103 nm^3 and periodic boundary conditions [28]. The box size should be calibrated so that the solvation shell extends at least 10 Å from the protein surface at all times. The long-range electrostatics should be treated using the particle-mesh Ewald summation method [29]. A Nose–Hoover thermostat [28] may be adopted to maintain the temperature at 300 K, and a Tip3P water model with optimized potential for liquid simulations force field should be adopted [27]. A barostat scheme should be maintained through a dedicated routine with the pressure held constant at 1 atm using a weak-coupling algorithm. After 1 ns equilibration, average fugacity may be determined in each osculating sphere for each of the 2661 proteins in the database. The θ-dependence of f may be established by computing average values over all osculating spheres with fixed curvature radius (at, say, 1/4 Å-resolution). Dispersions may be obtained to distinguish polar and nonpolar hydration patterns.

1.2. Identify an electrostatic quantity that evaluates the total departure from Debye polarization for a soluble protein along the entire protein-water interface.

1.3. Identify the dehydrons in the structure of the antifreeze protein from the snow flea (PDB.2PNE, N = 81, two disulfide bonds or X = 2.47). The twisted strands do not fit into any of the standard structural motifs for proteins. What motif do they belong to?

1.4. At 67.65 % discrepancy between the estimated (Y = 32.35) and the real (Y = 100) Y-value, the antifreeze protein from the snow flea (PDB.2PNE, N = 81, two disulfide bonds or X = 2.47) is the single major outlier from the (X, Y)-balance equation. This equation relates the normalized numbers of packing defects (dehydrons) and disulfide bonds in a soluble protein. Can you give a reason for this anomaly?

HINT: Consider the possibility of a backbone structure that gets significantly hydrated so that the backbone exposure is no longer a structure-destablizing factor. Overexposed dehydrons do not create interfacial tension or destabilize the protein structure.

1.5. Identify the dehydron patter for the beta-helix antifreeze protein from beetle *tenebrio molitor* (PSB.1EZG, N = 81, eight disulfide bonds, or X = 9.87)

1.6. Obtain the expected number of carbonaceous HB-wrapping groups per residue.

HINT: Consider the number of carbonaceous HB-wrapping groups for the amino acids and multiply them by the respective relative abundance (frequency) of the amino acids in a nonredundant set of the PDB (see, for example, [66]). Table 1.4 shows frequency and number of carbonaceous HB-wrapping groups for each amino acid type.

Table 1.4 Frequencies (relative abundancies) in a PDB nonredundant subset [66] and number of carbonaceous HB-wrapping groups for each amino acid type

Amino acid	Abundance in PDB subset (%)	Carbonaceous HB-wrapping groups
Ala	7.65	1
Arg	5.94	4
Asn	4.05	2
Asp	5.45	2
Cys	1.37	1
Gln	3.93	3
Glu	7.05	3
Gly	5.96	0
His	2.67	4
Ile	5.96	4
Leu	9.66	4
Lys	5.84	4
Met	2.42	3
Phe	3.86	7
Pro	4.70	3
Ser	6.56	1
Thr	5.34	2
Trp	1.08	8
Tyr	3.52	7
Val	6.87	3

References

1. Wernet P, Nordlund D, Bergmann U, Cavalleri M, Odelius M, Ogasawara H, Näslund L, Hirsch TK, Ojamäe L, Glatzel P, Pettersson LG, Nilsson A. The structure of the first coordination shell in liquid water. Science. 2004;304:995–9.
2. Head-Gordon T, Hura G. Water structure from scattering experiments and simulations. Chem Rev. 2002;102:2651–70.
3. Cheng YK, Rossky P. Surface topography dependence of biomolecular hydrophobic hydration. Nature. 1998;392:696–9.
4. Fernández A. The principle of minimal episteric distortion of the water matrix and its steering role in protein folding. J Chem Phys. 2013;139:085101.
5. Fernández A. Epistructural tension promotes protein associations. Phys Rev Lett. 2012;108:188102.
6. Giovambattista N, Lopez CF, Rossky P, Debenedetti P. Hydrophobicity of protein surfaces: separating geometry from chemistry. Proc Natl Acad Sci U S A. 2008;105:2274–9.
7. Debye P. Polar molecules. New York: Dover; 1929.
8. Strekalova EG, Mazza MG, Stanley HE, Franzese G. Large decrease of fluctuations for supercooled water in hydrophobic nanoconfinement. Phys Rev Lett. 2011;106:145701.
9. Tanizaki S, Feig F. A generalized Born formalism for heterogeneous dielectric environments: application to the implicit modeling of biological membranes. J Chem Phys. 2005;122:124706.

10. Stanley HE, Buldyrev SV, Kumar P, Mallamace F, Mazza MG, Stokely K, Xu L, Franzese G. Water in nanoconfined and biological environments. J Noncryst Solids. 2011;357:629–40.
11. Kumar P, Han S, Stanley HE. Anomalies of water and hydrogen bond dynamics in hydrophobic nanoconfinement. J Phys Condens Matter. 2009;21:504108.
12. Fernández Stigliano A. Breakdown of the Debye polarization ansatz at protein-water interfaces. J Chem Phys. 2013;138:225103.
13. Fernández A. Transformative concepts for drug design: target wrapping. Berlin: Springer; 2010.
14. Schutz CN, Warshel A. What are the dielectric "constants" of proteins and how to validate electrostatic models? Proteins Struct Funct Genet. 2001;44:400–8.
15. Fernández A. Fast Track communication: water promotes the sealing of nanoscale packing defects in folding proteins. J Phys Chem Condens Matter. 2014;26:202101.
16. Fernández A. Nanoscale thermodynamics of biological interfacial tension. Proc R Soc A. 2011;467:559–68.
17. Lee M, Salsbury F, Olson MA. An efficient hybrid explicit/implicit solvent method for biomolecular simulations. J Comp Chem. 2004;25:1967–78.
18. Allen MP, Tildesley DJ. Computer simulation of liquids. Oxford: Clarendon; 1989.
19. Bachmann M. Thermodynamics and statistical mechanics of macromolecular systems. Cambridge: Cambridge University Press; 2014.
20. Lee MS, Olson MA. Comparison of volume and surface area nonpolar solvation free energy terms for implicit solvent simulations. J Chem Phys. 2013;139:044119.
21. Fernández Stigliano A. Biomolecular interfaces: interactions, functions and drug design. Heidelberg: Springer; 2015.
22. Clackson T, Wells JA. A hot spot of binding energy in a hormone-receptor interface. Science. 1995;267:383–6.
23. Pal SK, Peon J, Zewail AH. Biological water at the protein surface: dynamical solvation probed directly with femtosecond resolution. Proc Natl Acad Sci U S A. 2002;99:1763–8.
24. Evans LC. Partial differential equations. Providence: American Mathematical Society; 2010.
25. Fernández A. Quantum theory of interfacial tension quantitatively predicts spontaneous charging of nonpolar aqueous interfaces. Phys Lett A. 2015;379:2405–8.
26. Matsumoto M, Tanaka K. Nano bubble—size dependence of surface tension and inside pressure. Fluid Dyn Res. 2008;40:546.
27. Jorgensen WL, Chandrasekhar J, Madura JD, Impey RW, Klein ML. Comparison of simple potential functions for simulating liquid water. J Chem Phys. 1983;79:926–34.
28. Wang J, Wolf RM, Caldwell JW, Kollman PA, Case DA. Development and testing of a general amber force field. J Comput Chem. 2004;25:1157–74.
29. Darden T, York D, Pedersen LJ. Particle Mesh Ewald-an N.Log(N) method for Ewald sums in large systems. J Chem Phys. 1993;98:10089–92.
30. Zhang L, Wang L, Kao Y, Qiu W, Yang Y, Okobiah O, Zhong D. Mapping hydration dynamics around a protein surface. Proc Natl Acad Sci U S A. 2007;104:18461–6.
31. Ashkenazi A, Presta L, Marsters S, Camarato J, Rosenthal K, Fendly B, Capon D. Mapping the CD4 binding site for human immunodeficiency virus by alanine-scanning mutagenesis. Proc Natl Acad Sci U S A. 1990;87:7150–4.
32. Castro MJ, Anderson S. Alanine point-mutations in the reactive region of bovine pancreatic trypsin inhibitor: effects on the kinetics and thermodynamics of binding to beta-trypsin and alpha-chymotrypsin. Biochemistry. 1996;35:11435–46.
33. Kuhlmann UC, Pommer A, Moore JR, James R, Kleanthous C. Specificity in protein-protein interactions: the structural basis for dual recognition in colicin endonuclease-immunity protein complexes. J Mol Biol. 2000;301:1163–78.
34. Bottger V, Bottger A, Garcia-Echeverria C, Chène P, Hochkeppel HK, Sampson W, Ang K, Howard SF, Picksley SM, Lane DP. Molecular characterization of the hdm2–p53 interaction. J Mol Biol. 1997;269:744–56.
35. Anfinsen CB. Principles that govern the folding of protein chains. Science. 1973;181:223–30.

36. Fernández A, Sosnick TR, Colubri A. Dynamics of hydrogen-bond desolvation in folding proteins. J Mol Biol. 2002;321:659–75.
37. Chandler D. Interfaces and the driving force of hydrophobic assembly. Nature. 2005;437:640–7.
38. Jewett A, Pande VS, Plaxco KW. Cooperativity, smooth energy landscapes and the origins of topology-dependent protein folding rates. J Mol Biol. 2003;326:247–53.
39. Scalley-Kim M, Baker D. Characterization of the folding energy landscapes of computer generated proteins suggests high folding free energy barriers and cooperativity may be consequences of natural selection. J Mol Biol. 2004;338:573–83.
40. Fernández A, Colubri A, Berry RS. Three-body correlations in protein folding: the origin of cooperativity. Phys A. 2002;307:235–59.
41. Fernández A, Kostov K, Berry RS. From residue matching patterns to protein folding topographies: general model and bovine pancreatic trypsin inhibitor. Proc Natl Acad Sci U S A. 1999;96:12991–6.
42. Fernández A, Colubri A, Berry RS. Topology to geometry in protein folding: beta-lactoglobulin. Proc Natl Acad Sci U S A. 2000;97:14062–6.
43. Fernández A, Kardos J, Goto J. Protein folding: could hydrophobic collapse be coupled with hydrogen-bond formation? FEBS Lett. 2003;536:187–92.
44. Fernández A. Conformation-dependent environments in folding proteins. J Chem Phys. 2001;114:2489–502.
45. Fernández A, Kardos J, Scott R, Goto Y, Berry RS. Structural defects and the diagnosis of amyloidogenic propensity. Proc Natl Acad Sci U S A. 2003;100:6446–51.
46. Fernández A. Keeping dry and crossing membranes. Nat Biotechnol. 2004;22:1081–4.
47. Pietrosemoli N, Crespo A, Fernández A. Dehydration propensity of order-disorder intermediate regions in soluble proteins. J Proteome Res. 2007;6:3519–26.
48. Fernández A, Scott R. Dehydron: a structure-encoded signal for protein interactions. Biophys J. 2003;85:1914–28.
49. Avbelj F, Baldwin RL. Role of backbone solvation and electrostatics in generating preferred peptide backbone conformations: distributions of phi. Proc Natl Acad Sci U S A. 2003;100:5742–7.
50. Krantz BA, Moran LB, Kentsis A, Sosnick TR. D/H amide kinetic isotope effects reveal when hydrogen bonds form during protein folding. Nat Struct Biol. 2000;7:62–71.
51. Fersht A. Transition-state structure as a unifying basis in protein-folding mechanisms: contact order, chain topology, stability, and the extended nucleus mechanism. Proc Natl Acad Sci U S A. 2000;97:1525–929.
52. Plaxco KW, Simmons KT, Baker D. Contact order, transition state placement and the refolding rates of single domain proteins. J Mol Biol. 1998;277:985–94.
53. Fernández A, Scott LR. Adherence of packing defects in soluble proteins. Phys Rev Lett. 2003;91:018102.
54. Fernández A, Zhang X, Chen J. Folding and wrapping soluble proteins: exploring the molecular basis of cooperativity and aggregation. Prog Nucl Acid Res Transl Sci. 2008;83:57–87.
55. Fernández A, Scheraga HA. Insufficiently dehydrated hydrogen bonds as determinants of protein interactions. Proc Natl Acad Sci U S A. 2003;100:113–8.
56. Fernández A, Berry RS. Proteins with H-bond packing defects are highly interactive with lipid bilayers: implications for amyloidogenesis. Proc Natl Acad Sci U S A. 2003;100:2391–6.
57. Deremble C, Lavery R. Macromolecular recognition. Curr Opin Struct Biol. 2005;15:171–5.
58. Ma B, Elkayam T, Wolfson H, Nussinov R. Protein-protein interactions: structurally conserved residues distinguish between binding sites and exposed protein surfaces. Proc Natl Acad Sci U S A. 2003;100:5772–7.
59. Fernández A. What caliber pore is like a pipe? Nanotubes as modulators of ion gradients. J Chem Phys. 2003;119:5315–9.
60. Despa F, Fernández A, Berry RS. Dielectric modulation of biological water. Phys Rev Lett. 2004;93:228104.

61. Demetri G. Efficacy and safety of imatinib mesylate in advanced gastrointestinal stromal tumors. N Engl J Med. 2002;347:472–80.
62. Fernández A, Sanguino A, Peng Z, Ozturk E, Chen J, Crespo A, Wulf S, Shavrin A, Qin C, Ma J, Trent J, Lin Y, Han HD, Mangala LS, Bankson JA, Gelovani J, Samarel A, Bornmann W, Sood AK, Lopez-Berestein G. An anticancer c-Kit kinase inhibitor is reengineered to make it more active and less cardiotoxic. J Clin Invest. 2007;117:4044–54.
63. Baldwin RL. In search of the energetic role of peptide hydrogen bonds. J Biol Chem. 2003;278:17581–8.
64. Powers ET, Deechongkit S, Kelly JW (2006) Backbone-backbone H-bonds make context-dependent contributions to protein folding kinetics and thermodynamics: lessons from amide-to-ester mutations. In: Baldwin RL, Baker D (eds) Peptide solvation and H-bonds. Adv Prot Chem 72:40–79
65. MacKinnon R, Reinhart PH, White MN. Charybdotoxin block of Shaker K^+ channels suggests that different types of K^+ channels share common features. Neuron. 1988;1:997–1001.
66. Fernández A, Berry RS. Golden rule for buttressing vulnerable soluble proteins. J Proteome Res. 2009;9:2643–8.
67. Kumar MD. ProTherm and ProNIT: thermodynamic databases for proteins and protein-nucleic acid interactions. Nucl Acid Res. 2006;34:D204–6.
68. Doig AJ, Williams DH. Is the hydrophobic effect stabilizing or destabilizing in proteins: the contribution of disulfide bonds to protein stability. J Mol Biol. 1991;217:389–98.
69. Betz SF. Disulfide bonds and the stability of globular proteins. Protein Sci. 1993;2:1551–8.

Chapter 2
Dielectric Structure of Aqueous Interfaces: From Classical Non-Debye Electrostatics to a Quantum Theory of Interfacial Tension

This chapter explores the dielectric structure of interfacial water from a classical and quantum electrostatics perspective. The protein-water interface is treated classically, while quantum effects are investigated only for simple featureless nonpolar interfaces due to the inherently higher complexities of the quantum approach. In a classical context, the most striking feature arising from the *epistructural physics* is the breakdown of the Debye *ansatz* that postulates the alignment of water polarization with the protein electrostatic field. The complexities of biological interfaces are shown to be in good measure due to this departure from the standard dielectric picture that has been historically—and incorrectly—extrapolated from a bulk interface. Accordingly, concepts like the dielectric permittivity coefficient are shown to be inadequate to describe interfacial electrostatics. The departure from bulk-like behavior is shown to enhance the physicochemical inhomogeneity of protein surfaces and to enable the chemical functionality of the aqueous interface. Epistructural physics of the protein-water interface identifies a structural defect known as *dehydron* as causative of anomalous polarization effects causing a breakdown of the Debye standard ansatz. The previous chapter revealed that interfacial tension is a central thermodynamic factor driving biomolecular events and may be alternatively stored as electrostatic energy associated with the non-Debye component of water polarization. This chapter substantively supplements this picture by showing that dehydrons induce chemical basicity in interfacial water as a consequence of the breakdown of Debye dielectrics. Thus, the relevance of dehydrons as catalytic elements is highlighted. We anticipate that this discovery will prompt a re-writing of vast mechanistic chapters of biological chemistry.

A quantum mechanical approach is shown to be essential to deal with the spontaneous negative charging of aqueous nonpolar interfaces. This phenomenon has eluded quantitative first-principle prediction, in part because it steadfastly challenges the Debye dielectric picture. In this chapter we also show that quantitative prediction requires not only a substantive revision of Debye's linear dielectric

A. Fernández, *Physics at the Biomolecular Interface*, Soft and Biological Matter,
DOI 10.1007/978-3-319-30852-4_2

ansatz to incorporate anomalous polarization yielding interfacial tension but also the implementation of a quantum mechanical theory of interfacial tension. The latter is necessary to compute the reduction in hydrogen-bond frustration at the interface that takes place upon hydroxide anion adsorption. The quantitative predictions are validated vis-à-vis experimental measurements of hydroxide adsorption uptake obtained using sum-frequency vibrational spectroscopy.

Unfortunately, the daunting complexities prevent us from extending this quantum mechanical approach to protein-water interfaces at this time. We anticipate that ionic adsorption at packing defect sites may indeed occur and play a role in enzymatic catalysis but cannot prove it at this point in time.

2.1 Interfacial Tension Stored as Non-Debye Polarization Energy

As noted in the previous chapter, the non-Debye orthogonal component $\vec{P^{\#}}$ of polarization is commensurate with the distortion in the structure of water due to sub-nanoscale confinement [1]. More precisely $\vec{P^{\#}} = \xi \vec{\nabla} g$, yielding an equivalence between electrostatic energy of orthogonal polarization and interfacial tension:

$$U^{\#} = \frac{1}{2}\varepsilon_0^{-1} \int \left\| \vec{P^{\#}} \right\|^2 d\vec{r} = \frac{1}{2}\lambda \int \left\| \vec{\nabla} g \right\|^2 d\vec{r} \tag{2.1}$$

This equation asserts the electrostatic origin of interfacial tension, prompting the question: Where do the most significant contributions to non-Debye polarization take place? Since dehydrons create the biggest drops in g-value, we may intuitively assert that dehydrons are the culprits of the breakdown of the Debye picture and a good portion of this chapter is devoted to prove this conjecture.

A measure of the local departure from the Debye scenario within the protein structure may be obtained for each residue generically denoted by n. This measure is furnished by a structure-dependent parameter ϑ_n that we term anomalous polarization fraction (APF) and define as

$$\vartheta_n = \left\langle U^{\#}{}_n / U_n \right\rangle \tag{2.2}$$

where the symbol "$\langle . \rangle$" denotes time average, and

$$U_n = \frac{1}{2}\varepsilon_0 \int \left\| \vec{\nabla} \varphi \right\|^2 d\vec{r}, \quad U^{\#}{}_n = \frac{1}{2}\varepsilon_0^{-1} \int \left\| \vec{P} - \left(\vec{P} . \vec{e} \right) \vec{e} \right\|^2 d\vec{r} \tag{2.3}$$

where electrostatic potential φ is defined through the relation: $-\vec{\nabla}\varphi = \vec{E} + \varepsilon_0^{-1}\vec{P}$. Integration in equations (2.3) extends over a neighborhood around residue n defined as a sphere of radius r centered at its α-carbon. To capture the environment of all side chains we adopted r = 6 Å, the approximate diameter of tryptophan (W), the largest side chain. We also evaluated the APF using a larger radius (r = 8 Å) to determine the dilution of the anomalous polarization effect as the bulk solvent region is approached. The centering of the n-residue sphere at the α-carbon is justified *a posteriori*, as shown subsequently. We expected and confirmed that anomalous polarization would be mainly related to poor packing of the protein backbone, while the packing defects are identified by introducing backbone solvation domains consisting of spheres of radius 6 Å centered at the α-carbons. Thus, by centering the residue spheres at the α-carbons we simultaneously interrogate the backbone and the side chains in search for anomalous dielectric patterns in interfacial water. A more obvious reason for choosing the α-carbon as opposed to other side chain carbon atom relates to the fact that we would otherwise need to treat glycine (G) as an exceptional case.

The parameter ϑ_n is computed at protein/water interfaces as a time average over a 5 ns-period beyond equilibration of the protein structural backbone with the solvent. Thus, water polarization for soluble natural proteins with structures reported in the Protein Data Bank (PDB) may be computed along molecular dynamics trajectories. Each 10 ns-trajectory is generated using as starting point the equilibrated structural coordinates that result after thermalization of the PDB-reported structure immersed in a pre-equilibrated solvent bath. The referenced computational details [2–9] are provided in the caption for Fig. 2.1. Simulations are performed within an isobaric/isothermal ensemble (1 atm, 298 K). The optimized systems are pre-equilibrated for 500 ps. The resulting structures become the starting point for the 5 ns-thermalization trajectories. A total of 100 interfacial solvent configurations, one every 50 ps, are used to compute the time average of ϑ_n. To this end, we recorded charge distribution $\rho\left(\vec{r},t\right)$, internal field $\vec{E}\left(\vec{r},t\right)$ and polarization $\vec{P}\left(\vec{r},t\right)$ for each intermediate structure/solvent configuration.

The structure/solvent system is considered equilibrated at time t_o if the RMSD of backbone atomic coordinates averaged over randomly chosen pairs of chain conformations within a time interval $[t_o, t_o + \tau]$ ($\tau \approx 1$ ns) is less than 1 Å. For all nine proteins in this study (PDB entries 1SRL, 1ESR, 1A8O, 1PIT, 1QGB, 1ATA, 1Q7I, 1PI2, 2PNE), this criterion was fulfilled for $t_o = 500$ ps. Solvent and side-chain conformations continue to vary significantly (i.e. RMSD > 2.25 Å) on the 1 ns timescale.

The APFs for individual residues for the natively folded SH3 domain (PDB.1SRL) are shown in Fig. 2.1. The context-dependence of APFs is evident since residues of the same type (i.e. serines S18 and S47, tryptophans W42 and W43) can have very different APFs depending on their location within the protein chain and therefore, within the structure. If we exclude the residues A12, I56 and

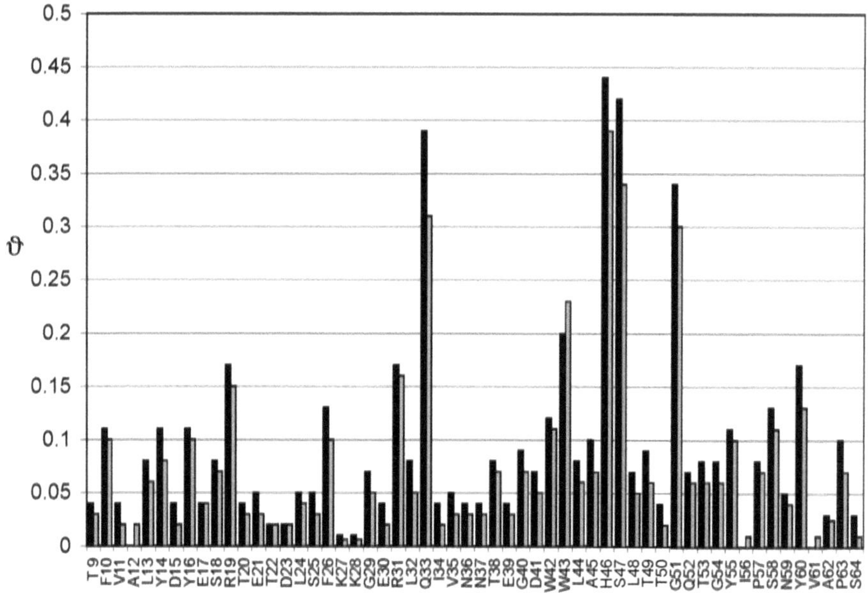

Fig. 2.1 Anomalous polarization fraction (ϑ) for each residue within the solvent-equilibrated folded protein chain for the soluble SH3 domain (PDB.1SRL). The PDB file numbering is followed in naming residues along the chain and the one-letter code for amino acids is adopted. *Black bars* represent protein surface interrogation with spheres of radius r = 6 Å centered at α-carbons, while *grey bars* were generated using radius r = 8 Å. The all-atom trajectories used to compute the time-averaged APF values thermalize the PDB structures in contact with a pre-equilibrated solvent bath consisting in a truncated octahedral cell of TIP3P water molecules that provide at least four water layers of solvent envelope [2]. Protein atoms are described with the parm99SB force field parameterization [3]. Water molecules extended at least 12 Å from the surface of the protein. Ewald sums [4] and an 8 Å-distance cutoff are used for treating long-range electrostatic interactions. A Shake scheme is employed to keep bonds involving hydrogen atoms at their equilibrium length [5] which allowed us to employ a 3 fs time step for the integration of Newton's equations. Constant pressure of 1 atm and a temperature of 298 K are maintained using the Berendsen coupling scheme [6]. An AMBER package [7] was adopted for these MD simulations, with charges on the molecules assigned according to the BCC charge model using AM1 optimized geometries and potentials [8, 9]. After protein/solvent equilibration (as defined in main text), the protein backbone coordinates are partially constrained according to the Shake scheme [5] and only side chains are allowed to explore conformation space, generating a gamut of local hydration patterns. Reprinted from [Fernández Stigliano A (2013) Breakdown of the Debye polarization ansatz at protein-water interfaces. J Chem Phys 138:225103], copyright 2013 with permission from AIP Publishing LLC

V61 that are fully buried within the structure, it is clear that the positively charge lysines (K27, K28) have the lowest APFs. As described below, this is expected since the ammonium cation ($-NH_3^+$) in lysine has the highest charge concentration of all amino acids and hence it is the most capable of organizing solvent in accord with its highest hydration requirements.

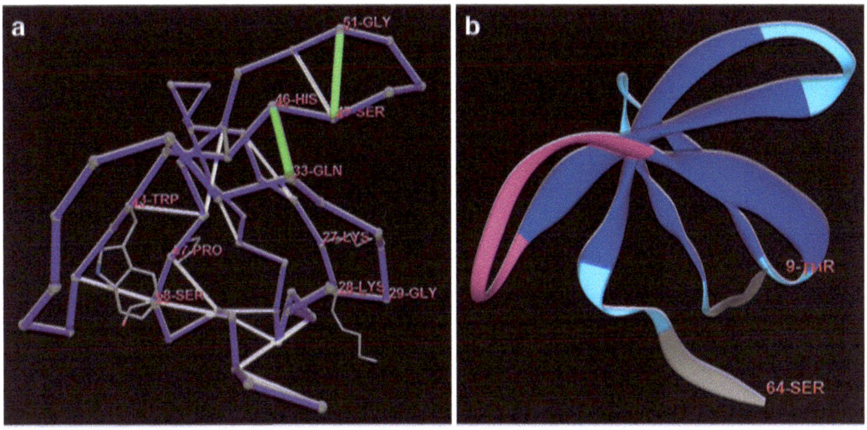

Fig. 2.2 (**a**) Dehydron pattern for the solvent-equilibrated structure of the soluble Src-SH3 domain (PDB. 1SRL). The backbone is represented as virtual bonds joining the α-carbons of residues along the chain, with well-shielded backbone hydrogen bonds and dehydrons shown as segments sustained between the paired residues in *grey* and *green*, respectively. Dehydrons are determined from the PDB structural coordinates following the protocol indicated in Chap. 1. In accord with this protocol, the under-wrapping of the backbone hydrogen bond due to a low number of surrounding nonpolar groups from the flanking side chains is considered to be a surrogate for the extent of solvent exposure. (**b**) Ribbon rendering of the structure of Src-SH3 domain

Intriguingly, a structural context becomes the determinant factor for the APF high values, superseding individual residue propensities. Thus, the residues with the highest APFs, Q33, H46, S47 and G51, are the only ones paired by *dehydrons* (marked in green in Fig. 2.2a; the ribbon rendering in Fig. 2.2b is an aid to the eye). Due to the nanoscale water confinement created by the packing defect that gives rise to the dehydron, a significant nonvanishing component $\vec{P}^{\#}$ is expected for residues paired by such hydrogen bonds (cf. Chap. 1). These packing defects expose the backbone polar groups amide (> N-H) and carbonyl (> C = O) to structure-disruptive effects of backbone hydration with the net effect of steering water dipoles into orientations that are not collinear with $\vec{E}$. The confined water molecules relinquish some of their hydrogen bonding possibilities in order to form hydrogen bonds with the backbone polar groups. This reduction in coordination represents a departure from the bulk water structure embodied in the tetrahedral lattice and the resulting water polarization becomes statistically independent of the internal field $\vec{E}$. This is so since water molecules with reduced water coordination tend to preserve their hydrogen-bond pattern thereby becoming impervious to the torque

$$\vec{E}\left(\vec{r}\right) \times \vec{\mu}_P\left(\vec{r}\right) = \vec{E}\left(\vec{r}\right) \times \int \left(\vec{r'} - \vec{r}\right) \vec{\nabla} \cdot \vec{P}\left(\vec{r'}\right) d\vec{r'} \qquad (2.4)$$

imposed by $\vec{E}=\vec{E}\left(\vec{r}\right)$ on the polarization-associated dipole with moment $\vec{\mu}_P(\vec{r}) = \int \left(\vec{r'} - \vec{r}\right) \vec{\nabla} . \vec{P}\left(\vec{r'}\right) d\vec{r'}$.

Thus, interfacial water polarization in this context is expected to contain and indeed contains (Fig. 2.1) a significant anomalous non-Debye contribution.

2.2 Non-Debye Dielectric Structure of the Aqueous Interface for a Soluble Protein

In Fig. 2.3a we show the individual propensities of the 20 residue types to align interfacial water along the electrostatic field by computing the APF of residues in nine PDB-reported proteins (specified in caption for Fig. 2.4). The expected APF for each residue type is obtained by averaging the APFs for that residue type in all nine proteins. The computation amounts to average over the structural contexts in the nine proteins where the particular residue type occurs. Due to the dominance of dehydrons as structural determinants of APF (Figs. 2.1 and 2.4), superseding individual propensities (Fig. 2.1), we have excluded dehydron-paired residues from the calculations in Fig. 2.3a.

As a class, the aromatic residues (H, F, W, Y) have the highest APF values due to their water-organizing power and their role as significant disruptors of the tetrahedral water structure. Their delocalized π-electron quadrupole promotes interactions with partial positive charges in vicinal interfacial water molecules. Furthermore, the side chains of such residues cannot be clathrated (surrounded without disrupting the tetrahedral water lattice) as it is the case with nonpolar aliphatic side chains (L, V, I, A). Thus, the resilient non-tetrahedral hydrogen-bond pattern of vicinal water explains the superior APF-boosting activity of aromatic residues when compared with nonpolar aliphatic ones. The sharp contrast between the lowest APF-booster lysine (K) and the highest APF-booster arginine (R), both in the same class of positively charged residues with aliphatic (methylene) linkages, is also striking, yet expected. The ammonium cation ($-NH_3^+$) in lysine has the highest charge concentration of all amino acids, therefore it strongly organizes hydration along electrostatic field lines, while the guanidinium cation ($\left[-NH = C(NH_2)_2\right]^+$) in arginine contains the most delocalized charge of all amino acids, hence the resulting local electrostatic field has the weakest water-organizing power.

From the above discussion it becomes apparent that the polarization steering power of individual residues is tightly related to the localization and concentration of their net charge. Thus, a pH dependence of the APF for an individual residue is expected in accord with the pKa of the residue within the protein structure. The titration of a residue removes a net charge and thereby increases the APF by curbing the polarization-steering capabilities of the residue. This titration effect becomes apparent as we compare the expected APFs of individual residue types at neutral pH

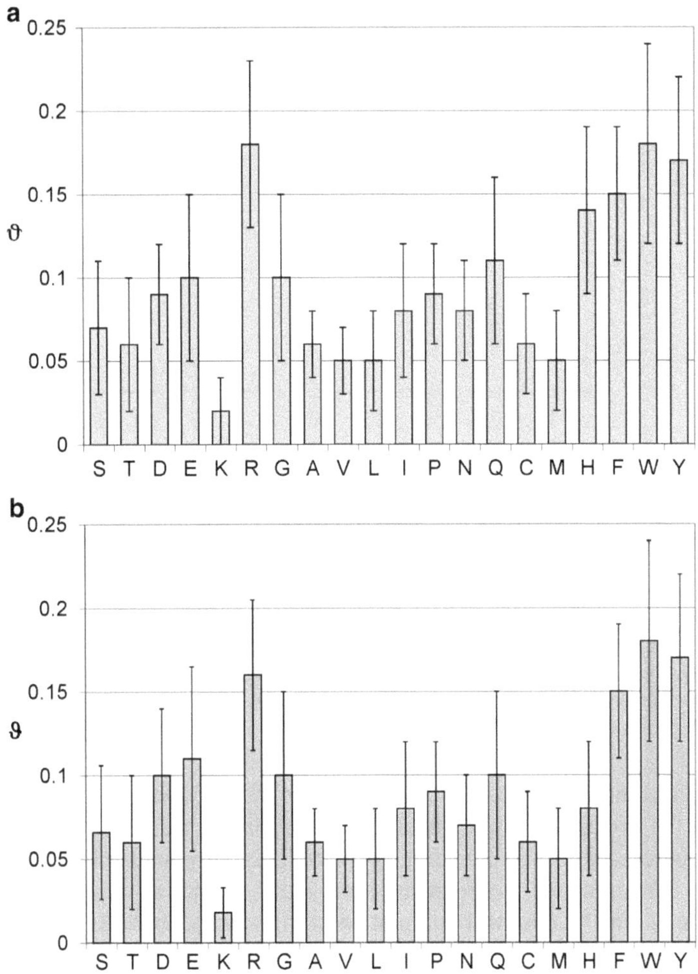

Fig. 2.3 APFs for each residue type averaged over all structural environments where the residue type occurs in 9 PDB-reported proteins described in the caption for Fig. 2.4. The radius r = 6 Å has been adopted and the error bars represent the dispersion in ϑ-values. (**a**) APFs at pH 7. (**b**) APFs at pH 5.5. Reprinted from [Fernández Stigliano A (2013) Breakdown of the Debye polarization ansatz at protein-water interfaces. J Chem Phys 138:225103], copyright 2013 with permission from AIP Publishing LLC

(Fig. 2.3a) and pH 5.5 (Fig. 2.3b). The pH window 5.5–7 apparent when contrasting Fig. 2.3a, b contains only the pKa $\approx$ 6.1 of histidine (H), and hence this residue is predicted and shown to undergo the most dramatic gain in polarization steering (decrease in expected APF) as pH is decreased from 7 to 5.5. The effects of titration on other residue types could not be assessed in this study since their pKa's dictate extreme pH values that would introduce denaturing conditions for the proteins studied.

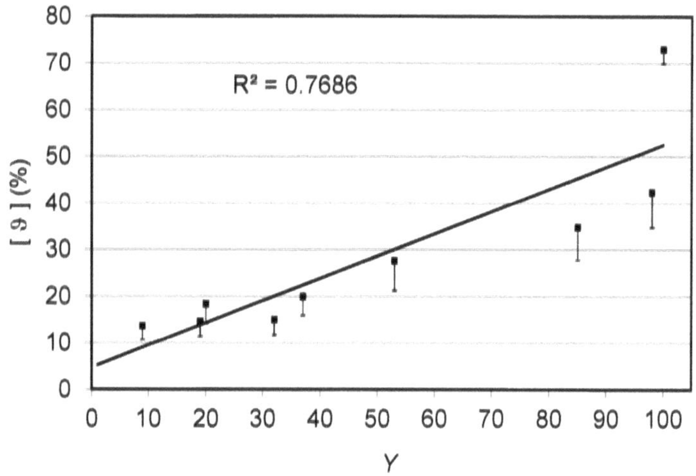

Fig. 2.4 Correlation between ϑ-value averaged over all residues in the protein ([ϑ]), and the dehydron-to-backbone-hydrogen-bond ratio Y for the protein. Both parameters are given as percentages. The proteins studied identified by their respective PDB files and Y-ratios (in brackets) are: 1SRL (14.3), 1ESR (27.9), 1A8O (32.1), 1PIT (40.0), 1QGB (48.6), 1ATA (57.7), 1Q7I (70.1), 1PI2 (98.0), 2PNE (100.0). The average APF values indicated by the *filled squares* correspond to $r = 6$ Å. The low error bar indicates the net decrease in APF as bulk solvent is approached when adopting $r = 8$ Å. Reprinted from [Fernández Stigliano A (2013) Breakdown of the Debye polarization ansatz at protein-water interfaces. J Chem Phys 138:225103], copyright 2013 with permission from AIP Publishing LLC

A significant correlation is established for PDB-reported proteins (Fig. 2.4) between the average APF ([ϑ]) over all residues in a protein and the protein ratio Y of dehydrons-to-backbone-hydrogen-bonds. This correlation validates the assertion that dehydrons are the main structural motif promoting anomalous polarization. At $Y = 100$ %, the antifreeze protein from snow flea in PDB entry 2PNE [10], with its anomalously high APF-boosting activity, is a significant outlier. This enhanced effect can be understood based on the extreme solvent exposure of its dehydrons, promoting a local backbone-hydrated state that persists on a 100 ns timescale, compared with the ~1 ns lifetime of the hydrated state typical of the dehydrons present in the other proteins studied. Thus, the water-organizing power of the antifreeze protein is due to nanoscale confinement and supersedes the Debye polarization tendencies, introducing a major supra-nanoscale perturbation of the water structure, in accord with its purported function as a disruptor of the ice nucleation.

This connection between disruption of ice nucleation and anomalous polarization suggests mutational studies aimed at removing dehydrons by improving the backbone protection in antifreeze proteins. We predict that the removal of dehydrons by backbone-protective valine (V) substitutions of poor backbone protector residues (S, T, G) should significantly impair the antifreeze potency of the snow flea protein reported in PDB.2PNE.

2.3 Epistructural Physics Reveals a Chemical Functionality for the Aqueous Interface

This section carries the analysis of the dielectric structure of the aqueous interface one step further by showing that dehydrons not only promote protein associations but also functionalize interfacial water by inducing basicity at the interface. These packing defects confine interfacial water molecules turning them into proton acceptors. This result has profound ramifications for bioengineering and drug design as it implies that dehydrons are actually involved in chemical events, acting as stimulators of enzymatic activity.

To make notation more agile, we introduce aligned ($\Gamma^{\parallel}$) and orthogonal ($\Gamma^{\#}$) polarization-induced charges defined as

$$-\vec{\nabla}\,.\,\vec{P}^{\parallel} = \Gamma^{\parallel};\ -\vec{\nabla}\,.\,\vec{P}^{\#} = \Gamma^{\#};\ \Gamma^{\parallel} + \Gamma^{\#} = \Gamma \tag{2.5}$$

Departures from bulk water structure (spatially measured by $\nabla g \neq 0$) induce orthogonal polarization. Given the relation $\vec{P}^{\#} = \xi\,\vec{\nabla}\,g$, the Poisson equation $\vec{\nabla}\,.\left(\vec{P}^{\#}\right) = \rho - \vec{\nabla}\,.\left(\varepsilon_0\,\vec{E} + \vec{P}^{\parallel}\right)$ may be written in terms of the curvature $\nabla^2 g$ of the scalar field g:

$$\xi\nabla^2 g = -\Gamma^{\#} \tag{2.6}$$

Equation (2.6) incorporates the nanoscale structure of water within an electrostatic relation, revealing that the curvature of the scalar field g is a measure of the departure from linear dielectrics.

We also introduce the scalar field $\phi\left(\vec{r}\right) = 4 - g\left(\vec{r}\right)$ that quantifies the extent of distortion from bulk-like water structure, with $\phi = 0$ representing no distortion. To obtain a partial differential equation for ϕ we first note that at each position $\vec{r}$, the quotient $\Gamma^{\#}/\Gamma$ measures the local deviation from a Debye scenario where polarization fully aligns the protein field (Fig. 2.5). Thus, we expect that a relation of the form $\frac{\Gamma^{\#}}{\Gamma} = c\phi$ must hold, where c is a proportionality constant. This relation is indeed valid with c = 0.191 as shown in Fig. 2.6a. The quotient $\Gamma^{\#}/\Gamma$ is computed at protein/water interfaces as a time average over a 10 ns-period beyond equilibration of the protein structure with the solvent. Thus, the episteric polarization $\vec{P}$ for nine soluble natural proteins with structures reported in the Protein Data Bank (Table 2.1, Protocol in legend for Fig. 2.1) is computed along thermalization molecular dynamics trajectories. Each 10 ns-trajectory is generated using as starting point the equilibrated structural coordinates that result after thermalization of the

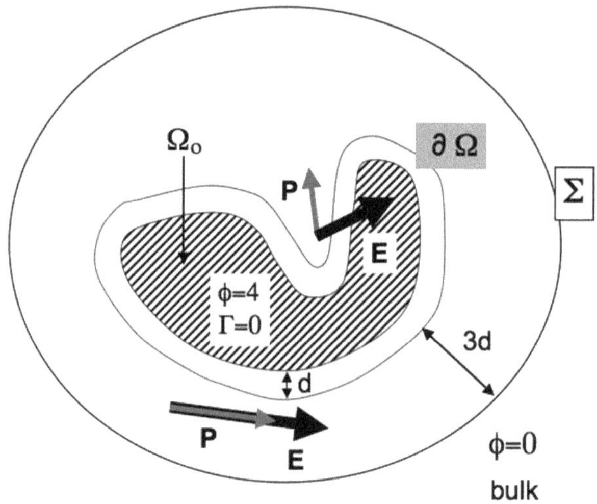

Fig. 2.5 Schematic depiction of the aqueous interface with boundary conditions that become constraints in determining the coarse-grained descriptor $\phi = 4 - g$ of water structure. The interface is defined by the water-smeared envelope $\partial\Omega$ of the solute. It is assumed that the solute surface has a topography endowed with nanoscale detail that may perturb the structure of interfacial water relative to bulk-like patterns. We get $\phi = 4$ (no water) in the "core" volume Ω_0 at distance d = 2r = 8 Å from each point on $\partial\Omega$, while $\phi = 0$ for points at distance larger than 3d = 24 Å from $\partial\Omega$. The latter condition holds since 3d $\gg$ 4 water layers ($\sim$ 13Å) from the interface and hence in this region, water structure is assumed to have recovered its bulk-like pattern. The regions where the structure of interfacial water is relatively undistorted show an alignment between episteric polarization $\vec{P}$ and the electrostatic field $\vec{E}$, whereas regions of high structural distortion likely depart from the linear dielectrics picture, as reflected by a lack of alignment between the fields $\vec{P}$ and $\vec{E}$. Reprinted from [1], copyright 2013 with permission from AIP Publishing LLC

PDB-reported structure immersed in a pre-equilibrated solvent bath. Simulations are performed within an isobaric/isothermal ensemble (1 atm, 298 K). A total of 100 interfacial solvent configurations, one per 100 ps along a 10 ns-thermalization trajectory, are used to compute the episteric polarization quotient as a time average using the relation

$$\frac{\Gamma^{\#}}{\Gamma} = \frac{\rho - \vec{\nabla}.\left[\varepsilon_0 \vec{E} + \vec{P^{\parallel}}\right]}{\rho - \vec{\nabla}.\varepsilon_0 \vec{E}} \tag{2.7}$$

To this end, we recorded charge distribution, internal field and polarization from the 100 snapshots that partition the 10 ns period in identical intervals.

The structure/solvent system is considered equilibrated at time t_o if the RMSD of backbone atomic coordinates averaged over randomly chosen pairs of chain

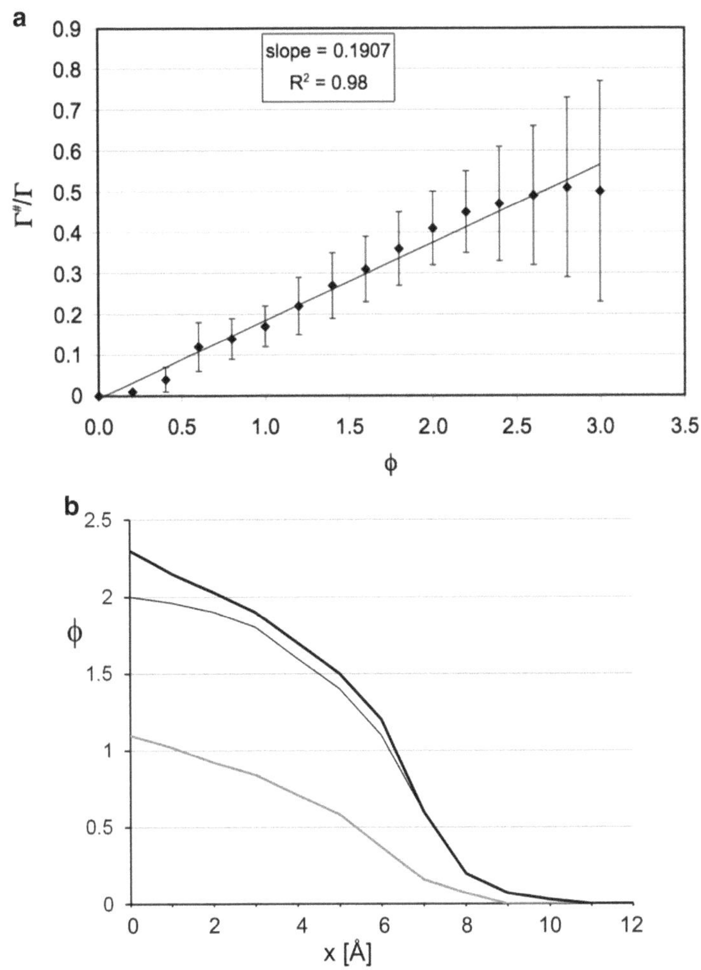

Fig. 2.6 (**a**) Local deviation from $\vec{E}$-aligned polarization measured as $\Gamma^{\#}/\Gamma$ and plotted against the structural function $\phi = 4 - g$. The quotient $\Gamma^{\#}/\Gamma$ is evaluated at protein/water interfaces as a time average over an interval beyond protein structure/solvent equilibration. To determine $\Gamma^{\#}/\Gamma$, the episteric polarization $\vec{P}$ for nine soluble natural proteins with structures reported in the Protein Data Bank (Table 2.1) is computed and averaged for each position in space along a thermalization molecular dynamics trajectory. The region spanning a distance of $3d = 24$ Å from $\partial\Omega$ (Fig. 2.5) is exhaustively interrogated by covering it with disjoint spheres of radius 4 Å centered at points $\vec{r}_n$(n = 1, 2,. . .) to determine the set of water-structure values $\phi\left(\vec{r}_n\right)$. The trend line and correlation coefficient were obtained by linear regression. (**b**) Behavior of $\phi(x)$ relative to the distance x to an interface. The interface is assumed nonpolar, consisting of a concave region of fixed curvature radius θ, with $x = 0 =$ center of curvature. The flat surface represents the macroscopic limit $\theta \gg 1$ nm, and the ϕ-values are shown in the *grey plot*. Other perturbations were obtained for $\theta = 3$ Å (*thin black plot*), and $\theta = 2.5$ Å (*thick black plot*). Reprinted from [1], copyright 2013 with permission from AIP Publishing LLC

Table 2.1 PDB accession codes and free energy changes (ΔG) for the thermal denaturation of soluble monomeric proteins with reported structures

PDB accession code	ΔG (kJ/mol)	T (C)
1BSQ	46.46	40.00
1RTB	42.28	25.00
4LYZ	37.76	26.85
1CX1	22.52	24.85
1QG5	36.84	40.00
2AIT	28.05	25.00
3SSI	17.04	20.00
1HIC	21.01	25.00
1PMC	4.60	20.00

Fernández A (2012) Communication: Nanoscale electrostatic theory of epistructural fields at the protein-water interface. J Chem Phys 137:231101, and references therein
The temperature (T) for thermal denaturation is indicated

conformations within a time interval $[t_o, t_o + \tau]$ ($\tau \approx 1$ ns) is less than 1 Å. For all nine proteins in this study, this criterion was fulfilled for $t_o = 500$ ps. Solvent and side-chain conformations continue to vary significantly (i.e. RMSD > 2.25 Å) on the 1 ns timescale.

By introducing the relation $\frac{\Gamma^{\#}}{\Gamma} = c\phi$, and defining the constant $k = c/\xi$, (2.6) becomes a linear homogeneous differential equation in $\phi\left(\vec{r}\right)$ of the Schrödinger type:

$$-\nabla^2\phi + k\Gamma\phi = 0 \tag{2.8}$$

The boundary conditions are described in Fig. 2.5 and are determined by the water-smeared envelope $\partial\Omega$ of the solute-protein interface. Thus, $\phi = 4$ (no water) in the "core" volume Ω_0 at distance $d = 2r = 8$ Å from each point on $\partial\Omega$, while $\phi = 4$ for points at distance larger than $3d = 24$ Å from $\partial\Omega$. The latter condition holds since $3d \gg 4$ water layers (~ 13 Å) from the interface and hence in this region, water structure is assumed to have recovered its bulk-like tetrahedral pattern.

The behavior of $\phi = \phi\left(\vec{r}\right)$ relative to the distance x to an interface is indicative of the propagation in space of the distortion of the water structure and is shown in Fig. 2.6b. The results were obtained by numerical integration of (2.8) for structural perturbations generated by confinement of water at the interface. To eliminate confounding factors, the interface was assumed nonpolar and physico-chemically featureless, consisting of a concave region of fixed curvature radius q, capable of partially confining water molecules, with $x = 0$ representing the center of curvature of the surface. The flat surface represents the macroscopic limit $\theta \gg 1$ nm, and the interfacial ϕ-value is expectedly close to 1 ($g \sim 3$), as revealed by the grey plot in Fig. 2.6b. Other perturbations were obtained for $\theta = 3$ Å (thin black plot), and $\theta = 2.5$ Å (thick black plot), generating $\phi(0)$-values 2.01 and 2.30, respectively. In all cases, the matrix distortion decays to zero for $x > 11$ Å (less than four water layers). No water molecule enters the cavity if doing so implies that the molecule retains on average less than 1.6 hydrogen bonds, making the range $\theta < 2.37$ Å, a forbidden region in real terms.

Equation (2.8) is the central result of this section and governs the interplay between episteric polarization and the nanoscale structure of interfacial water.

We now compute the energy increment ΔU_ϕ associated with spanning a protein/water interface. This energy is in fact an elastic contribution stored in the distortion of water structure as shown in Chap. 1, with $\nabla\phi\left(\vec{r}\right)$ measuring the local structural distortion at position $\vec{r}$ and the elastic integrand of $\int \frac{1}{2}\lambda \left\|\vec{\nabla}\phi\right\|^2 d\vec{r}$ ($\lambda = 9.0$ mJ/m) quantifying the energetic contribution of spanning a differential region $d\vec{r}$ centered at point $\vec{r}$. Using Gauss' divergence theorem we obtain:

$$\Delta U_\phi = \int \frac{1}{2}\lambda \left\|\vec{\nabla}\phi\right\|^2 d\vec{r} = -\frac{1}{2}\lambda \int \phi \nabla^2 \phi d\vec{r} \tag{2.9}$$

Using (2.8), we can substitute the integrand in the r. h. s. of (2.9) obtaining the alternative expression for the interfacial elastic energy:

$$\Delta U_\phi = -\frac{1}{2}c(\lambda/\varepsilon_0)^{1/2} \int \Gamma \phi^2 d\vec{r} \tag{2.10}$$

Since $\Delta U_\phi = \int \frac{1}{2}\lambda \left\|\vec{\nabla}\phi\right\|^2 d\vec{r} \geq 0\,(\lambda > 0)$, the r.h.s of (2.10) is a positive term, and thus (2.9) and (2.10) imply:

$$\int \Gamma \phi^2 d\vec{r} \leq 0. \tag{2.11}$$

Equation (2.11) implies that $\Gamma \leq 0$, *that is, the polarization-induced charge is negative around dehydrons known to promote interfacial tension.* This is a fundamental result of broad applicability and it establishes the following

Theorem *Interfacial water molecules tend to orient and organize in subnanoscale cavities leaving negative charges uncompensated* ($\Gamma < 0$) *when deprived of hydrogen-bonding opportunities* ($\phi > 0$).

This is a crucial result as it delineates the chemical basicity of dehydrons, or more properly, the dehydron-induced basicity of interfacial water. The theorem reveals that packing defects play a crucial role in enzyme catalysis and in biochemical events in general by functionalizing nucleophilic protein groups through dehydron-promoting proton accepting events. This type of catalytic stimulation requires that dehydrons be spatially close to catalytic nucleophilic groups and that the proton transfer event induced by the dehydron have a reasonable probability.

These striking aspects of the functionalization of biomolecular interfaces will be explored and ultimately established in Chap. 7.

The interfacial energy stored in the anomalous polarization or, equivalently, in the distortion of water structure, is readily evaluated using (2.10). This result has been contrasted against thermodynamic data on the spanning of aqueous interfaces with nanoscale detail. A suitable testing ground is provided by the aqueous interfaces for soluble monomeric proteins with a stable fold characterized by structural and thermodynamic information (Table 2.1) [1]. The reversible work performed on the system to span the protein-water interface is destabilizing of the native fold, thus facilitating thermal denaturation, and hence it should anticorrelate with the free energy change for protein denaturation, as it is indeed the case (cf. [1]). To compare interfacial thermodynamics with thermodynamic data on protein denaturation (Table 2.1), we introduce the entropic cost of solvent confinement at the interface $\Delta S_\phi = k_B \ln\left[\Pi_{j=1}^{L} g_j/4\right]$, where $k_B =$ Boltzmann constant, $g_j =$ time averaged number of hydrogen bonds for the jth-water molecule (L = total number of water molecules), and the dummy index j labels molecules within 3d = 24 Å from the solvent-smeared envelope of the protein (Fig. 2.5). Note that $-T\Delta S_\phi \geq 0$ and reinforces the trend defined by ΔU_ϕ. The reversible work $\Delta G_\phi = \Delta U_\phi - T\Delta S_\phi \geq 0$ performed on the system to span the protein-water interface is destabilizing of the native fold, thus facilitating thermal denaturation.

To quantitatively assess the folding-destabilizing effects of spanning the interface of the folded protein, we examined the same soluble monomeric proteins used to generate the data in Fig. 2.6 (Table 2.1). In thermodynamic terms, protein denaturation is facilitated proportionally to the reversible work required to span the interface, attesting to the folding-destabilizing effect of interfacial tension arising from the structural distortion of surrounding water. Thus, the computed reversible work for creating the interface measures the extent to which the "protein structure is at odds with the structure of surrounding water" since it quantifies the distortion of water structure around the protein. The tight anticorrelation between interfacial free energy and the stability of the protein structure (Fig. 2.7) provides experimental support to the underlying equation (2.8) since it reveals that protein destabilization is commensurate with the thermodynamic cost of creating its interface with water, computed using (2.10). This observation prompts us to formulate the principle of minimal episteric distortion (MED) that should govern conformational changes in the solute that generate concomitant changes in the interface. The validity of the MED principle in the context of protein folding is established in Chap. 3 that provides a semiempirical solution to the protein folding problem.

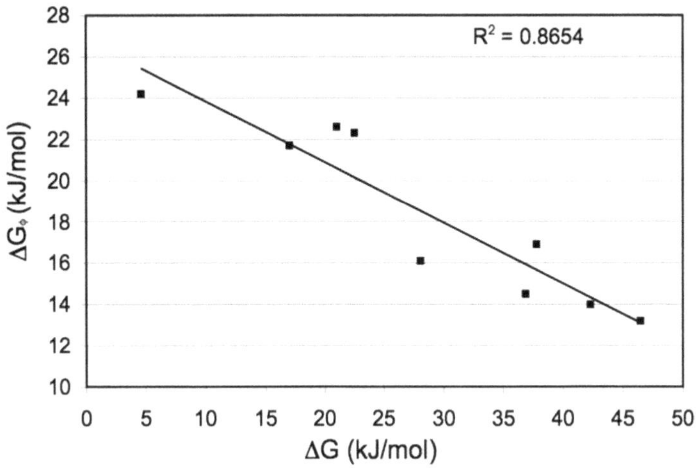

Fig. 2.7 Anticorrelation between reversible work required to span the protein-water interface, ΔG_{ϕ}, and free energy change for thermal denaturation, ΔG, for the nine monomeric soluble proteins indicated in Table 2.1. Reprinted from [1], copyright 2013 with permission from AIP Publishing LLC

2.4 Packing Defects as Catalytic Enablers

Enzyme catalysis is often viewed as a closed chapter where the core issues have already been dealt with [11, 12]. Yet, several open problems still stand on the way of progress in mechanistic understanding [13–15], and the design and optimization of enzyme catalysts [15–17] and drug-based enzyme inhibitors [17–19]. Especially opaque yet germane to these problems is the role of interfacial water in enzymatic reactions [13, 20, 21]. Because the protein aqueous interface is essentially sculpted by the protein structure [20, 21], the problem may be said to belong to the field of *epistructural biology*, as argued in Chap. 1. In this realm, as we have already noted, one structural feature of proteins stands out: the so-called *dehydron*, a packing defect that creates interfacial tension and thereby promotes protein associations that exclude surrounding water [21, 22]. Thus, the recently established fact that catalytic sites in enzymes are actually "decorated" with dehydrons [13] proves to be tale-telling and gives a significant spin to related biotechnologies. These observations will be properly delineated, expanded and validated in Chap. 7.

As implied by the theorem formulated in Sect. 2.3 (cf. [14]), besides promoting dehydration, dehydrons are also likely to be endowed with a biochemical role that may prove to be exquisitely complementary: they turn local interfacial water into a chemical base, a proton acceptor. Thus, if a catalytic group (hydroxyl in Ser, Thr or Tyr, thiol in Cys, amide in His) performs a nucleophilic attack on a substrate, the dehydrons nearby enhance its catalytic potential through a chemical functionalization of vicinal water that promotes deprotonation of the catalytic group.

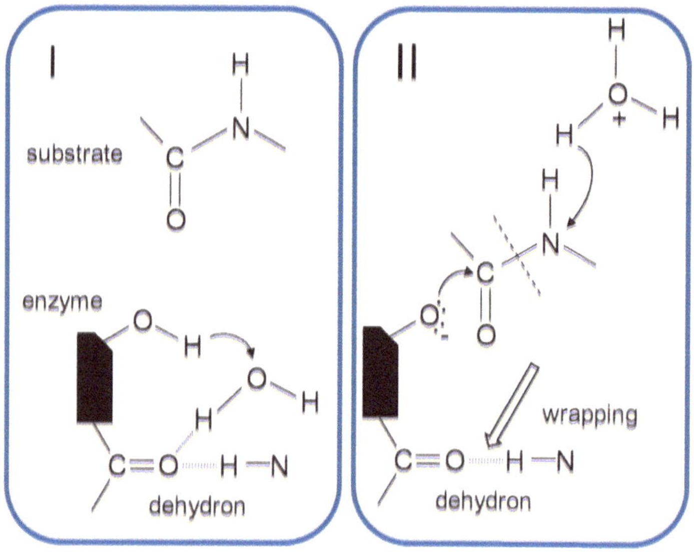

Fig. 2.8 Dehydron as two-stroke molecular engine sustaining enzyme catalysis by (**a**) functionalizing nano-confined water around the catalytic group and (**b**) promoting enzyme-substrate association, thereby correcting the nanoscale packing defect

The dual functional and interactive role of dehydrons combined with the fact that they are ubiquitously found at catalytic sites suggest a dual participation in catalysis: *first, dehydrons prepare the solvent for enzyme activity and, once the enzymatic nucleophilicity is enhanced and the solvent turns into a better leaving group (as hydronium), dehydrons promote enzyme-substrate association in consonance with their dehydration propensity*. This duality of functionality and interactivity makes dehydrons both enablers and stimulators of enzyme catalysis, an observation with paramount biotechnological implications, especially in regards to what we may term "epistructure-based enzyme design".

In more rigorous terms, as dehydrons activate nearby catalytic groups to perform a chemical (nucleophilic) attack on the substrate, causing trans-esterification, they turn the local water into hydronium (H_3O^+, a product of proton acceptance). In turn, the hydronium is easily removed from the active since it requires further hydration, thereby enabling enzyme-substrate association. This association process entails the exogenous "wrapping" of the dehydron, which is tantamount to the intermolecular correction of the structural defect [22]. Thus, *the dehydron may be regarded as a two-stroke molecular engine that agonizes and enables enzyme catalysis*, as described in Fig. 2.8.

This discovery heralds the advent of novel biomolecular design based on "dehydron enablers-stimulators" that may be created or removed though engineered mutations directed to fine-tune the protein structure. This finding makes it possible to activate or silence a catalytic site in a protein enzyme by respectively creating or annihilating a nearby dehydron through a change in the chemical composition of the protein. On the other hand, novel drug-based enzyme inhibitors will emerge as dehydron enablers-agonists are targeted through engineered protein-drug associations [19], as described in Chap. 10.

The newly established participation of dehydrons in enzymatic reactions will likely invite an extensive revision of the biochemical mechanistic literature, while novel molecular designs inspired by "epistructural catalytic stimulation" are expected to herald a new era in the optimization of enzyme catalysts and pharmaceuticals.

2.5 A Quantum Theory of Interfacial Tension and Its Experimental Verification

The complexities of the protein-water interface preclude us at this time from attempting the implementation of a quantum mechanical approach to study interfacial water frustration and the polarization effects resulting thereby. Thus, we shall content ourselves with studying the simplest nonpolar aqueous interfaces using a quantum approach. Be as it may, we are keenly aware that the results may be one day extrapolated *mutatis-mutandis* to the study of realistic biological interfaces.

Interfacial water on a nonpolar surface remains a subject of intense scrutiny due to its relevance in delineating the molecular forces that steer protein folding [1] and molecular associations [20], and drive biotechnological innovation [20, 23–26]. The spontaneous negative charging of such interfaces through transference of hydroxide ions (OH^-) from bulk water has been established [27–29], yet the topic still remains a subject of contention [25, 27, 29]. No first-principle inference of spontaneous interfacial charging has so far generated quantitative predictions amenable of experimental verification [28]. In this regard, the treatment of interfacial water through molecular dynamics (MD) appears to yield contradictory results, even predictions of acidic (proton donating) interfaces [27, 30]. This controversy stimulates the fundamental approach taken in this section, which involves a substantive revision of Debye's dielectric ansatz [24]. The latter would predict no polarization-induced charge since there is no intrinsic electrostatic field to speak of. Thus, we may state that the spontaneous charging of an aqueous nonpolar interface poses a challenge to the standard dielectric picture, demanding a reworking of the Debye ansatz starting from first principles.

In this section we argue that experimentally verifiable quantitative predictions of spontaneous negative charging of aqueous hydrophobic interfaces require a general treatment of interfacial water dielectrics that cannot be subject to the constraints imposed by Debye's ansatz regarding polarization alignment [1, 31] but needs to be further enriched by incorporating quantum effects. The first conceptual departure is essential to include the anomalous polarization component that arises as a result of hydrogen-bond frustration at the contact region between the two bulk phases. This revision could not have been reasonably envisioned at the time Debye's linear dielectric ansatz was formulated since details on water hydrogen-bonding structure were unknown [24]. As shown previously in this chapter and in Chap. 1, the non-Debye contribution generates interfacial tension that is stored as electrostatic energy, a tension shown in this section to be mitigated by the adsorption of hydroxide ions. To assess the level of interfacial frustration of hydroxide ions we need to incorporate quantum effects. By frustration in this context we refer to a hydrogen bonding opportunity that cannot materialize because the proton acceptor or nucleophilic group is not physically present, as it is the case with the water layer in contact with the nonpolar surface.

As previously noted, the departure from the Debye linear dielectric *ansatz* becomes imperative because no charge $\gamma = -\nabla.\mathbf{P}$ may be generated by water polarization ($\mathbf{P}$) if the latter is made proportional to the vanishing electrostatic field $\mathbf{E}$ at the nonpolar surface, as it is the case with the Debye assumption $\mathbf{P} = (\varepsilon - \varepsilon_o)\mathbf{E}$, where ε, ε_o denote dielectric and vacuum permittivity, respectively.

To predict quantitatively the spontaneous charging of the aqueous nonpolar interface, the more general dielectric scheme presented in Chap. 1 is required. The hydrogen-bond frustration of interfacial water, described by the scalar field $\phi = \phi(\mathbf{r})$, with $\phi(\mathbf{r})$ = expected number of unfulfilled hydrogen bonds for a water molecule at position $\mathbf{r}$, generates a non-Debye polarization component $\mathbf{P}^{\#} = \mathbf{P} - (\mathbf{P}.\mathbf{e})\mathbf{e}$ ($\mathbf{e} = \mathbf{E}/||\mathbf{E}||$). In turn, this "orthogonal" polarization induces a net charge $\gamma^{\#} = -\nabla.\mathbf{P}^{\#}$. In previous work [1, 31] we showed that distortions from bulk structure cause $\mathbf{P}^{\#}$ to be proportional to the frustration gradient, according to the equilibrium relation

$$\mathbf{P}^{\#} = -\xi\nabla\phi \tag{2.12}$$

The generation of net charge, and thereby of polarization, arises from the hydrogen-bond frustration at the interface (2.12) that leaves partial charges on the water atoms untitrated. The lack of interfacial charge neutralization implies that interfacial tension, which by definition arises from hydrogen-bond frustration, is also stored electrostatically in the anomalous polarization [1, 24]. Equation (2.12) will be subsequently corroborated in the context of this study. From (2.12), the net charge $\gamma^{\#}$ induced by $\mathbf{P}^{\#}$ becomes $\gamma^{\#} = \xi\nabla^2\phi$. By making interfacial tension proportional to electrostatic energy, the non-Debye treatment enables the computation of the net charge induced by the frustration at the water layer in contact with the nonpolar surface, and hence becomes adequate to predict quantitatively the spontaneous charging of the interface.

For simplicity, we consider water in contact with a featureless slab on nonpolar nonpolarizable material typically realized by hydrogenated graphene [29]. The frustration or net loss of hydrogen bonding opportunities at the hydrophobic interface generates interfacial tension quantifying the disruption of the bulk-like tetrahedral hydrogen-bond coordination of water, as shown in Chap. 1 [1, 20, 31]. The extent of frustration of a water molecule at a liquid/solid interface is expected to be higher than that of a hydroxide ion ($^{-}$OH) since the donated proton in the latter chemical species is expected to have a lower electrophilicity than the water proton, and therefore is expected to be less prone to form hydrogen bonds with a nearby nucleophile [29]. This conjecture will be corroborated in this work and is based on the fact that the hydroxide oxygen is less electrophilic than the water oxygen, since the former gained an electron while the latter shares an electron pair with a hydrogen atom. The argument leads us to postulate that the interfacial tension is lowered by adsorption of hydroxide ions or, in other words, the spontaneous negative charging of the nonpolar interface is expected to result from the relief of interfacial tension.

Given the above-noted observations, a quantum parameter, ζ, will be subsumed into the MD computations scaling a term accounting for interfacial energy in order to quantify the degree of frustration of interfacial hydroxide relative to interfacial water. The relative hydrogen bond frustration is given by the quotient $\zeta = \zeta(OH^-/H_2O) = F_h/F_w = 0.021/0.179 = 0.117\ (\pm 0.017)$, where F_h, F_w are the expected net quantum electrostatic fields (in e/Å^2 units) at the position or probe site of a putative hydrogen-bonding nucleophile (lone electron pair) that acts as putative acceptor of the proton donated by OH^- and H_2O, respectively [29]. The fields were obtained by molecular projection of the delocalized quantum charges in the aqueous condensed phase [32] along the direction of proton donation for hydrogen bonding [29]. The authors in [29] implemented a Car-Parrinello molecular dynamics scheme in a Wannier representation of the condensed phase [32], thus generating a unique set of maximally localized Wannier functions that realized the solid-state equivalent of localized molecular orbitals.

Based on the estimation of the quantum parameter ζ given above, we may assert that *the extent of frustration due to an unfulfilled hydrogen bond in the hydroxide ion at a nonpolar interface is 11.7 % of the hydrogen-bond frustration of water with a dangling OH group*. Conversely, $\zeta(H_3O^+/H_2O) = 1.922\ (\pm 0.102)$, (cf. [29]) implying that hydronium adsorption would entail almost twice as much frustration than water at the interface. To summarize, mitigation of interfacial tension dictates that hydroxide ion—and not hydronium—must be preferentially adsorbed.

To predict the spontaneous charging of the hydrophobic interface, we first incorporate into the MD simulations the interfacial energy per unit volume, Δu_ζ, associated with spanning a water interface that envelops a solid non-polarizable hydrophobic slab of macroscopic dimensions. The energy-density term Δu_ζ quantifies the distortion of the water hydrogen-bond matrix and is therefore given by the elastic integral [1]:

$$\Delta u_\zeta = (1/2)\lambda \int ||\nabla\phi||^2 d\mathbf{r}, \qquad (2.13)$$

where, as indicated previously, $\phi = \phi(\mathbf{r})$ denotes hydrogen-bond frustration of the solvent and the integration is carried over a spatial domain Ω large enough so that its border $\partial\Omega$ is fully contained in bulk water, where the following conditions are satisfied: $\phi(\mathbf{r}) = 0$, $\nabla\phi(\mathbf{r}) = 0\ \forall r \in \partial\Omega$. To describe the local distortion of water

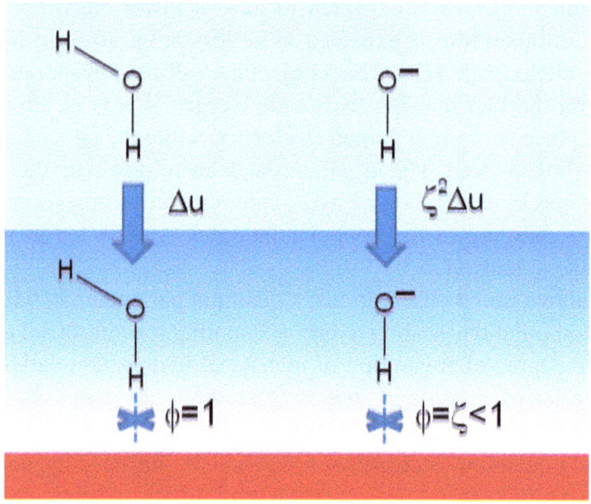

Fig. 2.9 Scheme of interfacial tension quenching due to hydroxide ion adsorption onto a nonpolar aqueous interface. The interfacial region is represented with color gradient, bulk water is in *solid blue* and the nonpolar slab is in *brown*. The frustrated hydrogen-bonding opportunity is symbolized by a *crossed dotted line*. The level of frustration at the interface with the nonpolar slab is reduced by a factor ζ (< 1) upon hydroxide ion adsorption, while the interfacial energy density change associated with transference from the bulk to the interface is reduced by a factor ζ^2 due to the square dependence on the frustration gradient of the elastic integrand in (2.13)

structure at spatial location $\mathbf{r}$ relative to the bulk hydrogen-bond pattern, we introduce the scalar field $\phi = \phi(\mathbf{r})$ that *in this context* represents the expected frustration of a water molecule or a hydroxide ion while any of them visits a sphere of radius $r = 4$ Å centered at position $\mathbf{r}$ for a 1 ps-timespan [1, 20, 31]. If the proton donation of the hydroxide ion leading to hydrogen bonding is frustrated due the absence of a proton acceptor, we get $\phi = \zeta$ (< 1) while the frustration of water hydrogen bonding yields $\phi = 1$.

In essence, we compute the interfacial tension arising from the hydrogen bonding frustration of interfacial water, and take into account the fact that this tension is minimized by the adsorbed hydroxide ions, whose dangling donated proton is 88.3 % less frustrated than the water counterpart (Fig. 2.9). Our theoretical treatment may be benchmarked and validated, as done in this work, since progress in spectroscopic methods based on second-order nonlinear optics enable direct examination of water structure at the interface [28].

The quantum-mechanical parameter ζ is thus incorporated in the MD simulations to account for the reduction in the frustration levels on the first layer directly in contact with the nonpolar surface due to hydroxide ion adsorption.

To validate (2.12), we observe that this equation dictates that the interfacial tension is actually stored at equilibrium as the non-Debye polarization energy density since

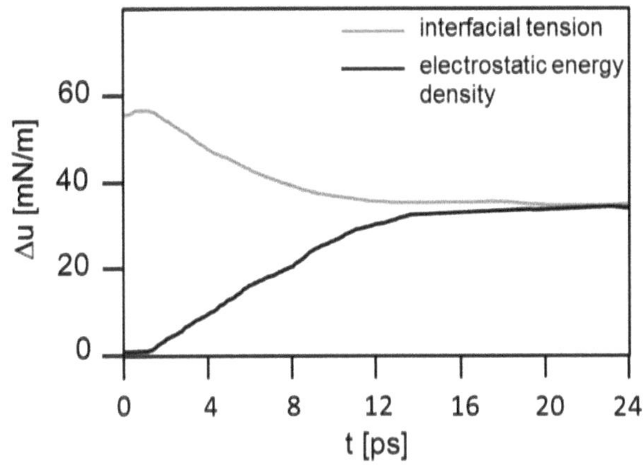

Fig. 2.10 Convergence of the interfacial tension (Eq. 2.13) and non-Debye electrostatic energy density as equilibrium is reached. The curves displayed are averages over 10 MD 24 ps-trajectories for water dynamics within an isobaric/isothermal ensemble (1 atm, 298 K) with 10,160 water molecules and a total of 160 ions, with 80 OH^- and 80 H^+ ($[OH^-] = 0.4$ M) packed against a featureless layer of hydrogenated graphene. The MD 24 ps-trajectories for water dynamics are generated as described in [1] except for the incorporation in the potential energy of an interfacial term to account for hydrogen bond frustration at the nonpolar region of contact. The solvent bath provides at least four water layers of solvent envelope with water molecules extending 18 Å from the graphene surface. The evaluation of the interfacial tension-generating frustration incorporates a quantum mechanical parameter ζ describing the relative electrophilicity of the donated hydroxide proton relative to the water proton, a measure of the hydrogen bonding propensity of the dangling O-H bond in hydroxide relative to water

$$\Delta u_\zeta = (1/2)\lambda \int ||\nabla\phi||^2 \mathbf{dr} = (1/2)\varepsilon_o^{-1} \int \mathbf{P}^{\#2} \mathbf{dr} \tag{2.14}$$

The direct MD computation of the time evolution of the electrostatic energy density and interfacial tension has been performed by incorporating the global term Δu_ζ in the potential energy stepwise evaluation. The convergence of the two quantities has been observed (Fig. 2.10), thus validating the equilibrium relation presented in (2.12).

Since the surface integral $\oiint[\phi\nabla\phi] \cdot d\sigma$ over $\partial\Omega$ vanishes as per the choice of Ω, integration of (2.13) by parts yields the usual relations, now specialized for an ionic solution at the interface:

$$0 \le \Delta u_\zeta = -(1/2)\lambda \int \phi\nabla^2\phi \mathbf{dr} = -(1/2)(\lambda/\varepsilon_o)^{1/2} \int \gamma^{\#}\phi \mathbf{dr} \tag{2.15}$$

Since $\phi \ge 0$, we obtain:

$$\gamma^{\#}{}_{\mathrm{MV}} = -(\lambda\varepsilon_{o})^{1/2} \int ||\nabla\phi||^{2}d\mathbf{r} / \int \phi d\mathbf{r} \leq 0, \tag{2.16}$$

where $\gamma^{\#}{}_{\mathrm{MV}}$ is the mean-value induced charge per unit volume. We have then proven the following *theorem*:

An aqueous featureless nonpolar interface causes hydrogen-bond frustration which is mitigated by hydroxide anion adsorption generating a net negative charge at the interface that may be estimated at $\gamma^{\#}{}_{\mathrm{MV}}$.

Thus, the net induced charge at the interface results from the adsorption of the negatively charged hydroxide ions. The net charge was estimated according to (2.16), requiring the construction of the field $\{\phi(\mathbf{r}), \nabla\phi(\mathbf{r})\}$ [1, 31] within an NPT-statistical ensemble for a featureless hydrophobic interface in the limit of infinite curvature. Thus, the calculation of $\gamma^{\#}{}_{\mathrm{MV}}$ following (2.16) yields $[\mathrm{OH}^-]^{\#} = \gamma^{\#}{}_{\mathrm{MV}}/\mathrm{Y} = 15.4$ mM, where $[\mathrm{OH}^-]^{\#}$ = hydroxyl molar concentration at interface and Y is the net charge per mole of hydroxyl ion. Thus, at pH 7 (neutral solution) we get the equilibrium constant

$$\mathrm{K} = [\mathrm{OH}^-]^{\#}/[\mathrm{OH}^-] = 153{,}832 \tag{2.17}$$

Let $\mathrm{M} = [H_2O] \gg [\mathrm{OH}^-]$ be the molar concentration of "free adsorption sites" $\approx$55.5 mole/liter at T = 298 K, then $\mathrm{K} = \mathrm{M}^{-1}\exp(-\Delta\mathrm{G}/\mathrm{RT})$, where ΔG is the free energy change associated with the transference of a hydroxide ion from bulk water to the interface. Thus, our predicted value $\Delta\mathrm{G} = -40.02$ kJ/mole compares satisfactorily within experimental error with the experimental value −45 kJ/mole obtained from the Langmuir adsorption isotherm monitored by phase-sensitive sum-frequency vibrational spectroscopy [28].

The spontaneous negative charging of the aqueous interface sustained at a flat featureless nonpolar surface is a fact that challenges and hence motivates a substantive revision of Debye's dielectric picture, whereby water polarization aligns with the intrinsic electrostatic field. This is so simply because the latter vanishes in the case under study. Thus, the hydrogen bonding frustration of water at a nonpolar interface generates a non-Debye polarization component that stores electrostatic energy as interfacial tension. As shown in this section, the reduction of the interfacial tension is caused by a net adsorption of hydroxide ions resulting in a spontaneous negative charging of the interface. The adsorption of the hydroxide ion is favored on quantum mechanical grounds as its hydrogen bond frustration at the interface is 88.3 % lower than that of water. The quantum mechanical basis for the induced interfacial effect within a non-Debye dielectric picture is quantitatively validated vis-à-vis surface measurements using second-order nonlinear optics.

The spontaneous negative charging of aqueous nonpolar interfaces has eluded quantitative first-principle prediction. We showed that quantitative prediction required a reworking of Debye's linear dielectric *ansatz* to incorporate non-Debye electrostatic energy stored as interfacial tension. To gain interfacial stability, the minimization of interfacial tension becomes operative promoting a reduction in hydrogen-bond frustration. The latter takes place upon hydroxide ion adsorption, as the quantum theory of interfacial tension predicts. The quantitative predictions were validated vis-à-vis experimental measurements of hydroxide adsorption uptake.

2.6 Problems

2.1. Using Gauss' divergence theorem, prove the following relation:

$$\Delta U_{\phi} = \int \frac{1}{2}\lambda \left\| \vec{\nabla}\phi \right\|^2 d\vec{r} = -\frac{1}{2}\lambda \int \phi \nabla^2 \phi d\vec{r}$$

2.2. Water promotes the sealing of nanoscale packing defects in proteins. This is hinted by the fact that during protein folding, a minimization of the departure from the Debye alignment is observed. How does the solvent promote the structure sealing? Assume a flexible chain that seeks to minimize interfacial free energy.

2.3. Using the protocol provided in this chapter, thermalize the monomeric soluble proteins with PDB-reported structure identified by entries 1SRL, 1ESR, 1A8O, 1PIT, 1QGB, 1ATA, 1Q7I, 1PI2, 2PNE, and validate the following empirical relation indicated in Chap. 1 concerning the density fluctuations: $P(N=0) = e^{-\left[\frac{N^2}{2\sigma^2}\right]} = 1 - g/4, \sigma^2 = \left\langle (N - \langle N \rangle)^2 \right\rangle$.

2.4. Assume you are able to monitor the extent of frustration of an interfacial water molecule at location and time t at any time along a trajectory in phase space. Provide an expression for the frequency-dependent orthogonal dielectric coefficient [15–17].

References

1. Fernández A. The principle of minimal episteric distortion of the water matrix and its steering role in protein folding. J Chem Phys. 2013;139:085101.
2. Jorgensen WL, Chandrasekhar J, Madura J, Impey RW, Klein ML. Comparison of simple potential functions for simulating liquid water. J Chem Phys. 1983;79:926–35.
3. Hornak V, Abel R, Okur A, Strockbine B, Roitberg A, Simmerling C. Comparison of multiple Amber force fields and development of improved protein backbone parameters. Proteins. 2006;65:712–25.
4. Darden T, York D, Pedersen L. Particle mesh Ewald: An N.Log(N) method for Ewald sums in large systems. J Chem Phys. 1993;98:10089–92.
5. Ryckaert JP, Ciccotti G, Berendsen HJC. Numerical-integration of Cartesian equations of motion of a system with constraints: molecular dynamics of N-alkanes. J Comput Phys. 1977;23:327–41.
6. Berendsen HJ, Postma JP, van Gunsteren WF, Di Nola A, Haak JR. Molecular dynamics with coupling to an external bath. J Chem Phys. 1984;81:3684–90.

7. Wang J, Wolf RM, Caldwell JW, Kollman PA, Case DA. Development and testing of a general Amber force field. J Comput Chem. 2004;25:1157–74.
8. Jakalian A, Bush BL, Jack DB, Bayly CI. Fast, efficient generation of high-quality atomic charges. AM1-BCC model: I. Method. J Comput Chem. 2000;21:132–46.
9. Jakalian A, Jack DB, Bayly CI. Fast, efficient generation of high-quality atomic charges. AM1-BCC model: II. Parameterization and validation. J Comput Chem. 2002;23:1623–41.
10. Fernández A, Berry RS. Golden rule for buttressing vulnerable soluble proteins. J Proteome Res. 2009;9:2643–8.
11. Frushicheva MP, Mills MJ, Schopf P, Singh MK, Prasad RB, Warshel A. Computer aided enzyme design and catalytic concepts. Curr Opin Chem Biol. 2014;21C:56–62.
12. Fersht A. Structure and mechanism in protein science: a guide to enzyme catalysis and protein folding. New York: Freeman; 1998.
13. Fernández A. Communication: chemical functionality of interfacial water enveloping nano-scale structural defects in proteins. J Chem Phys. 2014;140:221102.
14. Frauenfelder H. What determines the speed limit on enzyme catalysis? Nat Chem Biol. 2008;4:21–2.
15. Korendovych IV, DeGrado WF. Catalytic efficiency of designed catalytic proteins. Curr Opin Struct Biol. 2014;27C:113–21.
16. Privett HK, Kiss G, Lee TM, Blomberg R, Chica RA, Thomas LM, Hilvert D, Houk KN, Mayo SL. Iterative approach to computational enzyme design. Proc Natl Acad Sci U S A. 2012;109:3790–5.
17. Fernández A, Rogale K, Scott RL, Scheraga HA. Inhibitor design by wrapping packing defects in HIV-1 proteins. Proc Natl Acad Sci U S A. 2004;101:11640–5.
18. Walsh R, Martin E, Darvesh S. Limitations of conventional inhibitor classifications. Integr Biol. 2011;3:1197–201.
19. Fernández A. Transformative concepts for drug design: target wrapping. Berlin: Springer; 2010.
20. Kapcha LH, Rossky PJ. A simple atomic-level hydrophobicity scale reveals protein interfacial structure. J Mol Biol. 2014;426:484–98.
21. Fernández A. Epistructural tension promotes protein associations. Phys Rev Lett. 2012;108:188102.
22. Fernández A, Lynch M. Non-adaptive origins of interactome complexity. Nature. 2011;474:502–5.
23. Giovambattista N, Rossky PJ, Debenedetti PG. Computational studies of pressure, temperature, and surface effects on the structure and thermodynamics of confined water. Annu Rev Phys Chem. 2012;6:179–200.
24. Fernández Stigliano A. Biomolecular interfaces: interactions, functions and drug design, Chap. 2. Berlin: Springer; 2015.
25. Ball P. Nanobubbles are not a superficial matter. Chemphyschem. 2012;13:2173–7.
26. Ramsden JJ. Biomedical surfaces. Norwood: Artech House; 2008.
27. Iuchi S, Chen H, Paesani F, Voth GA. The hydrated excess proton at water-hydrophobic interfaces. J Phys Chem B. 2009;113:4017–30.
28. Tian CS, Shen YR. Structure and charging of hydrophobic material/water interfaces studied by phase-sensitive sum-frequency vibrational spectroscopy. Proc Natl Acad Sci U S A. 2009;106:15148–53.
29. Kudin NK, Car R. Why are water–hydrophobic interfaces charged? J Am Chem Soc. 2008;130:3915–9.
30. Buch V, Milet A, Vacha R, Jungwirth P, Devlin JP. Water surface is acidic. Proc Natl Acad Sci U S A. 2007;104:7342–7.
31. Fernández A. Chemical functionality of interfacial water enveloping nanoscale structural defects in proteins. J Chem Phys. 2014;140:221102.
32. Marzari N, Mostofi AA, Yates JR, Souza I, Vanderbilt D. Maximally localized Wannier functions: theory and applications. Rev Mod Phys. 2012;84:1419–75.

Chapter 3
Solution to the Protein Folding Problem

This chapter unravels a first-principle solution to the protein folding problem. The solution incorporates the dynamic interplay between the formation of packing defects and the interfacial tension created by such defects. Thus, the solution symbiotically combines a structural and epistructural approach to compute the dynamic entanglement between protein chain and solvent. The structural perspective explores the concept of wrapping, its intimate relation to cooperativity and its bearing on the expediency and reproducibility of the folding process. Wrapping refers to the environmental enhancement of intramolecular electrostatic interactions through an exclusion of surrounding water that takes place as the chain folds onto itself. In this way a many-body picture of the folding process emerges whereby the folding chain interacts with itself and at the same time shapes the microenvironments that stabilize or destabilize the intramolecular interactions. This picture reflects a dynamic competition between chain folding and backbone hydration, where ultimately, backbone hydrogen bonds prevail through cooperative wrapping, upholding the picture that "folding is a struggle for the survival of backbone hydrogen bonds".

On the other hand, a solvent-centric or epistructural analysis introduces a crucial component to the free energy of structural assemblage: the reversible work required to span the protein-water interface. Failures of cooperativity, i.e. wrapping deficiencies known as *dehydrons*, generate interfacial tension. Thus folding is steered by a variational principle of minimal interfacial frustration or, translated into structural terms, of minimization in the number of dehydrons. The interfacial contribution to the free energy complements and steers the many-body wrapping dynamics, leading to the semiempirical solution to the protein folding problem presented in this chapter.

To computationally validate the solution to the protein folding problem, we test the reciprocal influence of structural and epistructural events and their dynamic

---

The original version of the book was revised. An erratum can be found at DOI 10.1007/978-3-319-30852-4_18

A. Fernández, *Physics at the Biomolecular Interface*, Soft and Biological Matter,
DOI 10.1007/978-3-319-30852-4_3

entanglement at folding timescales of realistic relevance. Such timescales become accessible through a coarse graining of the chain torsional dynamics within a "modulo Ramachandran basin protodynamics", an underlying steering stochastic process developed by the author.

3.1 Protein Folding as a Many-Body Problem

The protein folding problem may be operationally defined as the inference of the 3D-structure based solely on the amino acid sequence or chemical composition of a protein chain with no additional information save for the basic principles of physics (Fig. 3.1). Unfortunately, in spite of the claims of many, such principles remain elusive or, rather, difficult to cast in a useful form that enables structure prediction and the inference of plausible folding pathways [1–10]. The possibility of inferring the folding pathways of a soluble protein solely from dynamics rooted in physical principles continues to elude major research efforts. This is entirely expected since protein folding in vitro is a spontaneous process, hence thermodynamically irreversible. Therefore, intermediate states of the system that comprises protein chain and solvent statistical bath are essentially irretrievable and the folding pathway is untraceable.

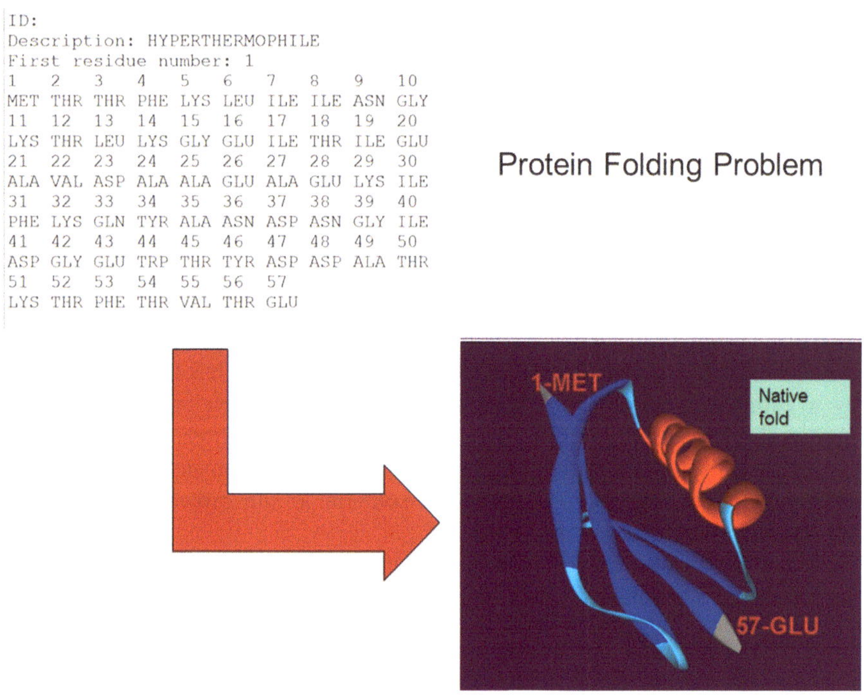

Fig. 3.1 Illustrative depiction of the protein folding problem. Inference of the 3D structure of the thermophilic variant of protein G domain from its primary sequence of amino acids

A major difficulty arises as we attempt to tackle the following problem: as a peptide chain folds onto itself, it also shapes the microenvironments of the intramolecular interactions, and hence the strength and stability of such interactions need to be rescaled according to the extent to which they become "wrapped" or surrounded by other parts of the chain. Thus, interactions between different parts of the peptide chain not only involve the units directly engaged in the interaction, but also the units involved in shaping their microenvironment, and the latter are just as important as they determine either the persistence or the ephemeral nature of such intramolecular interactions. This fact makes the folding problem essentially a many-body problem and points to the heart of cooperativity, a pivotal attribute of the folding process [4, 6]. Furthermore, it highlights the crucial identification of folding cooperativity with hydrogen-bond wrapping: *intramolecular hydrogen bonds prevail only if properly wrapped and this requires a cooperative process.*

To further explore the molecular basis of cooperativity, we need to examine the folding process from a physico-chemical perspective: With an amide and carbonyl group per residue, the backbone of the protein chain is highly polar and this molecular property imposes severe constraints on the nature of the hydrophobic collapse and on the chain composition of proteins capable of sustaining such a collapse [2, 9, 11]. Thus, the hydrophobic collapse entails the dehydration of backbone amides and carbonyls and such a process would be thermodynamically disfavored if it were not for the possibility of amides and carbonyls to engage in hydrogen bonding with each other. Hence, *not every hydrophobic collapse qualifies as being conducive to folding the protein chain: only a collapse that ensures the formation and protection of backbone hydrogen bonds is likely to ensure an expedient folding of the chain* [2]. On the other hand, polar-group hydration competes with intramolecular hydrogen bonds, compromising the structural integrity of proteins with a deficiently wrapped backbone [12]. Thus, the need for formation and protection of intramolecular hydrogen bonds from water attack imposes constraints on the chain composition of an efficient folder capable of sustaining a reproducible and expedient collapse.

In accord with this picture, it has been postulated that as water-soluble proteins fold, the hydrogen-bond pairing of backbone amides and carbonyls is concurrent with the hydrophobic collapse of the chain [11, 13]. This fact has been rationalized taking into account that the thermodynamic cost associated with the dehydration of unpaired polar groups is relatively high, and that the hydrophobic collapse hinders the backbone hydration by shielding it from water. On the other hand, the strength and stability of hydrogen bonds clearly depend on the microenvironment where they occur: The proximity of nonpolar groups to a hydrogen bond enhances the electrostatic interaction by de-screening the partial charges, and stabilizes it by hindering the hydration of the polar groups in the nonbonded state. *Thus, to guarantee the integrity of soluble protein structure and the expediency of the folding process, most intramolecular backbone hydrogen bonds must be surrounded or "wrapped" by nonpolar groups fairly thoroughly as to become significantly dehydrated at all times during the folding process.* On the other hand, if the intramolecular backbone hydrogen bonds fail to be properly shielded,

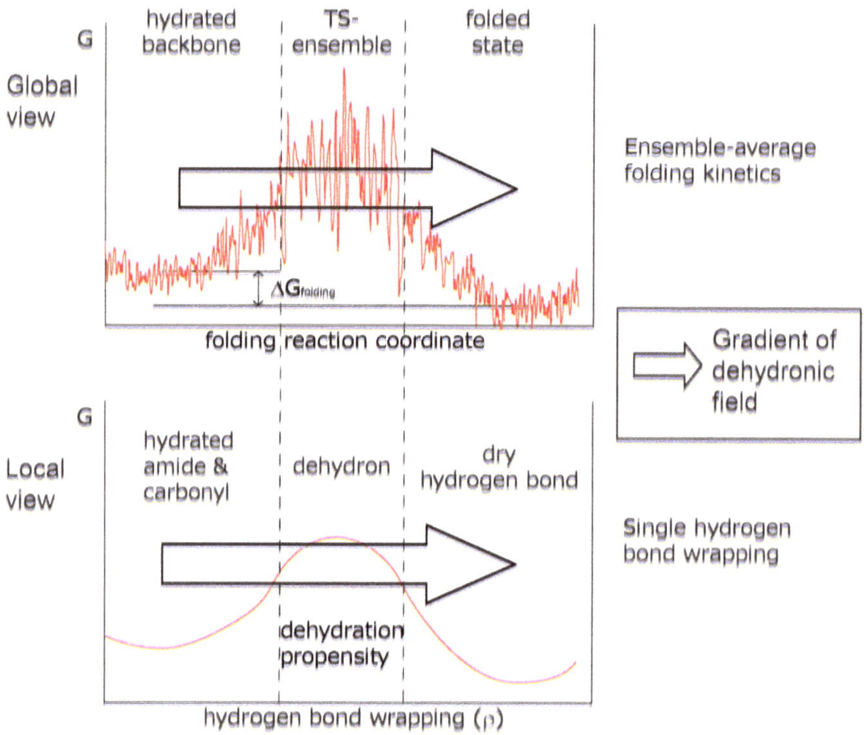

Fig. 3.2 The cross-over behavior from backbone hydration to dehydration of backbone hydrogen bonds triggered by hydrophobic collapse translates at the ensemble folding-kinetics level into a commitment of the chain to fold as the folding reaction progresses beyond the transition state

the backbone tends to maximize its hydration level, as shown subsequently. This observation has implications at an ensemble-average level accessible to experimentalists [13, 14]. It may help understand the fact that single-domain proteins are likely to be two-state folders, with a single kinetic barrier dominating the folding process at the ensemble-average level [13, 15].

The hydration propensity of amide and carbonyl and the dehydration-induced strengthening of their electrostatic association represent two conflictive tendencies, suggesting that there must be a crossover point in the dehydration propensity of a backbone hydrogen bond. If the bond is poorly wrapped by a few nonpolar groups that cluster around it, then hydration of the paired amide and carbonyl is likely to be favorable, but as the hydrogen bond becomes better wrapped intramolecularly, the surrounding water loses too many hydrogen bonding partnerships and thus, further removal of surrounding water is promoted. *Thus, the dehydronic field per hydrogen bond must significantly increase as hydrogen bonds get sufficiently dehydrated and this enhancement promotes further compaction of the chain and commits it fold. This switch-over behavior from backbone hydration to dehydration of backbone hydrogen bonds translated into an enhanced dehydronic field reflects the commitment to fold into a compact structure in which most backbone hydrogen bonds will be thoroughly dehydrated. This general picture is schematized in Fig. 3.2. The rationalization of the two-state folding kinetics of single-domain proteins is thus based on the cooperative nature of wrapping interactions.*

The crossover point in hydrogen-bond dehydration propensity may be regarded as representing a local characterization of the folding transition state if we adopt the backbone hydrogen bond dehydration as a generic folding coordinate. Once the folding process has progressed beyond the crossover point, further dehydration of the backbone is favored in consonance with the downhill nature of the folding process beyond the transition state [15]. Thus, *a transition-state conformation commits the chain to fold partly because the partially wrapped hydrogen bonds trigger their further desolvation, in turn promoting further chain compaction* (Fig. 3.2). This compaction is essential to augment the number of nonpolar groups within the hydrogen-bond microenvironments, thus protecting the bonds from water attack.

However compelling, the scenario described above requires a rigorous justification that entails an understanding of the cross-over behavior at a local level. This understanding prompts us to focus on the solvent environment of individual hydrogen bonds during the conformational exploration performed by the protein chain as the chain collapse takes place [12, 16, 17]. This scenario should be distinguished from the sterile "two-state folding scenario" often invoked to account for the difficulty in trapping protein folding intermediates. The latter are not difficult to find but simply irretrievable, as it is the case for any spontaneous process subject to the second law of thermodynamics.

3.2 Cooperativity Arises from Hydrogen-Bond Wrapping

The dynamics of hydrogen-bond wrapping concurrent with the folding process cannot be probed directly at this time, much like the conformational exploration defining the folding process still remains inaccessible to experimentalists [13–15]. To make progress in our understanding of the wrapping dynamics, we need to capture this process through *semiempirical* folding algorithms independently benchmarked and validated against PDB-reported structures and experimental data on the folding kinetics [2, 13, 15].

To properly describe folding cooperativity, we exploit a program that algorithmically reproduces folding dynamics by stochastically and coarsely representing transitions in the conformation space of individual residues and effectively incorporating wrapping 3-body interactions. To access realistic folding timescales beyond the μs range, the program builds on an implicit solvent model of cooperativity [6] and exploits a coarse-grained representation of the torsional Ramachandran (Φ, Ψ)-state of each residue, as described in Fig. 3.3 [10]. The implicit treatment of the solvent dictates that the program assess at each step the microenvironments of intramolecular hydrogen bonds that form as byproducts of the chain attempts at achieving a sustainable compaction.

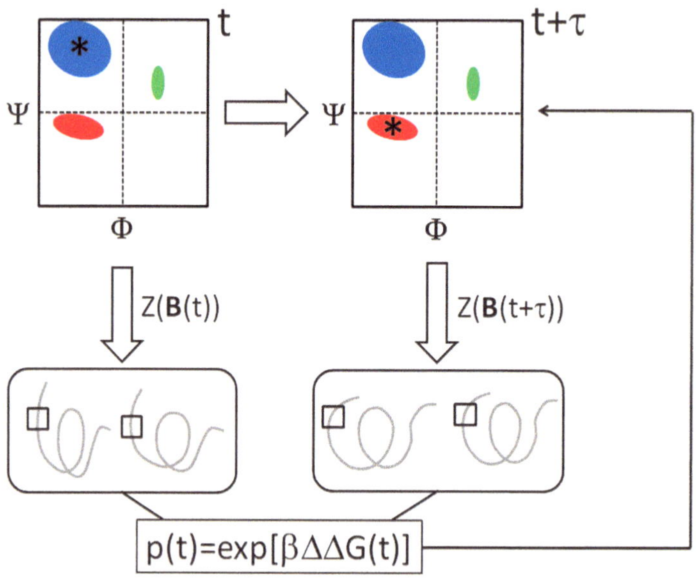

Fig. 3.3 Coarse grained stochastic process steering the generation of folding trajectories by the FM (Folding Machine), as developed by the author and coworkers [2, 6, 10]. The coarse grained stochastic process is provided by a "modulo-Ramachandran-basin metadynamics" whereby the local topology of the protein chain is specified by the basin of attraction within the (Ψ, Φ) Ramachandran torus for each individual residue. A basin assignment for a single generic residue is shown in the figure for the sake of illustration. The local topology of the generic residue at time t is marked by an *asterisk* and corresponds to the β-strand basin of attraction. The full chain topology at time t is specified by a vector **B**(t) of Ramachandran basin assignments, one for each residue, with color convention: blue for extended conformation, *red* for right handed helix and *green* for left-handed helix (a fourth conformation for glycine is marked in *grey*). Each Ramachandran vector spans Z(**B**(t)) microstates, geometrically specified by local (Ψ, Φ) torsional coordinates. The residue conformation corresponding to the Ramachandran basin choice is schematically marked by a square in the geometric microstate representation of the full chain. A coarse grained move $\mathbf{B}(t) \rightarrow \mathbf{B}(t+\tau)$ is determined by the lake areas of the putative destiny basins and by the energetic barriers in the respective Ramachandran plots. In the case of the individual residue monitored, a β-strand → α-helix basin transition (marked by the shifting location of the *asterisk*) has operated at time $t+\tau$. The coarse grained state $\mathbf{B}(t+\tau)$ is retained at time $t+\tau$ with probability p(t), determined by the free energy difference ΔΔG(t) between coarse states $\mathbf{B}(t+\tau)$ and **B**(t) with entropy contribution defined by the respective microstate realization numbers Z $(\mathbf{B}(t+\tau))$ and Z(**B**(t))

In order to reduce computation time and access relevant timescales, the micro-environmental dielectric within which intramolecular hydrogen bonds form is computed in a coarse manner at each step of the simulations. The strength and stability of intramolecular hydrogen bonds are thus rescaled with each computational step in consonance with the degree of dryness or dehydration of the bond determined by the over-all folding state of the chain. Thus, the extent of bond dehydration defines its net hydrophobicity (Chap. 1, Eq. 1.13), which in turn promotes further wrapping begetting further compaction of the chain.

The *ab-initio* folding algorithm reproduces the essential dynamical features of cooperativity while possessing predictive potential in regards to both end-point structure and folding pathway [2]. Thus, the algorithm appears to reproduce a basic feature of the folding kinetics of single-domain proteins: its purported two-stateness observed at an ensemble-average level of experimentation [13]. This dynamic feature is likely to arise as an over-all consequence of the cross-over behavior in hydrogen bond dehydration propensity, although a "hard proof" of this statement has not been produced to this day.

In spite of the difficulties inherent to a direct probing, the folding of a soluble protein is believed to be accompanied by a progressive structuring, immobilization and eventual exclusion of water surrounding backbone hydrogen bonds. This process turns hydrogen bonds into major determinants of the folding pathway and compensates for the thermodynamic penalties associated with desolvation of the backbone polar groups [18]. Thus, the need to wrap hydrogen bonds as a means to ensure their integrity has been shown to determine a constraint and hence define a constant of motion in the long-time limit behavior of coarse-grained *semiempirical* simulations. In this regard, an examination of one of the longest available (1 μs) all-atom simulations with explicit solvent [19] reveals that *the extent of wrapping is a constant of motion for the folding trajectory* [2]. It is well documented that such stabilization is best achieved by clustering five residues with nonpolar side-chain groups around the backbone hydrogen bonds, an arrangement that yields an optimal stabilization of the intramolecular hydrogen bond with a minimal conformational-entropy cost [2]. The results hereby reported thus support and clarify the view that hydrophobic surface burial should be commensurate with hydrogen-bond formation [13, 20] and enable us to introduce a basic wrapping motif inherent to the folding dynamics of soluble proteins. *Ultimately, the cross-over behavior outlined in Fig.* 3.2 *is reproduced, lending ample support to the two-state picture that emerged from ensemble-averaged kinetic studies and to the observed expediency of the folding process.*

The connection between hydrophobic interactions, water structuring and the strength of hydrogen bonds was first delineated by Scheraga and co-workers [21]. Thus, the inaccessibility of hydrogen bonds to solvent takes place as the protein strategically places hydrophobes around its backbone polar groups. This process induces hydrogen bond formation as a means to compensate for the unfavorable burial of the backbone polar groups. In this regard, natural questions arise and will be addressed through the examination of wrapping dynamics along the folding process:

(a) Does the wrapping or protection of backbone hydrogen bonds promote an expedient folding and if so, how?
(b) How can we identify the conformation or conformational ensemble that commits the chain to fold expeditiously into its native structure?
(c) Can we establish whether hydrophobic collapse precedes or is concurrent with secondary structure formation?

We approach these problems by establishing a relationship between the modulation of electrostatic shielding and the wrapping of the protein conformation along generated folding trajectories. This link is delineated based on statistical information on hydrophobic clustering around native backbone hydrogen bonds, information that leads us to establish a pervasive wrapping motif in native structures [22].

The analysis is carried out exploiting both a coarse-grained *ab initio* folding algorithm based on an implicit solvent model [18] and a long (1 μs) all-atom molecular dynamics simulation [19] of the type pioneered by the Karplus [23], Levitt [24] and Brooks [25] groups. Both algorithmic approaches reveal a nearly constant average extent of hydrogen-bond wrapping along the folding trajectories with relatively small dispersion, suggesting the existence of a constant of motion.

These results provide evidence supporting a dynamic picture of folding in which surface burial is commensurate with hydrogen bond formation or, phrased differently, hydrophobic collapse is concurrent with the formation of secondary structure [9, 13, 26]. Furthermore, the results add a new element to the picture: *hydrophobic collapse occurs productively in so far as hydrophobes can also exert a protective role* [26], *wrapping the backbone hydrogen bonds, and they do so to the same extent throughout the folding pathway, irrespective of the number of hydrogen bonds present at a given time*. This last assertion enables us to postulate a *constant of motion* underlying the folding process.

3.3 Cooperative Folding Pathways

To validate the previous scenario, the dynamics of backbone desolvation are studied using the so-called folding machine (FM), an *ab initio* wrapping-based algorithm designed to generate low-resolution folding pathways [2, 7, 8, 18], and contrasted against wrapping information extracted from all-atom explicit-solvent simulations [19] as well as experimental kinetics data [13].

The trustworthiness of the FM-generated folding pathways is guaranteed by an independent corroboration of the predictive value of the FM. This algorithm has been successful at predicting crucial dynamic features of complex folders that do not follow the two-state kinetic picture, such as β-lactoglobulin [8]. The native-like and non-native conformations occurring along the folding pathway in this presumed non-hierarchical folder as well as the productive role of non-native interactions in preventing misfolding or incorrect structure condensation were predicted through the FM and subsequently validated experimentally [27]. This corroboration added significant leverage to the wrapping-based FM algorithm, making it a powerful tool to study folding cooperativity.

Rather than encompassing all the structural detail for each step, the FM focuses on the time-evolution of backbone torsional constraints imposed by steric clashes with side-chains and rescales the chain's intramolecular potential according to the wrapping microenvironments around pairwise dielectric-dependent interactions. Thus, each pairwise interaction of the peptide chain, whether hydrophobic or electrostatic, is effectively enhanced or weakened according to the extent of burial of the paired groups. The net decrease of free energy associated with the backbone hydrogen bond desolvation is typically only in the range 0.5–1.2 kcal/mol, due to the opposing increase in the solvation free energy of the polar amide and carbonyl groups. The implicit treatment of the solvent requires that we introduce 3-body correlations to characterize the wrapping of pre-formed intramolecular interactions, and accordingly, rescale the internal energy terms with each iteration; that is, every time the pattern of 3-body correlations around an interaction has changed. Both the reduction in structural resolution and the implicit-solvent treatment are essential to make realistic folding time-scales (> 1 μs) accessible to the FM computations. Thus, the FM *ab initio* approach is geared to generate folding pathways with a coarse structural resolution needed to make folding time-scales computationally accessible. The FM algorithm utilizes no *a priori* information on target folds (unlike the so-called Go models which use the native fold as input for the simulator [28–30]), nor does it incorporate any energetic biases. The model focuses on the torsional Φ, Ψ constraints that are applied to backbone torsional state due to the steric hindrances imposed by the side-chains. Each residue is assigned to a region, or basin, in the Ramachandran map, and changes in configuration occur by hopping to a new basin.

By dealing with the evolution of constraints (*i.e. Ramachandran basins*) *rather than the backbone torsional coordinates themselves, the dynamics are judiciously simplified* [31]. The algorithm consists of a stochastic simulation of the coarsely resolved dynamics, simplified to the level of time-evolving Ramachandran-basin assignments, as described in Fig. 3.3. An operational premise is that steric restrictions imposed by the side-chains on the backbone may be subsumed into the basin-hopping dynamics. The side-chain constraints define regions in the Ramachandran map that can be explored in order to obtain an optimized pattern of non-bonded interactions.

The basin location of each residue coarsely defines the topology of the protein conformation. This string of basin locations, termed the local topology matrix or LTM(t), reflects the inherent geometrical constraints of a real polypeptide chain. The precise coordinates of the chain (i.e. the physical realization of an LTM) are defined by explicit Φ, Ψ angles determined by an optimization process that is turned on every ten hopping steps. To maintain structural continuity during a folding trajectory, the explicit dihedral angles are retained for each residue from one time-step to the next until that residue Ramachandran basin is scheduled to change.

To make torsional moves in 3D space, translating the "*modulo-basin topology*", a conformation is generated with a set of explicit Φ, Ψ angles compatible with the basin string. This explicit realization is used to identify the extent of structural involvement of each residue. As expected, the more structurally involved the residue, the less likely it may be engaged in a basin hopping, and the algorithmic rules do in fact reflect this fact. The degree of structural involvement is quantified energetically with a semi-empirical potential. This potential is used to determine which residues change their Ramachandran basin in the next step. Upon a basin transition, the new structure is energetically minimized by changing Φ, Ψ angles within the chosen basins.

The basin-hopping probability is dependent on the extent of structural engagement of the residue, which is defined by the energetic cost associated with the virtual move of changing basin, higher the more structurally engaged the residue is. On the other hand, the probability of hopping to a target basin (given that a hopping move is scheduled to occur) depends on the target-basin lake area or its microcanonical entropy, as described in Fig. 3.3. To fit experimental folding measurements [13, 15, 32], a free residue is assigned a basin-hopping rate fixed at 10^9 Hz. The basic tenets governing inter-basin hopping in the FM algorithm are: (a) interbasin hopping is slower than intrabasin exploration. This "adiabatic tenet" warrants a subordination of the backbone Φ, Ψ search to the LTM evolution or "*modulo-basin dynamics*" (Fig. 3.3); (b) side-chain torsional exploration occurs on a faster times-scale than backbone LTM dynamics. The last premise introduces a second adiabatic approximation, justifying the averaging of side-chain torsional motions in the stages of folding that precede a final side-chain fine tuning on the native backbone fold. This simplification is adequate to represent early stages of compaction and hydrogen-bond wrapping.

The FM captures the molecular basis of folding cooperativity by introducing an effective enhancement of dielectric-dependent two-body interactions according to the extent of wrapping of the interaction (see Chap. 1). The cooperative effect translates as a rescaling of the zero-order (in-bulk) pairwise contributions depending on the number of wrapping side-chain groups. The weakening of hydrophobic attractions depending on the extent of hydrophobic burial of the paired nonpolar groups is treated in a similar manner.

Three alternative and interchangeable representations of the folding state of a chain are simultaneously captured by the FM, as depicted in Fig. 3.2a–c: the modulo-basin Ramachandran torsional state (LTM), the 3-dimensional rendering, and the wrapping pattern. Figure 3.4 shows the endpoint conformation of a representative simulation for the thermophilic variant of protein G (PDB.1GB4) performed at RMSD~3.0 Å, pH 7 and consisting of 313 K steps [2]. The endpoint was largely reproduced in 66 of 91 runs and has 10^6 from PDB entry 1GB4.

The wrapping model subsumed into the FM algorithm reflects the fact that hydrogen bonds are extremely context-sensitive. The algorithm, however, treats the solvent implicitly. This simplification requires that we introduce 3-body correlations involving the wrapping residues (Fig. 3.4c) as an operational means to incorporate rescaling of the intramolecular potential according to the microenvironmental changes that take place during the course of folding. Explicitly, these correlations rescale the "zeroth-order" pairwise interactions by determining their extent of desolvation. For consistency, the wrapping of a hydrogen bond is also introduced in a coarse-grained manner in this analysis. Thus, in contrast with the detailed wrapping assessment (cf. Chap. 1, Fig. 1.10), here the wrapping parameter, denoted ρ_{coar}, indicates the number of residues contributing with nonpolar side-chain groups to the dehydration of the hydrogen bond. Each residue contributing to

a

	1	2	3	4	5	6	7	8	9	10
Aminoacid	MET-M	THR-T	THR-T	PHE-F	LYS-K	LEU-L	ILE-I	ILE-I	ASN-N	GLY-G
R-basin	red	red	blue	blue	blue	blue	blue	blue	blue	red
Phi-angle	−60.48	−53.47	−111.35	−146.29	−77.71	−142.19	−60.03	−136.53	−146.28	−83.41
Psi-angle	−50.21	−46.87	114.70	166.19	134.23	132.85	117.82	139.68	137.50	0.57
Omega-angle	180.00	180.00	180.00	180.00	180.00	180.00	180.00	180.00	180.00	180.00
	11	12	13	14	15	16	17	18	19	20
Aminoacid	LYS-K	THR-T	LEU-L	LYS-K	GLY-G	GLU-E	ILE-I	THR-T	ILE-I	GLU-E
R-basin	blue	blue	red	blue	blue	blue	blue	blue	blue	red
Phi-angle	−77.16	−130.78	−61.56	−110.69	−76.55	−128.50	−141.75	−129.74	−131.06	−57.29
Psi-angle	2.03	132.05	−25.00	104.99	86.44	106.57	175.45	137.67	129.69	−32.23
Omega-angle	180.00	180.00	180.00	180.00	180.00	180.00	180.00	180.00	180.00	180.00
	21	22	23	24	25	26	27	28	29	30
Aminoacid	ALA-A	VAL-V	ASP-D	ALA-A	ALA-A	GLU-E	ALA-A	GLU-E	LYS-K	ILE-I
R-basin	red	red	red	blue	red	red	red	red	red	red
Phi-angle	−63.64	−60.87	−62.93	−144.65	−61.10	−59.75	−61.47	−61.16	−54.21	−61.01
Psi-angle	−32.49	−47.50	−33.90	136.37	−43.25	−27.33	−50.01	−46.13	−43.25	−46.07
Omega-angle	180.00	180.00	180.00	180.00	180.00	180.00	180.00	180.00	180.00	180.00
	31	32	33	34	35	36	37	38	39	40
Aminoacid	PHE-F	LYS-K	GLN-Q	TYR-Y	ALA-A	ASN-N	ASP-D	ASN-N	GLY-G	ILE-I
R-basin	red	red	red	red	blue	red			grey	red
Phi-angle	−63.08	−54.65	−63.38	−58.81	−118.93	−67.19	−88.71	−78.66	80.98	−61.82
Psi-angle	−45.96	−44.88	−49.83	−27.40	114.72	−53.04	0.80	138.34	−94.77	−33.78
Omega-angle	180.00	180.00	180.00	180.00	180.00	180.00	180.00	180.00	180.00	180.00
	41	42	43	44	45	46	47	48	49	50
Aminoacid	ASP-D	GLY-G	GLU-E	TRP-W	THR-T	TYR-Y	ASP-D	ASP-D	ALA-A	THR-T
R-basin	red	green	red	blue	blue	blue	blue		red	blue
Phi-angle	−65.78	84.02	−66.69	−133.65	−64.78	−145.61	−76.24	−132.24	−68.87	−127.87
Psi-angle	−48.66	−3.46	−52.64	174.31	124.11	171.01	137.33	143.91	−54.06	114.38
Omega-angle	180.00	180.00	180.00	180.00	180.00	180.00	180.00	180.00	180.00	180.00
	51	52	53	54	55	56	57			
Aminoacid	LYS-K	THR-T	PHE-F	THR-T	VAL-V	THR-T	GLU-E			
R-basin	green	blue	blue	blue	blue	blue	green			
Phi-angle	62.05	−81.78	−73.09	−129.16	−141.35	−62.94	59.51			
Psi-angle	58.19	131.81	136.63	137.30	132.79	120.79	55.61			
Omega-angle	180.00	180.00	180.00	180.00	180.00	180.00	180.00			

Fig. 3.4 Three representations of the conformational state of thermophilic variant of protein G (PDB code: 1GB4) obtained after 10^6 FM iterations. (**a**) LTM or backbone torsional state represented "modulo Ramachandran basin". Each basin is indicated by the quadrant in the (Φ, Ψ)-torus where it occurs. Thus, *blue* stands for the extended residue conformations including the β-strand states, *red* corresponds to a manifold of conformations containing the right-handed α-helix, *green* denotes the basin containing the left-handed helical conformation, while *grey* represents the basin in the lower right quadrant which is only accessible to Gly. (**b**) 3D ribbon representation of the endpoint chain conformation. (**c**) Wrapping state of the chain in the endpoint conformation. The chain backbone is shown as virtual bonds joining α-carbons depicted in *pink*, hydrogen bonds are shown as *grey* segments joining α-carbons, and 3-body "wrapping" correlations (cf. Chap. 1) are shown as *thin blue lines* joining the α-carbon of the wrapping residue with the center of the wrapped hydrogen bond. A wrapping residue is defined as a contributor of nonpolar groups to the hydrogen-bond microenvironment. Thus, the FM keeps track of the pairwise interactions as well as of the evolving microenvironments of such interactions determined by the evolving chain conformation. Reprinted from Progress in Molecular Biology and Translational Science 83, Fernández A, Zhang X, Chen J, Folding and wrapping soluble proteins: Exploring the molecular basis of cooperativity and aggregation (Chapter 2), pages 53-88, copyright 2008 with permission from Elsevier

b c

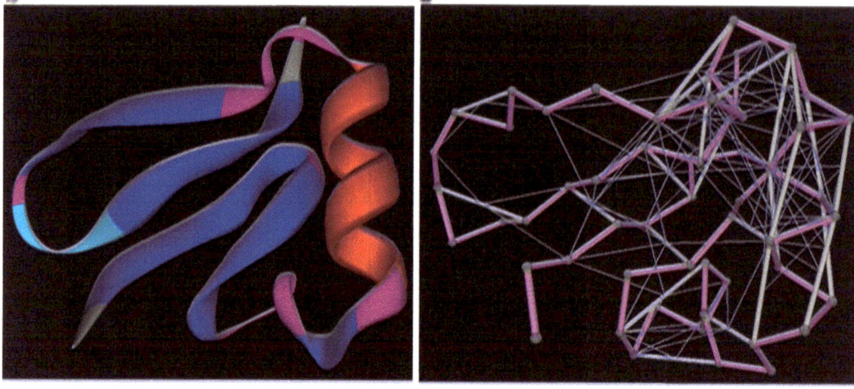

Fig. 3.4 (continued)

the dehydration of a hydrogen bond determines a 3-body correlation (cf. Fig. 1.9a). Notice that we have introduced a notational distinction absent in Chap. 1: a 3-body correlation denotes something coarser than a 3-body interaction. The latter refers to a wrapping group while the former refers to residues contributing with wrapping groups (often more than one) to shield a hydrogen bond. The overall number of 3-body correlations, denoted $C_3 = C_3(t)$ is invariably smaller than the number of 3-body interactions $I_3 = I_3(t)$, and both are roughly proportional for most natural proteins ($I_3 \approx 2.5C_3$).

The wrapping effect may be also cast in thermodynamic terms: due to their destabilizing effect on the non-bonded state, the hydrophobes surrounding a dielectric-dependent interactive pair become enhancers of the interaction. As folding progresses, the effective hydrophobic energy contribution becomes progressively transferred into an effective energy of backbone desolvation in which the amide–carbonyl hydrogen bonds become determinants of protein structure and compensate for the rise in self-energy of the paired groups.

To take into account this effect, we incorporate a phenomenological approach to wrapping electrostatics [10, 33] as a means of incorporating changes in permittivity associated with changes in the conformation-dependent environments that affect pairwise interactions. This procedure entails a rescaling of the intramolecular potential terms as folding progresses, an operation requiring keeping track of vicinal hydrophobe positions with each FM iteration (Fig. 3.4c).

3.4 Improving Wrapping Along Folding Pathways

To take into account the effect of wrapping on dielectric-dependent pairwise interactions, and assess its role in defining cooperativity, we have adopted a semi-empirical procedure to algorithmically keep track of the conformation-dependent microenvironments. Thus, the in-bulk potential energy contributions are regarded as

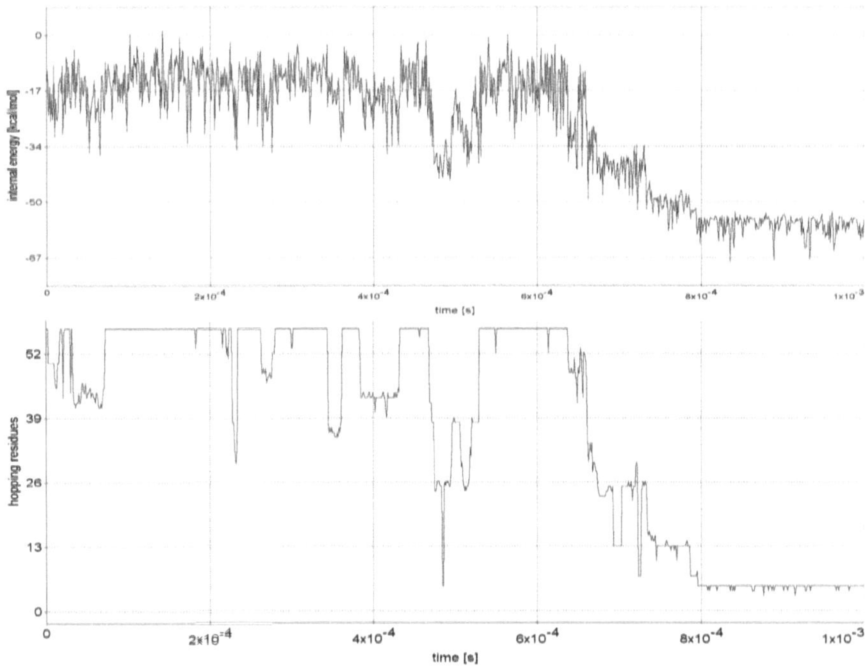

Fig. 3.5 Internal energy and number of hopping residues along a reproducible representative FM trajectory for the thermophilic variant of protein G. A hopping residue is defined as tagged for a Ramachandran basin transition in the coarse-grained stochastic process that underlies the FM torsional dynamics. Reprinted from Progress in Molecular Biology and Translational Science 83, Fernández A, Zhang X, Chen J, Folding and wrapping soluble proteins: Exploring the molecular basis of cooperativity and aggregation (Chapter 2), pages 53-88, copyright 2008 with permission from Elsevier

zero-order terms, while cooperative effects arise due to the wrapping of favorable interactions brought about by hydrophobic third-body participation (cf. Fig. 3.4c).

A basic question is addressed through this analysis: what is the dynamic relevance of optimal wrapping in regards to the protein's commitment to fold? To tackle this question, we analyze a representative simulation for the thermophilic variant of protein G. This simulation performed at 313 K, pH 7 consists of 10^6 steps and was essentially reproduced in 66 of 91 FM runs. All the runs generated a stationary fold within RMSD˜4 Å from the native structure (PDB entry 1GB44) and a dramatic decrease in potential energy around 0.6 ms (Fig. 3.5).

In accord with experimental tenets, direct examination of the time-dependent behavior of contact order suggests a *nucleation process*, whereby a sustainable large-scale organization is achieved only at 8×10^{-4} s after a relatively lengthy trial-and-error process (0–0.63 ms), followed by a critical regime (0.63–0.8 ms). Direct inspection of Fig. 3.6 reveals that the transition from local to large-scale organization is actually defined by a sudden burst in the number of 3-body correlations starting at $6.0 \times 10^{-4} - 7 \times 10^{-4}$ s.

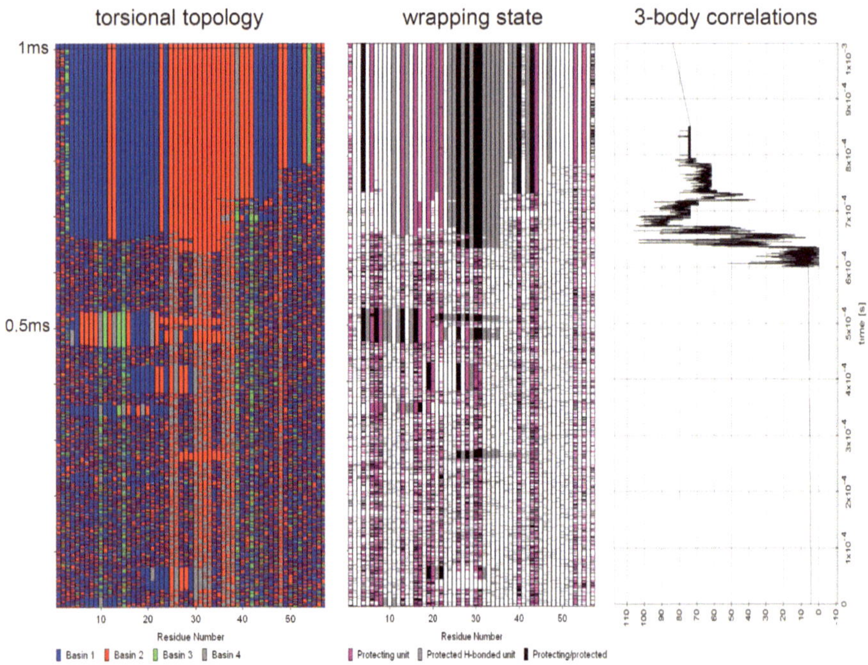

Fig. 3.6 Three views of the wrapping/folding dynamics for the thermophilic variant of protein G obtained from a representative expeditious FM trajectory. The *left panel* represents the time evolution of the local topology matrix (LTM) with the abscissas denoting residue numbers and the ordinates, the time axis. The Ramachandran basin assignment for each residue is given as a function of time and the topology of the entire chain is designated by a row in the histogram. Color convention is consistent with Fig. 3.4b. The *middle panel* represents the different roles exerted by residues along the chain at different times. Thus, a *pink* residue-time entry indicates a residue exerting a protecting or wrapping role at the particular time, a *grey* entry indicates a residue engaged in pairwise interaction which is being protected or wrapped by other hydrophobes, and a black entry indicates a dual role as protector or wrapper and also engagement in a hydrogen bond which is being exogenously protected. The *right panel* indicates the total number of 3-body correlations representing the wrapping dynamics concurrent with the expeditious folding process. *Notice that the burst phase in 3-body correlations coincides with the region of transition from a trial-and-error phase to a sustainable structure.* Reprinted from Progress in Molecular Biology and Translational Science 83, Fernández A, Zhang X, Chen J, Folding and wrapping soluble proteins: Exploring the molecular basis of cooperativity and aggregation (Chapter 2), pages 53-88, copyright 2008 with permission from Elsevier

The nucleation picture revealed by Figs. 3.5 and 3.6 has been further confirmed by examination of total internal energy of the peptide chain and solvent-exposed area plots. The energy experiences a sudden decrease in the $6.3 \times 10^{-4} - 8.0 \times 10^{-4}$ s region concurrent with a dramatic decrease in the solvent-exposed area. The point at which the protein is actually committed to fold (Fig. 3.7) can be inferred by performing runs with different starting conformations extracted from the $6.3 \times 10^{-4} - 8.0 \times 10^{-4}$ s time window. This commitment arises when a

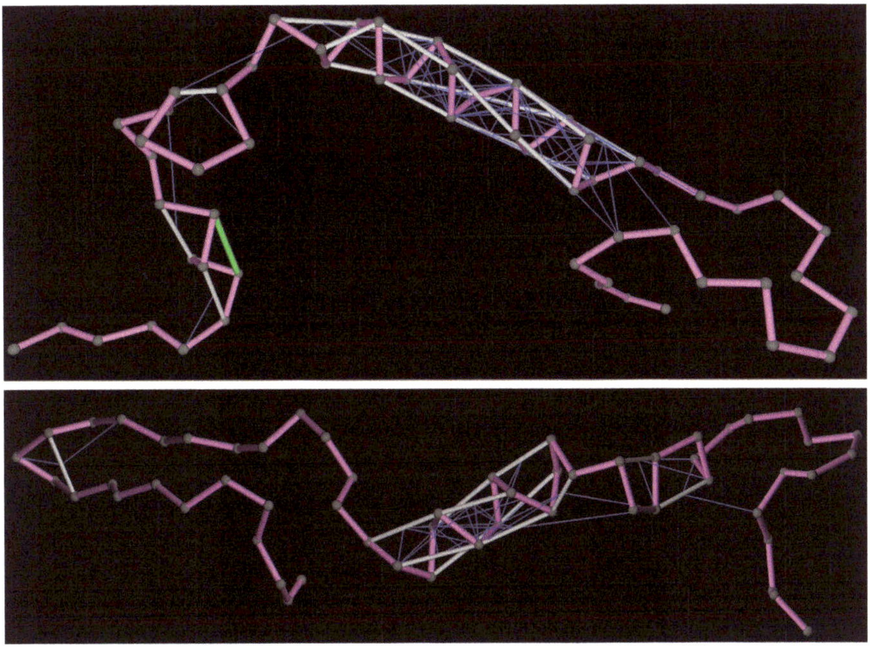

Fig. 3.7 Wrapping patterns for chain conformations occurring during the burst phase (cf. Fig. 3.4) of protein-G variant generated at 6.4×10^{-4}s (*upper panel*) and 6.5×10^{-4}s (*lower panel*). These conformations commit the chain to fold, are part of the "transition state ensemble", and *do not contain exclusively native interactions*, as it becomes apparent in the *upper panel*. Reprinted from Progress in Molecular Biology and Translational Science 83, Fernández A, Zhang X, Chen J, Folding and wrapping soluble proteins: Exploring the molecular basis of cooperativity and aggregation (Chapter 2), pages 53-88, copyright 2008 with permission from Elsevier

sustainable number of 3-body correlations (native or nonnative!) equal to or larger than the final almost stationary number is reached (Fig. 3.6). In the case of protein G, the burst time window is $6.3 \times 10^{-4} - 8.0 \times 10^{-4}$ s and a sustainable population of 3-body correlations is maintained in the region $7.0(\pm 0.2) \times 10^{-4}$ s.

A similar FM-computation was carried out for ubiquitin [32]. The exposed surface area at the transition state is estimated to be 7200 Å^2, while the random coil conformation exposes approximately 10,800 Å^2. Thus, we find that the transition state buries 3600 Å^2, approximately 60 % of the total area buried in the native fold, in good agreement with the experimental results [13].

The dominant folding pathway for the variant of protein G is coarsely displayed in Fig. 3.6, where the abscissas denote residue numbers and the ordinates, the time axis. The Ramachandran basin assignment for each residue is given as a function of time and the topology of the entire chain (cf. Fig. 3.3) is designated by a row in the histogram. The blue color (basin 1) designates the large basin containing the β strand and PP2 conformations, the red color (basin 2) designates the basin containing the right-handed helix conformation, green (basin 3) designates the

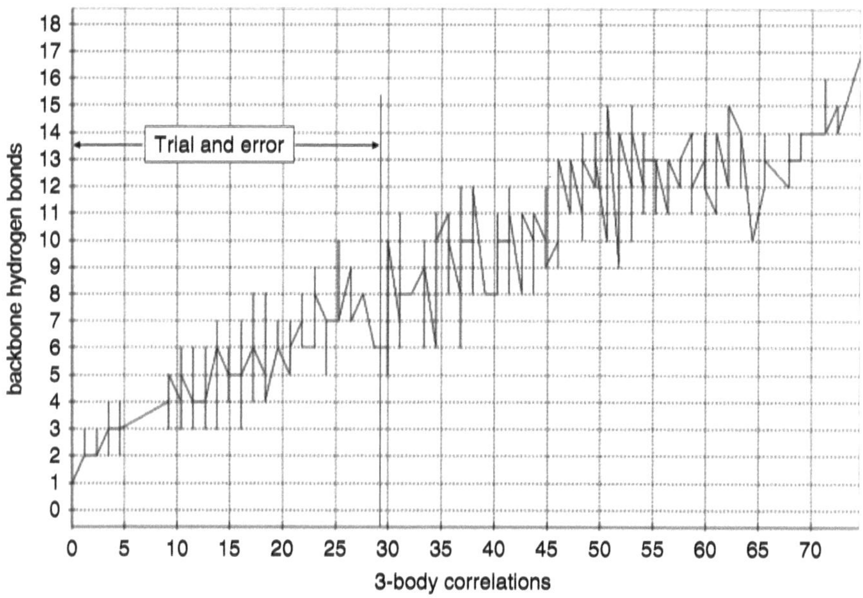

Fig. 3.8 Number of backbone hydrogen bonds plotted against number of 3-body correlations (C_3) extracted from representative FM trajectory capturing the wrapping/folding dynamics for the thermophilic variant of protein G. Reprinted from Progress in Molecular Biology and Translational Science 83, Fernández A, Zhang X, Chen J, Folding and wrapping soluble proteins: Exploring the molecular basis of cooperativity and aggregation (Chapter 2), pages 53-88, copyright 2008 with permission from Elsevier

basin containing the left-handed helix coordinates and gray (basin 4) corresponds to the fourth basin present only in glycine. The figure clearly reveals the emergence of a stable large-scale organization that prevails after a critical period located at the $6.3 \times 10^{-4} - 8.0 \times 10^{-4}$ s interval.

We have also reported the extent of protection of backbone hydrogen bonds along the folding pathway. Figure 3.8 shows that the average extent of hydrogen bond protection, $\rho_{coar}(t)$, converges to the value $\rho_{coar} = 5$ in the long-time limit that starts right after the trial-and-error period; that is, at the sharp burst in $C_3 = C_3(t)$. This regime is associated with the region $C_3 > 60$. The stationary native-like population of $15(\pm 1)$ backbone hydrogen bonds are protected by $75(\pm 5)$ 3-body correlations. We see that the $\rho_{coar} = 5$ value becomes an approximate constant of motion in the critical region and beyond, that is for $C_3 > 60$. A similar result holds for ubiquitin [2]: The native-like stationary population of $28(\pm 3)$ backbone hydrogen bonds is now wrapped by $140(\pm 5)$ 3-body correlations.

The wrapping results from Figs. 3.6 and 3.8 are more specific and informative than earlier attempts at establishing whether buried surface area is commensurate with hydrogen bond formation [13, 32]. It is difficult to infer from such studies whether hydrophobic collapse triggers hydrogen bond formation or whether the latter directs the former. However, Figs. 3.6 and 3.8 reveal that the productive build-up in hydrogen bond formation beyond the "trial-and-error folding regime" is invariably assisted by the nearly constant $\rho_{coar} = 5$ wrapping value.

To further validate our conclusions by contrasting them against all-atom molecular dynamics simulations, we have analyzed one of the longest all-atom simulations with explicit solvent: the 1 μs simulation of the villin headpiece by Duan and Kollman [19]. Their trajectory was examined using the FM's representation and visualization tools, and provides valuable information on the wrapping dynamics in the trial and error regime. *The analysis revealed an almost constant proportionality between C_3 and the number of backbone hydrogen bonds along the entire trajectory. The results reveal again that $\rho_{coar} = 5$ is a constant of motion for the folding trajectory.*

Testing the universal validity of this constant of motion may prove to be daunting, as the wrapping of most good folders would need to be investigated dynamically. Nevertheless, a theoretical analysis based on a nanoscale treatment of the solvent further supports this hypothesis. Thus, we now rigorously prove that *a hydrogen bond is embedded in the lowest dielectric when surrounded by five average-shaped hydrophobic residues, and this optimal wrapping arrangement represents a compromise between crowding and proximity to the hydrogen bond subject to the steric constraints determined by a minimum approach distance.*

This approximate law holds for both native structure [22] and folding dynamics [2]. In this regard, this wrapping motif may be regarded as a structural element that captures the basic component of energy transduction from hydrophobic association to structure formation. Furthermore, it implies that a fundamental constraint in protein architecture applicable to native structures applies also throughout the folding trajectory.

Kinetic amide isotope studies [13, 32] imply that helical hydrogen bond formation and surface area burial form to the same degree in the transition state for single-domain proteins that fold in a two-state manner. Extensive helix formation does not occur prior to significant hydrophobic association at the limiting step. The surveyed results on individual folding trajectories indicate that commensurate burial occurs both on the way up to the limiting barrier (pre-nucleation) as well as afterwards, in consistency with the ensemble-average evidence stemming from the kinetic folding experiments.

3.5 Dielectric Modulation by the Folding Protein

This section describes a semiempirical model of nanoscale solvation that captures the dielectric modulation brought about by the approach of a hydrophobe to a protein hydrogen bond. In essence, the model captures the solvent-ordering effect promoted by the hydrophobe and quantifies the effect of this induced organization on the electrostatics of a pre-formed amide-carbonyl hydrogen bond. *This model reproduces the crossover point in hydrogen-bond dehydration propensity that characterizes the folding transition state* (*Fig.* 3.2) *as we adopt the extent of backbone hydrogen-bond dehydration as generic reaction coordinate for the folding process* [9, 13]. *Strikingly, it also provides a rigorous justification of the* $n = 5$ *coarse-wrapping motif* (Fig. 3.8) *through an implicit-solvent model that reproduces the modulation of the dielectric environment in which intramolecular hydrogen bonds are formed. Finally, it establishes the fact that the progress of the folding process follows a reaction coordinate along the gradient of the dehydronic field, making the latter the dominant force driving chain compaction beyond the transition state.*

We start our theoretical treatment by defining a Cartesian coordinate system and placing the carbonyl oxygen atom effective charge q at the center of coordinates. We further define the x-axis as that along the carbonyl–amide hydrogen bond,

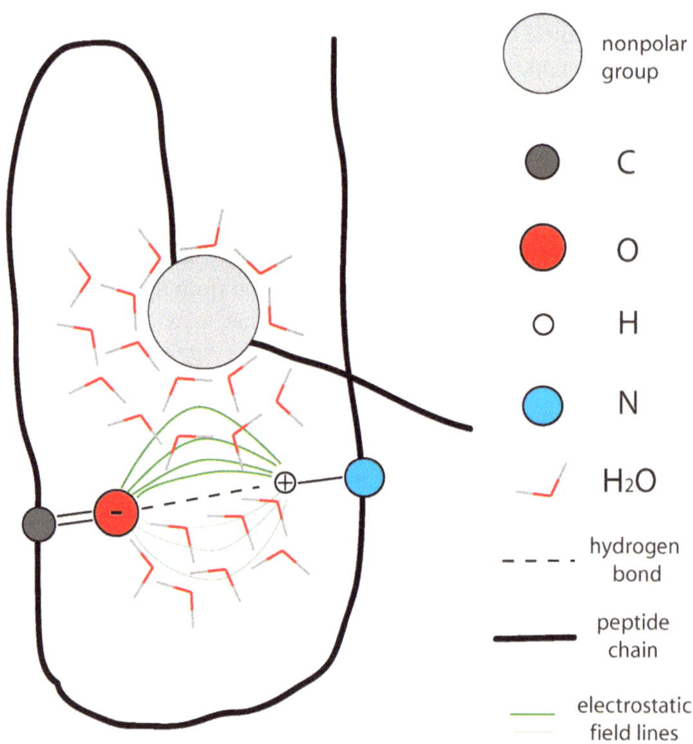

Fig. 3.9 Basic tenets of the nanoscale solvation theory. The solvent ordering promoted by the approaching hydrophobe enhances the electrostatics, an effect that must be captured by the model. The induced organization decreases the polarizability of the environment, preventing water dipoles from aligning with the electrostatic field lines. The *thicker lines* represent a stronger field. By contrast, the region exposed to bulk water facilitates dipole organization along the field lines, weakening the electrostatic field (*thin lines*). Reprinted from Progress in Molecular Biology and Translational Science 83, Fernández A, Zhang X, Chen J, Folding and wrapping soluble proteins: Exploring the molecular basis of cooperativity and aggregation (Chapter 2), pages 53-88, copyright 2008 with permission from Elsevier

and place the amide hydrogen atom at position **r**, 1.4–2.1 Å away along the positive x-axis. We assume the hydrogen bond to be surrounded by a discrete number of identical spherical hydrophobic units of radius $d/2$ (the parameter d is defined below) centered at fixed positions $\vec{r}_j, j=1,2,\ldots n$, This is an idealized picture but one that can be dealt with analytically.

Previously reported implicit-solvent approaches [33] take into account the solvent structuring induced by the solvent–hydrophobe interface (Fig. 3.9), translate this effect into a distance-dependence permittivity, $\varepsilon(\vec{r})$, and quantify the effect on the coulomb screening. A more heuristic, practical and phenomenological approach is rooted in two pivotal components: (a) perturbation of the diffraction structure of bulk water as hydrophobes are incorporated at fixed positions; and (b) recovery of their solvent-structuring effect by inverse Fourier transforming the previous result given in frequency space.

To propagate the solvent-structuring effect induced by the presence of the hydrophobic spheres, we replace the position-dependent dielectric by an integral kernel convoluted with the electric field at position $\vec{r}$ to represent the correlations with the field at neighboring positions $\vec{r'}$. This prompts us to replace the Poisson equation by the relation

$$\vec{\nabla}.\int K\left(\vec{r},\vec{r'},\{\vec{r}_j\}\right)\vec{E}\left(\vec{r'}\right)d\vec{r'} = 4\pi q\delta(\vec{r}) \tag{3.1}$$

where the kernel $K\left(\vec{r},\vec{r'},\{\vec{r}_j\}\right)$ is parametrically dependent on the fixed hydrophobe positions. In the absence of vicinal hydrophobic units, the correlations decay as $e^{-\|\vec{r}-\vec{r'}\|/\xi}$ (ξ=characteristic correlation length). In the limit $\xi \to 0$, we get: $K\left(\vec{r},\vec{r'}\right) \to \varepsilon_0\delta\left(\vec{r'}-\vec{r}\right)$, yielding the standard Poisson equation.

The correlation kernel reflects the relationship between diffraction and dielectric. For bulk water, we get $K\left(\vec{r},\vec{r'}\right) = K\left(\vec{r}-\vec{r'}\right)$, by inverse transforming its frequency $\vec{k}$-representation:

$$K\left(\vec{r}-\vec{r'}\right) = \int \exp\left\{i\vec{k}.\left(\vec{r}-\vec{r'}\right)\right\}L\left(\vec{k}\right)d\vec{k}, \tag{3.2}$$

In (3.2), $L(k) = \varepsilon_0 + (\varepsilon_w - \varepsilon_0)/(1 + \varepsilon_w\varepsilon_0{}^{-1}\|k\|^2\xi^2)$, $\xi \approx 5$ Å, denotes the characteristic length, and ε_w, ε_o are the permittivities of water and vacuum, respectively. To obtain the correlation kernel with n hydrophobic units, we need to incorporate their solvent-structuring effect:

$$K\left(\vec{r},\vec{r}',\{\vec{r}_j\}\right)=\left[\int \exp\left\{i\vec{k}.\left(\vec{r}-\vec{r}'\right)\right\}L\left(\vec{k}\right)d\vec{k}\right].\left[1+\sum_{j=1}^{n}\Gamma_j\left(\vec{r},\vec{r}'\right)\right] \tag{3.3}$$

On purely phenomenological grounds we get the following approximation [2, 33, 34]: $\Gamma_j\left(\vec{r},\vec{r}'\right)\sim\exp\left[-\frac{\|\vec{r}-\vec{r}_j\|+\|\vec{r}'-\vec{r}_j\|}{\Lambda}\right]$ for $\|\vec{r}-\vec{r}_j\|,\|\vec{r}'-\vec{r}_j\|>d/2$ with characteristic length Λ subsuming the range of the water-structuring effect. This parameter is typically fixed at 2.8 Å, the effective thickness of a single water layer (cf. Fig. 3.9).

We now solve (3.1) by Fourier transformation, obtaining the electric field $E(\vec{r})$ by inverse Fourier transformation of the solution to (3.1) in $\vec{k}$-representation:

$$\int E(\vec{r})d\vec{r}=-(4\pi q)\int e^{(i\vec{k}.\vec{r})}\|\vec{k}\|^{-2}K^{-1}(\vec{k},\{\vec{k}_j\})d\vec{k} \tag{3.4}$$

Direct residue evaluation at the first-order poles $k=\mp\left(\frac{\varepsilon_0}{\varepsilon_w}\right)^{\frac{1}{2}}\xi^{-1}$, $\vec{k}=\vec{k}_j\mp i\Lambda^{-1}$ yields the electric field by retaining only the real part in the calculation:

$$E(\vec{r})=\left(\frac{q}{r^2}\right)\left[\varepsilon_w{}^{-1}+(\varepsilon_0{}^{-1}-\varepsilon_w{}^{-1})\Omega(\{\vec{r}_j\})\left(1+\frac{r}{\xi}\right)e^{-\frac{r}{\xi}}\right] \tag{3.5}$$

where:

$$\Omega(\{\vec{r}_j\})=\prod_{j=1}^{n}\left[1+e^{-\|\vec{r}_j\|/\Lambda}\right]\left[1+e^{-\frac{\|\vec{r}-\vec{r}_j\|}{\Lambda}}\right] \tag{3.6}$$

Equations (3.5) and (3.6) describe the net effect of the wrapping hydrophobic arrangement on the electric field. The wrapping effect can be captured by replacing the permittivity constant ε_w for bulk water by an "effective permittivity" ε defined by:

$$\varepsilon^{-1}=\left[\varepsilon_w{}^{-1}+(\varepsilon_0{}^{-1}-\varepsilon_w{}^{-1})\Omega(\{\vec{r}_j\})\left(1+\frac{r}{\xi}\right)e^{-\frac{r}{\xi}}\right] \tag{3.7}$$

This permittivity tends to the bulk limit for long interaction distances:

$$lim_{r/\xi\to\infty}\varepsilon=\varepsilon_w \tag{3.8}$$

We now turn to the problem of finding the optimal wrapping arrangement and contrasting our implicit-solvent result with the phenomenological result emerging from the simulations (cf. Figs. 3.6 and 3.8). Since $\varepsilon_o{}^{-1}\gg\varepsilon_w{}^{-1}$, finding the wrapping cluster with the lowest dielectric in its interior is tantamount to finding the arrangement $(\{\vec{r}_j\})$ that maximizes the function $\Omega(\{\vec{r}_j\})$, in accord with (3.7). We name the maximum value Ω^*. First, we compute the maximum $\Omega^*(n)$ for each fixed n subject to the constraint of preserving a minimum distance d between any two hydrophobes. The d value is taken to be 5 Å in accord with typical minimal distances between α-carbon atoms in tertiary structure [22]. Our results are qualitatively invariant in the range 4.5 Å $\leq d \leq$ 6 Å. Using the Lagrange multipliers method to minimize the effective permittivity, we find that the optimal arrangement is invariably obtained by fixing $n-2$ hydrophobes at distance d from each other and equidistantly from the O and H atoms, and placing the remaining two along the x-axis at distance $(\Lambda+\eta)(1-n^{-2})$ (to first approximation) away from the C and N atoms, with $\eta = \mathrm{C}-\mathrm{O}$ distance in the carbonyl group. This gives for $n = 4$ (tetrahedron): $\Omega^*(4) = 3.419$; for $n = 5$ (trigonal bipyramid): $\Omega^*(5) = 4.144$; for $n = 6$ (square bipyramid), $\Omega^*(6) = 3.952$; and for $n = 7$ (pentagonal bipyramid), $\Omega^*(7) = 3.421$. Similar calculations for all n allow us to establish the following order relations:

$$\Omega^*(3) < \Omega^*(4) < \Omega^*(5) > \Omega^*(6) > \Omega^*(7) > \ldots \tag{3.9}$$

Thus, $\Omega^*(n)$ has a single maximum at $n = 5$. This maximum is expected on the basis of the two conflictive tendencies in the stabilization of a hydrogen bond: (1) bringing close to the hydrogen bond as many hydrophobes as possible and (2) bringing them as close to the hydrogen bond as possible. However, both demands start becoming mutually incompatible due to the steric hindrances implicit in the Lennard–Jones repulsive terms.

In full agreement with the result described in Fig. 3.8 ($\rho_{coar} = 5$ is a constant of motion along the folding trajectory), *a hydrogen bond is embedded in the lowest dielectric when surrounded by five wrapping residues, the optimal compromise between crowding and proximity to the hydrogen bond.* Thus, we have benchmarked the solvation theory against a rigorous dynamic result.

Equation (3.7) provides the first rigorous derivation of an effective permittivity that depends on the wrapping arrangement around the electrostatic interaction. Given the prior efforts by Warshel and coworkers to define a "protein permittivity" [34], this derivation becomes only of historical interest for structural biophysics. The rigorous non-Debye dielectric treatment of the aqueous interface presented in Chaps. 1 and 2 renders the scalar field ε a purely heuristic empirical concept.

3.6 The Dehydronic Field Steers the Folding Process

Equations (3.5)–(3.7) enable us to compute the dehydronic field, that is, the mechanical equivalent of the dehydration propensity of hydrogen bonds made along the folding process. This computation requires the evaluation of the gradient

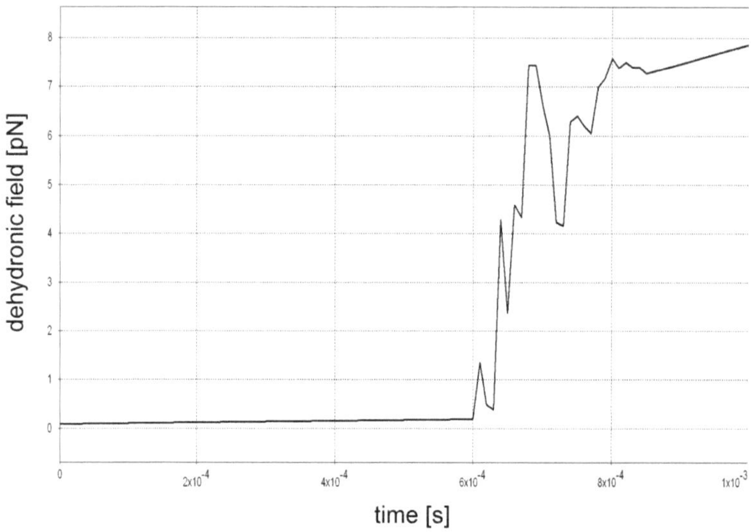

Fig. 3.10 Dehydronic field averaged over all backbone hydrogen bonds formed at time t for the protein G variant along the *ab initio* folding trajectory described in Figs. 3.5–3.8

$$\vec{\Phi}(\vec{\mathrm{R}}) = \frac{qq'}{r} \vec{\nabla}_{\vec{R}} \left[4\pi\varepsilon(\vec{R})\right]^{-1} \tag{3.10}$$

of the electrostatic energy with respect to the position vector $\vec{R}$ of the test hydrophobe (in the simulations we adopted methane as test hydrophobe). This analysis is motivated by the need to support the two-state kinetic picture outlined in Fig. 3.2. To normalize for the number of hydrogen bonds formed at any given time, we computed the dehydronic field per hydrogen bond as

$$\mathfrak{D}(t) = \left\langle \left\| \vec{\Phi}(\vec{\mathrm{R}}) \right|_{\|\vec{R}\| = 4\ \text{Å}} \right\| \right\rangle_t \tag{3.11}$$

where $< >_t$ denotes the average over the hydrogen bonds formed by the chain at time t. The results along the folding trajectory described in Figs. 3.4–3.8 are presented in Fig. 3.10.

Contrasting Fig. 3.10 with Figs. 3.5 and 3.6 clearly validates the two-state folding scenario depicted in Fig. 3.2: The dehydronic field starts as a very minor contributor during the hydrophobic collapse of the chain. *Yet, once hydrogen bonds are formed to compensate for the backbone burial and they become partially wrapped (transition state ensemble), the dehydronic force becomes the driving force behind chain compaction. Thus, the progress of the folding process follows a reaction coordinate dictated by the gradient of the dehydronic field, making the latter the dominant factor that commits the chain to fold. In plain terms, the*

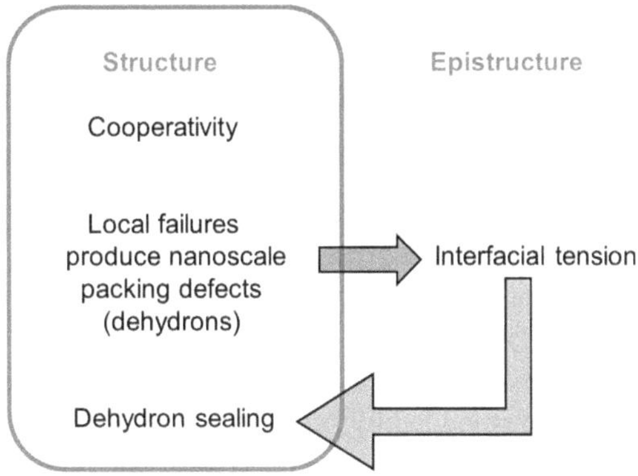

Fig. 3.11 Dynamic interplay between folding cooperativity and interfacial tension during the protein folding process

dehydronic field drags hydrophobes towards preformed hydrogen bonds and thus constitutes the mechanical realization of cooperativity in protein folding.

3.7 The Principle of Minimal Epistructural Distortion in Protein Folding

An epistructural approach to the protein folding problem entails incorporating interfacial free energy (reversible work to span the protein/water interface) in the algorithmic computational search in conformation space. *Since the interfacial free energy arises from failures of cooperativity, materialized as dehydrons, we can see how the minimization of the interfacial tension drives cooperativity, highlighting the epistructural feedback onto the conformational dynamics.* This interplay or dynamic entanglement between structural and epistructural events along protein folding pathways is clearly depicted in Fig. 3.11. As shown in Chap. 2, the interfacial tension arises from distortions of water hydrogen-bond structure as described by the scalar field $\phi(\vec{r}) = 4 - g(\vec{r})$. Since a folding protein changes its interface with water, the minimization of interfacial free energy actually becomes the thermodynamic translation of the principle of minimal epistructural distortion (MED) [35]. The steering role of the MED principle in the protein folding process will be investigated in this section.

The free energy associated with spanning the protein-water interface is

$$\Delta G_\phi = \Delta U_\phi - T\Delta S_\phi = \int \frac{1}{2}\lambda \|\vec{\nabla}\phi\|^2 d\vec{r} - Tk_B \ln\left[\prod\nolimits_{n=1,\ldots}(g_n/4)\right] \qquad (3.12)$$

where the interfacial energy term $\Delta U_\phi = \int \frac{1}{2}\lambda \left\|\vec{\nabla}\phi\right\|^2 d\vec{r}$ was introduced in Chap. 2 and the entropic cost of solvent confinement at the interface is evaluated as $\Delta S_\phi = k_B \ln\left[\Pi_{n=1,\ldots}(g_n/4)\right] \leq 0$, where k_B = Boltzmann constant, g_n = time averaged number of hydrogen bonds for the nth-water molecule, and the dummy index n labels molecules within a solvent-smeared envelope of the protein (cf. - Chap. 2). Beyond three water layers, the actual thickness of the envelope is immaterial as g = 4 for bulk solvent molecules. Since $-T\Delta S_{\phi} \geq 0$, the entropic contribution always reinforces the energetic term $\Delta U_{\phi} = \int \frac{1}{2}\lambda \left\| \vec{\nabla}\phi \right\|^2 d\vec{r} \geq 0$. The reversible work, $\Delta G_{\phi} = \Delta U_{\phi} - T\Delta S_{\phi} \geq 0$, performed on the system to span the protein-water interface is always *destabilizing* of the protein fold. This fact justifies the MED principle as an operative steer of the folding process [35].

To test this hypothesis, we run folding trajectories covering physically relevant timescales guided by "coarse moves" in the backbone (Φ, Ψ) dihedral torsions of the protein chain, as noted previously in this chapter (Fig. 3.3). Each coarse move is defined by transitions between basins of attraction (R-basins) in the Ramachandran (Φ, Ψ)-map for each residue along the chain. A Ramachandran map plots the internal energy of a residue versus the free dihedral backbone angles, and the R-basins are the allowed regions in (Φ, Ψ)-space. Thus, each coarse move represents a transition of the over-all topology of the protein chain where the string of (Φ, Ψ)-values is described in a coarse grained manner by identifying the R-basins where the torsional values lie. Since each residue is assigned an R-basin after a coarse move, the topology of the chain is in effect an ensemble of detailed conformations, with each conformation generated by selecting individual backbone torsional coordinates within the assigned R-basins. After each coarse move, the system is allowed to equilibrate with the solvent for 1 ns with backbone torsional angles constrained to remain within the pre-assigned R-basins. The equilibration is performed subject to the constraints in the torsional coordinates specified above. Other internal coordinates including side-chain torsional degrees of freedom and solvent coordinates are allowed to vary freely during equilibration.

To reach timescales of relevance to the folding process (> 10 μs), the folding process is steered by the coarse-grained stochastic process. The latter is defined in the time range [t, t + τ] (τ = 1 ns = fixed time step) by the probability p(t) of the transition **B**(t) → **B**(t + τ) for each vector of assigned R-basins **B**(t) = $(B_1(t), B_2(t) \ldots, B_N(t)$ representing the topology of the chain at time t, where N is the chain length and the dummy index denotes residue contour position (Fig. 3.3). The initial coarse state of the chain, **B**(0), is obtained by random assignment of R-basins for individual residues. All thermodynamic quantities are computed relative to the initial random coil ensemble or coarse state **B**(0). The transition probability p (t) is dependent on the overall free-energy difference between the two consecutive coarse states: $\Delta\Delta G(t) = \Delta\Delta H(t) - T\Delta\Delta S(t) = \Delta\Delta G(\boldsymbol{B}(t+\tau)) - \Delta G(\boldsymbol{B}(t)$, where the enthalpy contribution is in effect determined in the NPT ensemble (cf. Chap. 2) by the energy difference between the two equilibrated conformations belonging to topologies $\boldsymbol{B}(t+\tau)$, $\boldsymbol{B}(t)$, respectively. The entropy difference $\Delta\Delta S(t)$ is obtained from the Boltzmann formula:

$$\Delta\Delta S(t) = k_B \ln\left[Z\left(\boldsymbol{B}(t+\tau)/Z(\boldsymbol{B}(t)\right)\right] \tag{3.13}$$

where Z(**B**(t)) is the number of chain conformations subject to the constraints determined by the R-basin vector **B**(t). The possibility of subordinating or entraining the folding process to the coarse-grained stochastic dynamics hinges on an adiabatic scheme whereby intra-R-basin equilibration occurs faster that inter-basin transition. Since we adopt a Monte Carlo scheme [35, 36], we get:

$$p(t) = \exp\left[-\frac{\Delta\Delta G(t)}{k_B T}\right] \tag{3.14}$$

provided $\Delta\Delta G(t) > 0$ at T = 303 K and p(t) = 1 otherwise. If at time t, the coarse move **B**(t) = **B** → **B'** is rejected by the Monte Carlo procedure, the system remains

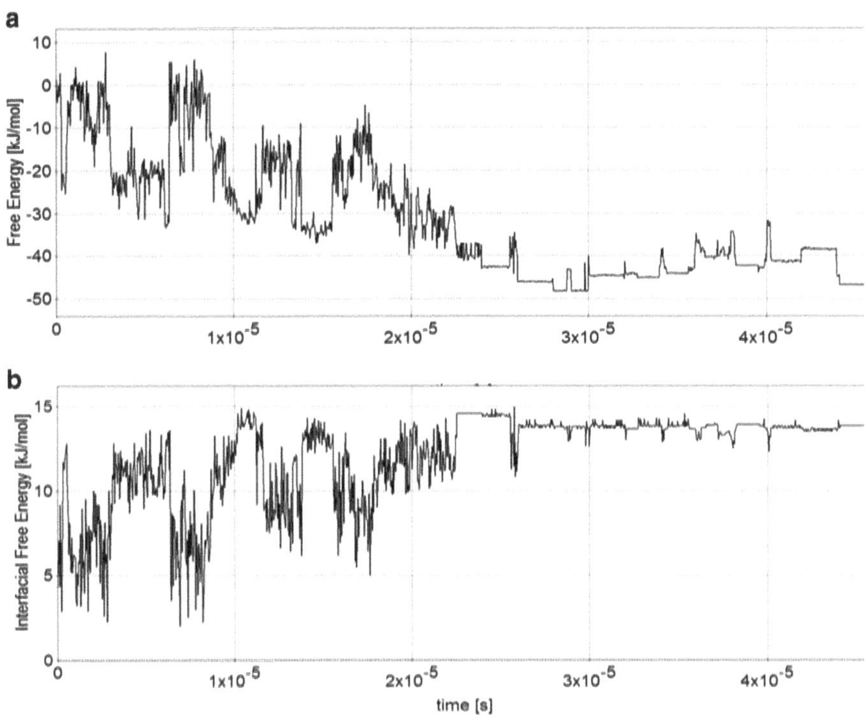

Fig. 3.12 Folding trajectory steered by an underlying coarse-grained stochastic process consisting of 5.2×10^4 coarse moves representing transitions and equilibrations of topological states of the chain for the thermophilic variant of the B1 domain of protein G, an autonomous folder with native fold reported in PDB entry 1GB4. The folding trajectory was generated adopting ΔGas potential determinant of coarse moves and equilibration. (**a**) Time-dependent behavior of the free energy relative to the random coil ensemble. (**b**) Time-dependent behavior of the interfacial free energy ΔG_ϕ. Reprinted from [35], copyright 2013 with permission from AIP Publishing LLC

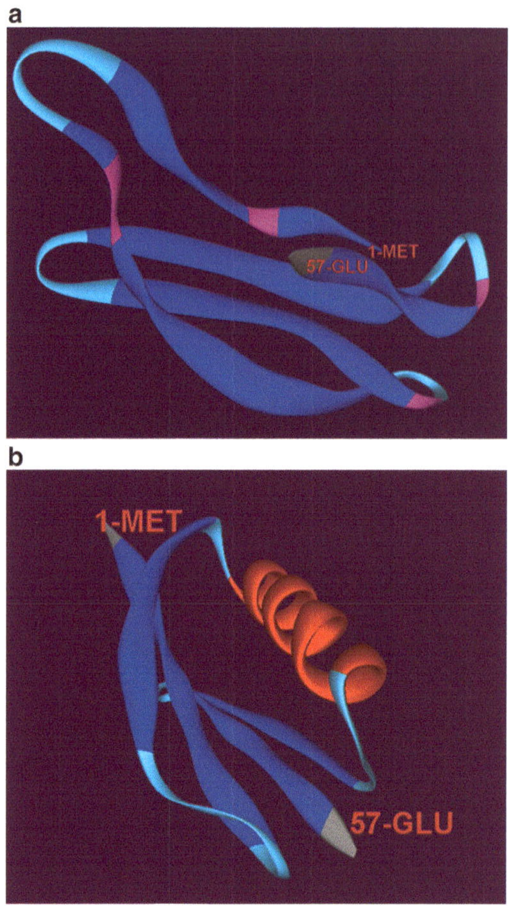

Fig. 3.13 (**a**) Ribbon representation of the destiny steady state conformation for the simulation described in Fig. 3.9. (**b**) Native fold in ribbon representation obtained from the structural coordinates reported in PDB.1GB4. Reprinted from [35], copyright 2013 with permission from AIP Publishing LLC

in topological state **B** for another 1 ns ($\mathbf{B}(t+\tau) = \mathbf{B}(t) = \mathbf{B}$) during which it undergoes a second round of equilibration with the solvent (cf. Fig. 3.3).

The AMBER package is adopted to equilibrate with the solvent (see Chap. 1 and 2 for details) with the constraint that backbone torsional coordinates remain within the R-basins that define the coarse state. Within the AMBER package, the energy stored in the orthogonal non-Debye polarization, or, equivalently, in the distortion of water structure, is not included in the thermodynamic potential ΔG.

To test the consequences of the MED principle, we generated eight folding MC trajectories for an autonomously folding protein within an NPT (isothermal/isobaric, $T = 303$ K) ensemble [35, 36]. The first four trajectories were generated with coarse states equilibrated and transitioned using the standard potential, ΔG while the remaining four trajectories contained coarse states equilibrated and transitioned using the potential $\Delta G_{tot} = \Delta G + \Delta G_{\phi}$ that incorporates the interfacial contribution. Each trajectory consists of 5.2×10^4 coarse moves, with each coarse state or chain topology generated by the underlying stochastic process. Two representative

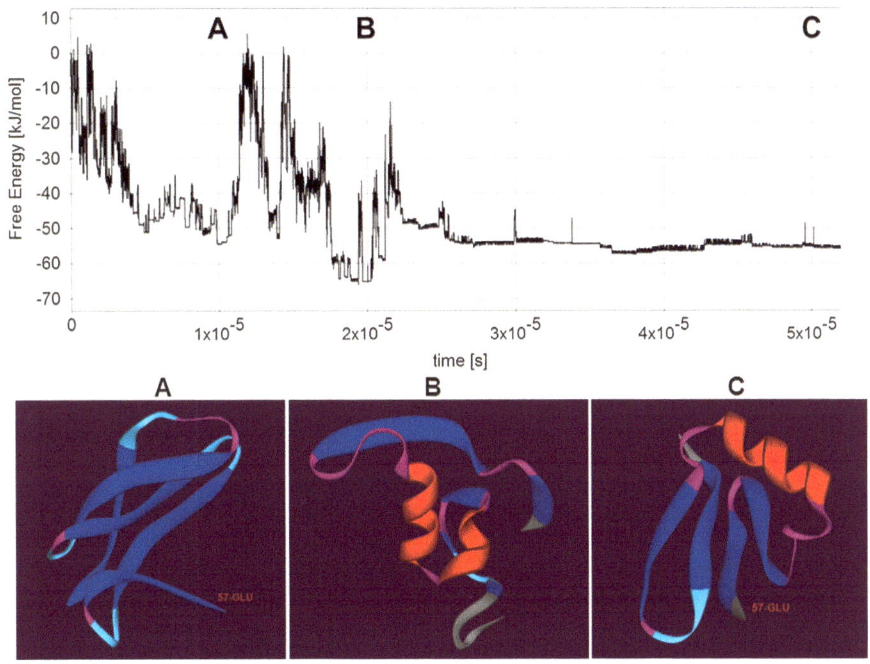

Fig. 3.14 Folding trajectory consisting of 5.2×10^4 coarse moves representing transitions and equilibrations of topological states of the chain for the thermophilic variant of the B1 domain of protein G. The folding trajectory was generated adopting the thermodynamic potential $\Delta G_{tot} = \Delta G + \Delta G_{\phi}$ as determinant of coarse moves and equilibration within coarse states. The trajectory is described by the time-dependent behavior of the free energy relative to the random coil initial state. Three equilibrated conformations at 10, 20 and 50 μs are shown in ribbon representation and denoted *A*, *B*, *C*, respectively. The steady-state conformation C is not the global minimum in free energy. Reprinted from [35], copyright 2013 with permission from AIP Publishing LLC

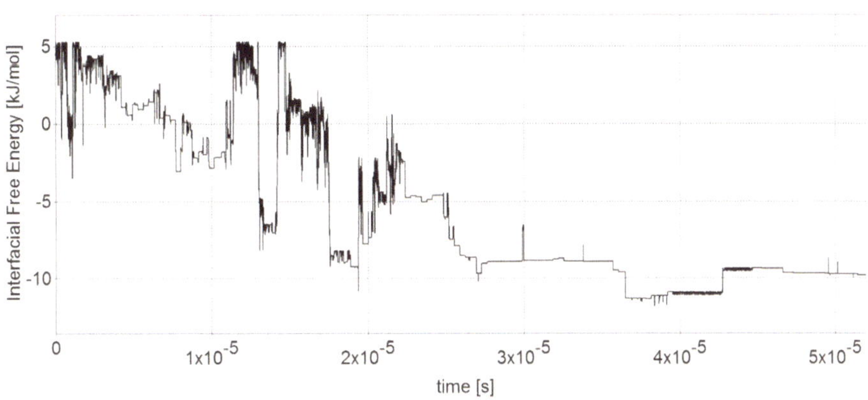

Fig. 3.15 Time-dependent behavior of the relative interfacial free energy ΔG_{ϕ} for the trajectory described in Fig. 3.14. The steady-state conformation C from Fig. 3.14 actually realizes the minimum of ΔG_{ϕ}. Reprinted from [35], copyright 2013 with permission from AIP Publishing LLC

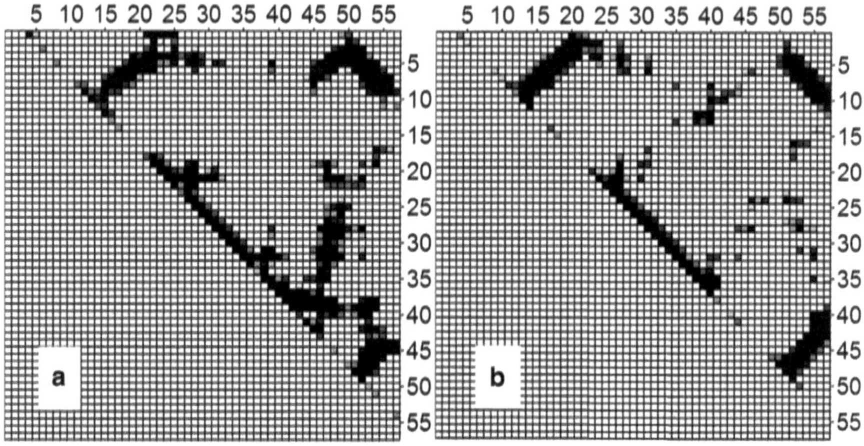

Fig. 3.16 Contact matrices associated with the equilibrated destiny conformation generated by the folding trajectory described in Figs. 3.14 and 3.15 (**a**) and with the native fold shown in ribbon representation in Fig. 3.13b (**b**), respectively. An (i,j) entry in the matrix is filled in *black* if the minimum Euclidean distance $d_{min}(i,j)$ between atoms in residues i and j is ≤ 4 Å, *grey* if 4 Å $< d_{min}(i,j) \leq 6$ Å, with the upper bound being the effective cut-off for long-range interactions, and *white* if $d_{min}(i,j) > 6$ Å. The two matrices reveal the same topological pattern of antiparallel and parallel β-sheets and α-helix. Reprinted from [35], copyright 2013 with permission from AIP Publishing LLC

trajectories are reported, the first (Figs. 3.12 and 3.13) adopting the potential ΔG, and the second (Figs. 3.14, 3.15, and 3.16), adopting the potential ΔG_{tot}. All trajectories in each group reproduce the destiny structures reported in this chapter to within 1 Å RMSD, as well as the free energy and interfacial free energy patterns hereby reported. In other words, the results are robust in the sense that the folding trajectories in the absence of the interfacial term consistently generate structures in the same topological class, and the same can be stated for trajectories that include the interfacial term. Due to our computational limitations, mitigated by the underlying coarse grained dynamics, we chose a relatively short ($N = 57$) protein chain capable of folding autonomously: the thermophilic variant of the B1 domain of protein G from *Streptococcus* (PDB.1GB4). The thermophilic variant was selected over the wild type due to its higher thermal stability. This higher stability should minimize the interfacial free energy cost beyond wild-type levels and hence should provide a better testing ground to validate the MED principle.

Figure 3.12 reveals that when the interfacial term ΔG_{ϕ} is excluded from the potential that controls the underlying stochastic process, the chain reaches a free energy minimum in about 27 μs but contains an interfacial free energy that is approximately 14 kJ/mol above the random coil value ($\Delta G_{\phi} = 14$ kJ/mol). The interfacial free energy has not been minimized along the trajectory but instead reaches its maximum at 27 μs. As expected, the destiny steady state has an equilibrated conformation (Fig. 3.13a) that is topologically different from the native state of the protein (Fig. 3.13b). This dynamic behavior, together with the incorrect destiny structure, suggest the need to incorporate the interfacial term ΔG_{ϕ} into the potential.

A representative folding trajectory incorporating the interfacial free energy in coarse grained transitions and structure equilibration is reported in Figs. 3.14–3.16. Thus, $\Delta U_{\phi}(t)$ was calculated as the difference in interfacial energy between equilibrated conformations within the ensembles **B**(t) and **B**(0) =random coil ensemble. This time, the trajectory converges at about 27 μs to a destiny steady state (structure C, Fig. 3.14) with a relative free energy ΔG_{ϕ} that is not the global minimum. In fact, the latter is achieved at 20 μs (structure B, Fig. 3.14), while another minimum is achieved at 10 μs (structure A, Fig. 3.14).

Strikingly, *although the destiny structure is not the global free energy minimum, it is the structure that realizes the minimum in interfacial free energy* ($\Delta G_{\phi} = -10$ kJ/mol approximately), *as shown in Fig.* 3.15. *It is also the correct structure, topologically equivalent to the native fold* (compare structure C in Fig. 3.14 with the native structure in Fig. 3.13b obtained from PDB.1GB4), *with a highly similar contact matrix* (*Fig.* 3.16) *and RMSD* = *1.32* Å. Furthermore, as shown in Fig. 3.15, leaving aside fluctuations, the interfacial free energy is consistently decreasing throughout the simulation, implying that the MED principle has been operative. The destiny structure that minimizes interfacial free energy (Figs. 3.14c and 3.16a) has a significantly higher number of tertiary (nonlocal) contacts (32, as shown in Fig. 3.16a) compared with those that represent local minima in ΔG, shown in Figs. 3.13 and 3.14a, b, with 21, 17 and 12, respectively. This observation suggests a structural signature of the MED principle (cf. Fig. 3.11).

The results reported in Figs. 3.14–3.16, jointly with the fact that the destiny steady state is topologically and geometrically very close to the native fold (PDB.1GB4), imply that protein folding seeks to minimize interfacial tension or the disruption of the water hydrogen-bond matrix, in accord with the MED principle.

3.8 Both Structural and Epistructural Perspectives Are Necessary to Solve the Protein Folding Problem

The results given in this chapter highlight the need to combine the structural and epistructural approach to cooperativity in order to solve the protein folding problem. The structural approach reveals the pivotal role of the dehydronic field to commit the chain to fold by promoting the cooperative wrapping of backbone hydrogen bonds, while the epistructural approach highlights the minimization of interfacial free energy as the steering factor governing the organized collapse of the protein chain. The mature picture that emerges from our analysis (Fig. 3.11) reveals a feedback mechanism whereby the epistructural tension generated by structural defects steers the conformational dynamics towards a correction or sealing of such defects. Thus, the structural-epistructural interplay leads to an iterative improvement of the quality of the protein fold that becomes progressively dryer in water.

It should be emphasized that this exquisite entanglement between protein folding and interfacial solvent dynamics (Fig. 3.11) is at this time not accessible to realistic computational resources. It requires a judicious coarse-grained stochastic process to steer the folding dynamics of the protein chain by lifting the modulo-Ramachandran-basin metadynamics invented by the author to the level of full geometric realization, as described in Fig. 3.3.

3.9 Discussion Forum: The Futile Search for Protein Folding Intermediates

Evidently, protein folding *in vitro* is a spontaneous process. That means it is irreversible in the thermodynamic sense. Surely the endpoint native and denatured states may be recovered by respectively restoring physiological or denaturing conditions but, according to the basic tenets of thermodynamics, the folding and unfolding pathways themselves are untraceable and irreproducible. Dissipative frictional forces intervene at any level of coarse-graining of the folding process [2, 18]. Therefore, the intermediates associated with the folding process are irretrievable, as it would be the case for any spontaneous process in nature regardless of the level of detail at which the system is specified.

Notwithstanding these basic thermodynamic considerations, and at odds with the rigorous thermodynamically consistent two-state picture of the folding process (Fig. 3.2) [2, 4, 5], a very active quest for folding intermediates continues [28–30] to this day. This search is futile, even meaningless. Real folding intermediates are not difficult to trap: they simply do not exist. True folding intermediates are irretrievable, and any claim to the contrary violates the second law of thermodynamics. By the same token, the so-called two-state picture is uninformative: it is simply the acknowledgement that the folding process is thermodynamically irreversible as all spontaneous processes are.

Perhaps even more worrisome are the efforts to characterize folding intermediates by denaturing the native state (however carefully). As the second law of thermodynamics tells us, the denaturation process is **not** the reverse of the folding process: only the endpoints are reversed. Perhaps to discourage further efforts in this direction, it would be more graphic to resort to an analogy. Imagine you need to delineate the way a standing house has been put together. No one in the right state of mind would address the problem by having the house demolished, taking snapshots at different stages of the demolition and then playing the movie backwards. Yet, the same preposterous reasoning has been assumed to hold *mutatis mutandis* in the context of the protein folding problem.

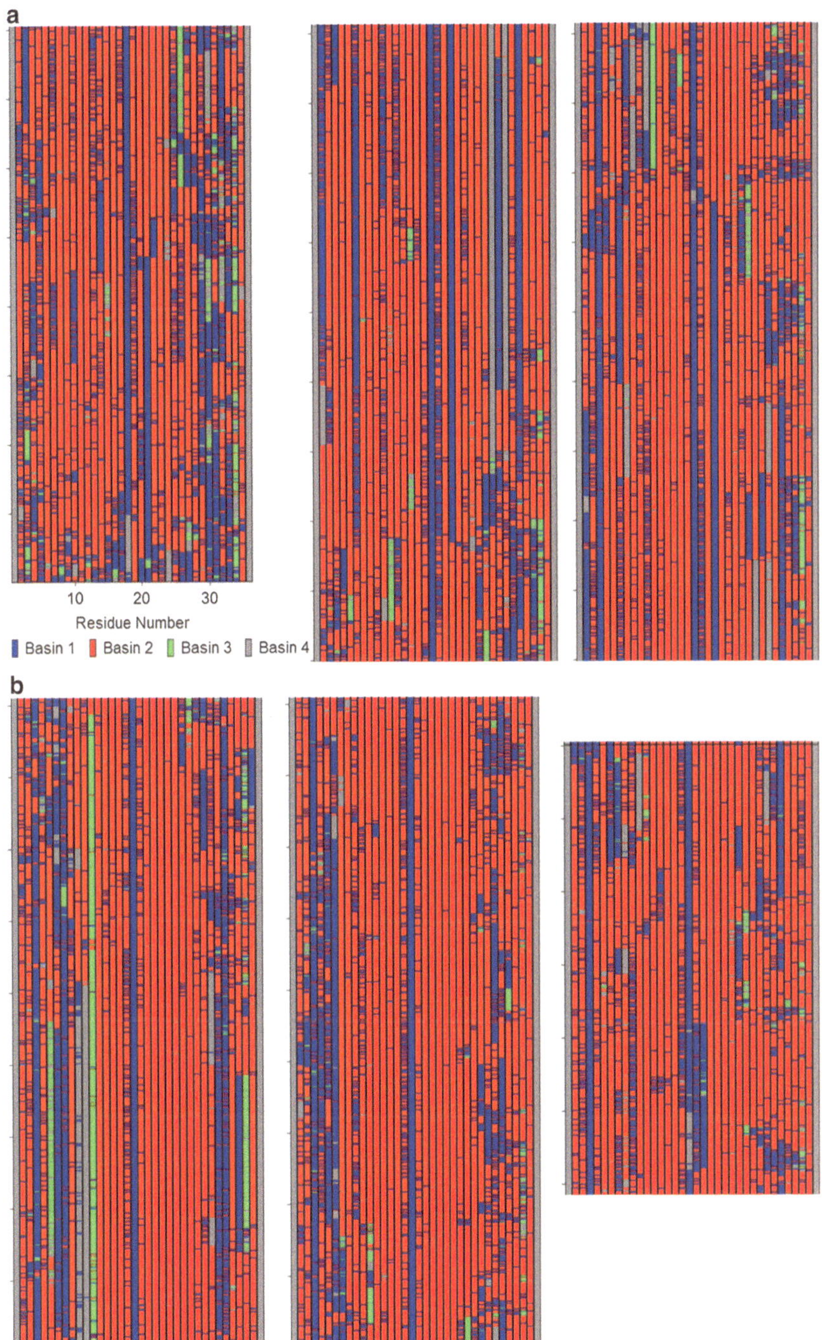

Fig. 3.17 (**a**, **b**) Coarse-grained (modulo Ramachandran basin) history of the conformational search of the villin headpiece subdomain simulated at the all-atom explicit solvent level for 1 μs with the AMBER package [19, 37]. The topological history of the trajectory must be read from *left* to *right* and from *bottom* to *top*

3.10 Problems

3.1. The "two-state folder" model for single-domain soluble proteins remains *en vogue* and there seems to be a seed of truth in it, regardless of the fact that it is uninformative as to the folding process itself and in regards to the factors involved in committing the chain to fold. Justify the two-state model based on the principle of minimization of the dehydronic field.

3.2. Describe structurally the "transition state" in the folding of the thermophilic variant of protein G.

3.3. Assessing the possibility of folding proteins with the AMBER simulation package [19, 37]. Figure 3.17 displays the coarse-grained (modulo Ramachandran basin) history of the conformational search of the villin headpiece subdomain simulated at the all-atom explicit solvent level for 1 μs using the AMBER package [37]. The simulation follows the pioneering effort by Duan and Kollman [19]. The topological history of the trajectory must be read from left to right and from bottom to top. Based on this test case, assess the potential of the AMBER simulation package to capture folding cooperativity and generate a viable folding pathway in all-atom-explicit-solvent simulations.

References

1. Anfinsen CB. Principles that govern the folding of protein chains. Science. 1973;181:223–30.
2. Fernández A, Sosnick TR, Colubri A. Dynamics of hydrogen-bond desolvation in folding proteins. J Mol Biol. 2002;321:659–75.
3. Chandler D. Interfaces and the driving force of hydrophobic assembly. Nature. 2005;437:640–7.
4. Jewett A, Pande VS, Plaxco KW. Cooperativity, smooth energy landscapes and the origins of topology-dependent protein folding rates. J Mol Biol. 2003;326:247–53.
5. Scalley-Kim M, Baker D. Characterization of the folding energy landscapes of computer generated proteins suggests high folding free energy barriers and cooperativity may be consequences of natural selection. J Mol Biol. 2004;338:573–83.
6. Fernández A, Colubri A, Berry RS. Three-body correlations in protein folding: the origin of cooperativity. Phys A. 2002;307:235–59.
7. Fernández A, Kostov K, Berry RS. From residue matching patterns to protein folding topographies: general model and bovine pancreatic trypsin inhibitor. Proc Natl Acad Sci U S A. 1999;96:12991–6.
8. Fernández A, Colubri A, Berry RS. Topology to geometry in protein folding: beta-lactoglobulin. Proc Natl Acad Sci U S A. 2000;97:14062–6.
9. Fernández A, Kardos J, Goto J. Protein folding: could hydrophobic collapse be coupled with hydrogen-bond formation? FEBS Lett. 2003;536:187–92.
10. Fernández A. Conformation-dependent environments in folding proteins. J Chem Phys. 2001;114:2489–502.
11. Avbelj F, Baldwin RL. Role of backbone solvation and electrostatics in generating preferred peptide backbone conformations: distributions of phi. Proc Natl Acad Sci U S A. 2003;100:5742–7.
12. Fernández A. Keeping dry and crossing membranes. Nat Biotechnol. 2004;22:1081–4.

13. Krantz BA, Moran LB, Kentsis A, Sosnick TR. D/H amide kinetic isotope effects reveal when hydrogen bonds form during protein folding. Nat Struct Biol. 2000;7:62–71.
14. Plaxco KW, Simmons KT, Baker D. Contact order, transition state placement and the refolding rates of single domain proteins. J Mol Biol. 1998;277:985–94.
15. Fersht A. Transition-state structure as a unifying basis in protein-folding mechanisms: contact order, chain topology, stability, and the extended nucleus mechanism. Proc Natl Acad Sci U S A. 2000;97:1525–929.
16. Fernández A, Scott LR. Adherence of packing defects in soluble proteins. Phys Rev Lett. 2003;91:018102.
17. Fernández A. What caliber pore is like a pipe? Nanotubes as modulators of ion gradients. J Chem Phys. 2003;119:5315–9.
18. Fernández A, Shen M, Colubri A, Sosnick TR, Freed KF. Large-scale context in protein folding: villin headpiece. Biochemistry. 2003;42:664–71.
19. Duan Y, Kollman PA. Pathways to a protein folding intermediate observed in a 1-microsecond simulation in aqueous solution. Science. 1998;282:740–4.
20. Baldwin RL. Making a network of hydrophobic clusters. Science. 2002;295:1657–8.
21. Nemethy G, Steinberg IZ, Scheraga HA. The influence of water structure and hydrophobic contacts on the strength of side-chain hydrogen bonds in proteins. Biopolymers. 1963;1:43–69.
22. Fernández A, Berry RS. Extent of hydrogen-bond protection in folded proteins: a constraint on packing architectures. Biophys J. 2002;83:2475–81.
23. Novotny J, Bruccoleri R, Karplus M. Analysis of incorrectly folded protein models. Implications for structure predictions. J Mol Biol. 1984;177:787–818.
24. Daggett V, Levitt M. A model of the molten globule state from molecular dynamics simulations. Proc Natl Acad Sci U S A. 1992;89:5142–6.
25. Brooks CL, Case D. Simulations of peptide conformational dynamics and thermodynamics. Chem Rev. 1993;93:2487–502.
26. Fernández A, Rogale K. Sequence-space selection of cooperative model proteins. J Phys A Math Gen. 2004;37:197–202.
27. Kuwata K, Shastry R, Cheng H, Hoshino M, Batt CA, Goto Y, Roder H. Structural and kinetic characterization of early folding events in beta-lactoglobulin. Nat Struct Biol. 2001;8:151–5.
28. Nymeyer H, Garcia AE, Onuchic JN. Folding funnels and frustration in off-lattice minimalist protein landscapes. Proc Natl Acad Sci. 1998;95:5921–8.
29. Onuchic JN, Luthey-Schulten Z, Wolynes PG. Theory of protein folding: the energy landscape perspective. Annu Rev Phys Chem. 1997;48:545–600.
30. Chan HS, Dill KA. From Levinthal to pathways to funnels. Nat Struct Biol. 1997;4:10–9.
31. Fernández A, Colubri A, Berry RS. Topologies to geometries in protein folding: hierarchical and nonhierarchical scenarios. J Chem Phys. 2001;114:5871–88.
32. Shi Z, Krantz BA, Kallenbach N, Sosnick TR. Contribution of hydrogen bonding to protein stability estimated from isotope effects. Biochemistry. 2002;41:2120–9.
33. Pietrosemoli N, Crespo A, Fernández A. Dehydration propensity of order-disorder intermediate regions in soluble proteins. J Proteome Res. 2007;6:3519–26.
34. Schutz CN, Warshel A. What are the dielectric "constants" of proteins and how to validate electrostatic models? Proteins Struct Funct Gen. 2001;44:400–8.
35. Fernández A. The principle of minimal episteric distortion of the water matrix and its steering role in protein folding. J Chem Phys. 2013;139:085101.
36. Fernández A. Fast track communication: water promotes the sealing of nanoscale packing defects in folding proteins. J Phys Condens Matter. 2014;26:202101.
37. Salomon-Ferrer R, Case DA, Walker RC. An overview of the Amber biomolecular simulation package. WIREs Comput Mol Sci. 2013;3:198–210.

Chapter 4
Epistructural Dynamics of Biological Water

Adopting the methods of epistructural physics previously developed, this chapter unravels dynamic singularities of the protein-water interface. It is shown that interfacial water in the vicinity of protein dehydrons is subject to a torque resulting from the breakdown of the Debye dielectric picture. This non-Debye torque accelerates interfacial water molecules beyond the kinetic energies characteristic of the bulk. The epistructural dynamic analysis highlights a link between the protein structural deficiencies and the defects of the aqueous interface. Thus, the tightness of the hydration shell is modulated by the mobility of interfacial water that varies widely becoming a maximum at dehydron locations. Interfacial water molecules subject to the non-Debye torque enhance the propensity for protein associations and raise their catalytic rate as proton acceptors by increasing the frequency of effective collisions. The most pronounced de-wetting propensity at the interface is promoted by dehydrons on the surface of soluble proteins. The result has crucial implications for the drug designer since ligands may be engineered to expel labile interfacial patches upon association. Furthermore, subtle differences in the interfacial dynamic singularities within a protein family may be exploited to promote drug specificity. This finding is of significance to the pharmaceutical industry since homologous proteins are known to share a common structural topology and therefore, discriminatory molecular recognition remains a daunting challenge.

4.1 Dynamic Singularities of Biomolecular Interfaces

The hydration shell relaxation of a soluble protein is generally assumed to slow down relative to bulk levels. Yet dynamic hydration of soluble proteins is not uniform, rather it is marked by vastly different mobilities of interfacial water molecules [1–5], with an expected tight correlation between water-density fluctuations and mobilities across the interface. The uneven distribution of mobilities of interfacial molecules affects protein folding dynamics and associations, since the

A. Fernández, *Physics at the Biomolecular Interface*, Soft and Biological Matter,
DOI 10.1007/978-3-319-30852-4_4

deprivation in hydrogen-bonding opportunities increases interfacial kinetic energy, as shown in this chapter, and is causative of interfacial tension. This tension steers the protein chain towards a conformational change that corrects the structural defect that produced the tension (Chap. 3, Fig. 3.11), or, if the conformational change is unfeasible, it promotes protein associations that exogenously correct the tension-generating structural defect.

The hydration shell gets tighter or less mobile at interfacial regions where the hydrogen-bond network of hydrating molecules is minimally disrupted, and becomes looser in concave or flat nonpolar regions of the protein surface [3–5]. Thus, dehydration patterns have been associated with surface topography [5] and with fold topology [6], both determinants of protein associations. However, the pronounced de-wetting propensity of interfacial regions remains difficult to assess in general, as evidenced by the difficulty in predicting binding sites [7], in inferring propensities for aberrant aggregation [8], and in rationally designing drugs or ligands, notwithstanding some level of success in ligand docking [9]. This chapter explores de-wetting patterns determined by under-wrapped backbone hydrogen bonds, the d*ehydrons* [10, 11] described in the preceding chapters. Since their levels of hydration may be modulated by protein associations, dehydrons are clearly shown to be implicated in molecular recognition [8–11]. More precisely, dehydrons may become stabilized and electrostatically strengthened by the attachment of a ligand or binding partner that further contributes to their dehydration. As indicated in Chap. 1, the net gain in Coulomb energy associated with wrapping or protecting a dehydron has been experimentally determined to be ~4 kJ/mol [12]. The adhesive force exerted by a dehydron on a hydrophobic group at 6 Å distance is ˜7.8 pN, a magnitude comparable to the hydrophobic attraction between two nonpolar moieties that frame unfavorable interfaces with water [12]. *Furthermore, the fact that dehydrons promote protein associations (Chaps. 1 and 2) strongly suggests that they may signal sites with a loose hydration shell, that is, sites where the work needed to remove interfacial water is minimal and amply compensated by the enhancement of the Coulomb energy.*

As shown in Chap. 8, this result will prove to be of paramount relevance for the rational drug designer as it introduces a blueprint to guide ligand engineering: the de-wetting propensities in the target protein may be sculpted into the ligand so that water is expelled from the interface upon association wherever the work needed to do so is minimal while the dehydronic force is maximal. In other words, the drug/ligand may be engineered to tightly fit against the hot spots of labile interfacial water.

In light of these considerations, a new picture of the interface dynamics is likely to emerge whereby the local physico-chemical parameters (kinetic energy, interfacial tension, non-Debye "orthogonal" polarization, cavity formation probability, density fluctuations, dehydronic field), all dependent on the water structural parameter $\phi(\vec{r}) = 4 - g(\vec{r})$, are expected to be tightly interrelated. Thus, the operational role of frustration in determining the local dynamics of episteric water becomes the focus of this chapter. The analysis naturally leads to the conclusion that frustration-inducing structural features of the protein introduce dynamic singularities that are at odds with the general slowing down of epistructural water relative to bulk levels.

4.2 Dehydrons Impact the Dynamics of the Aqueous Interface

In this section we introduce a descriptor of hydration tightness or de-wetting propensity in order to assess the functional role of dehydrons. Tightness is defined by the extent of mobility of hydrating molecules. Thus, the mean residence time of hydrating molecules within a domain around each residue on the protein surface serves as an adequate indicator. As an illustration, we describe in detail the interfacial water mobility for the autonomously folded SH3 domain (PDB.1SRL) [13], an all-beta protein involved in the regulation of cell-signaling. We also analyze ubiquitin (PDB.1UBI) [14], a quintessential α/β fold, and finally explore the relationship between extreme dehydration propensity and amyloidogenic aggregation [9].

Our results lead us to identify dehydrons as the structural feature causative of the most dramatic decrease in residence times (highest mobility) of hydrating molecules. Dehydrons promote local de-wetting because the Coulomb energy of the intramolecular hydrogen bonds becomes magnified upon water removal, and the net gain in stability upon dehydration offsets the work required to remove surrounding water [10, 15, 16].

In order to assess the local mobility of water molecules in the hydration shell, the local mean residence time, $\langle\tau\rangle_i$ of hydrating molecules around residue i is defined with respect to a microenvironment in the form of a spherical domain $D(i)$ of 6 Å-radius (~width of more than two water layers [12]) centered at the α-carbon of residue i (Fig. 4.1a). The computations are performed for a range of radii (see below). The residence time is obtained as follows:

$$\begin{aligned} &\langle\tau\rangle_i = \int \tau f_i(\tau)d\tau \Big/ \left(\int f_i(\tau)d\tau\right); \\ &\int_0^{\tau} f_i(\tau')d\tau' = P_i(0) - P_i(\tau); \\ &P_i(\tau) = \Theta^{-1}\int_0^{\Theta}\left[\sum_{\substack{v(t)\in U(i,t)\\ w(t+\tau)\in U(i,t+\tau)}} \delta(v(t), w(t+\tau))\right]dt; \end{aligned} \tag{4.1}$$

where $f_i(\tau)d\tau/\int f_i(\tau)d\tau$ is the expected fraction of water molecules that exit $D(i)$ within time interval $[\tau, \tau + d\tau]$; $P_i(\tau)$ is the expected number of water molecules remaining in $D(i)$ at time τ ($P_i(0)$-$P_i(\tau)$ = number of molecules that exited $D(i)$ in the time period $[0, \tau]$); $v(t), w(t+\tau)$ denote indexes labeling water molecules contained in $D(i)$ at times t and $t + \tau$, respectively; $U(i,t),\ U(i,t+\tau)$ denote the collection of indexes of water molecules contained in $D(i)$ at times t and $t + \tau$,

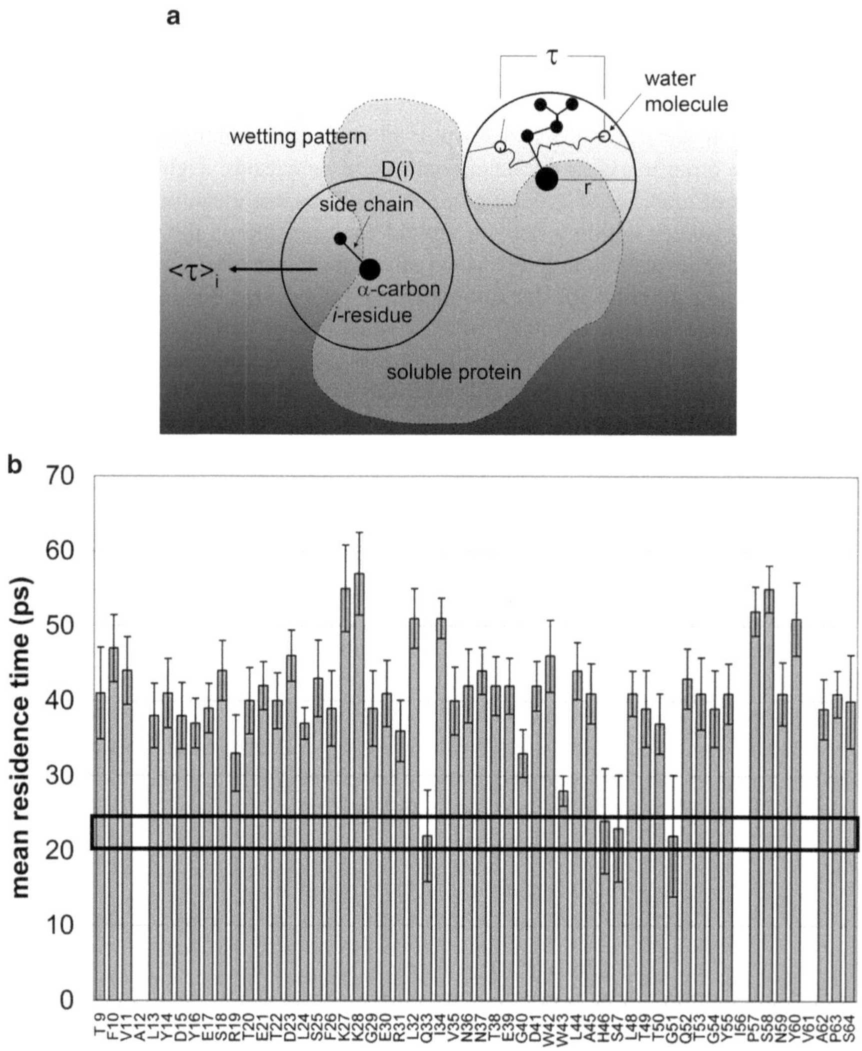

Fig. 4.1 (**a**) Basic scheme for the computation of de-wetting hot spots in soluble proteins. (**b**) Mean residence times of water molecules solvating the folded protein SH3 domain. Residue numbering follows PDB file 1SRL. The one-letter amino acid code was adopted for clarity. The mean residence times are computed using (4.1) and extracted from MD simulations of the hydration dynamics. The error bars represent variances and the thick rectangle represents the range of residence times for bulk water within a domain of the same dimensions of the one used to examine the protein surface. The range for bulk water serves as benchmark to single out hot spots of weak hydration. Three residues (Ala12, Ile56 and Val61) are fully buried in the folded structure and thus their hydration shells contain no water molecules. Figure (**b**) reprinted from [20], copyright 2007 with permission from AIP Publishing LLC

respectively; δ is the Kronecker symbol ($\delta(v(t), w(t+t)) = 1$ if $v(t) = w(t+\tau)$ and 0, otherwise); and the integration over variable t is carried out over the interval of sampled times ($t = 0$ to $t = \Theta = 10$ ns) after 5 ns of prior equilibration (the sampling is considered exhaustive since $< \tau ><< \Theta$ for all residues).

The mean residence times are obtained from classical trajectories generated by molecular dynamics (MD) simulations starting from the PDB structure embedded in a pre-equilibrated cell of explicitly represented water molecules and counterions [17, 18]. Computations are performed by integration of Newton's equations of motion with time step 2 fs using the GROMACS program [19] in an NPT ensemble with box size $8 \times 8 \times 8$ nm^3 and periodic boundary conditions maintained at 300 K and 1 atm. The box size is calibrated so that the solvation shell extended at least 12 Å from the protein surface at all times. Specific details on the generation of these classical trajectories may be found in [20].

4.3 De-wetting Propensities at the Protein-Water Interface

We start by providing a detailed analysis of the de-wetting propensities in the SH3 domain (N = 55, PDB.1SRL) and their correlation with dehydrons in the structure. The mean residence times of water molecules at the protein-water interface has an over-all average value of 44 ps, nearly twice the residence time for bulk water (~21 ps, Fig. 4.1b) [21]. The hydration shell reveals some standard features. Thus, short residence times correspond to de-wetting propensity associated with hydrophobic hydration at cavities with relatively large curvature radius (~6 Å) [22], like the hydrophobic pocket containing Trp43 (Figs. 4.1b and 4.2).

By contrast, exposed hydrophobic residues like Val35 may be effectively "clathrated" or accommodated within a water cavity that introduces a minimal perturbation of the tetrahedral hydrogen-bonding network of water. In fact, clathration actually tightens the hydration shell (Figs. 4.1b and 4.2).

Side chains with small exposed cations (Lys27, Lys28) have the longest residence times since they are tightly hydrated as are polar residues in cavities (like Ser58), since their hydrogen bonding to water is protected. By contrast, exposed residues with delocalized positive charges like Arg 19 have looser hydration [23].

The residues with the shortest residence times ($\langle\tau\rangle < 24$ ps) and also with the largest variances $\left(\left\langle(\tau - \langle\tau\rangle)^2\right\rangle^{1/2} > 6 \text{ ps}\right)$ are Gln33, His46, Ser47 and Gly51, indicating a highly fluctuating environment and a high de-wetting propensity. As shown below, these residues are paired by the only backbone amide-carbonyl hydrogen bonds which are incompletely wrapped or shielded from water by non-polar groups of the protein chain. In accord with their shortest water residence times (Fig. 4.1b), dehydrons become favorably dehydrated, a process that decreases the over-all polarizability of their microenvironments. This dehydration strengthens the electrostatic contribution and stabilizes the hydrogen bond by destabilizing the

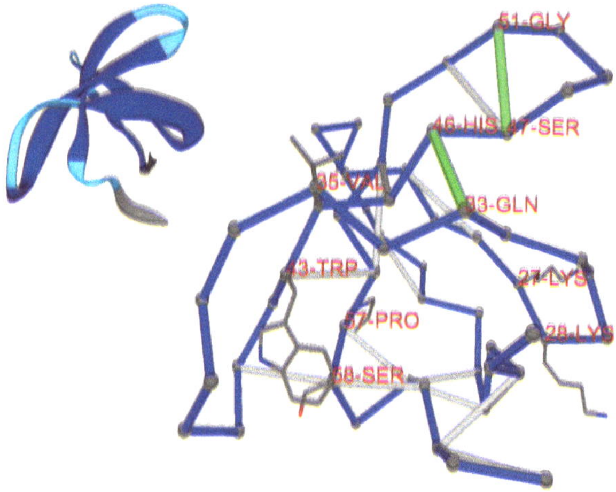

Fig. 4.2 PDB structure of SH3 domain indicating the packing defects in the protein. The backbone is displayed by *blue* segments representing virtual bonds joining a-carbons, well wrapped hydrogen bonds are shown as *white* segments joining α-carbons of paired residues, dehydrons are shown in *green*. Some residue labels and side chains are displayed to illustrate hydration patterns. Reprinted from [20], copyright 2007 with permission from AIP Publishing LLC

nonbonding or unbound state to the point of overcoming the work needed to remove surrounding water [8]. These packing defects and the decrease in medium polarizability or dielectric modulation they promote differ from packing-density variations in nonpolar interiors [24]: they relate to the exposure of the polar backbone due to incomplete wrapping by nonpolar side chains.

To explain the local weaknesses in the hydration shell of the protein (Fig. 4.1b), the dewetting propensity of dehydrons is quantified and compared with that of well wrapped hydrogen bonds. As shown in Chaps. 1 and 3, the dewetting propensity has a mechanical equivalent [12, 25]: the attractive force exerted on a nonpolar test group (i.e. methane, CH_4) by a preformed hydrogen bond that may be strengthened upon removal of surrounding water. As the nonpolar group approaches a dehydron, it displaces water thus decreasing the environmental polarity, thereby enhancing and stabilizing the bond. The wrapping microenvironment of a hydrogen bond may be described by the set of position vectors $\{\vec{R}_j\}_{j=1,\ldots,K} = \{\vec{R}_j(t)\}_{j=1,\ldots,K}$, relative to the hydrogen atom (origin of coordinates), of K nonpolar groups from the protein chain within the two spheres of radius 6 Å centered at the α-carbons of the paired residues. All time-dependent coordinates are taken from MD trajectories.

To compute the de-wetting of intramolecular hydrogen bonds, we take into account the modulation of local permittivity determined by a change in the "environmental coordinates": $\{\vec{R}_j\}_{j=1,\ldots,K} \to \{\vec{R}_j, \vec{R}\}_{j=1,\ldots,K}$ where $\vec{R}$ is the position vector of the test methane molecule that approaches the bond orthogonally to the

Coulombic field. The energy change ΔE associated with the change in wrapping is then given by [12, 25]:

$$\Delta E(t) = -\frac{qq'}{4\pi r}\left[\varepsilon^{-1}\left(\{\vec{R}_j, \vec{R}\}_{j=1,\ldots,K}\right) - \varepsilon^{-1}\left(\{\vec{R}_j\}_{j=1,\ldots,K}\right)\right] \tag{4.2}$$

where q, q′ are the effective charges at the hydrogen and oxygen atom of the amide-carbonyl hydrogen bond, r is the hydrogen-bond length $(r = \|\vec{r}\|,\ \vec{r} =$ position vector of carbonyl oxygen, origin of coordinates at the amide hydrogen), and the reciprocal permittivity ε^{-1} quantifying the Coulomb screening due to wrapping by nonpolar groups, is given by [25]:

$$\varepsilon^{-1}\left(\{\vec{R}_j\}_{j=1,\ldots,K}\right) = \left[\varepsilon_w{}^{-1} + (\varepsilon_0{}^{-1} - \varepsilon_w{}^{-1})\Omega\left(\{\vec{R}_j\}_{j=1,\ldots,K}\right)\left(1 + \frac{r}{\xi}\right)e^{-\frac{r}{\xi}}\right] \tag{4.3}$$

where ε_o, ε_w denote respectively the permittivity of vacuum and bulk water, ξ, fixed at 5 Å, is the characteristic length for water-dipole reorientation influence, Λ = 1.8 Å is the characteristic length for water structuring around a nonpolar group [25], and

$$\Omega\left(\{\vec{R}_j\}_{j=1,\ldots,K}\right) = \prod_{j=1}^{n}\left[1 + e^{-\|\vec{R}_j\|/\Lambda}\right]\left[1 + e^{-\frac{\|\vec{r}-\vec{R}_j\|}{\Lambda}}\right] \tag{4.4}$$

Applying (4.2–4.4) to the MD trajectories, we quantify the enhancement in the Coulombic contribution to the hydrogen bond resulting from the decrease in permittivity as the test hydrophobe approaches the protein surface (Fig. 4.3). Thus, the de-wetting field **Φ**(**R**) at position **R** of the test hydrophobe generated by a hydrogen bond wrapped by K nonpolar groups is given by:

$$\vec{\Phi}(\vec{R}) = \frac{qq'}{r}\vec{\nabla}_{\vec{R}}\left[4\pi\varepsilon\left(\{\vec{R}_j, \vec{R}\}_{j=1,\ldots,K}\right)\right]^{-1} \tag{4.5}$$

The time averages of the magnitude and variance of the de-wetting field were computed for each residue in the protein (Fig. 4.3).

Comparison of Figs. 4.1b and 4.3 prompt the following conclusions: (a) the tightness of the hydration shell of a soluble protein is not uniform, yielding an uneven distribution of mobilities for hydrating molecules; (b) dehydrons generate the most pronounced loosening of the hydration shell; (c) dehydrons are endowed with dehydration propensities.

A detailed analysis of dehydron dewetting propensities was conducted for other single-domain soluble proteins, unrelated evolutionarily or topologically to the study case SH3 domain and the same conclusions were shown to hold [20]. For example, we focused on the highly conserved ubiquitin (N = 76, 11 dehydrons, PDB.1UBI). In this case dehydrons also generate the most intense de-wetting field

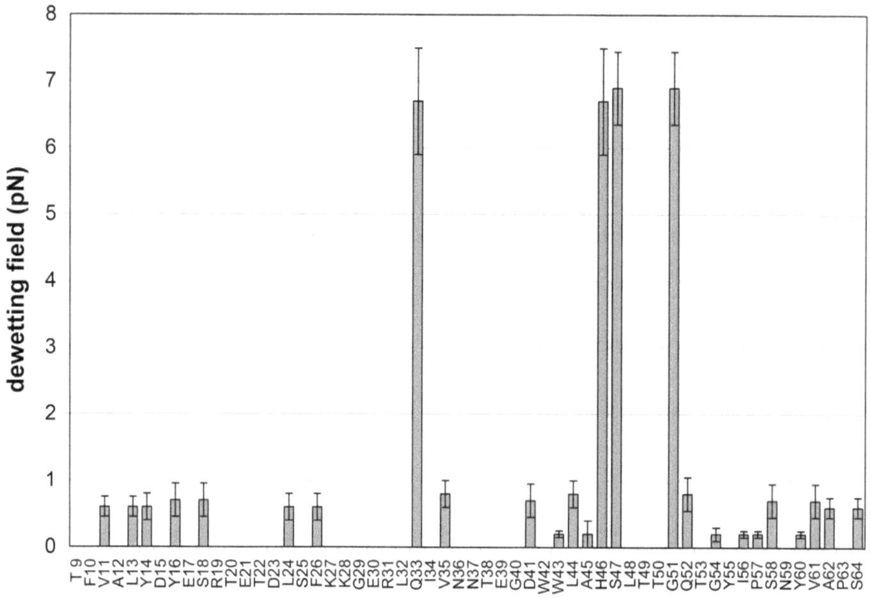

Fig. 4.3 De-wetting field exerted by residues paired by hydrogen bonds in the SH3 domain. The dehydration propensity of a hydrogen bond, $\|\mathbf{\Phi}(\mathbf{R})\|$, is determined at $\|R\| = 6$ Å, and is assigned to the two residues paired by the hydrogen bonds. The data displayed shows a time average and variance over the interval of sampled times ($t = 0$ to $t = \Theta = 10$ ns) after 5 ns of prior equilibration. Reprinted from [20], copyright 2007 with permission from AIP Publishing LLC

in the protein (Fig. 4.4), in accord with their role as structural markers for protein associations. The de-wetting hot spots account for 9 out of the 11 dehydrons identified for this structure, and only partially account for the remaining two. The two dehydrons which do not fully promote de-wetting are Glu24-Asp52 and Pro19-Ser57. The larger residence times in these cases may be attributed to the pronounced hydration demands of the two charged side chains in the former case and the special steric hindrance promoted by Pro19, which precludes water approach to the backbone in the latter.

The semiempirical approach conducted in this chapter (Eqs. 4.1–4.5) leads us to determine the net reversible work $\eta = \eta(\mathbf{\Phi})$ associated with the field $\mathbf{\Phi}$ and required to span the aqueous interface around a dehydron (cf. Eq. 1.13). The results for the dehydrons in the SH3 domain (PDB.1SRL), ubiquitin (PDB.1UBI) and lysozyme (PDB.2CDS) were obtained semiempirically. They compared favorably (Fig. 4.5, correlation coefficient $R^2 = 0.71$) with the variational computation of the interfacial free energy or reversible work to span the interface, $\Delta G_{if}(\phi)$,, presented in Chap. 1. The latter computation depends on the frustration scalar field ϕ and is conducted on the same domains, each containing an individual dehydron, in the proteins indicated. It still remains difficult to provide a rigorous first-principle derivation for $\varepsilon^{-1}\left(\{\vec{R}_j, \vec{R}\}_{j=1,\ldots,K}\right)$, as that would require the analytic form of the mapping

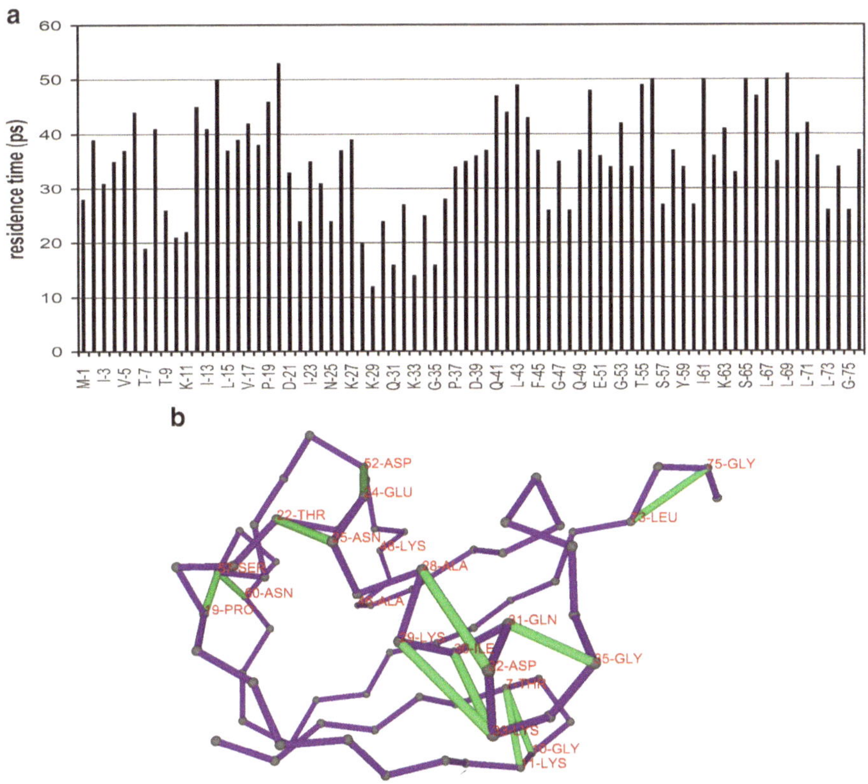

Fig. 4.4 (**a**) Mean residence times of water molecules solvating the folded protein *ubiquitin*. Residue numbering follows the file PDB.1UBI. (**b**) PDB structure of ubiquitin highlighting the dehydrons (displayed in *green*). Reprinted from [20], copyright 2007 with permission from AIP Publishing LLC

$\phi\left(\left\{\vec{R}_j, \vec{R}\right\}_{j=1,\ldots,K}\right)$ of the water-frustration field ϕ as a function of wrapping positions $\left\{\vec{R}_j, \vec{R}\right\}_{j=1,\ldots,K}$ for each dehydron. Yet, the relatively tight correlation in Fig. 4.5 suggests that this analytic derivation may be achieved one day.

4.4 Aqueous Interface as Blueprint for Drug Design

In conclusion, the identification of structural determinants of protein/water interfacial behavior is vital to understand protein associations and to design drugs/ligands with better affinity. The findings reported in this chapter represent a constructive step in this direction, since we have singled out structural singularities in soluble proteins—the dehydrons—that behave as de-wetting hot spots. The identification of these sites with defective and loose hydration opens up an engineering avenue to

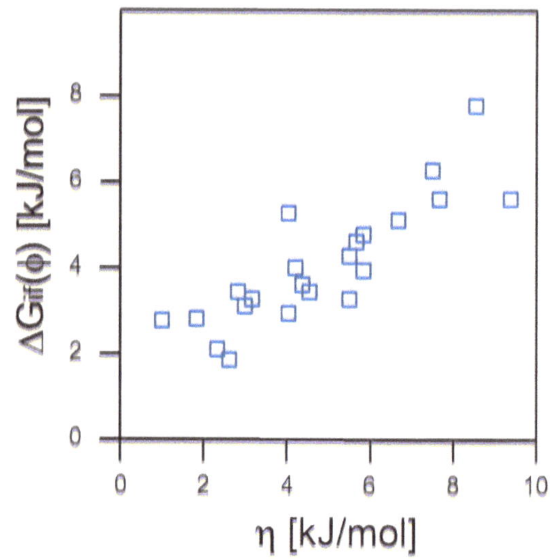

Fig. 4.5 Correlation between rigorous ($\Delta G_{if}(\phi)$) and semiempirical (η) estimation of the reversible work required to span the aqueous interface surrounding a dehydron. The data is for dehydrons in PDB entries 1SRL, 1UBI and 2CDS

understand and manipulate protein/ligand associations. Thus, by sculpting on the drug the de-wetting hot spots of the target, we take advantage of the minimal work associated with removal of dehydron-solvating water molecules, amply compensated by the resulting enhancement of the electrostatic interaction (see Chap. 3). In this way, we may enhance the affinity for the protein target by adopting the pattern of hydration defects in the target as a blueprint for molecular engineering (Fig. 4.6). This concept is brought to fruition in the reengineering of the powerful anticancer drug *imatinib* (Gleevec), as described in Chap. 9. Furthermore, it heralds the advent of a new generation of drugs based on an assessment of the architecture of the protein fold in relation to the solvating environment. As indicated, the implications of this concept for drug design will be explored in Chap. 8, where differences in de-wetting patterns across purported targets enable us to engineer a selective therapeutic impact.

4.5 Dehydron-Heated Interfacial Water

The preceding sections reveal that interfacial water mobility is maximized around dehydrons. A mechanistic explanation for this mobility rooted in a "semiclassical" treatment of interfacial dielectrics (Eqs. 4.2–4.5) would suggest that mobility is generated by the desolvation propensity of dehydrons, a propensity whose mechanical equivalent is provided by the dewetting or dehydronic field Φ. However, notwithstanding the efforts by Warshel and co-workers to define an interfacial permittivity [26], the treatment has serious shortcomings because a significant contribution to water polarization does not align with the electrostatic field of the protein, i.e. polarization does not follow the Debye *ansatz*, as shown in Chaps. 1 and 2.

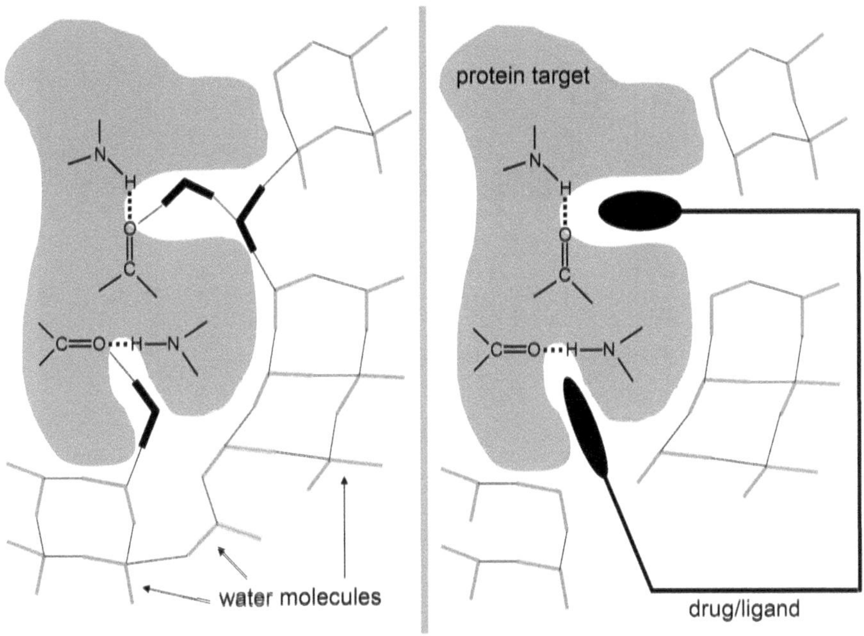

Fig. 4.6 The pattern of defects in the hydration shell of a target protein is a blueprint for drug design. Interfacial water molecules solvating dehydrons have short residence time and hence are easily displaced upon association with a purposely designed ligand. Their loss of hydrogen bonding partnerships results from their partial confinement. Fast interfacial water molecules are represented as *thick black angular lines*, while bulk-like water molecules are indicated as *grey angular lines*. Hydrogen bonds are shown as *thin black lines*, while dehydrons are indicated as *dotted lines*

A rigorous characterization of the water steering effect caused by dehydrons require that we first note that anomalous non-Debye polarization and water mobility are both tied up to the frustration scalar field $\phi(\vec{r}) = 4 - g(\vec{r})$ that measures the distortion in water hydrogen-bonded structure. Thus, we expect and indeed find a correlation between water mobility and the anomalous (non-Debye) polarization fraction ϑ defined in Chap. 2. As highlighted in Chaps. 1 and 2 the dominant structural motif promoting anomalous polarization is the dehydron. The separate study in this chapter revealed that the hydration patterns for residues paired by dehydrons are extremely ephemeral. The lifetime of a hydration pattern in picoseconds (ps) is given by the average residence time τ_n of a water molecule in the neighborhood of residue n defined in Fig. 4.1a. These considerations suggest an anticorrelation between the local parameters τ_n and ϑ_n, whereby a large anomalous polarization corresponds to a short residence time (high mobility). A complete analysis of the 9 PDB-reported proteins indicated in the caption for Fig. 4.7 (Table 2.1) reveals that such correlation is upheld. The tight τ-ϑ linear anticorrelation $\left(R^2 = 0.86\right)$ is displayed in Fig. 4.7 and can be rationalized by noting that anomalous polarization arises from nanoscale solvent confinement

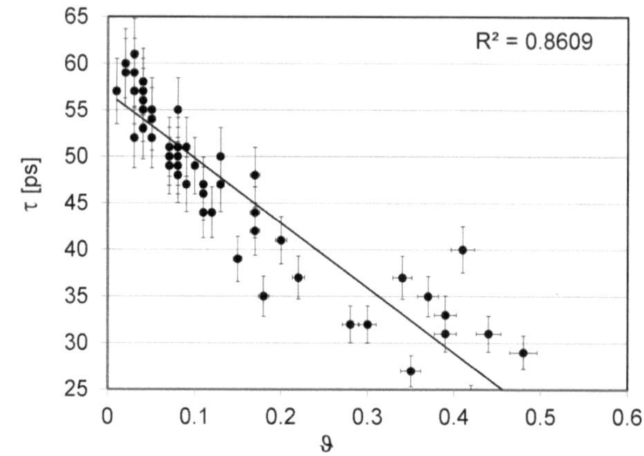

Fig. 4.7 Anticorrelation between water residence time τ and the anomalous non-Debye polarization fraction ϑ for all residues in the 9 PDB-reported proteins indicated in Table 2.1 (Chap. 2). The datapoints marked by *circles* correspond to the residues in PDB.1A8O. The error bars along both coordinates indicate Gaussian dispersions of datapoints over the remaining eight proteins

which deprives the water molecule of hydrogen-bond coordination possibilities when compared with bulk levels. The latter coordination deprivation enhances the kinetic energy, thereby decreasing the residence time.

The water steering factor can be elucidated upon inspection of Fig. 4.7. Interfacial water with high mobility has a significant collective dipole moment $\vec{\mu}_{P^{\#}}(\vec{r})$ due to anomalous non-Debye polarization

$$\vec{\mu}_{P^{\#}}(\vec{r}) = -\int (\vec{r'} - \vec{r})\vec{\nabla}.\vec{P^{\#}}(\vec{r'})d\vec{r'} \tag{4.6}$$

This polarization-associated dipole moment is subject to a significant torque $\vec{E}(\vec{r}) \times \vec{\mu}_{P^{\#}}(\vec{r})$ imposed by the orthogonal electric field $\vec{E}(\vec{r})$ created by the amide-carbonyl hydrogen bond:

$$\begin{aligned}\tau\# = \vec{E}(\vec{r}) \times \vec{\mu}_{P^{\#}}(\vec{r}) &= -\vec{E}(\vec{r}) \times \int (\vec{r'} - \vec{r})\vec{\nabla}.\vec{P^{\#}}(\vec{r'})d\vec{r'} \\ &= \vec{E}(\vec{r}) \times \int (\vec{r'} - \vec{r})\nabla^2\phi(\vec{r'})d\vec{r'}\end{aligned} \tag{4.7}$$

Thus, the dehydron steers interfacial water by imposing a torque $\tau^{\#}$ on the non-Debye polarization-induced dipole [27]. The emerging physical picture of dehydron-heated interfacial water is schematically represented in Fig. 4.8. The torque makes the hydration pattern around the dehydron ephemeral and destabilizes the protein structure by increasing the interfacial energy. This analysis inspires a new technology in drug-design, the wrapping technology, that targets dehydrons by ligands that displace "hot interfacial water" upon binding. The removal of highly mobile water molecules as a strategy to guide ligand optimization has been the

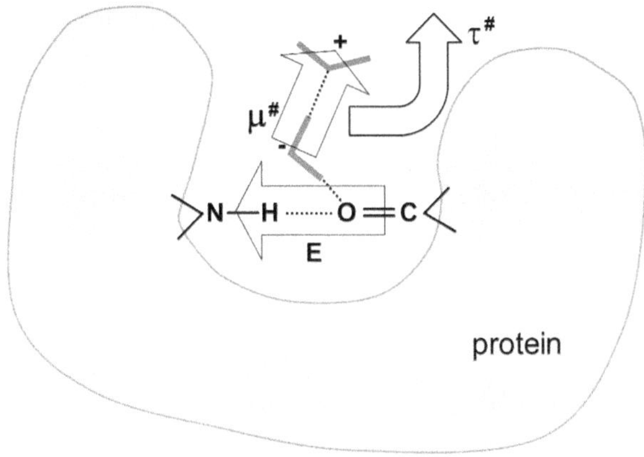

Fig. 4.8 Scheme of the mechanism by which dehydrons heat up interfacial water. The water molecules (grey angular segments) partially confined by the dehydron sub-nanoscale cavity become frustrated and hence polarized independently of the dehydron electrostatic field **E**. Thus, the latter exerts a torque $\tau^{\#}$ on the net dipole moment $\mu^{\#}$ corresponding to the non-Debye polarization-induced charge. The non-Debye torque accelerates the dehydron-confined water molecule, raising its kinetic energy beyond bulk-levels

premise of drug-design software like WaterMap® [28]. Such programs based on all-atom explicit-solvent dynamic analysis are far more costly than a dehydron identifier (see Appendix 1 for pyMol-based dehydron predictor), and do not reveal the structural feature that generates the dynamic singularity in the aqueous interface. Thus, the map of "dehydron-heated water molecules" serves as the best blueprint for drug designs based on the wrapping technology. We shall repeatedly return to this concept throughout Chaps. 8–17.

The catalytic activity of dehydron-bound water as proton acceptor is enhanced by the non-Debye torque to which the frustrated water molecule is subject. The torque-induced acceleration raises the frequency of effective collisions, which linearly increases the catalytic rate of proton transference to the dehydron-functionalized water molecule. This catalytic role of dehydrons, extensively explored in Chap. 7, has then a dynamic aspect which stands as a singularity, given the general slow-down experienced by epistructural water.

4.6 Problems

4.1. The results in this chapter suggest a very high mobility of interfacial water around dehydrons significantly at odds with the global slowing down of epistructural water. This mobility may be characterized in a number of ways, through density fluctuations, kinetic energy (K), etc. Derive an expression for the kinetic energy as a function of the non-Debye polarization created by the dehydron.

Hint: Water around dehydrons is subject to a significant torque exerted by the electrostatic field of the protein.

4.2. We adopt the coordination scalar field $g = g(\vec{r})$, a water-structure descriptor that assigns to each position vector $\vec{r}$ the expected value of hydrogen-bond coordination of a water molecule situated within a sphere centered at position $\vec{r}$ with radius 2.7 Å (thickness of a single water layer). Compared with bulk water ($g = 4$), interfacial water has reduced hydrogen-bonding opportunities ($g < 4$) that may be counterbalanced through interactions with polar groups on the protein surface or with induced electrostatic fields resulting from dipole alignments that in turn result from nanoscale confinement. Define g(j) as the expected g-value of a water molecule averaged over its residence time in domain D(j). Then, if residues i and j are paired with a backbone hydrogen bond, we define the g-value of the hydrogen bond environment as the average $\frac{1}{2}[g(i) + g(j)] = g(i,j)$.

4.2.1. Identify the g(i,j)-values for all hydrogen bonds in the free (uncomplexed) insulin subunits from the PDB entry 3E7Y. Show that $g(i,j) < 3$ only if residues i, j are paired by a dehydron. Thus, water is able to form hydrogen bonds with the paired polar groups of the backbone, the amide ($> NH$) and carbonyl ($> C = O$), only at the expense of reducing its g-value below bulk interface ($g = 3$) levels. This reduction in coordination results from partial confinement as the water molecule fills the cavity that results from the incomplete burial of the dehydron.

4.2.2. Show the spatial location of interfacial water molecules with a significant reduction of coordination ($g < 3$) in the two free (uncomplexed) subunits of human insulin (PDB.3E7Y). Identify the "hot" water molecules solvating dehydrons that become well-wrapped hydrogen bonds within the insulin complex. Thus, the hot water molecules are displaced upon association.

4.3. Analyze patterns of interfacial water exclusion resulting upon complex formation. Focus on complexes for which the residue contribution to the association free energy ΔG_a has been dissected through alanine scanning of the PP interface [29]. The residue substitution for alanine amounts to a truncation of the side chain at the alpha-carbon. By calorimetrically determining $\Delta\Delta G_a$ for the wild-type→mutant transformation, we can identify the residues that most contribute to the association free energy.

The contention of this problem is that the binding hot spots are residues that most significantly contribute to reduce the epistructural tension upon association by either displacing or becoming deprived of hot ($g < 3$) vicinal water upon association. Here $g = g(j)$ is computed for each individual hydration domain D(j) as indicated in problem 4.2.

4.3.1. Focus on alanine scanning analysis of PP (protein-protein) interfaces for well-studied complexes, and contrast the experimental results with the g-values of water excluded upon complexation. To identify the location of hot (g < 3) water molecules relative to hot-spot residues, define hydration vicinity D(j) of residue j as indicated in main text.

To contrast epistructural thermodynamics against alanine-scanning results for different complexes, group residues according to the $\Delta\Delta G_a$ for wild-type-to-mutant transformation and also according to the coordination quality of the interfacial water they displace or are deprived of upon protein-protein association. Classify hot-spot residues according to the ranges $\Delta\Delta G_a \geq 3$ kcal/mol, 1 kcal/mol $\leq \Delta\Delta G_a < 3$ kcal/mol and $\Delta\Delta G_a < 1$ kcal/mol. According to a second classifier, group residues according to the ranges for vicinal water: $g < 3$, $3 \leq g < 4$, $g = 4$.

4.3.2. Examine the PP interfaces for complexes with available alanine scanning data, classifying residues according to the $\Delta\Delta G_a$ and independently according to the g-value on free subunits. By mining the literature, gather alanine scanning data for the following complexes (PDB entries in brackets): human growth hormone/hGH receptor (3HHR), HIV-1-CD4/GP120 (1GC1), barnase and barstar in barnase/barstar complex (1BRS), P53/MDM2 (1YCR), trypsin inhibitor/beta-trypsin (2PTC), ribonuclease inhibitor/angiogenin (1A4Y), ribonuclease inhibitor/ribonuclease A (1DFJ), colicin E9 immuno-protein/colicin E9 DNase domain (1BXI).

4.3.3. Demonstrate that a statistically significant correlation exists between the $\Delta\Delta G_a$ classifier and the g-classifier of interfacial residues. *This correlation enables us to assert* $\left(P - value < 10^{-5}\right)$ *that protein association is driven by displacement of hot interfacial water that promotes a reduction in interfacial tension.*

References

1. Otting G, Liepinsh E, Wüthrich K. Protein hydration in aqueous solution. Science. 1991;254:974–80.
2. Qiu W, Kao Y, Zhang L, et al. Protein surface hydration mapped by site-specific mutations. Proc Natl Acad Sci U S A. 2006;103:13979–84.
3. Makarov V, Andrews BV, Smith PE, Pettitt BM. Residence times of water molecules in the hydration sites of myoglobin. Biophys J. 2000;79:2966–74.
4. Chandler D. Interfaces and the driving force of hydrophobic assembly. Nature. 2005;437:640–7.
5. Cheng Y, Rossky PJ. Surface topography dependence of biomolecular hydrophobic hydration. Nature. 1998;392:696–9.
6. Liu P, Huang X, Zhou R, Berne BJ. Observation of a dewetting transition in the collapse of the melitin tetramer. Nature. 2005;437:159–62.
7. Fernández A, Scheraga HA. Insufficiently dehydrated hydrogen bonds as determinants of protein interactions. Proc Natl Acad Sci U S A. 2003;100:113–8.

8. Fernández A, Kardos J, Scott R, Goto Y, Berry RS. Structural defects and the diagnosis of amyloidogenic propensity. Proc Natl Acad Sci U S A. 2003;100:6446–51.
9. Hetenyi C, van der Spoel D. Blind docking of drug-sized compounds to proteins with up to a thousand residues. FEBS Lett. 2006;580:1447–50.
10. Fernández A, Berry RS. Molecular dimension explored in evolution to promote proteomic complexity. Proc Natl Acad Sci U S A. 2004;101:13460–5.
11. Fernández A, Scott RL. Dehydron: a structurally encoded signal for protein interaction. Biophys J. 2003;85:1914–28.
12. Fernández A, Scott LR. Adherence of packing defects in soluble proteins. Phys Rev Lett. 2003;91:018102.
13. Yu H, Rosen MK, Schreiber SL. 1H and 15N assignments and secondary structure of the Src SH3 domain. FEBS Lett. 1993;324:87–92.
14. Vijay-Kumar S, Bugg CE, Cook WJ. Structure of ubiquitin refined at 1.8 angstrom resolution. J Mol Biol. 1987;194:531–44.
15. van der Spoel D, van Maaren P, Larsson P, Timneanu N. Thermodynamics of hydrogen bonding in hydrophilic and hydrophobic media. J Phys Chem B. 2006;110:4393–8.
16. Sheu S, Yang D, Selzle H, Schlag EW. Energetics of hydrogen bonds in peptides. Proc Natl Acad Sci U S A. 2003;100:12683–7.
17. Rizzo RC, Jorgensen WL. OPLS All-atom model for amines: resolution of the amine hydration problem. J Am Chem Soc. 1999;121:4827–36.
18. Jorgensen WL, Chandrasekhar J, Madura J, Impey RW, Klein ML. Comparison of simple potential functions for simulating liquid water. J Chem Phys. 1983;79:926–35.
19. Lindahl E, Hess B, Van der Spoel D. GROMACS 3.0: a package for molecular simulation and trajectory analysis. J Mol Model. 2001;7:302–17.
20. Fernández A, Chen J, Crespo A. Solvent-exposed backbone loosens the hydration shell of soluble folded proteins. J Chem Phys. 2007;126:245103.
21. Denisov V, Halle B. Protein hydration dynamics in aqueous solution. A comparison of bovine pancreatic trypsin inhibitor and ubiquitin by oxygen-17 spin relaxation dispersion. J Mol Biol. 1995;245:682–97.
22. Lum K, Chandler D, Weeks JD. Hydrophobicity at small and large length scales. J Phys Chem B. 1999;103:4570–7.
23. Mason PE, Neilson GW, Dempsey CE, Barnes AC, Cruickshank JM. The hydration structure of guanidinium and thiocyanate ions: implications for protein stability. Proc Natl Acad Sci U S A. 2003;100:4557–61.
24. Kocher JP, Prevost M, Wodak S, Lee B. Properties of the protein matrix revealed by the free energy of cavity formation. Structure. 1996;4:1517–29.
25. Fernández A, Sosnick TR, Colubri A. Dynamics of hydrogen-bond desolvation in folding proteins. J Mol Biol. 2002;321:659–75.
26. Schutz CN, Warshel A. What are the dielectric "constants" of proteins and how to validate electrostatic models? Proteins Struct Funct Gen. 2001;44:400–8.
27. Fernández A. Fast track communication: water promotes the sealing of nanoscale packing defects in folding proteins. J Phys Condens Matter. 2014;26:202101.
28. Pearlstein RA, Sherman W, Abel R. Contributions of water transfer energy to protein-ligand association and dissociation barriers: watermap analysis of a series of p38α MAP kinase inhibitors. Proteins. 2013;81:1509–26.
29. Weiss GA, Watanabe CK, Zhong A, Goddard A, Sidhu SS. Rapid mapping of protein functional epitopes by combinatorial alanine scanning. Proc Natl Acad Sci U S A. 2000;97:8950–4.

Chapter 5
Dehydron-Rich Proteins in the Order-Disorder Twilight Zone

As demonstrated in Chaps. 1 and 3, soluble proteins retain structural integrity by shielding their backbone hydrogen bonds from competing hydration of the paired polar moieties (amide and carbonyl). Thus, a *dehydron* constitutes a structural deficiency since it is prone to disruptive hydration. In this chapter we describe the physical and functional properties of dehydron-rich proteins with high dehydron density. We show that dehydron clusters represent structural singularities belonging to an order-disorder twilight zone that generates a sharp local quenching of the dielectric permittivity of the surrounding medium. The functional roles of these singularities are assessed for natural proteins. Special emphasis is placed on the molecular etiology of aberrant amyloidogenic aggregation arising in dehydron-rich soluble proteins with large deviations from the golden rule of molecular architecture established in Chap. 1. Our analysis of unstable aqueous interfaces requires a description of biological water that cannot be captured by conventional continuous models, where solvent degrees of freedom are averaged out.

The order-disorder twilight arising from high dehydron concentration at the protein surface is shown to be often a marker for dysfunctional misfolding. We also show how to predict such regions based solely on sequence information. The predictive tools are applied in the problems section to engineer a pharmaceutical disruptive of a protein-protein interaction as a therapeutic agent to treat heart failure. The problem is particularly challenging, not only because it involves inhibiting a protein-protein association, itself a major problem, but also because the structure of the target protein is unknown, forcing the drug designer to make inferences at the sequence level.

A. Fernández, *Physics at the Biomolecular Interface*, Soft and Biological Matter,
DOI 10.1007/978-3-319-30852-4_5

5.1 Dehydron Clusters and the Order-Disorder Twilight

As noted in the preceding chapters, the structural integrity of a soluble protein is contingent on its capacity to exclude water from its amide-carbonyl hydrogen bonds [1, 2]. Thus, under-wrapped intramolecular hydrogen bonds or dehydrons, constitute structural singularities representing packing defects that have been extensively characterized as implicated in protein associations and macromolecular recognition [2–13]. The strength and stability of dehydrons may be modulated by an external agent. More precisely, intramolecular hydrogen bonds which are not "wrapped" by a sufficient number of nonpolar groups may become stabilized and strengthened by the attachment of a ligand or binding partner that further contributes to their dehydration.

In this chapter we identify functional indicators of PDB-reported soluble proteins with clusters of dehydrons. Such regions, rich in structural vulnerabilities, may be characterized as belonging to a "twilight zone" between order and native disorder [4]. This characterization is already suggested by a strong correlation between wrapping of intramolecular hydrogen bonds (ρ) and propensity for structural disorder (f_d), as shown in Fig. 5.1. The correlation reveals that the inability to exclude water intramolecularly from pre-formed hydrogen bonds is causative of the loss of structural integrity.

The disorder propensity may be accurately quantified by a sequence-based score generated by the program PONDR-VLXT [14–16], a predictor of native disorder that takes into account residue attributes such as hydrophilicity, aromaticity, and their distribution within the window interrogated. The disorder score ($0 \leq f_d \leq 1$) is assigned to each residue within a sliding window, representing the predicted propensity of the residue to be in a disordered region ($f_d = 1$, certainty of disorder; $f_d = 0$, certainty of order). Only 6 % of 1100 nonhomologous PDB proteins gave false positive predictions of disorder in sequence windows of 40 amino acids. The strong correlation (over 2806 nonredundant nonhomologous PDB domains, Fig. 5.1) between disorder score of a residue and extent of wrapping of the hydrogen bond engaging the residue (if any) implies that clusters of dehydrons correspond to extended regions of structural vulnerability within a structured domain. Hence, the characterization of dehydron clusters as belonging to an order-disorder twilight zone appears to be warranted.

Both order and native disorder are well characterized structural attributes of protein chains [16]. However, the highly vulnerable regions in a soluble fold described in this chapter belong to the novel category of "tamed disorder" because they can acquire and maintain a structured state only upon association. Neither order nor disorder is an adequate category to describe such dehydron-rich protein regions.

In this chapter we explore the functional role of regions with the largest dehydron clusters. A cluster is defined as the maximal set of dehydrons with intersecting desolvation domains or overlapping microenvironments. These regions actually belong to an order-disorder twilight and are shown to be strong dielectric modulators,

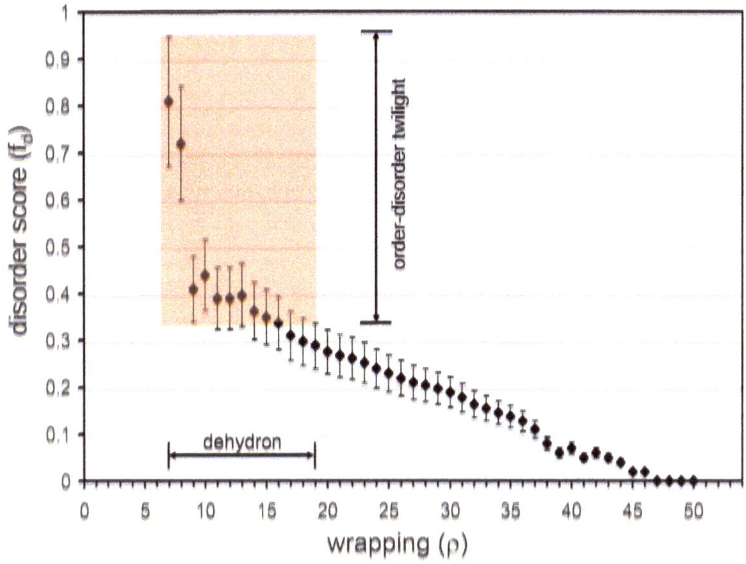

Fig. 5.1 Correlation between the disorder score of a residue and the extent of wrapping (ρ) of the backbone hydrogen bond engaging that particular residue (if any). The disorder score on each individual residue was obtained for 2806 nonredundant nonhomologous PDB domains. Residues have been independently grouped in 45 bins of 400 residues each, according to the extent of wrapping ($7 \leq \rho \leq 52$). The average score has been determined for each bin (*square*), and the error bars represent the dispersion of disorder scores within each bin. The strong correlation between the disorder score and the extent of wrapping and the dispersions obtained imply that dehydrons can be safely inferred in regions where the disorder score is $f_d > 0.35$. The *red rectangle* represents the region of order-disorder twilight where the existence of dehydrons may be inferred from disorder score with absolute certainty. No hydrogen bond in monomeric domains reported in PDB was found to possess less than seven wrappers, implying a threshold for structural sustainability in soluble proteins. Reprinted from [19], copyright 2007 with permission from the American Chemical Society

i.e. quenchers of the local dielectric permittivity, thus providing a local enhancement of Coulomb fields nearby. This property arises since clusters of packing defects promote local dehydration of the protein interface promoted by a significant reduction in hydrogen-bonding partnerships of solvating water molecules.

These discrete effects relating to local water structuring around packing defects cannot be captured properly by existing continuous models of the interfacial electrostatics [17]. This is mainly because such models are based on mean-force potential approximations to solvent interactions, where solvent degrees of freedom are averaged out, and this is clearly an inappropriate *ansatz* to deal with cavities of the size of solvent molecules themselves.

5.2 Semiclassical Dielectrics at Dehydron Sites on the Aqueous Interface

The dehydrons in a soluble protein ($\rho \leq 19$, for desolvation radius 6 Å) are partially exposed to solvent. As shown in the previous chapters, these bonds promote the removal of hydrating molecules as a means to enhance the underlying polar-pair electrostatics. Furthermore, the resulting bond stabilization overcomes the amount of work needed to remove solvating water molecules [3, 5]. To describe this dehydration propensity, we first compute the extent of constraint of interfacial water molecules. As shown in Chap. 1, this parameter is identified with the thermal average of g, here denoting the average number of hydrogen-bond partnerships involving water molecules within the desolvation domain of each residue along the chain ($0 \leq g \leq 4$, Fig. 5.2). As an illustration we focus on the DNA-binding domain of antitumor gene p53 [18]. This domain was selected because it contains three of the largest dehydron clusters to be found in PDB (Fig. 5.3). The functional significance of these clusters relates to their interplay with the electrostatics of DNA recognition by this transcription factor [19].

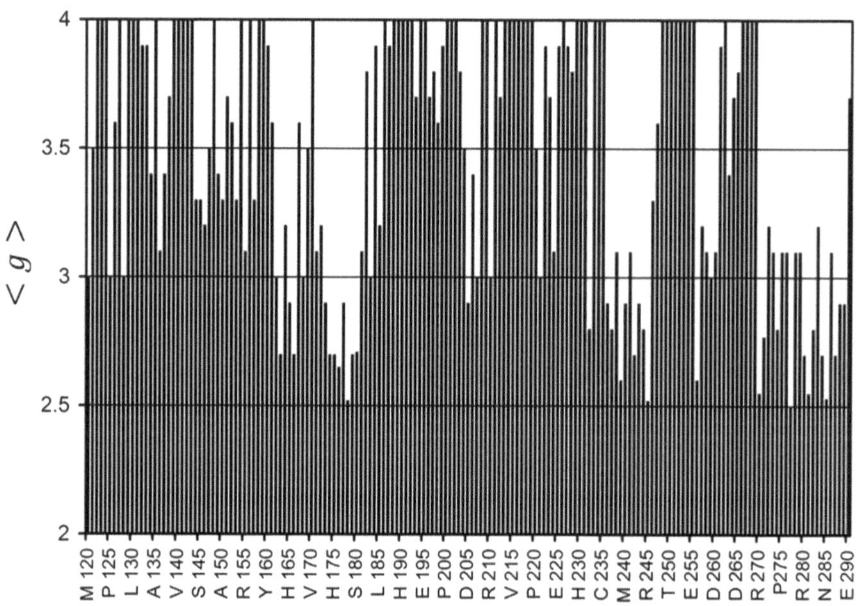

Fig. 5.2 Thermal average of the average number of hydrogen-bond partnerships, $\langle g \rangle$, for water molecules located within the desolvation domain of each residue in the DNA-binding domain of p53. If no water is found in the desolvation domain (buried residue), the bulk water value $g = 4$ is adopted. Reprinted from [19], copyright 2007 with permission from the American Chemical Society

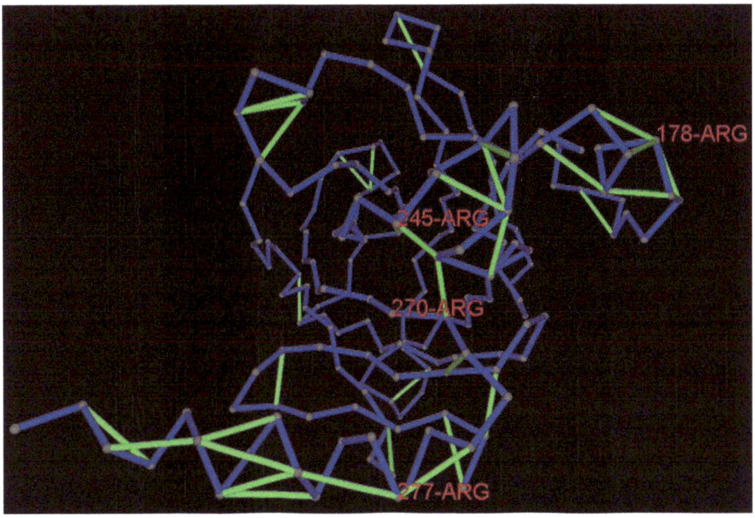

Fig. 5.3 Dehydrons for p53 DNA-binding domain. The backbone is indicated by *blue* virtual bonds joining α-carbons and dehydrons are shown as *green* segments joining the α-carbons of residues paired by backbone hydrogen bonds. Reprinted from [19], copyright 2007 with permission from the American Chemical Society

Figure 5.2 shows the thermal average, ⟨g⟩, of the average number of hydrogen-bond partnerships involving water molecules within the desolvation domain for each residue. Three dehydration hot spots are apparent, comprising residues 171–181, 236–246 and 270–289, respectively. The location of these hot spots corresponds to the three major dehydron clusters shown in Fig. 5.3.

The thermal average, ⟨g⟩, of the number of hydrogen-bond partnerships involving water molecules hydrating the p53 DNA-binding domain was obtained from classical trajectories generated by 5 ns molecular dynamics (MD) simulations (see Chap. 4). The g-values were determined for each water molecule within a 6 Å-radius spherical domain centered at the α-carbon of the residues. The adopted starting geometry is the monomeric structure from PDB entry 2GEQ. The starting conformation was embedded in a pre-equilibrated cell of explicitly represented water molecules and counterions [19]. The entire system was initially equilibrated for 5 ns. Computations were performed by integration of Newton's equations of motion with time step 2 fs using the GROMACS program [20] in the NPT ensemble with box size $8 \times 8 \times 8\,\text{nm}^3$ and periodic boundary conditions, as described in the preceding chapter.

This computation was repeated to include representatives of the three major fold topologies: all-β (SH3 domain, N = 55, two dehydrons, PDB.1SRL), all-α (λ-repressor, N = 86, 26 dehydrons, PDB.1LMB) and α/β (ubiquitin, N = 76, 16 dehydrons, PDB.1UBI). Consistency with the p53 calculation was obtained, as dehydrons proved to become the dehydration hot spots on the protein interface [19].

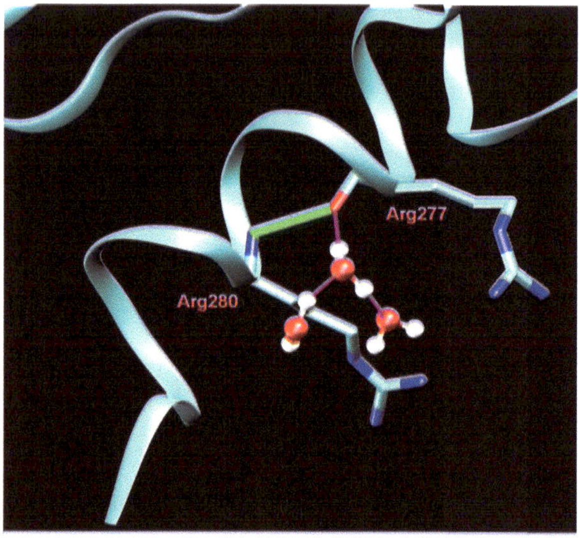

Fig. 5.4 Selected snapshot (after 1 ns of MD) of a solvating water molecule and its hydrogen bond partnerships (*purple* bonds) within the desolvation domain of Arg277 in the DNA-binding domain of tumor antigen p53 (ribbon representation, fragment). The backbone amide-carbonyl dehydron Arg277-Arg280 is shown in *green*. Reprinted from [19], copyright 2007 with permission from the American Chemical Society

The confinement of solvating water determined by a packing defect is illustrated in Fig. 5.4. The figure displays a snapshot of a water molecule within the desolvation domain of Arg277, a residue paired by a dehydron to Arg280. This water molecule has three hydrogen-bond partners, two with vicinal water and one with the Arg277 backbone carbonyl. A 3.6 Å-threshold is adopted for hydrogen bond distance between the heavy atoms. The snapshot was extracted after 1 ns of MD simulations equilibrating the protein chain with surrounding water. Due to the incomplete wrapping of dehydron (Arg277, Arg280) in the p53 domain, the closest water molecule is found at a distance 2.8 Å between carbonyl and water oxygen atoms. While electrostatically engaged with the Arg277 backbone carbonyl, this water molecule is deprived of one hydrogen bond partnership when compared with bulk water (Figs. 5.3 and 5.4).

In order to determine a generic relation between ρ and g, three additional single-domain folds representative of protein topologies were also analyzed: SH3-domain; ubiquitin, and λ-repressor, as indicated above.

The correlation between wrapping and dehydration propensity (Fig. 5.5a) has the following characteristics: (a) dehydrons ($\rho \leq 19$) generate g-values in the range $2 \leq g \leq 3.6$; (b) the upper wrapping bound, $\rho = 28$ corresponds to bulk-like water ($g = 4$) in the desolvation domain; and (c) all solvating water is excluded from the desolvation domain for $\rho > 28$.

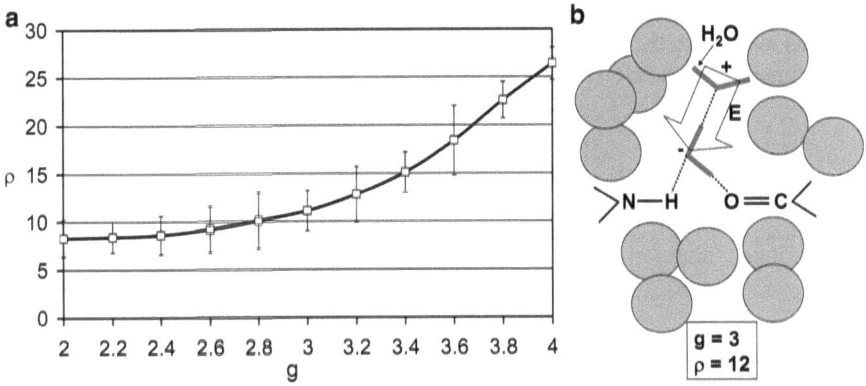

Fig. 5.5 (**a**) Correlation between hydrogen-bond wrapping ρ and wetting parameter g. Each residue is assigned a ρ-value averaged over all backbone hydrogen bonds in which it is engaged. The data was extracted from the wetting computation on the p53 DNA-binding domain and three additional folds: the SH3 domain (N = 55, 2 dehydrons, PDB.1SRL); ubiquitin (N = 76, 16 dehydrons, PDB.1UBI), and λ-repressor (N = 86, 26 dehydrons, PDB.1LMB). (**b**) Schematic representation of the relation between protein wrapping, interfacial water confinement, restrictions on hydrogen-bonding partnerships for interfacial water molecules and net internal electrostatic field arising from such restrictions. Figure (**a**) reprinted from [19], copyright 2007 with permission from the American Chemical Society

5.3 Semiclassical Treatment of Dielectric Modulation of Interfacial Water Around Dehydrons

In this section we show how dielectric modulation is promoted by discrete solvent effects arising from local imperfections in the wrapping of the protein backbone. These discrete effects cannot be captured properly by conventional continuous models, which should in principle be adapted to deal with local dielectric modulations. As demonstrated in this section, the dielectric modulation, i.e. the quenching in the local dielectric permittivity is caused by the local reduction in hydrogen-bonding partnerships of solvating water molecules [21]. To quantify this effect, we compute the polarizability associated with restricted interfacial water under the influence of a nonzero net internal electrostatic field **E** (**E** = **0** in bulk water as there is no preferred dipole orientation and therefore, no collective net effect). The relation between wrapping-based confinement of interfacial water, its hydrogen-bonding partnerships and the net internal field **E** resulting from water confinement is illustrated in Fig. 5.5b.

For convenience, we define the entropy loss of a water molecule associated to the reduction of hydrogen-bond exchange possibilities from 4 in bulk water (the parameter for the tetrahedral lattice of hydrogen bonds) to g at the interface [22] as: $\Delta s(g) = -k\,ln(g/4)$ (k = Boltzmann constant). In addition, we introduce the dimensionless parameter $\lambda(g) = T\Delta s(g)/(Ld)$, where d = dipole moment of a single water molecule and L = Lorentz field = $\eta d/3\varepsilon_o$ (η = bulk water density and ε_o = vacuum permittivity). The thermal average of the dipole moment vector **d** is given by: $\langle \mathbf{d} \rangle = d\langle \boldsymbol{\mu} \rangle$, where $\boldsymbol{\mu} = \boldsymbol{\mu}(\mathbf{E})$ is the dipole moment unit vector. The thermal average of the dipole projection μ_E onto the net internal field may be readily determined from the Boltzmann average [23]:

$$\langle \mu_E \rangle = \left[\coth(\beta Ed) - (\beta Ed)^{-1}\right], \tag{5.1}$$

where $\beta = (kT)^{-1}, E = \|\mathbf{E}\|$. The scattering $\sigma(\mathbf{E})$ of the net field produced by dipole interactions within volume V is the real part (Re) of

$$S = V - \int \exp\left[-iv(\mathbf{r})\boldsymbol{\rho} \cdot \langle \boldsymbol{\mu} \rangle\right] d\mathbf{r}, \tag{5.2}$$

where $v(\mathbf{r}) = d/(4\pi\varepsilon_o r^3)$ is the dipole interaction field and $\boldsymbol{\rho}$ = net field unit vector. Integration in (5.2) is carried out in the region of correlation: $r_w \le \|\mathbf{r}\| \le r_g$, with $r_g = r_w[1 - \lambda(g)]^{1/3}$, where r_w is the bulk dipole-dipole correlation distance. Thus, the field scattering is given as:

$$\begin{aligned}\sigma(E) &= \operatorname{Re} S = \lambda(g)L^2(|\langle \mu_E \rangle|^2)/18 \\ &= \eta kT[27\varepsilon_o]^{-1}\ln(4/g)\left[\coth(\beta Ed) - (\beta Ed)^{-1}\right]^2\end{aligned} \tag{5.3}$$

The probability distribution of net internal fields is then

$$P_g(\mathbf{E}) = [\pi\sigma^2(E)]^{-3/2}\exp[-E^2/\sigma^2(E)], \tag{5.4}$$

satisfying:

$$\lim_{g\to 4} P_g(\mathbf{E}) = \delta(\mathbf{E}), \tag{5.5}$$

in accord with the fact that there is no net internal field in bulk water. The dielectric susceptibility χ along a preferential direction z is directly computed as function of g by introducing a perturbation Δ_z of the net internal field $\tilde{\mathbf{E}} = \mathbf{E} + \Delta_z$.

By definition:

$$\chi_z = \lim_{\|\Delta\|\to 0} \partial \Xi_z / \partial \Delta_z, \tag{5.6}$$

where $\Xi_z = \eta d \int \left[\coth(\beta \tilde{E}_z d) - (\beta \tilde{E}_z d)^{-1}\right] P_g(\mathbf{E}) d(\mathbf{E})$ is the polarizability along the z-direction. This gives:

$$\chi_z = \chi_z(g) = \eta d^2\beta/3\varepsilon_o - \eta d^2\beta/3\varepsilon_o \int \left[\coth^2(\beta E_z d) - (\beta E_z d)^{-2}\right] P_g(\mathbf{E}) d(\mathbf{E}), \tag{5.7}$$

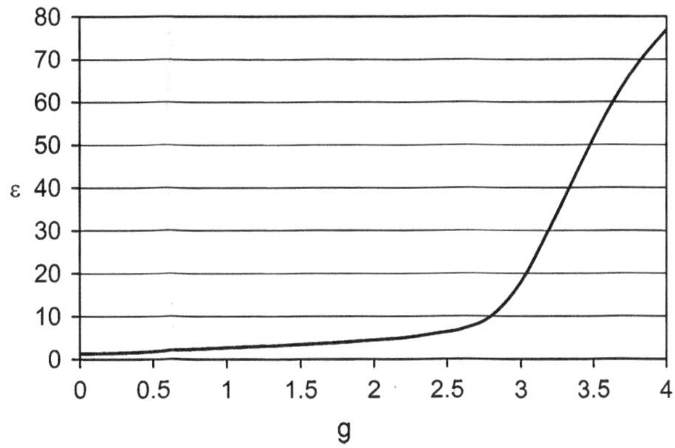

Fig. 5.6 Analytically dependence of the dielectric permittivity ε on the parameter g. Reprinted from [19], copyright 2007 with permission from the American Chemical Society

where $\eta d^2\beta/3\varepsilon_o = \chi_{zw}$ is the bulk water susceptibility. The bulk limit is obtained substituting (5.5) into (5.7) in the limit g → 4:

$$\begin{aligned}\lim_{g\to 4}\chi_z &= \eta d^2\beta/3\varepsilon_o - \eta d^2\beta/3\varepsilon_o\int\left[\coth^2(\beta E_z d) - (\beta E_z d)^{-2}\right]\delta(\mathbf{E})d(\mathbf{E}) \\ &= \eta d^2\beta/3\varepsilon_o = \chi_{zw},\end{aligned} \quad (5.8)$$

since $\lim_{\xi\to 0}\left[\coth^2(\xi) - (\xi)^{-2}\right] = 0$.

Figure 5.6 displays the rigorously derived g-dependence of the dielectric permittivity $\varepsilon = 1 + \chi(g)$ (the subindex z denoting a generic direction has been dropped). The dielectric quenching is extreme upon moderately small losses in hydrogen-bond partnerships. Thus, the most dramatic decrease is marked by a drop in ε-values from 50 to 7 as g is reduced from 3.5 to 2.5.

The combination of Figs. 5.5 and 5.6 leads us to the conclusion that *clusters of packing defects act as dramatic enhancers of the electric fields generated at the protein interface*. The typical loss in hydrogen bonding partnerships associated with dehydron solvation puts g in the range: $2 \le g \le 3.6$. This range contains the region of most dramatic dielectric quenching, decreasing the permittivity by an order of magnitude with respect to bulk water. In turn, this effect translates in an order of magnitude increase in electrostatic interactions, hence rationalizing the functional role of these dehydron-rich regions.

5.4 Dielectric Modulation by Dehydrons in the p53 DNA-Binding Domain

The functional significance of the three dielectric modulators in the DNA-binding domain of p53 may be understood by examining its dimeric state and its role as transcription factor. Thus, a major cluster involving the five dehydrons (173, 176),

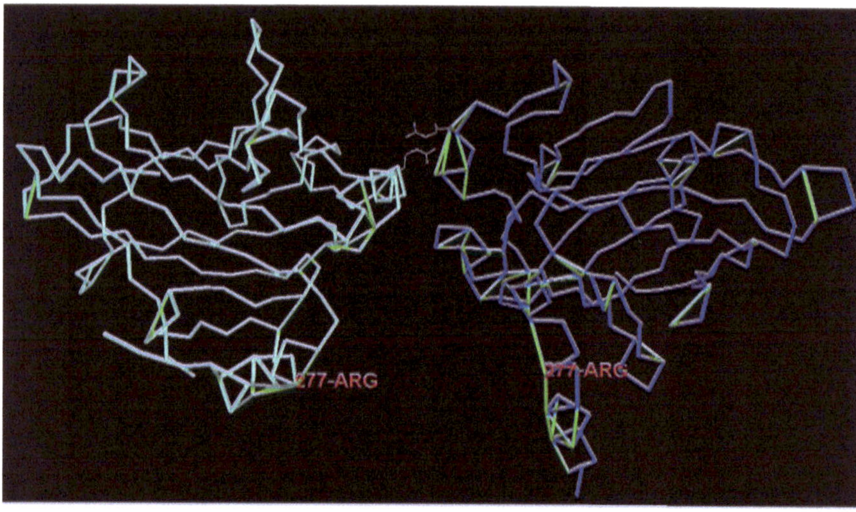

Fig. 5.7 Backbone/dehydron representation of the dimmer interface for the DNA-binding domain of p53 (PDB.2GEQ). The side chains of the Arg178 of each monomer involved in a resonance pair are shown. Reprinted from [19], copyright 2007 with permission from the American Chemical Society

(174, 178), (175, 178), (176, 179) and (178, 180) is found at the dimer interface (Figs. 5.3 and 5.7). This cluster fosters dimerization in accord with the dehydration propensity of dehydrons and their role as promoters of protein associations [2, 5, 10–13]. The dimerization involves a resonant pairing of the Arg178 from each monomer (Fig. 5.7) likely, to promote supramolecular charge delocalization with distal charge separation at all times. Significantly, ‹g› reaches a minimum precisely at Arg178 (Fig. 5.2), in accordance with the low dehydration penalty for the guanidinium ion [24] and with the fact that the Arg-Arg resonant association requires guanidinium dehydration (cf. Fig. 5.7).

Besides Arg 178, there are three additional minima in $\langle g \rangle$, corresponding to residues Arg245, Arg270 and Arg277 (Fig. 5.2). The latter three arginines play a pivotal role in DNA recognition [18], as discussed below. Residue Arg245 is engaged in the dielectric quenching region 236–246 and is part of the dehydron cluster involving pairs (236, 239), (237, 245), (237, 271), (239, 242), (239, 244), (240, 242) (Fig. 5.3). Residues 270 and 277 lie within the dielectric quenching region 270–289 and are part of the dehydron cluster involving pairs (237, 271), (274, 277), (277, 281), (280, 284), (281, 285), (282, 285), (285, 288) (Figs. 5.3 and 5.8). The latter cluster is one of the largest to be found in a PDB-reported soluble protein, as discussed below.

Direct examination of the protein-DNA complex (PDB.2GEQ), reveals that the three residues directly implicated in DNA recognition are precisely Arg245, Arg270 and Arg277 (Fig. 5.8). Residue Arg277 acts as intra-base intercalator, while the other two interact with the negatively charged backbone phosphates (Fig. 5.8). The electrostatics of protein-DNA recognition is not merely the result

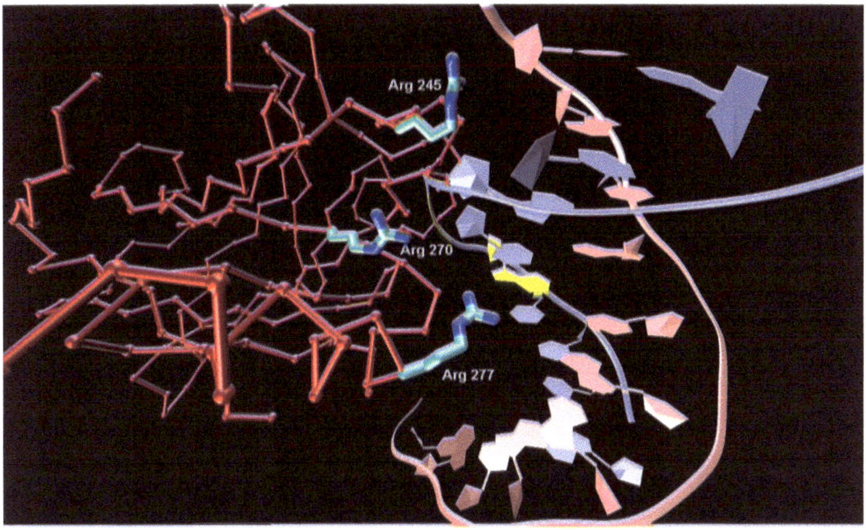

Fig. 5.8 Protein-DNA complex of the DNA-binding domain of p53 (PDB.2GEQ). Side chains of the key residues directly implicated in DNA recognition, Arg245, Arg270 and Arg277 are shown. The pyridine base recognized by Arg277 is shown in *yellow*, while the individual DNA strands are shown in *lilac* and *light magenta*. Reprinted from [19], copyright 2007 with permission from the American Chemical Society

of matching charges along the geometrically compatible interfaces, otherwise ion pairs would prevail in water. Rather, electrostatic recognition of the DNA polyelectrolyte requires a device to promote dehydration at the protein-nucleic acid interface. The large dehydron clusters surrounding the three arginines directly implicated in the protein-DNA association (Figs. 5.3 and 5.8) provide such an expedient, as they quench the local dielectric (Fig. 5.2), thus enhancing the electrostatic recognition. Thus, the fact that the three arginines involved in DNA recognition are also dehydration hot spots satisfies a functional imperative for the transcription factor.

5.5 Proteins with Dehydron Clusters

A curated PDB-derived database was constructed free of redundancy for single-domain proteins with contour length $N > 50$. The database is comprised of 2982 entries with <50 % pairwise identity in aligned sequences. Proteins were grouped according to n, the size of the dehydron cluster in the structure. The groups intersect to a considerable extent [19].

Each n-group has been lumped into five nondisjoint functional categories: biosynthesis, enzymology, cell signaling, cytoskeleton and cancer. The contribution of each category to each n-group is quantified through suitable normalization,

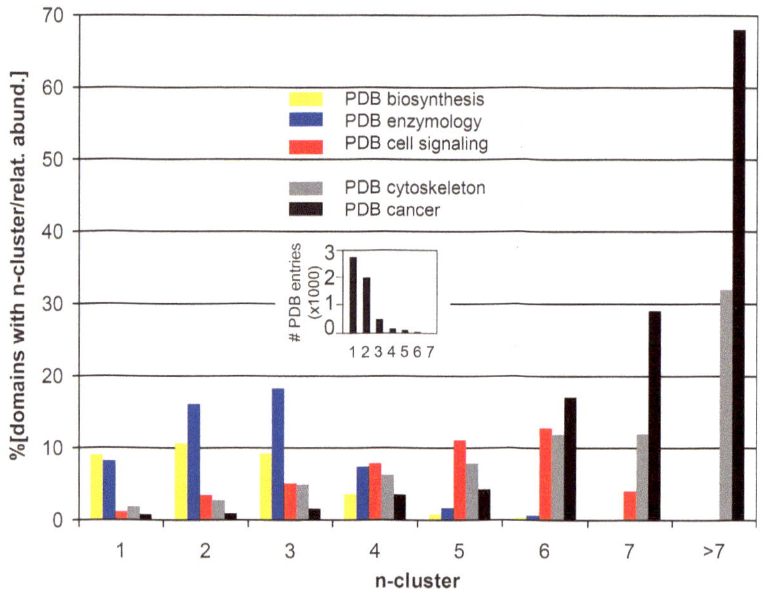

Fig. 5.9 Percentages of PDB-domains in functional categories binned into groups determined by dehydron-cluster size n. Each cluster-size group is divided into five nondisjoint functional categories: biosynthesis, enzymology, cell signaling, cytoskeleton and cancer. The number of PDB domains in each group is normalized to the relative abundance of the functional category. Thus, the number of PDB-domains in a cluster-size group and functional category is divided by the total number of PDB domains in the category. The *inset* indicates the number of domains in each cluster-size group. Reprinted from [19], copyright 2007 with permission from the American Chemical Society

taking into account the relative abundance of each category in the curated database. Thus, the number of PDB-domains within an n-cluster in a functional category is divided by the total number of PDB domains in the category. After normalization, percentages indicating representativity from each functional category are computed for each n-group (Fig. 5.9).

The distribution of dehydron clusters becomes a marker of the functional category. Thus, biosynthesis peaks at 2, enzymology at 3, cell signaling at 6, while cytoskeleton and cancer are monotonically increasing with dehydron cluster size. *Intriguingly, the dominance of the cancer category becomes apparent for the poorest packed proteins*. The dearth of domains for $n \geq 7$ facilitates individual identification and assessment. Thus, there are only five domains with $n = 7$: calmodulin (cell signaling, Fig. 5.10) [25], actin (cytoskeleton) [26], p53 DNA-binding domain (cancer, Figs. 5.3 and 5.7) [18], BRCT, the terminal repeat domain of breast cancer gene BRAC1 [27], and the cellular prion protein (Fig. 5.11) [28]. The group with $n > 7$ only contains 3 members: severin (cytoskeleton) [29] and two oncogenic transcription factors with DNA-stabilizing induced fit, jun/fos (Fig. 5.12, [30]) and myc/max [31].

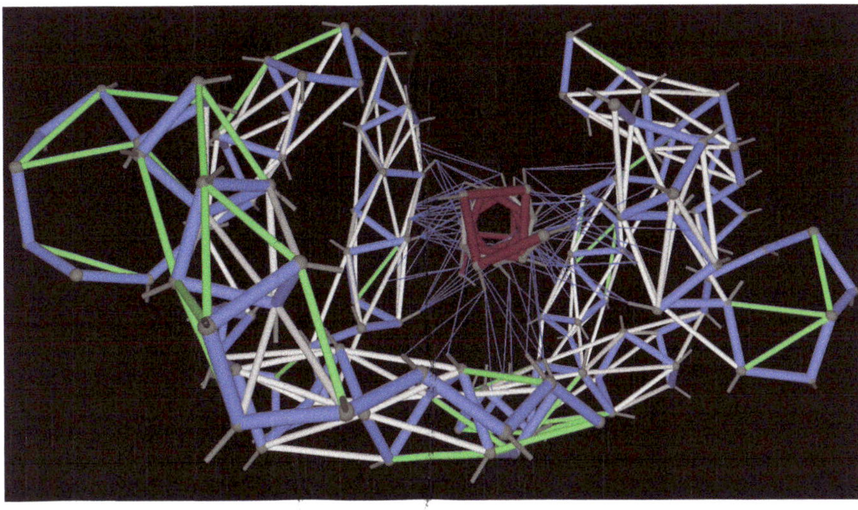

Fig. 5.10 Dehydron pattern of calmodulin (PDB. 1CDM) complexed with the calmodulin-binding domain of calmodulin-dependent protein kinase II. The backbone is indicated by *blue* virtual bonds joining α-carbons and dehydrons are shown in *green*. The intermolecular wrapping of calmodulin hydrogen bonds by the ligand side chains is indicated by *thin blue lines*. The extremities of these lines are the β-carbon of the residue contributing to the desolvation shell of the hydrogen bond and the midpoint of the hydrogen bond that is being wrapped intermolecularly. Reprinted from [19], copyright 2007 with permission from the American Chemical Society

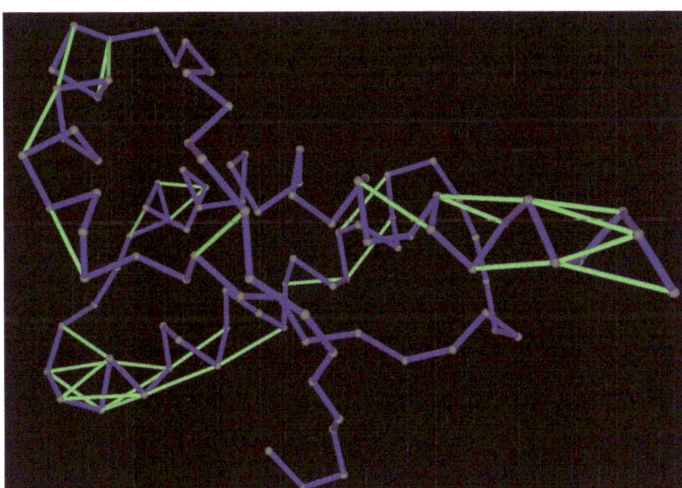

Fig. 5.11 Dehydron pattern of the cellular prion protein (PDB.1QM0). Reprinted from [19], copyright 2007 with permission from the American Chemical Society

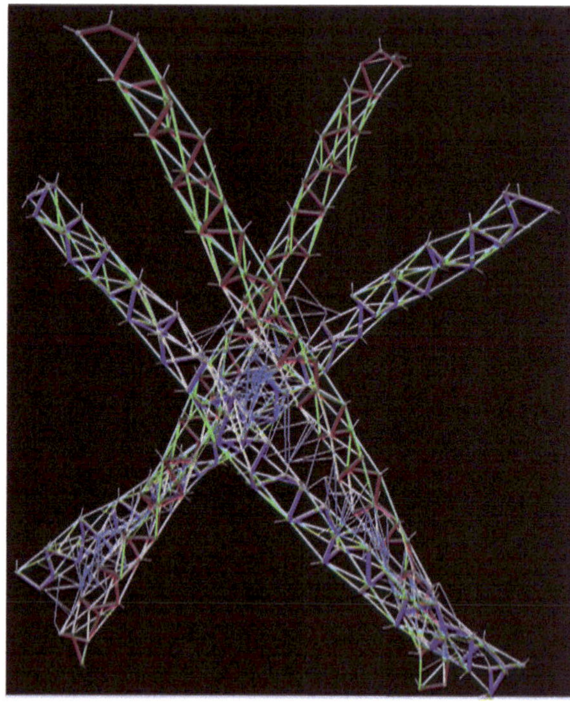

Fig. 5.12 Dehydron pattern of oncogenic transcription factor jun/fos (tetramer, PDB.1FOS). Intermolecular wrapping is displayed as in Fig. 5.10. Monomeric *jun* chains are in *blue*, *fos* chains, in *dark red*. Reprinted from [19], copyright 2007 with permission from the American Chemical Society

The eight protein domains with unusually large dehydron clusters are highly interactive proteins. In spite of their functional diversity, a common functional motif is discerned: as soluble proteins, they all possess a major weakness in the hydration shell. Thus, a dehydron cluster plays at least three interrelated roles: (a) promoter of protein associations (calmodulin, actin, severin), (b) dielectric modulator enhancing intermolecular electrostatic interactions (cancer-related transcription factors), and (c) a structural weakness promoting water attack on backbone hydrogen bonds with concurrent refolding leading to aggregation (cellular prion protein).

To summarize, proteins with severe weaknesses in their hydration shell resulting from extended regions with poor wrapping are capable of creating a significant dielectric quenching of interfacial water. This property becomes most apparent for proteins possessing the largest dehydron clusters in the PDB (seven or more dehydrons). We found that there are eight such proteins reported in PDB, possessing structural regions in the order-disorder twilight. These regions belong to cancer-related (oncogenic or antitumor) proteins and also to highly interactive proteins and to a cellular prion that promotes misfolding and aberrant aggregation.

5.6 Inferring Dehydrons from Protein Sequence: Water-Exposed Backbone and Disorder Propensity

When a protein has no reported structure, a dehydron prediction from protein sequence is required. This imperative becomes particularly pressing for drug designers that often face the problem of a lack of structural information on the target protein. As noted in Sect. 5.1, dehydron-rich regions entail a significant exposure of the protein backbone that translates into a significant propensity for native structural disorder. Thus, sequence-based predictors of disorder can provide the signal for identifying dehydrons in the missing structure. The publicly available Predictor for Native Disorder PONDR® is a tool of choice to infer dehydrons from protein sequence. At the time when this book is being composed, PONDR may be freely accessed at the URL: http://pondr.com.

PONDR provides a coarse resolution, generating a smeared-out plot resulting from prediction of disorder propensity on a sliding sequence window. We need to deconvolute the smeared-out signal to obtain individual amino-acid values of disorder propensity, as described below. When using PONDR, one can only report averaged trends over sequence windows, not individual disorder propensities. This is so because the dehydron prediction with PONDR is only a statistical inference over a sliding window. The broad disorder regions ($f > 0.5$) without disorder certainty usually have a fine-grained saw-like structure embedded in them with many order-disorder twilight zones rich in dehydrons. This fine structure is usually missed in a PONDR reading because the latter provides a coarse-grained picture which is adequate to predict large disorder regions (PONDR's intent) but not adequate for obtaining detailed dehydron prediction at the residue-level resolution that is necessary for the drug designer. In PONDR's broad signals we are looking at the envelope of a saw-like structure, and in the reported envelope we are missing the dehydron-rich twilight regions.

A protocol of how to resolve PONDR signals to the level of fine graining needed to infer specific dehydrons is provided subsequently. The raw data uses PONDR to infer dehydrons but uses also structural information (templates in a training set for supervised learning) to get the level of resolution needed to generate dehydron predictions. In other words, the output from a structure-based dehydron calculator (Appendix A) is needed to infer dehydrons from PONDR. The dehydron calculator enables one to construct a training set that is used to "learn to infer dehydrons" from PONDR plots through learned resolution enhancement (LRE). Once a smeared-out (window-averaged) PONDR signal is obtained, the signal has to be resolved at the amino acid level in order to identify the twilight regions rich in dehydrons. This process of LRE is implemented by developing a computational resource, hereby named *Twilighter*, based on supervised learning technology trained with [sequence/PONDR-score/(# of dehydrons)] windows extracted from the PDB and from the PONDR plots generated for PDB-reported proteins. Once trained, the Twilighter maps the number of dehydrons onto a sequence/PONDR-score compound window. A goal is to generate the wrapping parameter $\nu = D/H$ (D = number of dehydrons, H = number of backbone hydrogen bonds) subsumed in the smeared-out PONDR plot.

Since dehydrons are local features and disorder predictions cannot be resolved to the level of locality required to infer the presence of individual dehydrons, a scale or resolution problem arises when PONDR is used for dehydron inference. The twilight regions corresponding to dehydrons are often smeared out as the window is slid along the sequence, and often the dehydron microstructure embedded in the PONDR signal is lost due to poor resolution.

This problem is akin to probing reaction kinetics at different temperatures and finding different orders for the chemical reaction depending on the temperature. Fine-grained features of the potential energy surface that are smeared out at high temperature become important when thermal fluctuations scale down, dramatically altering the observable kinetics of the reaction.

In our particular context of interest, a helical structure with "frayed ends", rich in Dehydrons (cf. helix 25–35 in ubiquitin, Fig. 1.2c), will be typically read by PONDR as a broad disorder signal with no certainty of disorder, unless the length of the helical region is incommensurately larger than the length of PONDR sliding window. The resolution problem is solved exploiting unsupervised learning technology to interpret and de-convolute the broad signals generated by PONDR. Here we describe the supervised learning algorithm that we have aptly named Twilighter that de-convolutes PONDR-signals to identify dehydrons by learned resolution enhancement (LRE) [4].

Twilighter is easily implemented and adopts as training set the direct dehydron identification obtained using Dehydron Calculator (Appendix A) for a large proportion (typically 66 %) of the proteins reported in the PDB combined with the PONDR plot output on the protein sequences for the same PDB-entries. Each PONDR signal is discretized as a 9-tuple of disorder propensities corresponding to a typical PONDR-sliding window and the training data is represented by combining the outputs of PONDR and Dehydron Calculator into an 19-tuple: nine entries for amino acid sequence window, nine entries for PONDR scores, and one entry for number of dehydrons obtained from Dehydron Calculator. In this way, the learning machine is trained to interpret each PONDR signal from a PDB reported protein as containing a well determined number of dehydrons that would not be identifiable from direct reading of the PONDR plot. This training is then used to make sequence-based dehydron inferences based on 9-tuples of PONDR scores for 9-amino acid sequence windows.

The testing set for this machine is the rest (approx. 33 %) of PDB-reported proteins, where we can directly contrast a PONDR-based learned dehydron prediction assigning # dehydrons to each 18-tuple (window sequence identity + the 9 respective PONDR scores) with the actual identification of dehydrons obtained using Dehydron Calculator.

A training set amounting to roughly 66 % of PDB (40,719 entries) gives a prediction accuracy of 93 %, with most outliers being small floppy peptides ($N < 35$) with structure determined from NMR.

This tool is needed to extract wrapping information from sequence-based disorder prediction, since the latter is an attribute averaged over a sliding sequence window, while dehydrons pair two specific residues, thus requiring fine resolution at the single amino acid level. Thus, a learned resolution enhancer (LRE) was required to translate PONDR signals into dehydron patterns. A feedforward neural network resource can be built trained with the wrapping and PONDR-based disorder scores from PDB-represented proteins in order to infer wrapping patterns of foldable proteins with unknown or unreported structure.

The network may be trained with information on proteins with PDB representation. The training data is represented as a vector $(\mathbf{s}, \mathbf{w}, \mathrm{F}) = (\mathbf{X}, \mathrm{F})$, where **s** is the protein sequence window of length $\mathrm{N} = 9$, **w** is the 9-vector of "raw" PONDR scores in the window, $\mathbf{X} = (\mathbf{s}, \mathbf{w})$, and F is the number of residues within the window that are paired by dehydrons. The F-values for proteins in PDB (training set) are computed using Dehydron Calculator.

The inference is then defined by input **X** and output $\mathrm{F} = \mathrm{F}(\mathbf{X})$ for a protein with unknown structure, for which **X** is determined using PONDR. Thus, learned resolution enhancement (LRE) is simply an *in silico* lens used to get information from a smeared signal. The exact work flow of the LRE is described in Fig. 5.13.

The LRE uses the standard learning strategy. The idea is to generate the function $\mathrm{F}(\mathbf{X}) = $ (# residues paired by dehydrons) from a data representation $\mathbf{X} = (\mathbf{s}, \mathbf{w})$. For simplicity, we discretized the single amino acid score f in three ranges: low (1/3) $0 < \mathrm{f} < 0.35$; medium (2/3): $0.35 < \mathrm{f} < 0.8$ ("dehydron range") and high (1): $\mathrm{f} > 0.8$.

We defined the training set H as the set of **X**-vectors obtained from PDB-reported proteins and their PONDR scores. We define the standard metric d

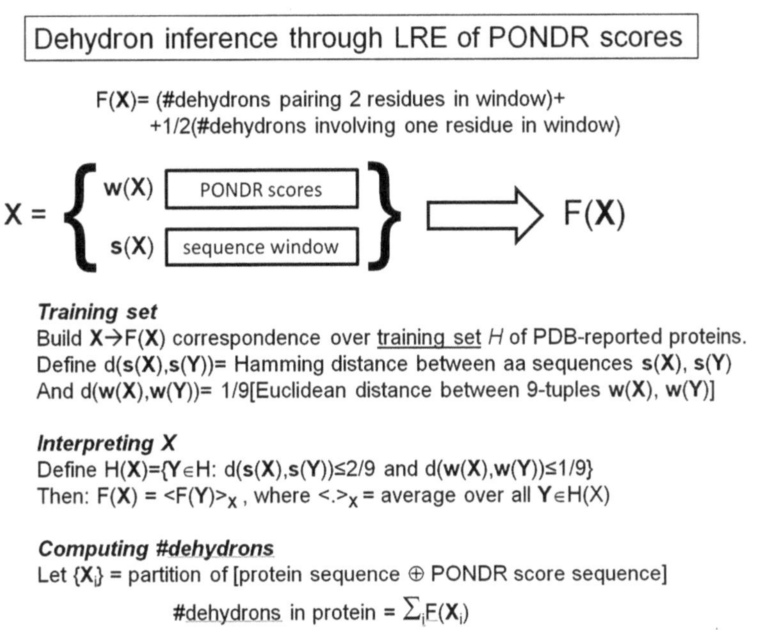

Fig. 5.13 Work flow of the learned resolution enhancement (LRE) machine "Twilighter" used in [4] to infer dehydrons from sequence-based predictions of disorder propensity

in sequence space and the standard metric d in the space of ternary w-vectors. We of course know the F(**Y**) values for all **Y**s in *H*, because we can compute them directly from structure using the Dehydron Calculator.

For crude inferences, given an **X** not belonging to *H*, we infer the value F(**X**) by defining the set *H*(**X**) as the set of vectors in the training set *H* that are "closest" to **X**. The set *H*(**X**) is constructed using the BLAST tool (available at the US National Library of Medicine site http://blast.ncbi.nlm.nih.gov/Blast.cgi). Then, we estimate F(**X**) as the [integral part] of the average of F(**Y**) over all **Y**'s in *H*(**X**).

To get the exact result F(**X**), we exploit a basic property of the network: For any two vectors ($\mathbf{X}_1$, F_1) and ($\mathbf{X}_2$, F_2) in the training set, F satisfies the strong "continuity" relation:

$$F_1 - F_2 = q[d_w(\mathbf{s}_1, \mathbf{s}_2) + (3/2)d_{PONDR}(\mathbf{w}_1, \mathbf{w}_2)], \tag{5.9}$$

where d_w is the wrapping quasi-distance between the two sequence windows, defined as the difference in the number of carbonaceous side-chain nonpolar groups; d_{PONDR} is the compound difference in the PONDR scores (residue-by residue) in the window, and $q = 0.16$ is a constant. The value $q = 0.16$ (approx. $1/(4 + 2)$) is obtained by noting that $\underline{1}$ new dehydron is created when the number of wrappers in the window decreases by $\underline{4}$ and $\underline{2}$ residues qualitatively raise their discretized PONDR score ($1/3 \rightarrow 1$) as defined above. Thus, F(**X**) is "continuous" in the sense that $F(\mathbf{X}_2) - F(\mathbf{X}_1) = q\Delta(\mathbf{X}_2, \mathbf{X}_1)$, where

$$\Delta(\mathbf{X}_2, \mathbf{X}_1) = [d_w(\mathbf{s}_1, \mathbf{s}_2) + (3/2)d_{PONDR}(\mathbf{w}_1, \mathbf{w}_2)] \tag{5.10}$$

Then the rigorous way to infer the # dehydrons in a window takes advantage of the "continuity" relation for the neural network output F. Thus, for **X** not in *H* we get:

$$F(\mathbf{X}) = F(\mathbf{X}^*) + q\Delta(\mathbf{X}, \mathbf{X}^*), \tag{5.11}$$

where $\mathbf{X}^*$ belongs to *H* (hence F($\mathbf{X}^*$) can be computed with certainty using Dehydron Calculator) and is obtained by blasting **X** in *H* ($\mathbf{X}^*$ realizes the minimum distance between **X** and *H*).

To summarize, dehydron predictions can be adequately generated from sequence-based disorder propensity inference using supervised learning technology trained with PDB-reported structures, their structure-based dehydron pattern and their PONDR plots. The computational toolbox presented in this section is extremely useful for the design of drugs targeting proteins with unreported or unknown structure and to build pharmacoinformatics platforms, as shown in Chaps. 10 and 16.

5.7 Misfolding and Aggregation: Flagrant Violation of the Architectural Golden Rule for Protein Structure

In Chap. 1 we investigated the architectural constraints imposed by the extent of under-wrapping of a soluble protein. In view of the characterization of proteins with large dehydron clusters given in the previous section, time is ripe to pose the question: *What is physically the fate of a soluble protein with a large dehydron cluster and with* $Y \gg 5X + 20$, *that is, with poor disulfide-bridge buttressing relative to the high level of structural deficiency*? The answer obviously depends on the extent to which the protein is capable of recruiting binding partners to maintain its structural integrity. As noted in Chap. 1, the worse wrapped proteins are potassium channel neurotoxins but their buttressing is actually adequate for an extracellular protein with such a level of structural deficiency.

On the other hand, the cellular form of the human prion (PDB.1QM0) is very poorly buttressed for its extensive under-wrapping, belongs to the reducing cytosolic space, contains spots of large dehydron density and makes no obligatory complex. We know that prions misfold and aggregate aberrantly into highly organized amyloidogenic fibrils. Is this then a generic behavior of proteins sharing the same molecular attributes? This question prompts us to investigate the amyloid phase more closely. Amyloidogenic aggregation is essentially an epistructural (interfacial) tension effect (Fig. 5.14a). The folded state for a highly "deformed" soluble protein satisfying $Y >> 2X+20$ (Chap. 1) is severely underwrapped. Hence, the internal free energy benefit of intramolecular interaction is insuficient to offset the thermodynamic cost of spanning the dehydron-rich interface. The interface thus becomes severely frustrated, with $g<4$, destabilizing the folded state. In the unfolded state, interfacial frustration vanishes ($g=0$) and there is hardly any contribution from internal free energy. In the amyloid phase, there are no dehydrons, no interfacial frustration and hence there is virtually no thermodynamic cost associated with spanning the interface. Thus, the extensive free energy benefit of forming well wrapped intermolecular interactions makes the amyloid the most favored state. This free energy balance makes amyloid formation a spontaneous, and therefore thermodynamically irreversible process.

There is significant evidence supporting the view that amyloidogenic aggregation is a generic phase of peptide chains [32–34]. The term generic phase refers to a three dimensional organization dominated by main-chain interactions which is essentially insensitive to the information encoded in the primary sequence. Such intermolecular associations appear to be dominated by a basic structural motif: the cross-β structure [35], an intermolecular sheet-pleated pattern ubiquitous in fibrillogenic aggregation. This assertion remains conjectural, as no crystal of the fiber for natural prions has been obtained. By contrast, the folded state of the peptide chain is relatively well understood: natural soluble proteins tend to adopt single-molecule conformations of marginal stability.

While the folding process and its final stable outcome are crucially dependent on the amino-acid composition of the chain, the amyloid state appears to be fairly insensitive to this information: At first sight, amyloidogenic aggregation does not seem to require an "aggregation code", as some recent claims would have it [32–34]. But further analysis reveals that it must place severe constraints on the primary sequence, as some proteins tend to be prone to aggregate under physiological conditions [8], while others require extreme conditions to do so or simply do not aggregate reproducibly [32, 36]. In addition, negative-design features of the folded state purposely encoded in the protein primary sequence may avert aggregation [37]. *Thus, it is not entirely correct to characterize the aberrant aggregation as a "polymer physics phase", shared by polypeptides with arbitrary, suboptimal or random sequence, in contrast with the folded state, determined unambiguously by the primary sequence.*

Proteins are subject to selection pressure and adapt to become good folders, i.e., expeditious structure seekers with a reproducible and stable soluble structure. This optimization is needed to prevent the functionally competent fold from reverting to a primeval amyloid phase. On the other hand, certain sequences are better optimized to escape aggregation than others even under conditions known to sustain the native fold [8].

While amyloidogenic aggregation has been shown to be always plausible provided sufficiently stringent denaturation conditions are applied [32–36], a marked amyloidogenic propensity has been detected on a number of proteins even under physiological or near-physiological conditions, particularly if the monomeric folding domain is deprived of its natural interacting partners [8]. *Such findings imply that not all soluble structures have been optimized to the same degree in order to avert aggregation, and that the more reliant the structure is on binding partnerships or complexations, the more vulnerable it becomes in regards to reverting to the primeval phase.* Thus, an over-expression of a folding domain with high complexation requirements *in vivo*, or the modification of its binding partners as a result of genetic accident, or any factor that distorts its natural interactive context is likely to bolster a transition to an amyloidogenic state.

These observations lead us to the following question: *What type of deficiency in the native fold constitutes a signal for aberrant aggregation?* A recent assessment of the *wrapping* of soluble structure might prove critical to address this problem. For highly under-wrapped proteins (~50 % dehydrons or more), densities higher than four dehydrons per 1000 $Å^2$ on the protein surface become inducers of protein aggregation [8].

Thus, the condition of "keeping the structure dry in water" becomes a requirement to preserve the structural integrity of soluble proteins and imposes a severe building constraint (and thereby an evolutionary pressure) on such proteins. It is expected that the optimization of the structures resulting from this type of evolutionary constraint would be uneven over a range of soluble proteins, resulting in marked differences in aggregation propensity.

This wrapping analysis clarifies the physical picture put forth by Dobson [32–34], in which amyloidogenic propensity depends crucially on the fact that main-chain interactions become dominant in detriment of the amino-acid sequence that encodes the folded state. Precisely, main-chain interactions may dominate as the main chain of the folded state is not properly protected from water attack.

It is instructive to compare this statement with the local analysis of Avbelj and Baldwin [38] in the sense that backbone solvation is a determinant of β-sheet propensity. Thus, an over-exposed backbone hydrogen bond in the native fold is an indicator of a failure in folding cooperativity, as it reveals an inability to remove water from an interactive polar pair by means of a many-body correlation, and at the same time, it is a signal enabling the diagnosis of amyloidogenic propensity. Thus, the wrapping concept enables us to discern why some soluble proteins may have been better optimized to avoid amyloidogenic aggregation than others.

Direct inspection of the pattern of desolvation of the main chain clearly reveals that the cellular fold of the human prion [39] (Fig. 5.11) is too vulnerable to water attack and at the same time too sticky to avert aggregation. Clearly, its sequence has not been optimized to "keep the backbone hydrogen bonds dry" in the folded state. In fact, *their extent of exposure of backbone hydrogen bonds is the highest among soluble proteins in the entire PDB*, with the sole exception of some toxins whose stable fold is held together by a profusion of disulfide bridges, as shown in Chap. 1. This discussion leads us to identify proteins with amyloidogenic propensity as significant outliers in flagrant violation of the X-Y architectural golden rule described in Fig. 5.14b.

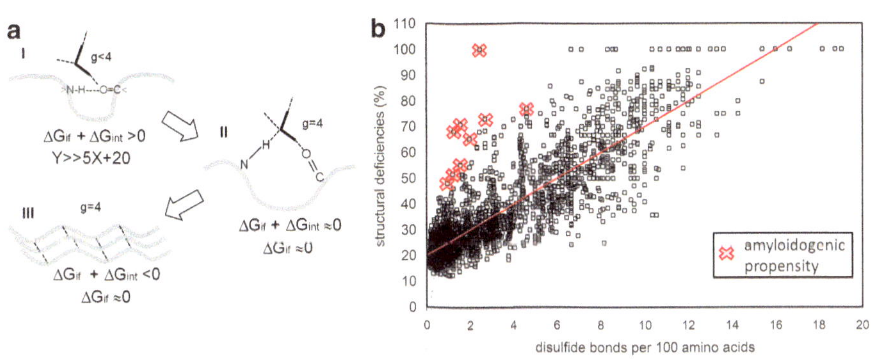

Fig. 5.14 (**a**) Amyloidogenic aggregation as an interfacial tension effect. The folded state (I) for a protein with Y>>2X+20 (Chap. 1) is severely underwrapped and hence, the internal free energy benefit of intramolecular interaction is insuficient to offset the thermodynamic cost of spanning the interface. The interface thus becomes severely frustrated. In the unfolded state (II) interfacial frustration vanishes and there is hardly any contribution from internal free energy. In the amyloid phase (III) there are no dehydrons and hence there is virtually no thermodynamic cost associated with spanning the interface. Thus, the extensive free energy benefit of intermolecular interactions makes III the most favored state. (**b**) Distribution of proteins with known or established amyloidogenic propensity in the X-Y plane (Y = ordinates = %dehydrons, X = abscissas = number of disulfide bridges every 100 amino acids) with datapoints linearly fitted to the architectural golden rule $Y = 5X + 20$ established by Ariel Fernández and R. Stephen Berry and described in Chap. 1

It is suggestive that an inability to protect the main chain is precisely the type of deficiency that best correlates with a propensity to reverse to a primeval aggregation phase determined by main-chain interactions. The actual mechanism by which such defects induce or nucleate the transition is still opaque, although the inherent adhesiveness of packing defects obviously plays a role.

Recently, an atomic-detail structure of a fibrillogenic aggregate, with its β-sheets parallel to the main axis and the strands perpendicular to it, was reported and revealed a tight packing of β-sheets [40]. The cross-β spine of the fibrillogenic peptide GNNQQNY reveals a double parallel β-sheet with tight packing of side chains leading to the full dehydration of intra-sheet backbone-backbone and side-chain-side-chain hydrogen bonds. However, *there is not a single pair-wise interaction between the β-sheets, no hydrogen bond and no hydrophobic interaction.* Instead, a direct examination of the crystal structure reveals that the association is driven by the dehydration propensity of pre-formed intra-sheet dehydrons, as depicted in Fig. 5.15.

This discussion naturally leads us to some evolutionary considerations. A paradigmatic discovery in biology revealed that folds are conserved across species to

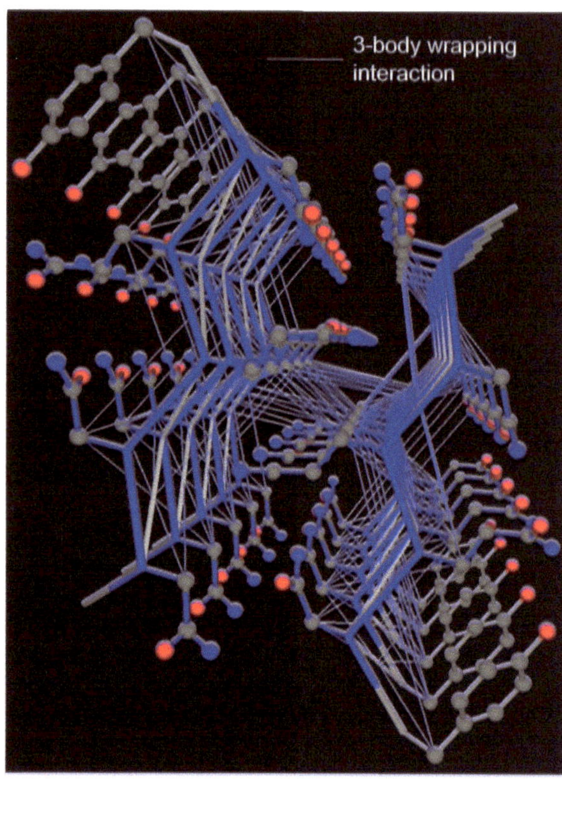

Fig. 5.15 Pattern of inter-sheet wrapping of backbone-backbone hydrogen bonds in the fibrillogenic state of peptide GNNQQNY. The strand backbone is represented as virtual bonds (*blue*) joining α-carbons, and backbone hydrogen bonds are displayed as *light grey lines* joining the α-carbons of the paired residues. A *thin blue line* from the β-carbon of a residue to the barycenter of a hydrogen bond indicates wrapping of the bond by the residue: at least one non-polar group from the residue is contained in the desolvation sphere of the bond

perform specific functions. However, the wrapping of such folds is clearly not conserved [4]. This fact suggests how complex physiologies may be achieved without dramatically expanding genome size, a standing problem in biology. Considerable network complexity may be achieved by actually fostering a higher level of complexation or binding partnership, as promoted by a more precarious wrapping of the isolated protein domains. According to our previous analysis, such complex design also entails an inherent danger: the reversal of highly under-wrapped folding domains to an amyloidogenic phase even under physiological conditions. Amyloidosis is thus likely to be a consequence of high complexity in proteomic connectivity, as dictated by the structural fragility of highly interactive proteins (cf. Fig. 5.14).

Prion-like aggregation has been widely recognized as a causative of pathogenic dysfunction [39], but recent work [41] has revealed that there may be also epigenetic consequences to prion-like aggregation. These novel insights lead to a significant extension of the prion hypothesis. Furthermore, while a connection between protein under-wrapping and propensity for aberrant amyloidogenic aggregation has been unraveled, no misfolding inference can be made for cases when the extent of under-wrapping is so severe that no soluble structure can be maintained by the monomeric chain. This case appears to be of biological interest in light of recent research [41] that focused on nonpathogenic yeast prions. These prions are capable of promoting phenotypic polymorphism through a transmittable conformational change that they regard as having epigenetic consequences, thus defining a protein-based element of inheritance. The yeast prions studied, especially *PSI* in yeast gene Sup35, are unlikely to sustain a monomeric structure.

From the wrapping perspective, this is so because the prion sequences contain large windows (>30 residues) containing mostly under-wrapping residues (especially G, N, Q, S, A, P). In turn, these residues are under-wrappers because they contain few nonpolar side-chain groups, thereby exposing the backbone to the solvent, while some (i.e., S, N, Q) even prevent other nonpolar groups from clustering around the backbone hydrogen bonds. The other under-wrapping residue, D, is unlikely to be found in such prions as its charge promotes hydration, thus introducing a negative design element for aggregation.

These observations are in accordance with the structural characterization by Krishnan and Lindquist [41], which places yeast prions in the twilight between order and disorder, in consonance with the wrapping-based characterization of the prion described in this chapter.

Thus, the epigenetic consequences associated with misfolding of severely under-wrapped proteins incapable of sustaining monomeric structure should be mandatory subjects of any research agenda built on the premises of this chapter. The focus of such endeavors are proteins endowed with large windows (>30) of under-wrapping residues. When combined with tyrosine (probably needed for stacking), such motifs are likely promoters of self-aggregation leading to pathology [39] or epigenetic prion-based phenotype polymorphism [41]. Thus, future research efforts should be directed at identifying highly under-wrapped human genes containing the sequence motifs indicated, and assessing the epigenetic consequences of their transformation into sequestered aggregates. A preliminary scanning of circa 16,000 human proteins in Swiss-Prot reveals 13 genes with the severely under-wrapped motif. Among them, RNF12, AF9, MAML2 are implicated in cancer promotion, while seven are involved in transcription regulation, hence with likely epigenetic consequences associated with a conformational switching into sequestered aggregate states. An epistructural comparison of orthologous proteins across distant species (Chap. 6) reveals that these aberrant states prone to aggregation are likely to arise evolutionarily in organisms where natural selection is inefficient, like humans or higher vertebrates with low effective population. Such "fitness catastrophes" are shown to result from an enrichment in structural disorder which appears to be a hallmark of ineffcient selection, as revealed by direct epistructural analysis of established protein orthologies.

5.8 Problems

5.1. Show that proteins that aggregate aberrantly must fulfill two conditions: (a) large ratio of dehydron-to-well-wrapped-hydrogen-bonds, and (b) significant departure from the (X, Y)-architectural golden rule. Using PONDR®, examine soluble and aggregation-prone sequences [42] in order to validate the postulate.

5.2. *Disrupting a protein-protein interaction as a therapeutic interference to cure heart failure*. Myosin binding protein C (MyBP-C) has been shown to be a central regulator of cardiac contraction (US patent 9,051,387, Inventors: Richard L. Moss and Ariel Fernández). In murine models, it has been observed that genetic ablation or phosphorylation of MyBP-C by PKA or CAMKII accelerates the kinetics of contraction and increases the force of contraction in cardiac muscle. Most recently, it has been shown that CAMKII phosphorylation of MyBP-C at residues S282 and S302 in the mouse and S284 and S304 in the human underlies the increase in myocardial force of contraction as heart rate is increased, the so-called staircase phenomenon [43]. Proof for this conclusion was obtained by observing that (1) phosphorylation at these two residues is increased when stimulus frequency is increased, but other potential phosphorylation sites are not, and (2) the staircase phenomenon is absent in hearts in which these residues are replaced with non-phosphorylatable residues. In terms of molecular mechanism, it is believed that MyBP-C normally depresses the speed and strength of contraction by means of its interaction with the contractile protein myosin and thereby reduces the probability of myosin binding to actin [43]. Ablation of MyBP-C or phosphorylation of MyBP-C by PKA or CAMKII disrupts this interaction and relieves the repression of myosin by MyBP-C. Once phosphorylated, MyBP-C no longer binds to myosin, myosin moves closer to actin, the probability of myosin binding to actin increases, and the speed and strength of contraction both increase.

In heart failure, MyBP-C is phosphorylated minimally or not at all by either PKA or CAMKII due to down-regulation of β-adrenergic receptors. Since phosphorylation of MyBP-C improves contraction, it is proposed to target the CAMKII site(s) on MyBP-C with a pharmaceutical to disrupt its interaction with myosin and thereby improve cardiac contraction and pump function. This problem involves designing an optimal MyBP-C-derived peptide. The peptide will contain the motifs responsible for interaction with myosin and hence will disrupt the MYBP-C-myosin interaction as required for therapeutic action, as schematically illustrated in Fig. 5.16. The 3D-structure of MYBP-C is unknown, and so the only tools to design the peptide are PONDR® and Twilighter described in Sect. 5.6.

The myosin-binding peptide-based therapeutic agent may be identified based on the premise that disruption of the myosin-cMyBP-C interface would release a molecular brake on cardiomyocyte contractility imposed by the inhibitory activity of the unphosphorylated form of cMyBP-C on myosin. Since there is no PDB-reported structure for the 11-domain protein cMyBP-C, the molecular design of the peptide must be developed on a sequence-based approximate prediction of a putative myosin-binding site. To predict the sequence of the peptide, seek a region between the C1 and C2 domains of cMyBP-C (which binds to myosin) in the twilight zone between order and disorder, using PONDR®, the predictor of native disorder (see Fig. 5.17 and Table 5.1). The version used, PONDR-VLXT®, assigns a disorder propensity

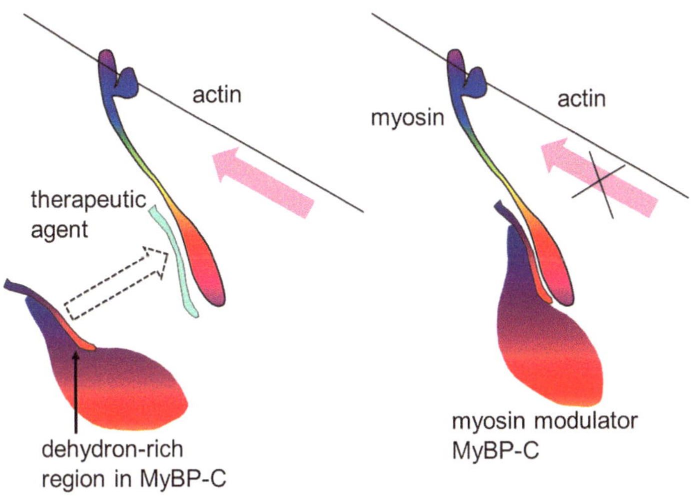

Fig. 5.16 Scheme of the therapeutic disruption of the MyBP-C/myosin interface to treat heart failure. The therapeutic agent consists of a small peptide mimicking the dehydron-rich region in MyBP-C that binds to myosin. To prevent detachment, the peptide is made to be non-susceptible to phosphorylation. As the peptide attaches to myosin, it displaces the myosin modulator MyBP-C, enabling the motor to slide along the actin filament as needed for myocyte contractility. Thus, the therapeutic impact reflects a "molecular brake release" mechanism

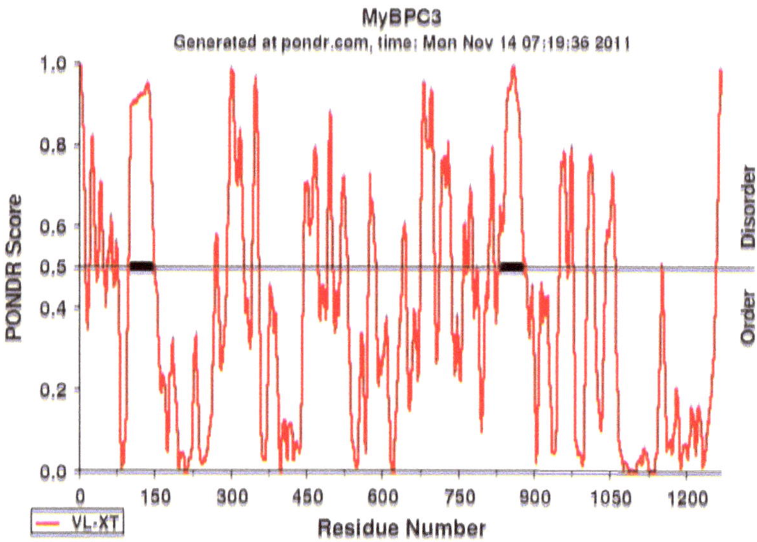

Fig. 5.17 Order-disorder prediction plot for MYBP-C3 obtained from PONDR-VLXT®

score D to each amino acid along the chain, with D = 0 corresponding to certainty of order and D = 1 corresponding to certainty of structural disorder. On the other hand, as established in this chapter, regions in the twilight between order and disorder, with $0.35 < D < 0.8$, are rich in dehydrons. These regions are inherently sticky since dehydrons enhance their stability by promoting dehydration.

Identify the twilight region(s) containing phosphorylation sites S302, S307 in the motif region intercalated between domains C1 and C2 of cMyBP-C (see Fig. 5.17 and Table 5.1 for PONDR order/disorder plot based on protein sequence) and determine the sequence of the optimal peptide that may be used as therapeutic agent to treat heart failure.

Table 5.1 PONDR-VLXT disorder prediction for MYBP-C3

```
                                 PREDICTOR OUTPUT
          "D" = Disordered                                 " " = Ordered

1       MPEPGKKPVS AFNKKPRSAE VTAGSAAVFE AETERSGVKV RWQRDGSDIT
VLXT    DDDDDDDDDD DDDD        DDDDDDDDD DDDD    DD DDDDDDDDDD

51      ANDKYGLAAE GKRHTLTVRD ASPDDQGSYA VIAGSSKVKF DLKVTEPAPP
VLXT          DDDD DDDDDDDD     DDDDD                       DD

101     EKAESEVAPG APKEVPAPAT ELEESVSSPE GSVSVTQDGS AAEHQGAPDD
VLXT    DDDDDDDDDD DDDDDDDDDD DDDDDDDDDD DDDDDDDDDD DDDDDDDDDD

151     PIGLFLMRPQ DGEVTVGGSI VFSARVAGAS LLKPPVVKWF KGKWVDLSSK
VLXT

201     VGQHLQLHDS YDRASKVYLF ELHITDAQTT SAGGYRCEVS TKDKFDSCNF
VLXT

251     NLTVHEAIGS GDLDLRSAFR RTsLAGAGRR TsDsHEDAGT LDFSSLLKKR
VLXT                          DDDDD                     DDDDDDDD

301     DsFRRDsKLE APAEEDVWEI LRQAPPSEYE RIAFQHGVTD LRGMLKRLKG
VLXT    DDDDDDDDDD DDDDDDDDDD DDDDDD                  DDDDDDDD

351     MKQDEKKSTA FQKKLEPAYQ VNKGHKIRLT VELADPDAEV KWLKNGQEIQ
VLXT    DDDDDD

401     MSGSKYIFES VGAKRTLTIS QCSLADDAAY QCVVGGEKCS TELFVKEPPV
VLXT                                                    DDDDDD

451     LITRSLEDQL VMVGQRVEFE CEVSEEGAQV KWLKDGVELT REETFKYRFK
VLXT    DDDDDDDDDD DDDDDDDDDD DDDD                  DDDDDDDDDD

501     KDGRKHHLII NEATLEDAGH YAVRTSGGQS LAELIVQEKK LEVYQSIADL
VLXT    DD               DDDD DDDDDDDDDD

551     AVGAKDQAVF KCEVSDENVR GVWLKNGKEL VPDNRIKVSH IGRVHKLTID

VLXT                             DDDDDDD DDDDDDD
```

```
601       DVTPADEADY SFVPEGFACN LSAKLHFMEV KIDFVPRQEP PKIHLDCPGS
VLXT                                                DDDDDDDD

651       TPDTIVVVAG NKLRLDVPIS GDPAPTVVWQ KTVTQGKKAS TGPHPDAPED
VLXT                                 DDDDD DDDDDDDDDD DDDDDDDDDD

701       AGADEEWVFD KKLLCETEGR VRVETTKDRS VFTVEGAEKE DEGVYTVTVK
VLXT      DDD           DDDDDDD DDDDDDDDDD DDDDDD

751       NPVGEDQVNL TVKVIDVPDA PAAPKISNVG EDSCTVQWEP PAYDGGQPVL
VLXT               D DDDDDD  DD DDDDDDDDD        D

801       GYILERKKKK SYRWMRLNFD LLRELSHEAR RMIEGVAYEM RVYAVNAVGM
VLXT               D DDDDDDDDDD D        D DDDDDDDDDD DDDDDDDDDD

851       SRPSPASQPF MPIGPPGEPT HLAVEDVSDT TVSLKWRPPE RVGAGGLDGY
VLXT      DDDDDDDDDD DDDDDDDDDD DDDDDDDDDD DD D

901       SVEYCQEGCS EWTPALQGLT ERTSMLVKDL PTGARLLFRV RAHNVAGPGG
VLXT                                                         D

951       PIVTKEPVTV QEILQRPRLQ LPRHLRQTIQ KKVGEPVNLL IPFQGKPRPQ
VLXT      DDDDDDDDDD DDDDD  DDD DDDDDDDD

1001      VTWTKEGQPL AGEEVSIRNS PTDTILFIRA ARRTHSGTYQ VTVRIENMED
VLXT           DDDDD DDDDDDDDD                       DDDDDDDDD

1051      KATLILQIVD KPSPPQDIRI VETWGFNVAL EWKPPQDDGN TEIWGYTVQK
VLXT      DDDDDDDDDD

1101      ADKKTMEWFT VLEHYRRTHC VVSELIIGNG YYFRVFSHNM VGSSDKAAAT
VLXT

1151      KEPVFIPRPG ITYEPPKYKA LDFSEAPSFT QPLANRSIIA GYNAILCCAV
VLXT        D

1201      RGSPKPKISW FKNGLDLGED ARFRMFCKQG VLTLEIRKPC PYDGGVYVCR
VLXT

1251      ATNLQGEAQC ECRLEVRVPQ
VLXT                  DDDDDDDDD
```

References

1. Fernández A, Sosnick TR, Colubri A. Dynamics of hydrogen-bond desolvation in folding proteins. J Mol Biol. 2002;321:659–75.
2. Fernández A, Scheraga HA. Insufficiently dehydrated hydrogen bonds as determinants of protein interactions. Proc Natl Acad Sci U S A. 2003;100:113–8.
3. Fernández A, Scott LR. Adherence of packing defects in soluble proteins. Phys Rev Lett. 2003;91:018102.
4. Fernández A, Lynch M. Nonadaptive origins of interactome complexity. Nature. 2011;474:502–5.
5. Fernández A, Scott RL. Dehydron: a structurally encoded signal for protein interaction. Biophys J. 2003;85:1914–28.
6. Fernández A. Keeping dry and crossing membranes. Nat Biotechnol. 2004;22:1081–4.
7. Fernández A, Scott RL, Berry RS. Packing defects as selectivity switches for drug-based protein inhibitors. Proc Natl Acad Sci U S A. 2006;103:323–8.
8. Fernández A, Kardos J, Scott R, Goto Y, Berry RS. Structural defects and the diagnosis of amyloidogenic propensity. Proc Natl Acad Sci U S A. 2003;100:6446–51.
9. Fernández A, Berry RS. Proteins with hydrogen-bond packing defects are highly interactive with lipid bilayers: Implications for amyloidogenesis. Proc Natl Acad Sci U S A. 2003;100:2391–6.
10. Deremble C, Lavery R. Macromolecular recognition. Curr Opin Struct Biol. 2005;15:171–5.
11. Ma B, Elkayam T, Wolfson H, Nussinov R. Protein-protein interactions: structurally conserved residues distinguish between binding sites and exposed protein surfaces. Proc Natl Acad Sci U S A. 2003;100:5772–7.
12. Ma B, Pan Y, Gunasekaran K, Venkataraghavan RB, Levine AJ, Nussinov R. Comparison of the protein-protein interfaces in the p53-DNA crystal structures: towards elucidation of the biological interface. Proc Natl Acad Sci U S A. 2005;102:3988–93.
13. Rajamani D, Thiel S, Vajda S, Camacho CJ. Anchor residues in protein-protein interactions. Proc Natl Acad Sci U S A. 2004;101:11287–92.
14. Dunker AK, Cortese MS, Romero P, Iakoucheva LM, Uversky VN. Flexible nets: the roles of intrinsic disorder in protein interaction networks. FEBS J. 2005;272:5129–48.
15. Iakoucheva LM, Dunker AK. Order, disorder, and flexibility: prediction from protein sequence. Structure (London). 2003;11:1316–7.
16. Dunker AK, Obradovic Z. The protein trinity-linking function and disorder. Nat Biotechnol. 2001;19:805–6.
17. Wade RC, Gabdoulline RR, Lüdemann SK, Lounnas V. Electrostatic steering and ionic tethering in enzyme-ligand binding: insights from simulations. Proc Natl Acad Sci U S A. 1998;95:5942–9.
18. Ho WC, Fitzgerald MX, Marmorstein R. Structure of the p53 core domain dimer bound to DNA. J Biol Chem. 2006;281:20494–502.
19. Pietrosemoli N, Crespo A, Fernández A. Dehydration propensity of order–disorder intermediate regions in soluble proteins. J Proteome Res. 2007;6:3519–26.
20. Lindahl E, Hess B, Van der Spoel D. GROMACS 3.0: a package for molecular simulations and trajectory analysis. J Mol Model. 2001;7:302–17.
21. Fernández A. What caliber pore is like a pipe? Nanotubes as modulators of ionic gradients. J Chem Phys. 2003;119:5315–9.
22. Ben-Naim A. Hydrophobic interactions. New York: Plenum Press; 1980.
23. Debye P. Polar molecules. New York: Dover; 1929.
24. Mason PE, Neilson GW, Dempsey CE, Barnes AC, Cruickshank JM. The hydration structure of guanidinium and thiocyanate ions: implications for protein stability in aqueous solution. Proc Natl Acad Sci U S A. 2003;100:4557–61.
25. Kabsch W, Mannherz HG, Suck D, Pai EF, Holmes KC. Atomic structure of the actin: DNase I complex. Nature. 1990;347:37–44.

26. Meador WE, Means AR, Quiocho FA. Modulation of calmodulin plasticity in molecular recognition on the basis of X-ray structures. Science. 1993;262:1718–21.
27. Williams RS, Green R, Glover JN. Crystal structure of the BRCT repeat region from the breast cancer-associated protein BRCA1. Nat Struct Biol. 2001;8:838–42.
28. Zahn R, Liu A, Luhrs T, et al. NMR solution structure of the human prion protein. Proc Natl Acad Sci U S A. 2000;97:145–50.
29. Schnuchel A, Wiltscheck R, Eichinger L, Schleicher M, Holak TA. Structure of severin domain 2 in solution. J Mol Biol. 1995;247:21–7.
30. Glover JN, Harrison SC. Crystal structure of the heterodimeric bZIP transcription factor c-Fos-c-Jun bound to DNA. Nature. 1995;373:257–61.
31. Lavigne P, Crump MP, Gagne SM, et al. Insights into the mechanism of heterodimerization from the 1H-NMR solution structure of the c-Myc-Max heterodimeric leucine zipper. J Mol Biol. 1998;281:165–81.
32. Dobson CM. Protein misfolding, evolution and disease. Trends Biochem Sci. 1999;24:329–32.
33. Dobson CM. The structural basis of protein folding and its links with human disease. Philos Trans R Soc Lond B. 2001;356:133–45.
34. Fändrich M, Dobson CM. The behavior of polyamino acids reveals an inverse side chain effect in amyloid structure formation. EMBO J. 2002;21:5682–90.
35. Sunde M, Blake CCF. From the globular to the fibrous state: protein structure and structural conversion in amyloid formation. Q Rev Biophys. 1998;31:1–39.
36. Dobson CM. Protein misfolding diseases: getting out of shape. Nature. 2002;418:729–30.
37. Richardson JS, Richardson DC. Natural β-sheet proteins use negative design to avoid edge-to-edge aggregation. Proc Natl Acad Sci U S A. 2002;99:2754–9.
38. Avbelj F, Baldwin RL. Role of backbone solvation and electrostatics in generating preferred peptide backbone conformations: distributions of phi. Proc Natl Acad Sci U S A. 2003;100:5742–7.
39. Prusiner SB. Prions. Proc Natl Acad Sci U S A. 1998;95:13363–83.
40. Nelson R, Sawaya M, Balbirnie M, et al. Structure of the cross-beta spine of amyloid-like fibrils. Nature. 2005;435:773–8.
41. Krishnan R, Lindquist SL. Structural insights into a yeast prion illuminate nucleation and strain diversity. Nature. 2005;435:765–72.
42. Fang Y, Fang J. Discrimination of soluble and aggregation-prone proteins based on sequence information. Mol Biosyst. 2013;9:806–11.
43. Moss RL, Fitzsimons DP, Ralphe JC. Cardiac MyBP-C regulates the rate and force of contraction in mammalian myocardium. Circ Res. 2015;116:183–92.

Chapter 6
Dehydron as a Marker for Molecular Evolution: Lessons for the Drug Designer

This chapter explores the significance of protein dehydron patterns as markers for molecular evolution. Since, as described in Chap. 4, dehydrons may be targeted by wrapping ligands that may eventually become therapeutic drugs, the evolutionary insights hereby obtained naturally have ramifications for the drug designer. Such ramifications will be exhaustively investigated in the remaining chapters of the book. Proteins with common ancestry (homologs) typically share a common fold, but this structural similarity introduces major problems for molecular targeted therapy since it may lead to hazardous off-target effects and prevent the control of specificity. As shown in this chapter, while the topology of the native fold is highly similar across homologs, the wrapping patterns tend to be different, enabling the wrapping drug to funnel its impact solely on clinically relevant targets. The evolutionary root of the differences in dehydron pattern across homologous proteins is dissected in this chapter both across species and within the human species. As first hinted in this chapter and further developed in the subsequent ones, wrapping variations across homologs can be exploited in drug design to considerable advantage as we aim at engineering target-specific and species-specific therapeutic agents and build meaningful animal models for disease and malignancy. In establishing the evolutionary forces that promote differences in the dehydron patterns across orthologous proteins, we discovered that random genetic drift plays an operational role in promoting dehydron enrichment. This type of structural degradation enhances the propensity for protein interactivity and becomes more pronounced in species with low population, such as humans, where the mildly deleterious mutations typically resulting from random drift have a higher probability of getting fixed in the population. The fitness consequences of this evolutionary strategy adopted by nature are assessed for humans, and reveal the high exposure of the human species to fitness catastrophes resulting from aberrant protein aggregation.

This chapter also brings to fruition the insights from epistructural biology as we tackle the problem of the origin of multicellularity, a holy grail in molecular evolution. We identify a structural signature for the origin of multicellularity by assessing global properties in certain unicellular proteomes that enable a type of

A. Fernández, *Physics at the Biomolecular Interface*, Soft and Biological Matter,
DOI 10.1007/978-3-319-30852-4_6

developmental cooption of basic unicellular processes needed to promote multicellularity. Clues on the evolution of multicellularity have not revealed significant innovation at the genomic or protein domain level, but have identified a causative plexus in the co-option of unicellular functionalities into genes and pathways responsive to developmental cues. We argue that without proteomic innovation, this co-option could only materialize with a degree of functional cooperativity well above unicellular levels. What raw properties of a unicellular proteome are harnessed by such functional exaptations? This chapter addresses this crucial question by examining the structural dependence of proteins (Q) across species. Thus, Q becomes a proxy for the dehydron content of orthologous proteins. The Q parameter differentiates unicellular and multicellular species by globally characterizing their proteome interdependence, and identifies the unicellular volvocine alga *Chlamydomonas reinhardtii* as the sole outlier, with a level of structural dependence closer to that of a multicellular species. This result is significant since volvocine algae provide a model for the evolution of multicellularity and is revealing since structural dependence implies a lack of functional autonomy that enables co-opting without the need to innovate at the proteomic level. This process would not be supported by a proteome with a higher degree of structural independence, typical of most unicellular species. By singling out the volvocine algae taxon, our epistructural analysis upholds prevailing models for the evolution of multicellularity and encourages further studies on the epistructural basis of functional co-option.

6.1 Molecular Evolution for the Drug Designer

From a structural biology perspective, one of the most striking observations regarding protein evolution is the uncanny similarity of the 3D structure of proteins with a common ancestry. This similarity across homologous proteins is often quantified by a RMSD < 1.5 Å for the aligned backbone atoms and holds even at low levels of sequence identity (~25 %), that are nevertheless sufficient to establish protein homology.

From the perspective of a drug designer, this structural similarity across homologs of the protein target is particularly troublesome. Major goals in drug design are related or entangled with this evolutionary aspect. They are: (a) the control of drug specificity; (b) the engineering of species-selective therapeutic impact in the treatment of infectious diseases; and (c) the building of insightful animal models to properly assess drug efficacy. In all three contexts, the existence of homologs of the protein target within the species (paralogs) or across species (orthologs) can introduce undesired cross reactivities due to their high level of structural similarity with the target. These cross reactivities, in turn, can result in adverse and even health-threatening side effects or can render inconsequential or irrelevant the animal models for a particular disease or malignancy.

The fold may be highly conserved across homologs, but as shown in this chapter, the wrapping or dehydron pattern is not conserved, and neither is the regulation of the expression pattern across homologs. As described below, we should distinguish orthologs from paralogs when assessing the evolutionary origins of these dissimilarities. Thus, in the case of orthologs, the variations arise mostly from differences in the level of efficiency of natural selection across different species. On the other hand, in paralogs the dissimilarity is rooted in the imperative to curb dosage imabalances that would arise if two copies of the same gene would be retained undifferentiated. As shown in the subsequent chapters, these subtle variations of proteins that essentially share the same 3D-fold have paramount consequences to engineer drugs that can fulfill the three goals described above.

The picture that emerges from this chapter is essentially that wrapping constitutes a biological dimension utilized by evolution to promote complexity and at the same time, that wrapping differences within a fold offer escape routes to the fitness consequences introduced by paralog retention. Both evolutionary aspects will be harnessed in Chaps. 8–13 as we herald the new generation of wrapping-based drugs.

As we explored the evolutionary forces that sculpt differences in the dehydron patterns across orthologous proteins, we came to realize that random genetic drift is the main factor causing dehydron enrichment in species with low population. Because dehydrons are markers for protein association (Chaps. 1, 2, and 4), this type of structural degradation promotes higher protein interactivity which becomes more pronounced in species with low population, such as humans. This is because dehydron-generating mildly deleterious mutations resulting from random genetic drift have a higher probability of getting fixed in the relatively low population of the human species. As shown in this chapter, the fitness consequences of nature's "evolutionary gambit" reveal the high exposure of the human species to fitness catastrophes resulting from aberrant protein aggregation.

6.2 Dehydron Patterns Across Orthologous Proteins: Hallmarks of Non-adaptive Traits

Proteins with common ancestry tend to share the same fold or at least the same structural topology [1, 2]. Thus, essentially the same molecular machinery is operative in organisms that diverged from each other billions of years ago. For example, the metabolic enzymatic function "dehydrofolate reductase" (DHFR) is exerted by a molecular machinery that has hardly changed in structure as we compare species in different kingdoms of life, like *archea* (i.e. *Haloferax volcanii*), bacteria (i.e. *Escherichia coli*) and higher eukaryotes (i.e. *Homo sapiens*). When closely examined, the wrapping of the DHFR has gotten worse (richer in dheydrons) in species with small population size, that is, those where natural selection became less efficient [3] (Fig. 6.1a). This trend is apparent even when examining highly conserved protein domains, such as ubiquitin: *There is a progressive enrichment in dehydrons as the species population decreases* (Fig. 6.1b).

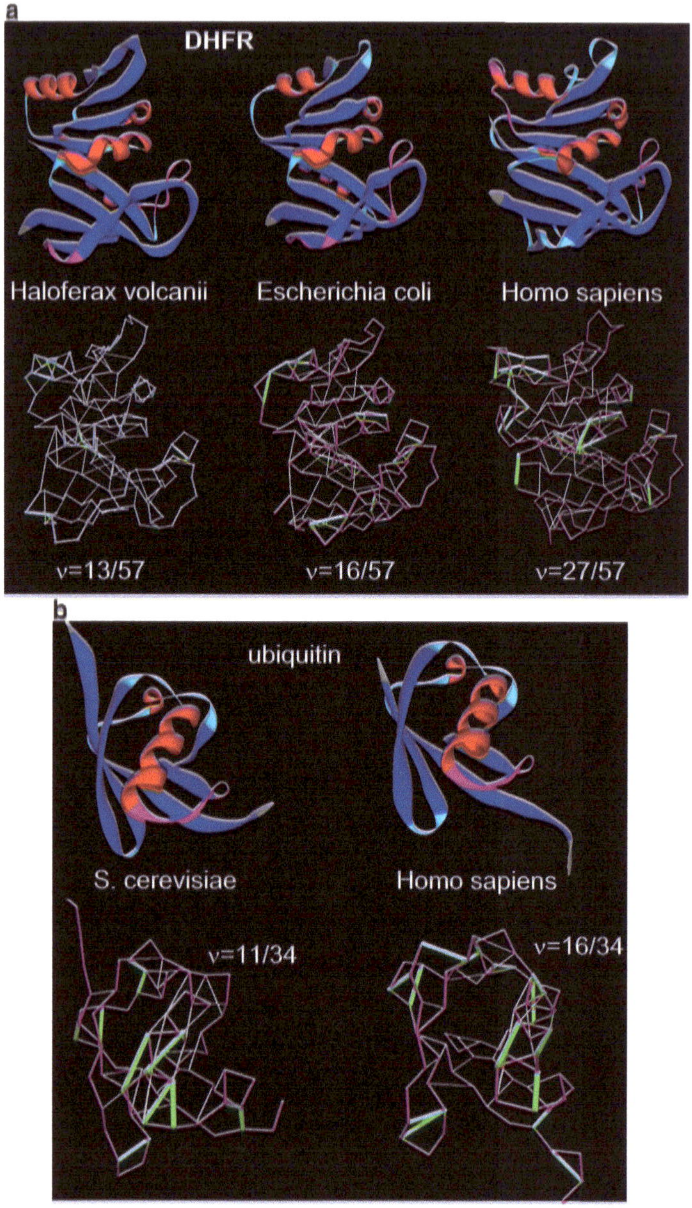

Fig. 6.1 (**a**) Under-wrapping of three orthologous versions of dehydrofolate reductase (DHFR) from three kingdoms of life archea, bacteria and eukarya, represented by species *Haloferax volcanii*, *Escherichia coli* and *Homo sapiens*, respectively. (**b**) Wrapping patterns of ubiquitin from *Saccharomyces cerevisiae* (*yeast*) and *Homo sapiens*. (**c**) Wrapping patterns of haemoglbin in human and of its ortholog leghaemoglobin in pea (*Pisum sativum*). In human, the monomer within the tetramer (*upper left panels*) contains less dehydrons that the monomer in isolation (*lower left panels*) due to intermolecular wrapping. As the monomer is taken in isolation, seven additional dehydrons (marked by *asterisks*) occur. These dehydrons promote the quaternary structure and become well wrapped hydrogen bonds in the homomeric complex. By contrast the pea leghaemoglobin ortholog is better wrapped and lacks the dehydrons that would trigger oligomerization and promote allostery

c

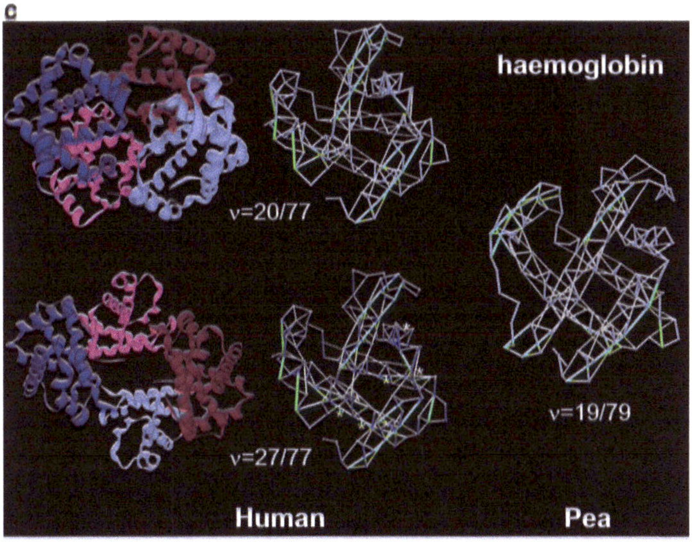

Fig. 6.1 (continued)

For operational reasons, the under-wrapping or packing deficiency of a protein may be defined as $\nu = \%$ dehydrons in the set of backbone hydrogen bonds. As we focus on any functionally competent fold, the anectdotal examples we harvested suggest a clear trend: ν(archea) < ν(bacteria) < ν(unicellular eukaryotes) < $\cdots$ < ν (higher eukaryotes). This trend probably follows from the fact that mildly deleterious mutations have a better chance to prevail and get fixed in species with smaller populations [3]. In this regard, the reader should note that most mutations are detrimental and hardly ever beneficial. Thus it is more likely that a mutation will loose up the fold (create a dehydron), rather than tighten it. These mildly detrimental mutations that prevail manage to degrade the wrapping of the protein, enriching its dehydron content, thereby making the protein more reliant on binding partnerships to maintain its structural integrity [1, 2]. *Thus, as protein interactivity, regulation and allostery increases with decreasing species population, one may say that complexity is in this sense promoted by nonadaptive forces.*

An illustration of the probable nonadaptive origin of allostery is furnished by the wrapping of haemoglobin across orthologs in species with vastly different population size. Thus, this protein becomes richer in dehydrons and more prone to oligomerization in species with smaller population, with the majority of the new dehydrons located at the interface that promotes the quaternary structure of the homomer (Fig. 6.1c).

6.3 Natural Selection and Dehydron Patterns

Dosage imbalances occur when protein concentration levels at specific locations in tissues or metabolic/developmental phases do not fit the stoichiometry of the complexes in which the proteins are involved [4–6]. The complexes may be transient, adventitious or obligatory with regards to maintaining the structural integrity of the protein [7] and hence the effects of the imbalances may vary widely. Therefore, dosage sensitivity, that is, the impact of dosage imbalances on fitness, must be influenced not only by whether the protein is part of a complex but also by the extent of reliance of the protein on its binding partners to maintain structural integrity and functional competence.

While overexpression, gene duplication, misfolding and self-aggregation may all cause dosage imbalance, the structural or molecular properties determining the magnitude of the resulting effects remain largely unknown. For example, as we focus on gene duplication, we notice that paralog proteins, identical when they initially diverge, are subject to higher or lower selection pressure depending on their dosage sensitivity [5].

Cross examination of genetic and structural information revealed that the under-wrapping or packing deficiency of a protein, ν, is a molecular quantifier of dosage sensitivity [5] precisely because it constitutes a measure of the reliance of the protein on binding partnerships to maintain the integrity of the native fold. As expected, a deficiently packed protein is more likely to be engaged in an obligatory complex [7] and its concentration imbalances relative to the complex stoichiometry are more likely to impact fitness than those of a well packed protein.

In unicellular organisms, the packing quality $(100 - \nu)$ of soluble gene products correlates with the number of paralogs or family size [5]. That is, the better the packing quality, the less likely that gene duplication would introduce a fitness disadvantage, hence the higher the chance that the duplicate gene would be retained. However, this correlation becomes less significant in higher eukaryotes, as shown in Fig. 6.2. Thus, paralog survival is dependent on the packing quality of protein structure with $P < 10^{-16}$ in *Escherichia coli* and $P < 6.7 \times 10^{-3}$ in human [5]. This contrast between simple and complex organisms is hard to interpret due to wide differences at the proteome level. However, alternative measures point to a similar trend. For example, the average difference in packing deficiency between singletons and duplicate genes is 18 % in *E. coli*, 6 % in worm (*Caenorhabditis elegans*), and ~2 % in human (*Homo sapiens*).

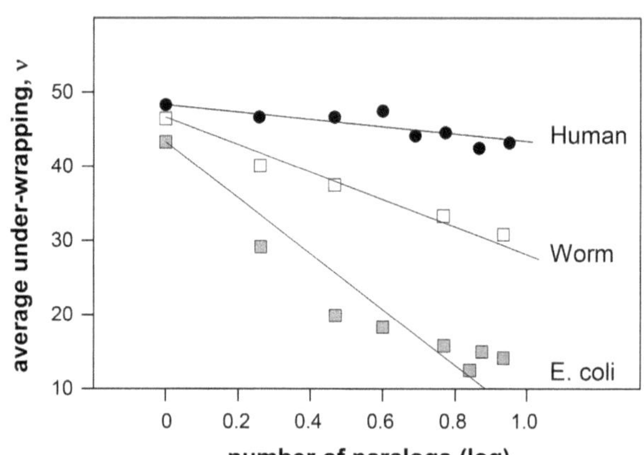

Fig. 6.2 Negative correlation between protein under-wrapping, ν, and gene duplicability or family size represented by the number of paralogs in the gene family. Structural and genetic information was obtained for different species and ν-values were averaged over all families with a fixed number of paralogs, as described in [5]

In human, this insensitivity to dosage imbalance may be attributed in part to selection inefficiency arising from smaller population size [3], implying that the selection pressure exerted on paralogs of deficiently packed proteins has simply not become operative. Alternatively, the higher complexity of expression regulation in higher eukaryotes may introduce a tolerance to dosage imbalance not found in unicellular organisms. This chapter explores this latter possibility, focusing on evolution-related dosage imbalances and the mechanisms that humans possess to cope with the troubling aspects of selection inefficiency.

6.4 Inefficient Selection in Humans: An Epistructural View of the Coping Mechanism

If selection is indeed inefficient in humans and this inefficiency enabled duplicate genes to stay around awaiting further specialization, significant dosage imbalances must have occurred during human evolution. How did we cope with such imbalances, which are likely to be significant when the gene products are very underwrapped? The key to this problem lies in the ability of humans to regulate the expression of both gene copies dissimilarly through a plausible process that involves random deleterious mutations.

To address this problem, we first assess the selection pressure on gene duplicates exerted as paralogs are co-expressed at the mRNA (messenger RNA) level and hence are likely to compete for their interactive partners. Then we relate packing deficiency (ν) with differences in post-transcriptional regulation patterns within families. Thus, we investigate how differences in miRNA-target patterns [8], telling apart paralogs through different patterns of translational repression, impinge on the selection pressure on duplicate genes by mitigating dosage imbalances. In human these patterns are significantly dissimilar across paralogs of poorly packed proteins while nearly coincident across paralogs of well packed proteins [9], thus underscoring a means to buffer dosage imbalance effects arising from gene duplication.

This miRNA-based capacitance is not expected to be nearly as significant in species with larger effective population size due to the higher efficiency of evolutionary forces in such organisms when compared with human [3]. Thus, the selection pressure affecting the retention of gene duplicates is likely to be more efficient in these organisms promoting adaptation through functional innovation or purifying selection.

The next two sections are devoted to provide further evidence in support of the claim that protein wrapping quality is a determinant of dosage sensitivity while upholding the view that resilience to dosage imbalance is achieved in human by diversifying miRNA-regulatory patterns across paralogs.

6.4.1 Expression Patterns Segregating Paralog Proteins

As indicated in Chaps. 1, 2, and 4, soluble proteins may further protect their backbone hydrogen bonds and improve packing quality through binding partnerships by increasing the number of nonpolar groups in their hydrogen-bond microenvironments [7, 10]. Hence, the extent of intermolecular protection determines whether the complex is obligatory, ephemeral or adventitious, and thus packing quality may be recognized as an important factor in determining dosage sensitivity [5].

Gene duplication introduces dosage imbalance and the resulting selection pressure on paralogs [6, 11, 12] appears to depend on the packing deficiency of the parental gene [5]. This trend is clear in *E. coli* and *S. cerevisiae*, but not so apparent in higher eukaryotes. This observation suggests that expression dissimilarities at the mRNA level and at post-transcriptional levels may be exploited to separate paralogs and avoid competition for the binding partners of the parental gene. Thus, to study human capacitance to dosage imbalance arising from gene duplication, we examined families with paralog coexpression at the mRNA level [13] and assessed post-transcriptional microRNA (miRNA) regulation patterns in relation to the packing quality of the proteins in the family.

To assess the role of miRNA regulation in the human capacitance to dosage imbalance, we selected human genes from an exhaustive set of 583 non-singleton families for which genetic [14], evolutionary [15], structural [5], expression [13] and post-transcriptional [16, 17] data is available for at least two paralogs (Table 6.1). We obtained human gene information from the Ensembl Genome Database (NCBI36). Using the Ensembl gene family annotation [14], 22,357 human genes were grouped into 12,394 families [9]. Gene expression data were obtained from Novartis Gene Expression Atlas [13] in the form of expression levels across a panel of 73 normal human tissues. We then mapped the putative miRNA-target patterns for each classified gene. Putative conserved target sites in the 3'UTR (untranslated region) of each gene for 156 conserved microRNA families were identified using TargetScanS (version 5.1). Thus, to determine co-expression and co-regulation patterns across paralogs, each gene i is represented by two vectors:

1. A normalized mRNA-expression vector $\Phi_i/\|\Phi_i\|$, where the vector Φ_i has 73 entries indicating mRNA expression levels in 73 normal tissues [13] and $\|\Phi_i\|$ is the norm of the vector.
2. A normalized miRNA vector $\Psi_i/\|\Psi_i\|$ of 156 entries representing the pattern of miRNA-related repression efficacy on gene i, with $\|\Psi_i\|$ = vector norm. This representation is in accord with the identified target sites for the 156 conserved miRNA families (broadly conserved, intermediately conserved and mammalian specific) in 17,444 human genes [8, 16]. The nth entry in Ψ_i is $\Psi_i(\mathrm{n}) = 1 - 2^{\mathrm{s(i,n)}}$, where $\mathrm{s(i,n)} \leq 0$ is the context score of conserved miRNA-binding site n in the 3'UTR of gene i [18]. Thus, $\Psi_i(\mathrm{n}) = 1$ indicates full repressive efficacy of the nth miRNA conserved site on gene i ($\mathrm{s(i,n)} = -\infty$), while $\Psi_i(\mathrm{n}) = 0$ (or $\mathrm{s(i,n)} = 0$) indicates absolute lack of repressive regulatory power. The context score is

Table 6.1 PDB representation of selected human gene families of different sizes

Ensembl ID	Family size	Swiss prot ID	Gene name	PDB ID
>ENSF00000000393	7	Q05066	SRY	1HRY_A
>ENSF00000000393	7	P48431	SOX2	1O4X_B
>ENSF00000000784	4	P08048	ZFY	1KLR_A
>ENSF00000002256	4	Q8IUE1	TGIF2LX	2DMN_A
>ENSF00000000640	6	Q9HC56	PCDH9	2EE0_A
>ENSF00000000504	6	P38919	EIF4A3	2HXY_A
>ENSF00000000661	16	Q01105	SET	2E50_A
>ENSF00000001155	23	Q15414	RBMY1B	2FY1_A
>ENSF00000001155	23	Q14011	CIRBP	1X5S_A
>ENSF00000000702	5	P17612	PRKACA	2GU8_A
>ENSF00000000530	19	P13501	CCL5	1B3A_A
>ENSF00000000530	19	O00175	CCL24	1EIG_A
>ENSF00000001027	4	Q92565	RAPGEF5	1WGY_A
>ENSF00000000301	14	Q05923	DUSP2	1M3G_A
>ENSF00000000301	14	Q16690	DUSP5	2G6Z_A
>ENSF00000001036	2	O00763	ACACB	2DN8_A
>ENSF00000002731	2	P61923	COPZ1	2HF6_A
>ENSF00000000164	18	Q9Y3D6	CLDN15	1NZN_A
>ENSF00000000212	11	Q9UHX3	EMR2	2BO2_A
>ENSF00000002211	2	Q8WUA7	TBC1D22A	2QFZ_A
>ENSF00000002844	2	Q8IWR0	ZC3H7A	2D9M_A
>ENSF00000001386	3	Q9UQB8	BAIAP2	1WDZ_A
>ENSF00000001211	4	P01100	FOS	1A02_F
>ENSF00000000285	19	Q8NHL6	LILRB1	1G0X_A
>ENSF00000001565	4	P50616	TOB1	2D5R_B
>ENSF00000002661	2	Q6XZF7	DNMBP	1UG1_A
>ENSF00000000647	6	P55201	BRPF1	2D9E_A
>ENSF00000001800	2	O96006	DHRSX	2CT5_A
>ENSF00000000849	6	Q9H8S9	MOBK1B	1PI1_A
>ENSF00000000030	2	P03999	OPN1SW	1KPN_A
>ENSF00000000192	21	P15157	TPSB2	1LTO_A
>ENSF00000000192	21	P20231	TPSAB1	1A0L_A
>ENSF00000000254	9	Q14832	GRM3	1S8M_A
>ENSF00000000521	7	Q05586	GRIN1	2NR1_A
>ENSF00000000521	7	Q13224	GRIN2B	1S11_A
>ENSF00000001623	2	P30291	WEE1	1X8B_A
>ENSF00000000001	458	Q8WTR7	ZNF473	2EMB_A
>ENSF00000000001	458	Q96SE7	ZNF347	2EMA_A
>ENSF00000000001	458	Q8NHY6	ZFP28	2EM2_A
>ENSF00000000001	458	Q9H4T2	ZSCAN16	2COT_A
>ENSF00000000822	3	Q86UL8	MAGI2	1UEP_A
>ENSF00000001433	3	Q9H2H8	PPIL3	1XYH_A

(continued)

Table 6.1 (continued)

Ensembl ID	Family size	Swiss prot ID	Gene name	PDB ID
>ENSF00000000610	5	P84022	SMAD3	1MHD_A
>ENSF00000000610	5	Q15797	SMAD1	1KHU_A
>ENSF00000000153	12	Q96JP2	MYO15B	2DLP_A

Human gene information is from Ensembl Genome Database (NCBI36)

known to correlate tightly with the post-transcriptional downregulation efficacy, $2^{s(i,\ n)}$, of the predicted binding site for the nth miRNA family within the 3'UTR of gene *i* [18]. Thus, $2^{s(i,\ n)} \approx g(i)/g(i,n)$, where g(i) is the translation level for gene i and g(i, n) is the i-translation level with knockout of the nth miRNA family. If the n-site in not predicted in the 3'UTR of gene i, $g(i) = g(i,n)$ and therefore $s(i,n) = 0$.

Only paralogs that are significantly co-expressed are likely to produce dosage imbalances if the genes have not diverged significantly. Thus, similarities between mRNA expression profiles of two genes *i*, *j* will be assessed by the Pearson correlation coefficient $\eta(i,j)$ of their expression vectors Φ_i and Φ_j. In general, for two expression vectors ***X*** and ***Y***, the Pearson coefficient is given by

$$\eta(\mathbf{X},\mathbf{Y}) = \frac{\langle (X - \langle X \rangle)(Y - \langle Y \rangle) \rangle}{\sqrt{\langle X^2 \rangle - \langle X \rangle^2}\sqrt{\langle Y^2 \rangle - \langle Y \rangle^2}}, \tag{6.1}$$

where *X*, *Y* are generic coordinates in the vectors ***X*** and ***Y*** respectively, and ‹ › indicates mean over cell types.

For paralogs with significant coexpression, a tolerance to dosage imbalance may still arise through differences in translational repression patterns. Thus, orthogonal miRNA-repression patterns for paralogs with high dosage sensitivity may introduce an escape route to the selection pressure introduced by the dosage imbalance. To test this hypothesis, we introduce the extent of miRNA-target coincidence $\tau(i,j)$, defined as the scalar (dot) product of the two miRNA-target vectors: $\tau(i,\ j) = \Psi_i/||\Psi_i|| \cdot \Psi_j/||\Psi_j||$.

To determine the dosage sensitivity we calculated the packing deficiency of each gene-encoded protein based on its PDB coordinates, if available. Otherwise, packing deficiency was determined based on homology-threaded structure coordinates adopting as templates PDB-reported paralogs [19]. The input for the computation consists of the set of constraints applied to the spatial structure of the amino acid sequence to be modelled and the output is the 3D structure that best satisfies these constraints.

The resulting homology model was validated by comparing its inferred dehydron pattern with the dehydron pattern predicted from a sequence-based computation of disorder score, as described in the previous chapter [9]. The inability of an isolated protein fold to protect specific intramolecular hydrogen bonds from water attack may lead to a structure-competing backbone hydration with concurrent local or global dismantling of the structure. This view of underwrapping implies a strong correlation between the degree of solvent exposure of intramolecular hydrogen bonds and the local propensity for structural disorder. Hence, the latter parameter was used to validate the former.

6.4.2 Dehydron Enrichment Enhances Dosage Imbalance

To assess the selection pressure imposed by dosage imbalance we first consider an exhaustive set of 457 non-singleton human gene families with paralog co-expression at the mRNA level: $\langle\eta\rangle > 0$, $\langle\ \rangle$ = family average. This condition is essential since paralogs expressed in different cell types cannot introduce dosage imbalance, regardless of their extent of identity. The families with paralog co-expression are selected to discern the factors that buffer dosage imbalance caused by gene duplication. Thus, the selection pressure may be assessed at the post-transcriptional level in terms of dissimilarities in miRNA-targeting patterns across paralogs.

The families with significant mRNA co-expression were deemed likely to generate dosage imbalance. To assess how these imbalances impinge on the degree of divergence in post-transcriptional repression patterns across paralogs, we must compare families with similar divergence time of gene duplicates. This is so since significant regulatory dissimilarities across paralogs may simply result from long divergence times. Thus, we adopt K_s, the synonymous nucleotide divergence [15], as a proxy for divergence time [20] and bin human families with $\langle\eta\rangle > 0$ according to their respective maximum K_s over paralog pairs. Each class contains families whose duplicate divergence is located in time vis-a-vis particular speciation events. Thus, we construct four classes of human families with significantly coexpressed paralogs [9]: class I: $K_s > 2.98$ (378 families); class II: $2.98 > K_s > 1.48$ (68 families); class III: $1.48 > K_s > 1.11$ (6 families) and class IV: $K_s < 1.11$ (5 families), in accord with the K_s values between human and orangutan (*Pongo pygmaeus*) ($K_s = 2.98$), human and gorilla (*Gorilla gorilla*) ($K_s = 1.48$) and human and chimpanzee (*Pan troglodytes*) ($K_s = 1.11$) [21]. All K_s values are given as percentages and the data for each orthology class is compiled in Table 6.2.

The conservation-based reliability of miRNA site prediction [8] is the highest in class I and decreases with lower divergence times for duplicate genes. This is so since the condition: K_s (duplicate genes) > K_s (speciation) implies that orthologs of the paralog human genes are likely to be found in the diverging species [22]. Thus, paralogs for families in class I are likely to have orthologs in orangutan, gorilla and chimpanzee, those in class II, only in gorilla and chimpanzee, etc.

Human families with paralog coexpression and the most reliable miRNA site inference (class I) exhibit a tight anticorrelation ($R^2 = 0.697$) between packing deficiency and miRNA-target coincidence (Fig. 6.3): *paralogs with deficient packing are more likely to be localized separated from each other as dictated by their dissimilar miRNA-target patterns of post-transcriptional regulation:* $\langle\tau\rangle \rightarrow 0$ *as*

Table 6.2 Dataset of evolutionary, expression, miRNA-based regulation and structural information on human families binned in different K_s-classes

Ensembl ID	Size	Max. K_s	Expression correlation	miRNA-target coincidence	Packing deficiency
Class I. Representative families from 378 with $K_s > 2.98$					
>ENSF00000000529	7	4.7368	0.092	0	58.611
>ENSF00000000100	20	4.3361	0.105	0	57.905
>ENSF00000000139	4	3.5898	0.009	0	57.905
>ENSF00000001722	3	4.5429	0.112	0	57.235
>ENSF00000001167	4	4.257	0.031	0.100156612	49.166
>ENSF00000000095	26	4.8626	0.091	0.100505579	41.711
>ENSF00000001308	3	4.5669	0.137	0.100723554	53.574
>ENSF00000000711	4	4.0037	0.058	0.100919215	43.801
>ENSF00000000137	10	5.618	0.054	0.146230992	40.942
>ENSF00000000284	10	4.2759	0.099	0.146538636	41.175
>ENSF00000001426	4	3.7137	0.026	0.183940909	31.682
>ENSF00000000564	5	4.4951	0.052	0.184436777	38.162
>ENSF00000002030	3	4.5383	0.032	0.203391736	44.664
>ENSF00000001001	2	3.1208	0.201	0.204847727	41.439
>ENSF00000001947	3	4.5458	0.355	0.206692562	44.336
>ENSF00000000469	6	4.2126	0.369	0.207894326	47.726
>ENSF00000001542	3	4.9462	0.088	0.20806405	33.63
>ENSF00000001394	3	4.4438	0.185	0.215967149	40.265
>ENSF00000000676	4	5.1652	0.052	0.220155165	32.679
>ENSF00000000750	4	4.555	0.032	0.249786777	31.144
>ENSF00000000662	6	4.1415	0.148	0.252284504	33.377
>ENSF00000000678	3	4.398	0.077	0.259919421	29.562
>ENSF00000001387	4	4.6489	0.132	0.265183678	33.705
>ENSF00000001289	2	3.2296	0.234	0.289453099	37.129
>ENSF00000002412	2	3.5859	0.504	0.291411364	34.575
>ENSF00000000546	7	4.4044	0.314	0.29621157	37.76
>ENSF00000001974	3	5.605	0.116	0.301903926	31.719
>ENSF00000001090	4	3.8111	0.092	0.30237686	38.952
>ENSF00000002334	3	4.0352	0.117	0.305980165	26.024
>ENSF00000001936	2	5.4117	0.076	0.309117355	26.752
>ENSF00000000747	3	4.4318	0.239	0.326588636	39.481
>ENSF00000001530	3	4.1787	0.1	0.326981198	36.141
>ENSF00000001820	2	3.007	0.048	0.337169628	34.925
>ENSF00000001225	3	4.5539	0.042	0.436954752	35.533
>ENSF00000004001	2	3.5857	0.042	0.467541529	21.877
>ENSF00000001160	3	3.8174	0.393	0.746814463	19.7721
Class II. The 68 human families with $2.98 > K_s > 1.48$					
>ENSF00000001825	2	2.9144	0.041	0	58.601
>ENSF00000001266	2	2.6079	0.155	0	57.999
>ENSF00000000708	2	1.7241	0.157	0	58.999

(continued)

Table 6.2 (continued)

Ensembl ID	Size	Max. K_s	Expression correlation	miRNA-target coincidence	Packing deficiency
>ENSF00000003280	2	2.9426	0.201	0	57.981
>ENSF00000001173	3	2.6772	0.23	0	56.613
>ENSF00000001409	3	2.0711	0.473	0	56.132
>ENSF00000004229	2	2.0857	0.062	0	55.729
>ENSF00000000656	5	2.736	0.471	0	55.62
>ENSF00000001587	2	2.6153	0.452	0	55.192
>ENSF00000000881	2	2.4164	0.12	0	54.574
>ENSF00000000675	5	2.4562	0.009	0	54.434
>ENSF00000001679	2	2.3438	0.48	0	54.33
>ENSF00000001375	4	2.2318	0.217	0	54.074
>ENSF00000001081	2	2.864	0.392	0	54.036
>ENSF00000000178	7	1.9737	0.146	0	53.456
>ENSF00000000572	2	2.1986	0.045	0	53.121
>ENSF00000003334	2	1.7844	0.035	0	52.671
>ENSF00000002313	2	1.8049	0.018	0	52.216
>ENSF00000000436	5	2.8531	0.29	0	51.945
>ENSF00000000074	21	2.7213	0.061	0	51.689
>ENSF00000003192	3	2.3561	0.4	0	51.646
>ENSF00000001401	3	2.8501	0.041	0	51.446
>ENSF00000000271	16	2.2647	0.292	0	51.438
>ENSF00000001153	4	2.6993	0.155	0	51.215
>ENSF00000001694	3	2.0918	0.258	0	51.198
>ENSF00000003861	2	1.6126	0.243	0	50.846
>ENSF00000001802	2	1.6984	0.537	0	50.768
>ENSF00000000847	4	2.8478	0.376	0	50.687
>ENSF00000001880	2	1.9662	0.219	0	50.687
>ENSF00000003481	2	1.8106	0.446	0	50.687
>ENSF00000003653	2	2.2285	0.183	0	50.332
>ENSF00000001645	3	2.2403	0.206	0	50.135
>ENSF00000000637	8	1.9319	0.041	0	50.062
>ENSF00000002038	2	1.8469	0.109	0	52.797
>ENSF00000002767	2	2.2562	0.07	0	52.795
>ENSF00000001162	5	2.0596	0.097	0	53.704
>ENSF00000001747	2	1.7805	0.019	0	52.532
>ENSF00000000570	6	2.1163	0.013	0	52.377
>ENSF00000000671	3	2.2928	0.178	0	52.279
>ENSF00000000389	5	2.4549	0.22	0	53.107
>ENSF00000002053	2	1.8579	0.411	0	58.107
>ENSF00000001405	2	2.2754	0.042	0	58.055
>ENSF00000002395	4	2.5819	0.365	0	57.152
>ENSF00000000402	5	2.2446	0.561	0.056	57.012
>ENSF00000001931	5	2.3206	0.035	0.091	45.743

(continued)

Table 6.2 (continued)

Ensembl ID	Size	Max. K_s	Expression correlation	miRNA-target coincidence	Packing deficiency
>ENSF00000003072	2	2.0413	0.093	0	54.973
>ENSF00000000492	6	1.9765	0.195	0.141	44.787
>ENSF00000002120	2	2.6622	0.02	0	54.427
>ENSF00000004883	4	2.7168	0.14	0	53.174
>ENSF00000002772	4	1.7219	0.132	0.094	51.132
>ENSF00000001313	3	2.5443	0.099	0.064	50.351
>ENSF00000001227	3	2.1405	0.285	0.17	33.99
>ENSF00000001199	2	2.9308	0.411	0.098	38.351
>ENSF00000001760	2	1.8274	0.123	0.088	37.322
>ENSF00000001365	3	1.5327	0.112	0.162	37.295
>ENSF00000000850	3	1.6986	0.193	0.335	22.644
>ENSF00000001242	3	1.8311	0.58	0.204	36.633
>ENSF00000002009	3	2.1064	0.339	0.267	31.05
>ENSF00000003693	2	2.8605	0.132	0.256	34.582
>ENSF00000000931	3	1.9106	0.07	0.35	24.055
>ENSF00000001181	3	2.5267	0.093	0.279	28.222
>ENSF00000000463	4	1.9925	0.134	0.152	32.358
>ENSF00000004286	2	1.869	0.564	0.385	29.439
>ENSF00000001127	2	2.7223	0.688	0.367	29.419
>ENSF00000000265	14	2.0125	0.699	0.213	28.91
>ENSF00000001157	4	2.3618	0.064	0.169	27.24
>ENSF00000002286	2	2.4737	0.046	0.41	17.284
>ENSF00000000715	9	2.4495	0.237	0.707	24.718
Class III. The six human families with $1.48 > K_s > 1.11$					
>ENSF00000001699	2	1.4313	0.223	0.408	21.703
>ENSF00000001755	4	1.4269	5.00E−001	0.136	31.305
>ENSF00000001091	4	1.4047	0.112	0.333	21.279
>ENSF00000000239	5	1.3547	0.442	0.298	34.019
>ENSF00000001198	3	1.3243	0.335	0	55.438
>ENSF00000000360	6	1.3127	0.18	0.343	27.938
Class IV. The five human families with $1.11 > K_s$					
>ENSF00000000399	13	1.102	0.138	0	43.174
>ENSF00000002309	2	1.0081	0.551	0	49.846
>ENSF00000000530	19	0.615	0.23	0	49.235
>ENSF00000002885	2	0.4883	0.012	0	47.305
>ENSF00000000786	6	0.28	0.107	0	43.32

$\langle v \rangle \rightarrow maximum \approx 58\,\%$. These disjoint localization patterns reduce paralog competition for binding partners, thereby buffering the evolution-related dosage imbalance. This result highlights the role of miRNA regulation as a capacitor for dosage imbalance.

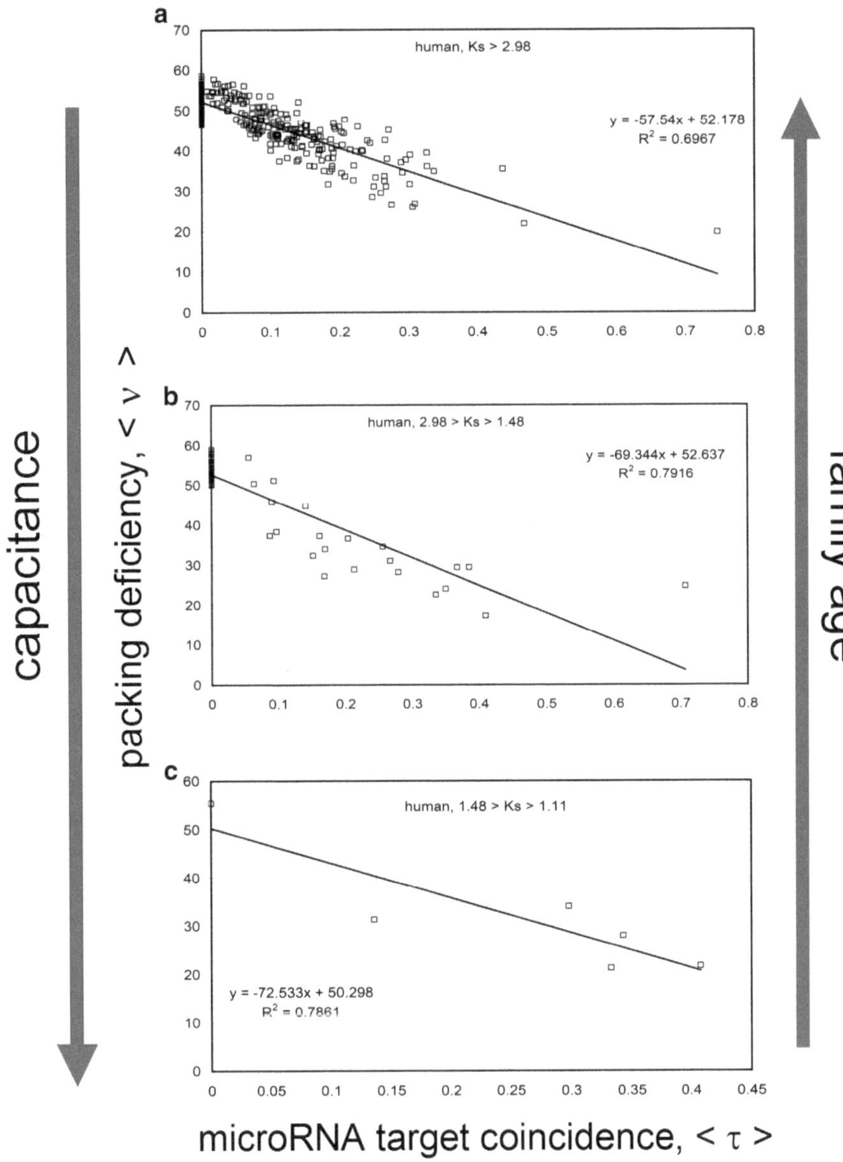

Fig. 6.3 Negative correlation between under-wrapping or packing deficiency (ν) and miRNA target coincidence (τ) for human families in K_s-classes I (**a**), II (**b**), III (**c**). The linear fits were obtained by least-squares linear regression

An even tighter anticorrelation between packing deficiency and miRNA target coincidence is found for family class II ($R^2 = 0.792$, Fig. 6.3). The slope of the linear fit obtained by the least squares linear regression is now significantly larger in magnitude (−69.34 versus −57.54 for class I). This implies that for a fixed level of packing deficiency, a more effective buffer (lower miRNA target coincidence) is needed for the newer families (K_s-class II) than for the older ones (K_s-class I). This result is expected since a longer exposure of surviving paralogs to the selection pressure promoted by dosage imbalance is likely to promote higher level of adaptation through functional divergence, and hence, as older paralogs become more differentiated, a capacitance to dosage imbalance becomes less necessary. The same trend is apparent as we examine class III (slope -72.53, $R^2 = 0.786$), although the scarcity of the data precludes a reliable statistical analysis. Class IV consists of only five families and hence no trend can be established, except that all families have zero miRNA target coincidence irrespective of their packing deficiency. This fact is clearly indicative of a pressing need to buffer dosage imbalances arising from duplicates that have not yet undergone sufficient functional differentiation.

The trends in terms of tighter $\nu-\tau$-anticorrelation and steeper slope as classes with lower K_s are considered (Fig. 6.3) implies that a miRNA-based capacitance to dosage imbalance is more operative for younger families (classes II–IV versus class I). This result is compatible with the fact that selection pressure on more recent paralogs has had comparably less time to promote adaptation through functional divergence and hence duplication-related dosage imbalances are more significant than those in older families.

These results reveal that the human capacitance to dosage imbalance is in part required due to the inefficiency of the selection pressure on duplicate genes, precluding sufficient differentiation over the evolutionary times of the latest speciations, thereby maintaining an evolutionarily related dosage imbalance.

The results of Fig. 6.3 imply that miRNA target dissimilarity across paralogs may be assimilated to a capacitance to dosage imbalance effects arising from gene duplication. The severity of such effects is in turn quantified by packing deficiency: Dosage imbalances are less tolerated for deficiently packed proteins forcing paralogs to be localized separately from each other.

6.5 Human Capacitance to Cope with the Impact of Dosage Imbalances

In this section we examined dosage imbalances that have an evolutionary origin. Thus, gene duplication events generate dosage imbalances that impose selection pressure on paralogs, and the magnitude of the effects of this pressure depend on the extent of under-wrapping or packing deficiency of the gene product. However, this dependence varies widely from unicellular to higher eukaryotes, with human being particularly insensitive to dosage imbalances. In human, there is a significant amount of genes with packing deficiency which are nevertheless extensively duplicated. This suggests that humans are resilient to evolution-related dosage imbalances, a capacitance that may be rationalized in terms of escape routes available to human but not to unicellular organisms, where dosage imbalances have clear deleterious effects.

In the absence of expression dissimilarity, the initially identical paralogs of deficiently packed proteins are subject to high selection pressure because they compete for binding partners needed to maintain structural integrity. Conversely, tight protein packing reduces dosage sensitivity, thereby curbing selection pressure. Cross examination of genetic and structural data reveals that humans have a built-in resilience or capacitance to dosage imbalances. The determinant of this human capacitance is traced in this chapter to the paralog-discriminatory power of miRNA regulatory patterns. In this way, dissimilar paralog localization governed by post-transcriptional regulation of protein levels mitigates the competition of paralogs for common binding partners that become obligatory for proteins of low packing quality. In other words, dissimilarity in paralog localization operative through miRNA control offers an escape route to dosage imbalances created by gene duplication, and this escape route becomes more necessary as protein packing deficiency makes these dosage imbalances less tolerable.

If selection is indeed inefficient in human as Lynch and coworkers proposed [3], one may wonder how miRNA-based capacitance could be achieved through random genetic drift. The removal of a miRNA binding site is readily achievable through a single deleterious mutation in one paralog and is unlikely to occur at the same binding site in another paralog. For instance, if M non-overlapping miRNA-binding sites are present in the 3'UTR of two paralogs ($1 << M < 156$), the probability that a pair of mutations (one in each paralog) will discriminate paralogs is $(M - 1)/M$, while the probability that they both occur at the same site in the two paralogs and hence do not contribute to increase the capacitance is $1/M$. Thus, dissimilarity and even orthogonality in post-transcriptional repressive patterns is readily achievable through widespread random mutation and hence may be the result of the nonadaptive forces prevalent in human.

6.6 Fitness Catastrophes for the Human Species Are the Consequence of Interactome Complexity

Human biological functions are exquisitely regulated and resilient to external variations, owing to complicated webs of interactions. Yet, we may be doomed as a species precisely because of the way in which our complexity arose [23]. Paraphrasing science writer Philip Ball, nature seems to have activated a time bomb, and our complexity is only a short-term fix [24, 25].

To grasp the nature of the problem, we need to examine orthologous proteins across vastly different species [23].

As previously said, it has been generally recognized that the basic "fold" of a protein is conserved across species, because there is a tight correspondence between structure and function. Proteins that retain the same function across different species—generally the case with orthologs—are expected to keep the same fold. But the sequence of amino acids that make up the protein chains in these orthologs can vary significantly. Sometimes the extent of sequence identity between two orthologs can be as low as 25 %, and yet their folds remain strikingly similar, attesting to the robustness of function to evolutionary change.

This conservation of the protein fold across species makes the origin of our complexity even more puzzling, as it is well known that the number of human genes is deceptively small, merely one order of magnitude larger than that of, say, rice. If the structure of the proteins is conserved across species, where is our complexity coming from? Better still, in what sense are we more complex?

We have discovered subtle structural variations occurring in orthologs from species that diverged from each other billions of years ago. In these structures, a feature subtler than overall topology changes across orthologs [23]. The structure in some seems "looser" than in others—less well wrapped, with surface regions that enable surrounding water to penetrate and disrupt the structure by interacting favorably with the protein backbone. These structural vulnerabilities known as *dehydrons* have been extensively studied in the preceding chapters.

As we examine orthologs, the proteins become more structurally degraded, or richer in dehydrons, in species with a lower effective population—a somewhat elusive indicator inversely related to the size and reproductive complexity of the organism (Fig. 6.1a–c). Thus, humans have significantly smaller (ten or more orders of magnitude) population than bacteria [23].

The observation that structural degradation is a reflection of decreasing species population (Fig. 6.1a–c) resonates in the field of evolution, because natural selection is known to be more inefficient as the population gets smaller [23–25]. Structural degradation is thus an indicator of the species' exposure to random genetic drift: mildly deleterious mutations that would typically degrade the protein structure are more likely to be selected against in bacteria before they can become fixed in the entire population (estimated in trillions of individuals), whereas such a mutation has a far better chance of prevailing in humans.

A protein that is richer in dehydrons than its ortholog is more vulnerable to becoming disrupted by surrounding water. Precisely for this reason, it becomes more reliant on binding partners to maintain its structural integrity. Furthermore, dehydrons are known to be sticky (Chaps. 1 and 2), so structurally degraded proteins are more likely to promote protein-protein associations than orthologs with a lower dehydron content. Thus, *protein-protein interactions, a hallmark of organismic complexity, are actually promoted by random genetic drift, the evolutionary force behind the protein degradation process.*

So, it seems, complexity is not really naturally selected, but instead arises as a short-term fix to the effects of selection inefficiency. At first reading, this assertion seems counterintuitive, but the root of the paradox is simply our dogmatic way of thinking, where complex traits are expected to be an outcome of natural selection.

And where is nature's gambit taking us? As shown in Chap. 5, the proteins with the largest accumulation of uncompensated structural defects are the prions, soluble proteins so poorly wrapped that they relinquish their functionally competent fold and form aberrant aggregates that may cause degenerative neuropathies.

This extreme case of an "aberrantly needy protein" illustrates the high level of genetic risk to which we are exposed as a species due to our small population. The prion is a "fitness" catastrophe that gives us clues as to where nature's gambit might take the human species. Perhaps the long-term evolutionary cost of our complexity is simply too high. Our survival as a species will ultimately depend on our ability to mitigate the fitness cost of our complexity through increasingly arduous therapeutic solutions.

Other fitness catastrophes seem to be in the horizon for the human species as dehydron enrichment caused by random genetic drift creates significant deviations from the architectural constraints required to maintain the structural integrity of soluble proteins (Chaps. 1 and 5). If we monitor over species within established orthologies [23] the deviation $\Delta(X, Y) = Y - (5X + 20)$ from the $Y = 5X + 20$ rule ($Y =$ number of dehydrons per 100 backbone hydrogen bonds; $X =$ number of disulfide bonds per 100 aminoacids) averaged over proteins for each species, we can see that the human species is ostensible more vulnerable to fitness catastrophes than species with significantly larger population (Fig. 6.4).

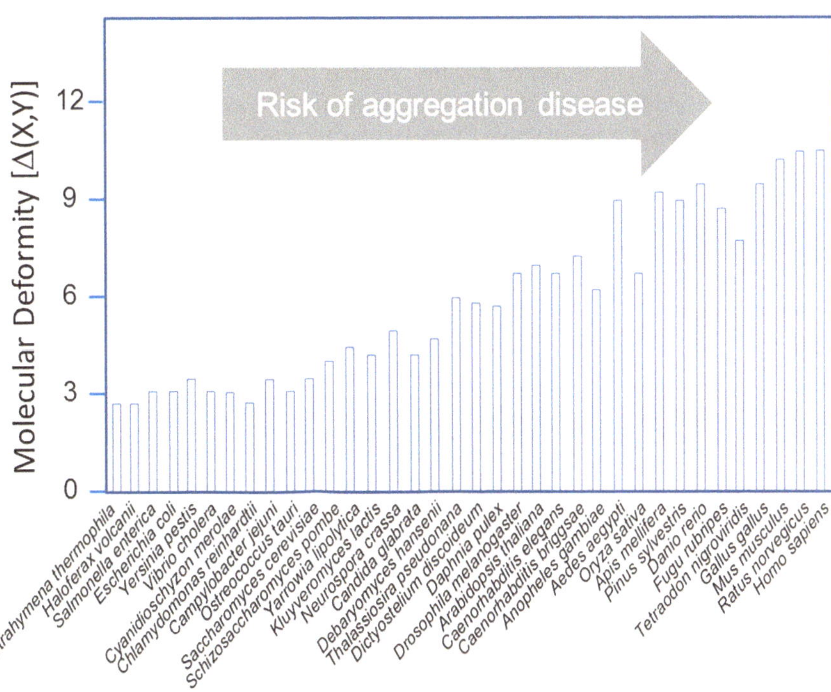

Fig. 6.4 Averaged deviation Δ(X, Y), named molecular deformity, for 36 species with established protein orthologies [23]

6.7 Molecular Evolutionary Insights for the Drug Designer: An Epistructural Perspective

As surveyed in this chapter, in spite of the uncanny structural similarity arising from common ancestry, both ortholog and paralog proteins exhibit vast differences in their wrapping or dehydron patterns. However, the root of these variations is very different depending on whether orthologs or paralogs are considered. As indicated in the previous sections, differences in selection efficiency across species appear to be responsible for wrapping differences across orthologs. On the other hand, in the case of paralogs, significant dosage imbalances would arise if paralogs were identical, simply because paralogs coexist in the same individual. To avert such imbalances and their fitness consequences, sufficient divergence of the original duplicate genes must take place through dissimilar expression regulation. Thus, in the previous sections we showed how paralog segregation is evolutionarily achieved.

An alternative evolutionary route to differentiate paralogs as needed to buffer the dosage imbalances they create is through dissimilar wrapping of their otherwise highly similar 3D-structures (Fig. 6.5). *As demonstrated in the subsequent chapters, this phenomenon has profound consequences for drug design because it enables designers to discriminate paralogs guided by their different dehydron patterns, exploiting the idea that a drug may function as an external wrapper of the packing defects in the target protein.*

Differences in wrapping patterns may be significant even for close paralogs with high levels (~88 %) of sequence identity, such as the kinase domains of the insulin receptor (IR) and that of the insulin-like growth factor receptor (IGF1R), as shown in Fig. 6.5. From the perspective of molecular targeted cancer therapy, there is a vested interest in engineering drugs able to discriminate between these two

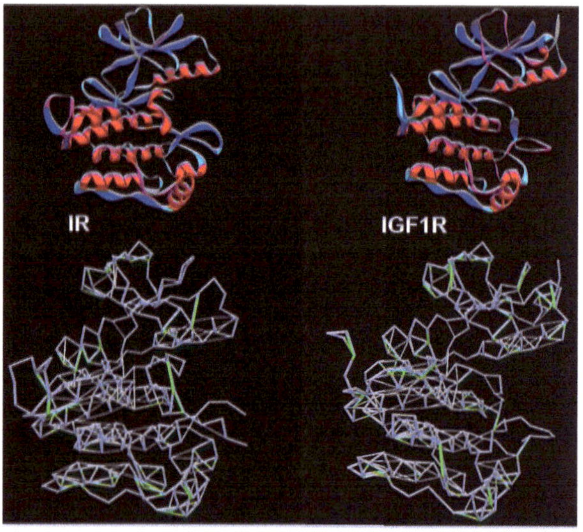

Fig. 6.5 Comparison of structures (ribbon representation) and wrapping patterns of the human paralog kinases for the IR (insulin receptor) and the IGF1R (insulin-like growth factor 1 receptor)

paralogs: IGF1R is probably a major target in the treatment of breast cancer [26], while inhibition of the IR kinase is likely to have devastating consequences, even the risk of inducing a diabetic coma in the patient. Yet, harnessing on the wrapping design concept, it should be possible to target differences in the wrapping patterns of both kinases (Fig. 6.4) to enhance therapeutic efficacy against breast cancer.

On the other hand, the observation that there are wrapping differences across orthologous proteins should also have a significant impact for the drug designer. *For example, the therapeutic efficacy of drugs is routinely tested in animal models through xenografts and other assays, always under the unwarranted assumption that the affinity and specificity of the drug for its target in human would be the same or very close to its affinity and specificity for the animal ortholog of the targeted protein. In light of the wrapping differences across orthologs, there is simply no assurance that this scenario is indeed valid. This observation could potentially undermine the very foundation on which animal testing is predicated.*

Yet, there is a positive implication of the observed wrapping differences across orthologs, and it relates to the drug-based treatment of infectious diseases such as AIDS (see Chap. 14). It is indeed highly advantageous for the drug designer to detect wrapping differences between the target protein in the infecting species and its human ortholog, if any: *A drug purposely designed to wrap specific dehydrons in the protein of the infecting species that are not present in the human ortholog is likely to be far less toxic than a drug that cannot tell apart the two orthologs.* Species selectivity is hardly a concern in an AIDS treatment geared at targeting the reverse transcriptase of the HIV-1 retrovirus, simply because this protein is not encoded in the human genome. On the other hand, a molecular design rooted in wrapping differences would be very helpful if the goal is to target a retroviral protein that is also represented in the human proteome. Species selectivity matters in this last context.

6.8 Dehydron-Based Non-adaptive Evolutionary Technology

Perhaps the most common mistake by those who misread Charles Darwin is the idea that evolution equates natural selection. By examining evolutionary patterns at the molecular level, mostly in the sequence of gene products, Motoo Kimura demonstrated that natural selection is only one of the intervening forces that shapes evolutionary outcome, and possibly not even the most important one [27]. While most scientists acknowledge this towering achievement, at some level they also distrust the iconoclastic nature of the proposal. Kimura's "Neutral Theory of Evolution" contains a scientific theory where the importance of random drift is clearly delineated as a major determinant of evolutionary change. In recent work described in the preceding sections of this chapter [2, 23], we have shown that random drift is likely to have a constructive role in promoting and enabling biological complexity at the molecular level. This complexity is cast in terms of the richness of associations or interactions of gene products, the hallmark of biological sophistication and subtlety.

This emerging picture built on Kimura's picture surely will have its detractors, if it does not have them already: Much evolutionary change arising from random drift is mildly deleterious, weakening the structure of gene products, thereby compromising their functionality. Yet, as shown in Chaps. 1 and 2, these structural weaknesses are inherently sticky, meaning that they promote or invite protein associations as the means to correct inter-molecularly the structural defects that arise in the free-standing molecule as a result of the degradation brought about by random drift [2, 23]. In other words, mildly degraded proteins, exposed to the vagaries of random drift, function within complexes and not as free-standing functional entities. These complexations enable allosteric relations and high regulation of protein function, hallmarks of sophistication and functional robustness rather than structural weakness.

Evolutionary technology (ET), that is, the directed control of test-tube evolutionary change to optimize biological function, is deeply rooted in Darwin's paradigm [28], or, rather, in the ET practitioners' own reading of Darwin's paradigm. ET largely discards Kimura's picture, perhaps because of the misguided perception that it would not lead to constructive improvement. While commendable, ETs have so far only offered results of marginal importance, mostly based on slight improvements of pre-existing biological functionalities and catalytic activity. ETs have not delivered the kind of dramatic functional innovation that arises from novel allostery, new cooperative and synergistic functions or novel functional regulation based on ephemeral protein complexation, the true hallmarks of organismic sophistication. By and large, we dare say that the achievements of ETs are not all that impressive. Nature continues to outdo us in this field, and by an ample margin. Real progress in ET will surely begin to emerge when scientists cease to voluntarily confuse evolution with natural selection.

The exciting new ET that can potentially emerge from the understanding of the constructive role of random drift at particular evolutionary junctures will require modulations of selection efficiency implemented at the test-tube level. Such modulations are necessary to open up the possibility of dramatic functional innovation of the kind that, paradoxically, random drift can bring about. Scientists are yet to incorporate selection *inefficiency* as a strategy to steer functional innovation in evolutionary technology.

6.9 Dehydron Richness Above Unicellular Levels Creates a Structural Dependence Exploited to Evolve Multicellularity

Clues on the evolution of multicellularity are emerging from comparative genomics and functional genetics studies of monophyletic unicellular/multicellular species pairs with recent divergence [29, 30]. Such studies have not revealed significant innovation at the genomic or protein domain level, but have identified a causative plexus in the co-option of unicellular functionalities into genes and pathways responsive to developmental cues [31, 32]. We argue that without proteomic innovation, this co-option could only materialize with a degree of functional cooperativity well above unicellular levels. What raw properties of a unicellular proteome are harnessed by such functional exaptations? This section addresses the question by examining the structural dependence of proteins (Q), a proxy for dehydron content (cf. Chap. 1), across species. As shown subsequently, this parameter differentiates unicellular and multicellular species by globally characterizing their proteome interdependence, and identifies the unicellular volvocine alga *Chlamydomonas reinhardtii* [31] as the sole outlier, with a level of structural dependence closer to that of a multicellular species. This result is significant since volvocine algae provide a model for the evolution of multicellularity [29, 33] and is revealing since structural dependence implies a lack of functional autonomy that enables co-opting without the need to innovate at the proteomic level.

The evolution of multicellularity poses a challenge to biologists [29, 33, 34]. Key insights are emerging from comparative studies of pairs of species that have diverged recently ($\approx 2 \times 10^8$ years ago) from a common ancestor, one of them being unicellular and the other, multicellular [29, 35]. Thus, the unicellular *Chlamydomonas reinhardtii* and the multicellular *Volvox carteri*, both *volvocine algae* in the *Chlorophyta* phylum, are uniquely positioned to provide clues on the origin of multicellularity [29, 30, 32, 35, 36]. Comparative genomics and functional genetics studies have not revealed significant innovation at the protein domain level ($\approx$14,500 gene products in both species) or significant differences in the protein-coding capacities [30]. Rather, the origin of multicellularity has been traced to the co-option of unicellular functionalities [31, 32, 35, 37–39]. This type of innovation is aptly illustrated by the functional adaptation of the unicellular gene pathway that represses cell growth in response to environmental stress. This function appears to have been co-opted into a pathway responsive to developmental cues and required for asymmetric cell division in the multicellular species [32].

To support this type of functional innovation, a high level of proteomic interdependence is required. Autonomous functions exerted by single protein units are not susceptible to be co-opted without significant innovation at the protein domain level, and there is no evidence favoring this type of innovation at the onset of multicellularity [29, 32]. Thus, two generic scenarios (I, II) for co-option can in principle be delineated (Fig. 6.6). Within scenario I, effector protein EFF originally recruits protein FUN1 to exert function 1 in the unicellular species and this pathway can be co-opted through proteomic innovation once another gene product FUN2 is expressed and gets competitively recruited by EFF to exert function 2. As said, extant evidence [29, 32] does not favor this scenario. By contrast, scenario II does not require innovation of coding capabilities but instead entails a higher degree of functional/structural interdependence. In this scenario, EFF recruits protein FUN to exert function 1 if FUN is cooperatively or allosterically activated in response to an environmental cue through interaction with the environmental receptor ENV. FUN may also be recruited by EFF in response to a developmental

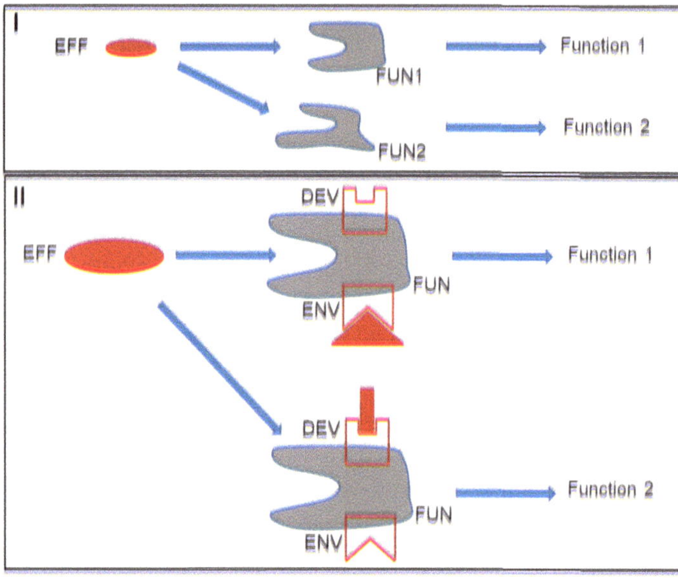

Fig. 6.6 Generic scenarios (I, II) for co-option of unicellular function 1 into multicellular function 2. Both functions implicate effector gene product EFF and are exerted once this protein is recruited into specific pathways. Within scenario I, EFF originally recruits FUN1 to exert function 1 and this pathway can be co-opted into function 2 through proteomic innovation once another gene product FUN2 is generated and competitively recruited by EFF. Within scenario II, EFF recruits FUN to exert function 1 if FUN is activated in response to an environmental cue (*red triangle*) through receptor ENV. FUN may also be recruited by EFF in response to a developmental cue (*red rectangle*) to exert function 2, since FUN is also coupled to receptor DEV. Thus, the higher levels of functional interdependence in scenario II enable functional cooption without proteomic innovation

cue to exert function 2, since FUN is also coupled to developmental receptor DEV. This duality in the recruitment capacity of FUN mediated by its binding partners ENV and DEV enables the cooption of the unicellular environment-responsive pathway into a developmental pathway without the need to innovate at the protein domain level.

In light of these observations, this section is devoted to globally characterizing a unicellular proteome endowed with a capacity to sustain functional exaptations without having to resort to innovation at the protein domain level. To tackle this problem we adopt a structural biology perspective and introduce the *structural dependence of a soluble protein* as a parameter q defined by $q = (h - h')/h$, where h is the structural deficiency (dehydron content) of the protein in its free (uncomplexed) state, and h' is the average structural deficiency of the protein within its PDB-reported complexes [26]. The parameter h is defined as $h = w/w_o$, where w is the number of dehydrons and w_o is the total number of backbone hydrogen bonds (BHBs) in the protein structure (Fig. 6.7) [40, 41]. Thus, q quantifies the structural protection gained by the soluble protein as it forms complexes relative to

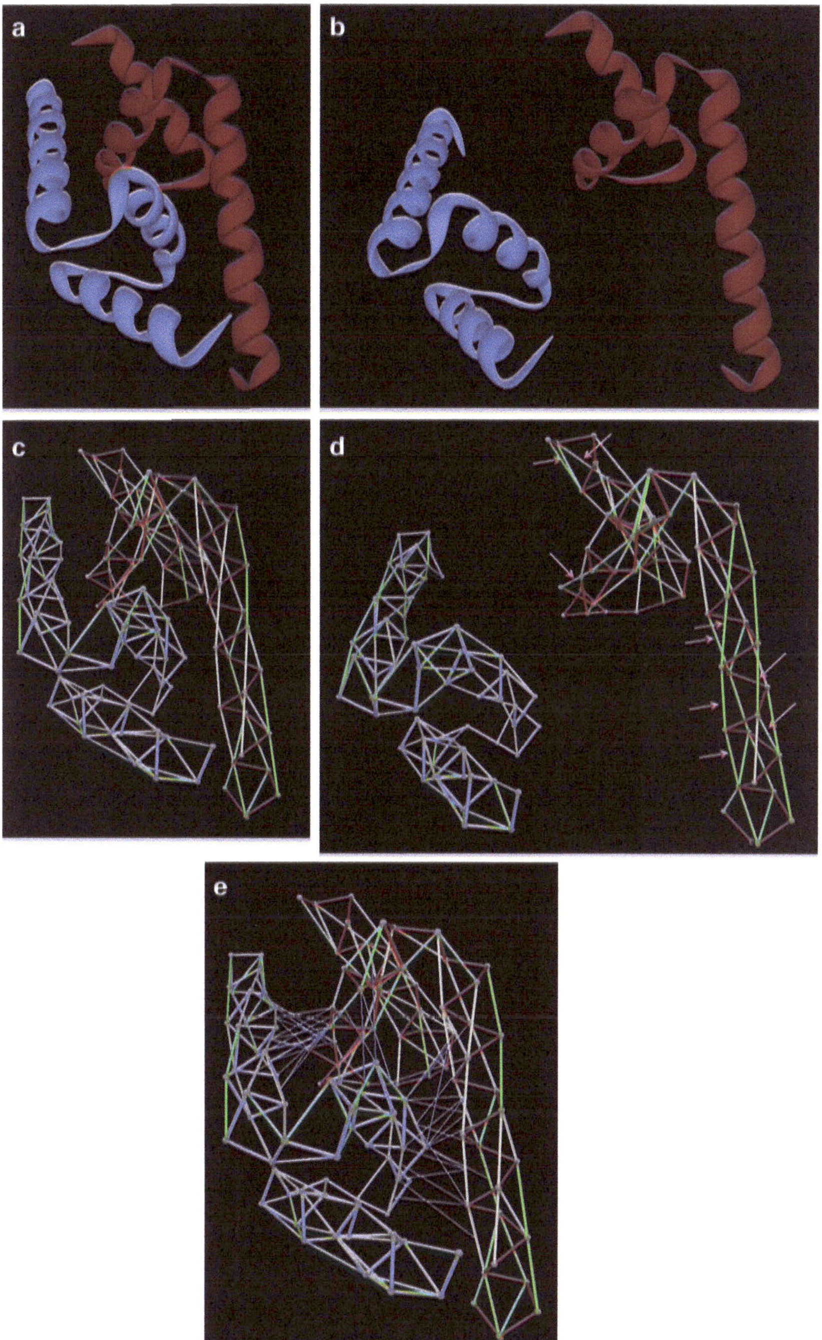

Fig. 6.7 Structural dependence of protein domains upon association illustrated by the dimerization of *doublesex* domain from *Drosophila melanogaster* (PDB.1ZV1). (**a**) Ribbon representation

its structural deficiency in the free state. The motivation for introducing q arises from the early observation that under-protected proteins ($h > 33\ \%$) are reliant on binding partnerships to maintain their structural integrity [26], but goes beyond that observation since the extent of structural dependence had not been quantified before.

A Q-versus-H plot of proteome-wide averages $Q = \langle q \rangle$ and $H = \langle h \rangle$ for the 38 species with adequate PDB representation (>30 non-homologous domains) distinguishes unicellular from multicellular species (Fig. 6.8). With the exception of *Chlamydomonas reinhardtii* (H = 41.4, Q = 25), unicellular species are contained in the sector $20\ \% < H < 41\ \%$, $5\ \% < Q < 20\ \%$, while multicellular species belong to the sector $41\ \% < H < 80\ \%$, $25\ \% < Q < 40\ \%$. The H-distribution is consistent with the degree of inefficiency of natural selection across species (prokaryotes < unicellular eukaryotes < invertebrates < vertebrates and land plants) [26]. Selection efficiency is commensurate with effective population size (highest in prokaryotes and archea, lowest in vertebrates) and is quantified by the probability that detrimental evolutionary change arising from random genetic drift become selected out from the species population. The trend for the H-values is expected since structural deficiency directly correlates with the probability of fixation of a mildly deleterious (structure-degrading) mutation in the species population [26]. On the other hand, the Q-distribution follows a monotonic trend towards increasing structural dependence with increasing structural deficiency, but the trend is not linear and its inflexion point at H = 41, Q = 23 is the closest point in the curve to the H, Q-coordinates of *Chlamydomonas reinhardtii*. The inflexion region signals a broad transition from unicellularity towards multicellularity with the unicellular species *Chlamydomonas reinhardtii* closely matching multicellular coordinates while lying outside the unicellular cluster.

Fig. 6.7 (continued) of dimer structure (individual chains are colored *red* and *blue*). (**b**) Ribbon representation of separated monomers. (**c**) Simplified representation of backbone and backbone hydrogen bonds (BHBs) of the protein chains within the complex. The backbone is depicted by virtual bonds joining α-carbons. The hydrogen bonds are shown as lines joining the α-carbons of the paired residues, with buried BHBs in *light grey* and dehydrons in *green*. We count $w = 22$ and $w_o = 50$ for the monomer within the dimeric complex, yielding $h' = 22/50$. (**d**) Structural deficiency of the monomers in the free state. Direct counting of dehydrons and BHBs gives $h = 31/50$. Nine dehydrons present in the free state (*pink arrows*) become protected upon association undergoing the wrapping transition $\zeta \leq 19 \rightarrow \zeta > 19$, and yielding $q = 9/31 = 0.29$ or 29.03 %. (**e**) Intermolecular protection within the dimer, describing the extent of structural dependence. Each chain confers protection to the BHBs formed by the other chain by contributing upon association with side-chain nonpolar groups to the desolvation domain of its binding partner. Such three-body interactions are shown as thin blue lines joining the α-carbon of the protective residue with the center of the BHB that is being intermolecularly protected

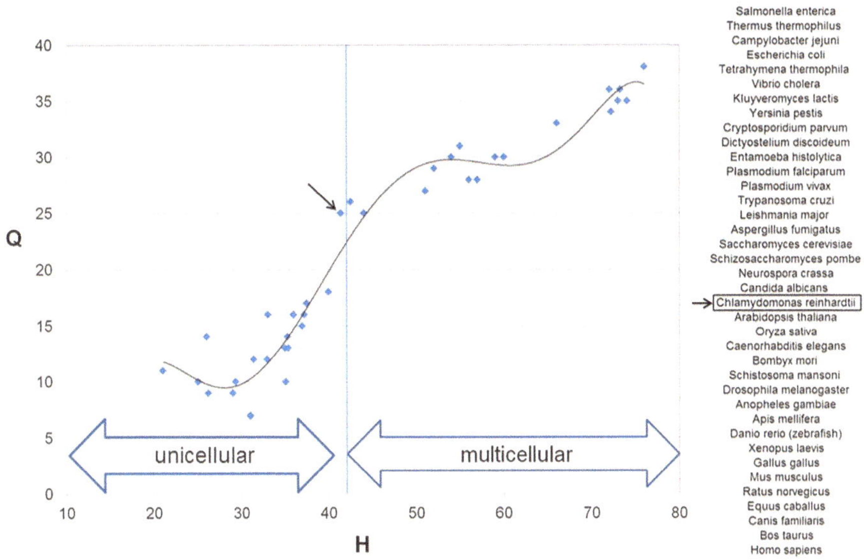

Fig. 6.8 Thirty-eight species with proteome PDB-representation (>30 structurally reported nonhomologous domains) distributed in the H-Q plane according to their respective H, Q values. The list to the right indicates the species ordered from top to bottom by increasing value of H. The fit with a sixth-order polynomial generates the lowest dispersion of datapoints relative to a trendline obtained by polynomial regression with correlation coefficient $R^2 = 0.95$. The structural information was extracted from the *wrapperome*, the universe of wrapping patterns for structurally reported proteins

The species distribution in the H-Q-plane suggests that higher levels of structural dependence for functionally competent complexes are required in multicellular species compared with unicellular species. Furthermore, the H-Q distribution identifies *Chlamydomonas reinhardtii* as having an anomalously high level of proteomic interdependence for a unicellular species. This singularity signals a striking lack of functional autonomy of protein domains when compared with other unicellular species. These unique properties turn the closest ancestor of *Chlamydomonas reinhardtii* into a suitable candidate for the co-optive hijacking of unicellular functions needed to fit the developmental demands of multicellular organisms. This process would not be supported by a proteome with a higher degree of structural independence, typical of most unicellular species (cf. Fig. 6.8). By singling out the *volvocine algae* taxon, our structural analysis upholds prevailing models for the evolution of multicellularity and encourages further studies on the structural basis of functional co-option.

6.10 Problems

6.1 As discussed, evolutionary technology, i.e. the implementation of directed evolution in the test-tube, may benefit in creative ways from conditions that promote selection inefficiency under controlled conditions for periods of time in the evolution reactor. Such inefficiencies promote the enrichment of the structure in dehydrons, which in turn, invite protein associations that would create allosteric conditions and higher-level regulation. Design a means of achieving and controlling such selection inefficiency in the test tube.

References

1. Fernández A, Scott R, Berry RS. The nonconserved wrapping of conserved folds reveals a trend towards increasing connectivity in proteomic networks. Proc Natl Acad Sci U S A. 2004;101:2823–7.
2. Fernández A, Berry RS. Molecular dimension explored in evolution to promote proteomic complexity. Proc Natl Acad Sci U S A. 2004;101:13460–5.
3. Lynch M, Conery JS. The origins of genome complexity. Science. 2003;302:1401–4.
4. Kondrashov FA, Koonin EV. A common framework for understanding the origin of genetic dominance and evolutionary fates of gene duplications. Trends Genet. 2004;20:287–90.
5. Liang H, Rogale-Plazonic K, Chen J, Li WH, Fernández A. Protein under-wrapping causes dosage sensitivity and decreases gene duplicability. PLoS Genet. 2008;4:e11.
6. Papp B, Pal C, Hurst LD. Dosage sensitivity and the evolution of gene families in yeast. Nature. 2003;424:194–7.
7. Fernández A, Scheraga H. Insufficiently dehydrated hydrogen bonds as determinants for protein interactions. Proc Natl Acad Sci U S A. 2003;100:113–8.
8. Bartel D. MicroRNAs: target recognition and regulatory functions. Cell. 2009;136:215–33.
9. Fernández A, Chen J. Human capacitance to dosage imbalance: coping with inefficient selection. Genome Res. 2009;19(12):2185–92.
10. Fernández A. Keeping dry and crossing membranes. Nat Biotechnol. 2004;22:1081–4.
11. Veitia RA. Exploring the etiology of haploinsufficiency. Bioessays. 2002;24:175–84.
12. Veitia RA. Gene dosage balance: deletions, duplications and dominance. Trends Genet. 2004;21:33–5.
13. Su AI, Wiltshire T, Batalov S, et al. A gene atlas of the mouse and human protein-encoding transcriptomes. Proc Natl Acad Sci U S A. 2004;101:6062–7.
14. Birney E, Andrews D, Caccamo M, et al. Ensembl 2006. Nucleic Acids Res. 2006;34: D556–61.
15. Yang Z, Nielsen R. Estimating synonymous and nonsynonymous substitution rates under realistic evolutionary models. Mol Biol Evol. 2000;17:32–43.
16. Friedman RC, Farth KK, Burge CB, Bartel DP. Most mammalian mRNAs are conserved targets of microRNAs. Genome Res. 2009;19:92–105.
17. Lewis B, Burge C, Bartel D. Conserved seed pairing, often flanked by adenosines, indicates that thousands of human genes are microRNA targets. Cell. 2005;120:15–20.
18. Grimson A, Farth KK, Johnston WK, et al. MicroRNA target specificity in mammals: determinants beyond seed pairing. Mol Cell. 2007;27:91–105.
19. Aloy P, Ceulemans H, Stark A, Russell RB. The relationship between sequence and interaction divergence in proteins. J Mol Biol. 2003;332:989–98.

20. Gu Z, Nicolae D, Lu HH, Li W-H. Rapid divergence in expression between duplicate genes inferred from microarray data. Trends Genet. 2002;18:609–13.
21. Chen F, Li W-H. Genomic divergences between humans and other hominoids and the effective population size of the common ancestor of humans and chimpanzees. Am J Hum Genet. 2001;68:444–56.
22. Gao L, Innan H. Very low gene duplication rate in the yeast genome. Science. 2004;306:1367–70.
23. Fernández A, Lynch M. Nonadaptive origins of interactome complexity. Nature. 2011;474:502–5.
24. Ball P. The Achilles' heel of biological complexity. Nature. 2011;474:502–5. doi:10.1038/news.2011.294.
25. Ball P. Why are you so complex? Complicated protein interactions evolved to stave off mutations. Scientific American. http://www.scientificamerican.com/article/complicated-protein-interactions-evolved-to-stave-off-mutations/. Accessed 18 May 2011.
26. Surmacz E, Bartucci M. Role of estrogen receptor alpha in modulating IGF-I receptor signaling and function in breast cancer. J Exp Clin Cancer Res. 2005;23:385–94.
27. Kimura M. The neutral theory of molecular evolution. Cambridge: Cambridge University Press; 2005.
28. Arnold FH, Meyerowitz JT. News and views: evolving with purpose. Nature. 2014;509:166–7.
29. Kirk D. Volvox: the molecular genetic origins of multicellularity and cellular differentiation. Cambridge: Cambridge University Press; 1998.
30. Prochnik SE, Umen J, Nedelcu AM, Hallmann A, Miller SM, et al. Genomic analysis of organismal complexity in the multicellular gree alga Volvox carteri. Science. 2010;329:223–8.
31. Kirk D. A twelve-step program for evolving multicellularity and a division of labor. Bioessays. 2005;27:299–310.
32. Miller SM, Kirk DL. GlsA, a *Volvox* gene required for asymmetric division and germ cell specification, encodes a chaperone-like protein. Development. 1999;126:649–58.
33. King N. The unicellular ancestry of animal development. Dev Cell. 2004;7:313–25.
34. Herron MD, Jeremiah D, Hackett JD, Aylward FO, Michod RE. Triassic origin and early radiation of multicellular volvocine algae. Proc Natl Acad Sci USA. 2009;106:3254–8.
35. Cheng Q, Fowler R, Tam LW, Edwards L, Miller SM. The role of GlsA in the evolution of asymmetric cell division in the green alga *Volvox carteri*. Dev Genes Evol. 2003;213:328–35.
36. Kirk DL, Nishii I. *Volvox carteri* as a model for studying the genetic and cytological control of morphogenesis. Dev Growth Differ. 2001;43:621–31.
37. Herron MD, Michod RE. Evolution of complexity in the volvocine algae: transitions in individuality through Darwin's eye. Evolution. 2008;62:436–51.
38. Nedelcu AM. Environmentally induced responses co-opted for reproductive altruism. Biol Lett. 2009;5:805–8.
39. Ueki N, Nishii I. *Idaten* is a new cold-inducible transposon of *Volvox carteri* that can be used for tagging developmentally important genes. Genetics. 2008;180:1343–53.
40. Pietrosemoli N, Crespo A, Fernández A. Dehydration propensity of order-disorder intermediate regions in soluble proteins. J Proteome Res. 2007;6:3519–26.
41. Fernández A, Berry RS. Golden rule for buttressing vulnerable soluble proteins. J Proteome Res. 2010;9:2643–8.

Chapter 7
Catalytic Role of Dehydrons in Soluble Proteins: Biological Chemistry of Frustrated Interfacial Water

This chapter unravels the catalytic role of dehydrons at biomolecular interfaces. It prompts a significant revision of the mechanisms of Biological Chemistry to encompass the chemical functionality of frustrated interfacial water. We first establish the chemical basicity of interfacial water that is frustrated in its hydrogen-bonding opportunities due to partial confinement at sub-nanoscale structural defects (*dehydrons*). Through the functionalization of vicinal water, the dehydron is shown to be an enabler and promoter of enzymatic activity. Through multiple steering molecular dynamics with a pulling coordinate spanning the proton-transference trajectory, we show that the vast majority of transesterification reactions, ubiquitous in Biological Chemistry, are actually enabled by nearby dehydrons that deprotonate the pro-nucleophiles that may potentially intervene in such reactions. The theoretical results are validated against experimentally measured pKa decreases at functional sites and by experimental corroboration of the aberrant deregulation of catalytic activity arising from dehydron-creating mutations.

The proton-acceptor role of dehydrons, or rather, of interfacial water enveloping dehydrons, is likely to revolutionize the mechanistic aspects of Biological Chemistry. Most if not all transesterification reactions requiring the activation of a nucleophilic group will need to be rewritten to incorporate the catalytic enablement triggered by nearby dehydrons. A new quasi-reactant ubiquitous in Biological Chemistry has been discovered.

The ramifications of this discovery within the field of Biological Chemistry are virtually unfathomable, since a new frustration-based chemistry has been invented, and the aqueous interface is now shown to play an active role in biochemical reactions. For instance, the general basicity of frustrated water enables it to titrate side-chain ammonium and guanidinium cations. Frustration-based chemistry is captured by a quantum mechanical treatment of proton transference and shown to remove same-charge uncompensated anticontacts at the interface found in the crystallographic record and in other spectroscopic information on the aqueous interface. Such observations are untenable within classical arguments since

A. Fernández, *Physics at the Biomolecular Interface*, Soft and Biological Matter,
DOI 10.1007/978-3-319-30852-4_7

hydronium is a stronger acid than ammonium or guanidinium. Furthermore, frustration enables a directed Grotthuss mechanism for proton transference stabilizing same-charge anticontacts. With these examples we portend to show that a window of opportunities to explore novel chemical phenomenology of interfacial water is now open. Here is a summary of the findings described in this chapter:

- We physically characterize frustrated water molecules at the protein interface.
- We show that frustrated water molecules exhibit a chemistry that prompts the revision of mechanistic manuals for Biological Chemistry.
- Frustrated water is shown to enable a directed intramolecular Grotthuss mechanism for proton transference.

7.1 Catalytic Dehydrons

The dielectric behavior of *episteric* ("around the solid") water in soluble proteins and in other chemically complex solutes endowed with nanometer-level detail is a subject of intense scrutiny [1–4]. The understanding and manipulation of such interfaces require a multiscale approach to water dielectrics, an unsettled matter at this time. As recently argued, such an approach demands a conceptual departure from the so-called "Debye *ansatz*" [1] intensely advocated by Warshel and coworkers [5] in their extrapolation from macroscopic behavior. The Debye ansatz has been incorrectly taken to mean that water polarization $\vec{P}$ at the protein-water interface must align with the internal electrostatic field $\vec{E}$ of the solute: $\vec{P} = \vec{P^{\parallel}}$, where $\vec{P^{\parallel}}$ is the Debye polarization component. In contrast with the bulk hydrogen-bond pattern, water molecules under nanoscale confinement are forced to relinquish hydrogen bonding opportunities to fit in the nanometer cavity, generating a net polarization component $\vec{P^{\#}}$ orthogonal and statistically independent of $\vec{E}$. While the Debye polarization introduces the well-known rescaling (screening) of the inherent solute charge [5], a consequence of the resilience of water structure to variations in the electrostatic field, the orthogonal polarization component induces an $\vec{E}$-independent net charge $\gamma^{\#} = -\vec{\nabla} \cdot \vec{P^{\#}}$ and is responsible for the dielectric anisotropy of the interface (Chap. 2, problem 2.4). This charge contribution is of paramount importance in assessing chemical functionalities related to topographical features of protein interfaces as it generates charges that are not accounted for by the protein chemical composition. As shown in this chapter, this net induced charge endows nano-confined water with a chemical functionality in accord with the sign of $\gamma^{\#}$, i.e. proton donor if $\gamma^{\#} > 0$ and proton acceptor if $\gamma^{\#} < 0$. Thus, the inequality $\gamma^{\#} > 0$ indicates that at least one water hydroxyl proton is on average forced by the interface geometry to relinquish its hydrogen-bonding capability, while $\gamma^{\#} < 0$ implies that at least one water oxygen atom remains on average unpaired to a neighboring hydroxyl proton (its orbital-localized lone electron pair is unutilized).

Building on the non-Debye treatment of the water interface, we identify the episteric chemical functionalities arising from nanoscale structural defects in soluble proteins, establishing their role as promoters or enhancers of catalytic activity [6]. To identify the chemical role of interfacial water subject to partial nanoscale confinement, we resort to the non-Debye treatment described in Chaps. 1 and 2.

Within this framework, the frustration-related interfacial energy, ΔU_ϕ, associated with spanning the solute-water interface is given by the elastic integral

$$\Delta U_\phi = \int \frac{1}{2}\lambda \|\vec{\nabla}\phi\|^2 d\vec{r} \tag{7.1}$$

where the integration is carried over a spatial domain Ω large enough so that its border $\partial\Omega$ is fully contained in bulk water. Thus, Ω is subject to the boundary condition

$$\{\phi(\vec{r}) = 0,\ \nabla\phi(\vec{r}) = 0\}\ \forall \vec{r} \in \partial\Omega \tag{7.2}$$

Integration of (7.1) by parts yields

$$0 \leq \Delta U_\phi = -\frac{1}{2}\lambda \int \phi \nabla^2 \phi d\vec{r} = -\frac{1}{2}(\lambda/e_0)^{1/2} \int \gamma^{\#} \phi d\vec{r} \tag{7.3}$$

since $\gamma^{\#} \propto \nabla^2\phi$ (cf. Chaps. 1 and 2) and

$$\oiint \phi \vec{\nabla}\phi \cdot d\vec{\sigma} = 0 \text{ on } \partial\Omega \tag{7.4}$$

Since $\phi(\vec{r}) = 4 - g(\vec{r}) \geq 0$, the mean value theorem of integral calculus applied to (7.3) yields:

$$\gamma^{\#}_{MV} = -\frac{(\lambda e_0)^{\frac{1}{2}} \int \|\vec{\nabla}\phi\|^2 d\vec{r}}{\int \phi d\vec{r}} \leq 0 \tag{7.5}$$

where $\gamma^{\#}_{MV}$ is the mean-value constant yielding:

$$\gamma^{\#}_{MV} \int \phi d\vec{r} = \int \gamma^{\#} \phi d\vec{r}. \tag{7.6}$$

We have then proven the following <u>*Theorem 1*</u>: *Interfacial water under local nanoscale confinement with $\phi \geq 0$ yields a non-Debye polarization-induced charge $\gamma^{\#}{}_{MV} \leq 0$. The net negative charge begets a proton-acceptor role compatible with unfulfilled hydrogen-bond coordination of water oxygen (where partial negative charge is located), turning the local interfacial water into a chemical base.*

Theorem 1 may appear counterintuitive at first. Why should there be asymmetry in the sign of the induced charge? Yet, the result is rigorous and may be derived

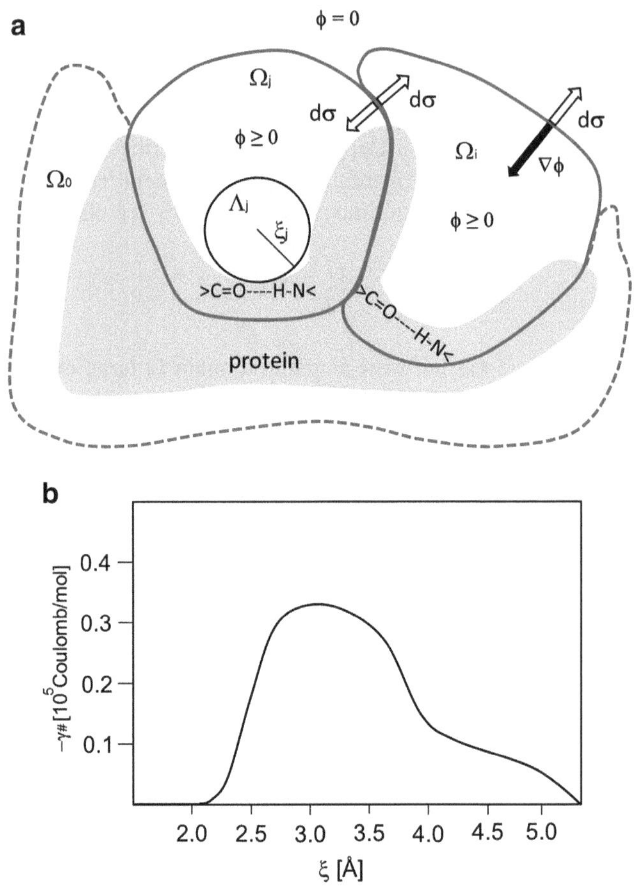

Fig. 7.1 (**a**) Schematic representation of the partition of domain Ω into subdomains Ω_n n = 0, 1, 2, . . . , M, where a single subdomain with n > 0 is required to contain each $(\phi > 0)$-dehydron in the protein structure and $\Omega_0 = \overline{\Omega/\cup_{n=1}^{M}\Omega_n}$ where the over-bar denotes closure. The generic subdomain Ω_j contains the amide-carbonyl dehydron *j* and the first-order contact (osculating) sphere of the water-smeared envelope of the protein surface at dehydron *j*. (**b**) Net non-Debye polarization-induced charge $\gamma^{\#}$ as a function of the local curvature radius at a dehydron site. The charge units for $\gamma^{\#}$ are scaled by the proton charge 1.6×10^{-19} Coul or ~10^5 Coul/mol. Reprinted from [6], copyright 2014 with permission from AIP Publishing LLC

alternatively from intuitive spatial models of the favorable presentation of a water molecule within a dehydron cavity (cf. Fig. 7.8). As noted in Chaps. 1 and 2, the interfacial tension is generated by dehydrons. While the physico-chemical properties of dehydrons have been characterized in previous work [1, 3] their chemical role has not been assessed so far. To determine the chemical functionality of *individual* dehydrons in the protein structure we consider a covering of the spatial domain Ω made up of M closed convex subsets Ω_j j = 0, 1, 2, . . . , M fulfilling the following conditions (cf. Fig. 7.1a):

1. $\Omega = \cup_{j=1}^{M} \Omega_j$
2. For $i \neq j$, $\Omega_i \cap \Omega_j = \partial\Omega_i \cap \partial\Omega_j$ (closed subsets only overlap along their borders).
3. Let the dummy index j = 1, . . . , M label the dehydrons in the protein structure, then the subset Ω_j j = 0, 1, 2, . . . , M contains dehydron *j* and also satisfies $\Lambda_j = \Lambda_j\left(\xi_j\right) \subseteq \Omega_j$, where the set Λ_j is the osculating (first-order contact) sphere of the water-smeared envelope of the protein surface at dehydron *j* with local curvature ξ_j [7].
4. $\forall j = 0, 1, \ldots; \forall \vec{r} \in \Omega_j$ one of the two conditions holds:

4a. $\left\{\phi(\vec{r}) = 0,\ \vec{\nabla}\phi(\vec{r}) = 0\right\}$

4b. if $\left\{\phi(\vec{r}) \neq 0\ or\ \vec{\nabla}\phi(\vec{r}) \neq 0\right\}$, $\exists\Omega_j : \vec{r} \epsilon \partial\Omega_i \cap \partial\Omega_j$, and the respective differential area normal vectors $d\vec{\sigma}(\vec{r})$ for $\vec{r}$ on the two subsets cancel each other out (cf. Fig. 7.1a).

Given conditions 1–4, the vanishing integral $\oiint \phi \vec{\nabla}\phi \cdot d\vec{\sigma} = 0$ over $\partial\Omega$ may be written as

$$0 = \oiint \phi \vec{\nabla}\phi \cdot d\vec{\sigma} = \sum_{j=1}^{M} \oiint \phi \vec{\nabla}\phi \cdot d\vec{\sigma}_j \tag{7.7}$$

where $\oiint \phi \vec{\nabla}\phi \cdot d\vec{\sigma}_j$ denotes the surface integral extended over $\partial\Omega_j$. Thus, we obtain:

$$0 \leq \Delta U_\phi = \frac{1}{2}\sum_{j=1}^{M} \left\{\lambda \oiint \phi \vec{\nabla}\phi \cdot d\vec{\sigma}_j - \left(\frac{\lambda}{\varepsilon 0}\right)^{\frac{1}{2}} \gamma_{MV}^{\#}(j) \int \phi d\vec{r}_j\right\} \tag{7.8}$$

where the integral $\int \phi d\vec{r}_j \geq 0$ extends over Ω_j and the mean value constant $\gamma_{MV}^{\#}(j)$ satisfies the relation: $\gamma_{MV}^{\#}(j) \int \phi d\vec{r}_j = \int \gamma^{\#} \phi d\vec{r}_j$. Thus, the individual contribution of dehydron *j* to the protein interfacial tension is

$$0 \leq \frac{1}{2} \oiint \phi \vec{\nabla}\phi \cdot d\vec{\sigma}_j - \frac{1}{2}\left(\frac{\lambda}{\varepsilon 0}\right)^{\frac{1}{2}} \gamma_{MV}^{\#}(j) \int \phi d\vec{r}_j \tag{7.9}$$

Since either $\vec{\nabla}\phi(\vec{r}) = 0$ or the vectors $\vec{\nabla}\phi(\vec{r})$ and $d\vec{\sigma}_j(\vec{r})$ point in opposite directions for $\vec{r} \in \partial\Omega_j$ (cf. Fig. 7.1a), the following inequality holds for all j's:

$$\frac{1}{2} \oiint \phi \vec{\nabla}\phi \cdot d\vec{\sigma}_j \leq 0 \tag{7.10}$$

Combining (7.9) and (7.10), it follows that $\gamma_{MV}^{\#}(j) \leq 0$. Thus we have proven the following *Theorem 2*:

As a generator of interfacial tension, a dehydron yields a negative non-Debye polarization-induced charge. Thus, interfacial water enveloping a dehydron constitutes a chemical base. This catalytic role of dehydrons is amenable of an intuitive corroboration that relates to the way a water molecule positions itself within the dehydron cavity (cf. Fig. 7.8).

The proton-acceptor role of dehydrons, or rather, of interfacial water enveloping dehydrons, will probably revolutionize our understanding of biochemical mechanism. It is likely that many if not all intermolecular transesterification reactions in biochemistry requiring the activation of a nucleophilic group will need to be rewritten to incorporate the catalytic enablement provided by nearby dehydrons. In plain terms, *dehydrons have been the missing reactants in the chemical equations of biological catalysis.*

From (7.4) it follows that the net contribution to interfacial energy from dehydron j is simply:

$$\Delta U_\phi(j) = \int \frac{1}{2}\lambda \|\vec{\nabla}\phi\|^2 d\vec{r}_j = -\left(\frac{\lambda}{\varepsilon_0}\right)^{\frac{1}{2}} \gamma^{\#}_{MV}(j) \int \phi d\vec{r}_j \tag{7.11}$$

yielding the net non-Debye polarization-induced charge

$$\gamma^{\#}_{MV}(j) = -\frac{(\lambda\varepsilon_0)^{\frac{1}{2}} \int \|\vec{\nabla}\phi\|^2 d\vec{r}_j}{\int \phi d\vec{r}_j} \tag{7.12}$$

The disruption of bulk water structure quantified by the field $\phi(\vec{r})$ is computed using the molecular dynamics protocol previously described for dehydron cavities of variable curvature radius 1–6 Å [7]. The results are integrated according to (7.12) to yield a net non-Debye polarization-induced charge $\gamma^{\#}_{MV}(j)$ shown in Fig. 7.1b.

The basicity of dehydrons is in consonance with their abundance near catalytically active side chains involved in nucleophilic attacks in enzymatic reactions, as hereby shown. We may assert that the concentration of $(\phi > 0)$-dehydrons in the vicinity of side chains involved in intermolecular transesterification attests to their proton-acceptor role as promoters of active-site nucleophilicity. Dehydrons vicinal to the catalytic site stabilize the polarized deprotonated state that empowers nucleophilic activity. The paramount importance of this finding will likely invite a major revisiting of biochemical mechanism, as we have already emphasized.

7.2 Dehydrons as Quasi-Reactants in Biological Chemistry

To illustrate the chemical functionality of dehydrons we first examine the dehydron distribution of the aspartic (Asp) protease from HIV-1 virus (PDB.4DJP) [8], a functionally competent homodimer. Each monomer within the complex has two dehydrons located in the vicinity of the active site (Asp25) and at the highly flexible

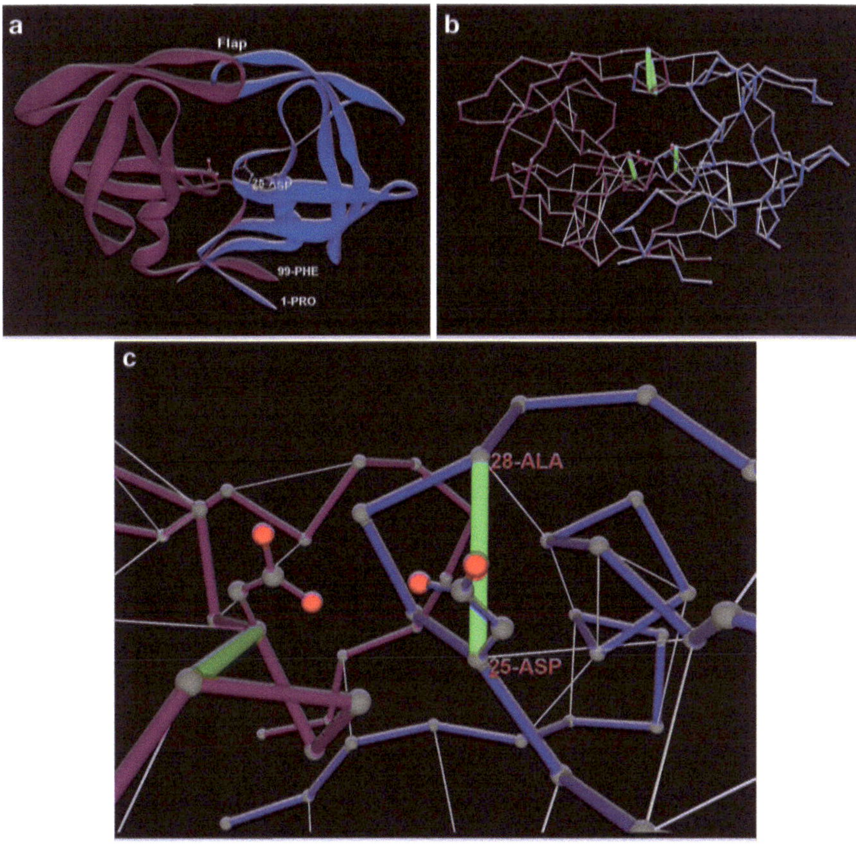

Fig. 7.2 (**a**) Ribbon representation of the functional homodimeric HIV-1 protease (PDB.4DJP), with active site Asp25 with displayed side chain. The monomeric chains in the complex are depicted in *magenta* and *blue*, respectively. (**b**) Dehydron distribution in HIV-protease. Dehydrons are shown as *green* segments joining the α-carbons of the paired residues, while solvent-shielded (well wrapped) backbone hydrogen bonds are shown in *grey*. (**c**) The ($\phi > 0$)-dehydron Asp25-Ala28 flanking the active site Asp25 in the dimeric HIV-1 protease. Reprinted from [6], copyright 2014 with permission from AIP Publishing LLC

flap (Fig. 7.2), with a single ($\phi > 0$)-dehydron involving the pair Asp25-Ala28. "Vicinal" is hereby defined as being within 6 Å of the α-carbon (desolvation domain of a dehydron-paired residue) of the catalytically active residue. Examination of an exhaustive nonredundant set of 198 Asp proteases with PDB representation at resolution better than 1.5 Å and relational Uniprot [9] sequence annotation reveals the same localization pattern of ($\phi > 0$)-dehydrons found exclusively in the vicinity of the active site.

A similar localized concentration of ($\phi > 0$)-dehydrons is identified at phosphorylation sites that involve the participation of a nucleophilic group (usually hydroxyl and less frequently histidine amide) from a side chain (Ser, Thr, Tyr, His) in a trans-phosphoesterification of the ligand ATP [10]. The reaction promotes the cleavage of the terminal (gamma) phosphoryl group from ATP and its covalent attachment to the protein through a phospho-ester linkage. The net non-Debye polarization-induced charge at such sites is significantly larger in magnitude than that for a control residue of the same type but not susceptible to phosphorylation (Fig. 7.3a, b). This assertion has been validated by examining phosphorylation sites

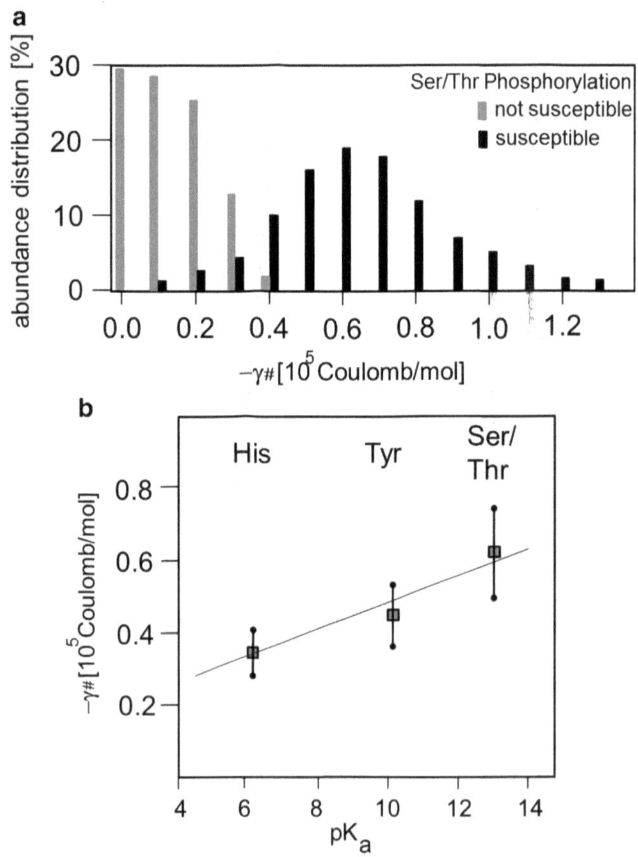

Fig. 7.3 (**a**) Abundance distribution of phosphorylation-susceptible Ser/Thr residues according to their vicinal non-Debye polarization-induced charge $\gamma^{\#}$. The $\gamma^{\#}$-distribution for phosphorylation-impervious Ser/Thr residues is displayed as control. (**b**) Dispersion ranges (error bars) of $\gamma^{\#}$-values as a function of bulk pK_a of the phosphorylation-susceptible side-chain group. (**c**) Chemical reaction of trans-phosphoesterification involving a nucleophilic attack on the terminal phosphoester linkage of ATP by a dehydron-functionalized Tyr (the phosphorylation-susceptible residue). The nucleophilicity of the Tyr phenolic hydroxyl is significantly enhanced through polarization induced by the proton-acceptor water molecule that envelops the vicinal dehydron with $\gamma^{\#} < 0$. (**d**) Cluster of five $(\phi > 0)$-dehydrons (maximum concentration) vicinal to the catalytic residue Ser195 in bovine trypsin (PDB.4I8G). The structure is rendered following the convention in Fig. 7.2 with backbone in magenta. The chemical basicity of these structural defects enables the proton relay mechanism from Ser to His in the Ser-His-Asp catalytic triad of this enzyme, which would otherwise be impaired due to the significant gap in pK_a value (~7 log units) between adjacent side chains Ser and His. Reprinted from [6], copyright 2014 with permission from AIP Publishing LLC

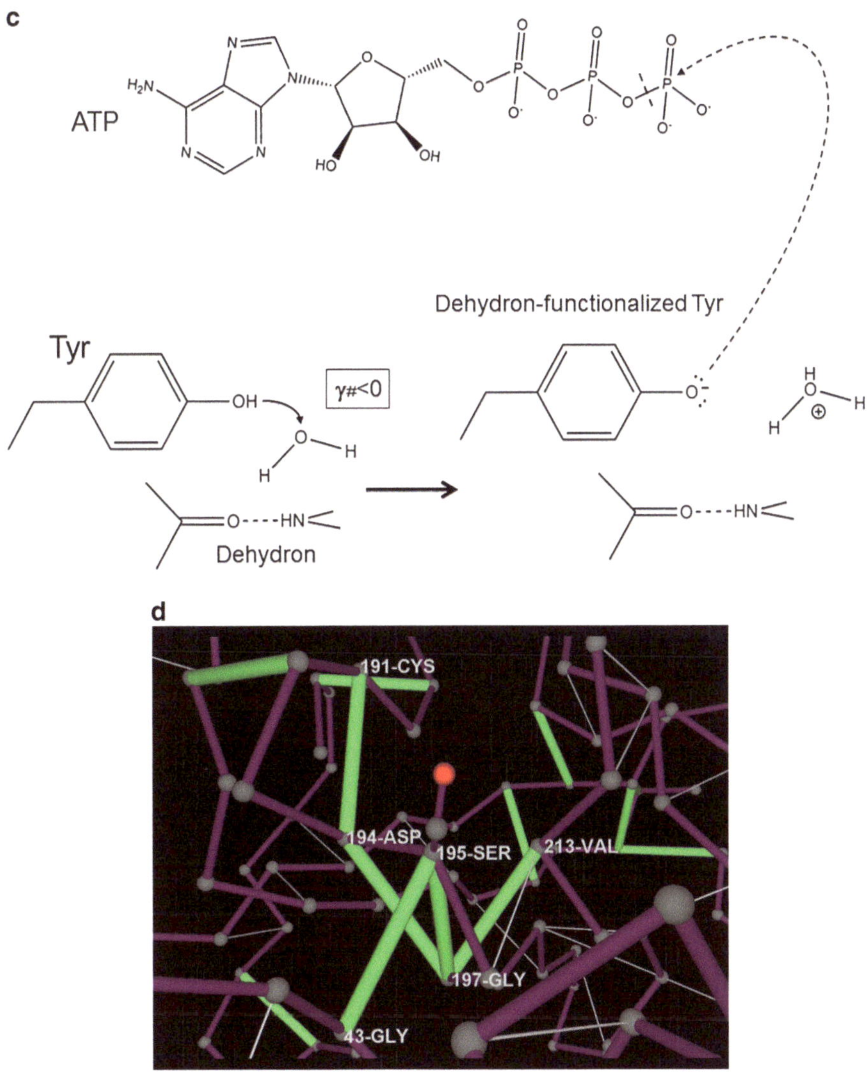

Fig. 7.3 (continued)

with reported local structure [11] in exhaustive nonredundant sets of PDB-reported kinases at resolution better than 1.5 Å with relevant sequence annotation on phosphorylation sites obtained from Uniprot. Thus, we examined 507 Ser/Thr kinases, 214 Tyr kinases and 19 His kinases. The tight correlation ($R^2 = 0.82$) between $\gamma^{\#}$ and bulk pK_a of the phosphorylation-susceptible residue is indicative of the de-protonation requirements of the side-chain group to enhance its nucleophilicity and thereby its susceptibility to phosphorylation. The basicity of the vicinal ($\phi > 0$)-dehydrons as measured by net non-Debye polarization-induced charge functionalizes the residue group implicated in the nucleophilic attack on the ATP terminal phospho-ester linkage (Fig. 7.3c).

Due to the extremely high pK_a (≈13), the functionalization of Ser/Thr as nucleophile requires a very large vicinal $\gamma^{\#}$ -value, which in turn requires a substantial dehydron concentration. This situation is illustrated in the active Ser195 of the serine protease trypsin (PDB.4I8G [12]) that contains five vicinal dehydrons (Fig. 7.3d), the maximum concentration found in PDB [13]. Since dehydron clusters are unstable and expose the backbone to hydration, they become disruptors of protein structure and so an extreme concentration (>5) of vicinal dehydrons yields a natively disordered region [13], as shown in Chap. 5. Such disordered regions are often found around nucleophilically functional and extreme base-demanding Ser/Thr and may be regarded as "ephemerally basic" since they visit conformations so rich in dehydrons that can only be temporarily sustained in water [13]. Yet, while the dehydrons are present in a transient conformation they act as proton acceptors and functionalize the Ser/Thr by stabilizing the polarized state, switching on their nucleophilic nature.

In the case of serine proteases [14], the high concentration of serine-vicinal ($\phi > 0$) -dehydrons becomes adjuvant to the inherent Ser-His-Asp proton-relay mechanism of such enzymes, facilitating proton transfer from serine to the nearby histidine. It should be noted that such transference is unlikely to occur without the assistance of vicinal basic dehydrons due to the large gap in pK_a (approximately 7 log units) between serine and the nearby histidine in the catalytic triad.

This section reveals the chemical functionality of interfacial water enveloping nanoscale structural defects in soluble proteins or other nano-materials and asserts its role as a quasi-reactant in biochemical reactions. Many such reactions require the activation of protein groups that perform or promote a nucleophilic attack leading to transesterification. *The typical biochemistry textbook would indicate a deprotonation reaction where a side-chain proton is transferred to a nearby water molecule, but this mechanism is severely challenged by the fact that hydronium is usually a stronger acid than the side-chain group that actually gets deprotonated.* This section shows that interfacial water enveloping a dehydron under nanoscale confinement acts as a chemical-base effector, enhancing the nucleophilicity of the adjacent active site. The concept of functionalized episteric water is introduced in this chapter and the results invite a substantive revision of the purported elementary steps in biochemical reactions. On the other hand, novel biomolecular engineering is also likely to emerge from the physico-chemical foundations delineated (cf. Chaps. 8–15), as dehydron-based enzymatic effectors may be created or removed though site-directed mutation altering side-chain packing.

7.3 The Dehydron as a Catalytic Engine

The biochemical role of dehydrons as promoters of basicity mechanistically complements their dehydration propensity [1, 3, 7] in the context of enzymatic reactions. Thus, water enveloping a dehydron becomes a better leaving molecule (hydronium seeking full hydration) as it functionalizes the nucleophilic moiety of the enzyme, while the dehydration propensity of the packing defect induces the

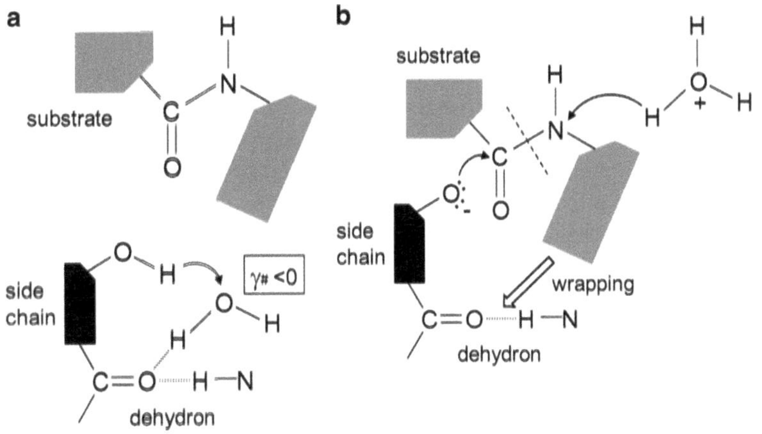

Fig. 7.4 The dehydron as a two-step catalytic engine. (**a**) The dehydron functionalizes the nucleophilic group by inducing basicity in vicinal water. (**b**) The dehydron promotes substrate binding and concurrent hydronium expulsion to enable intermolecular trans-esterification

expulsion of the hydronium as it promotes the binding of the substrate to the enzyme. This migration of the dehydron-enveloping hydronium is entropically favored due to a gain in translational and conformational freedom as the hydronium is transferred to the bulk, and is also enthalpically favored, as the transference enables the fulfillment of the hydration demands of the hydronium. Thus, the thermodynamic cost of transferring the proton from the pro-catalytic group to the dehydron-functionalized water molecule is defrayed by the subsequent stabilization of the dehydron that results from its wrapping or shielding upon substrate-protein association and by the free-energy gain associated with the transferring of confined ionized water to the bulk region. This established role of the dehydron as a two-step catalytic engine (Fig. 7.4) will surely prompt further research effort, as its impact in biotechnology is predicted to be of vast proportions.

7.4 Quantum Mechanics of Proton Transfer Events Involving Dehydrons

The following sections explore the participation of dehydrons in biochemical events through quantum mechanical (QM) calculations (for further computational details see Appendix A, Sect. A1.2). We delineate the enabling role of dehydrons as activators of nucleophilic groups. As previously shown, this activation results from the induction of chemical basicity in interfacial water molecules, an event that triggers deprotonation of adjacent nucleophiles. Through multiple steering molecular dynamics with pulling along the proton-displacement coordinate, we now show that nucleophilic groups are functionally enabled by nearby dehydrons that promote proton transference. The computations are validated against experimentally determined pKa decreases at functional sites and biochemical probes of deregulated catalytic activity arising from dehydron-generating mutations.

Dehydrons are markers for protein associations [15] and have been conjectured in the previous sections to induce chemical reactivity in interfacial water [16]. These two properties are interrelated, and a chain of reasoning starting by assuming the validity of one property takes us to assert the validity of the other, as rigorously shown in Sect. 7.1. Thus, dehydrons are known introduce structure-destabilizing nanoscale cavities on the protein surface. This interfacial tension is released upon protein associations that in effect displace the restricted interfacial water, turning dehydrons into promoters of protein interactions [17]. A simple computation of the interfacial tension leads us to the conclusion that water molecules enveloping dehydrons may act effectively as proton acceptors, as shown in Sect. 7.1. This chemical behavior arises from a non-Debye polarization-induced negative charge arising as nanoscale confinement hampers the alignment of water dipoles with the electric field.

In the forthcoming sections we unravel the mechanism of functional enablement promoted by dehydrons in their role as inducers of chemical basicity of the aqueous interface. To that effect, we investigate the chemical event of proton transference prompted by water molecules at dehydron interfaces. Specifically, we compute the shifts towards lower values in pKa [18] of groups functionalized through dehydron-promoted deprotonation. To assess this activity we need to consider chemically active dehydrons in the proximity of the functional site, i.e. those for which the water oxygen in the dehydron cavity is within 6 Å of the α-carbon of the functional residue. The method of choice to investigate the chemical event of dehydron-induced proton transference is multiple steering molecular dynamics computation [19]. The nucleophilic group and nearby proton-receptive water molecules at the dehydron interface are treated within a quantum mechanical (QM) scheme while the rest of the molecule and explicit solvent are treated using a classical molecular mechanics (MM) package, in accord with a QM-MM hybrid approach [20, 21]. The results are validated against experimentally determined pKa shifts [18] and functional studies of constitutively active mutant enzymes [22] whose aberrant deregulation is shown to arise from the creation of dehydrons not present in the wild type.

The study cases are selected so that the dehydron-promoted lowering of pKa value is significant and cannot be properly captured by current estimators of pKa shift. The latter are typically based on an empirical evaluation of pairwise interactions within a protein environment that favors a particular ionization state. Such estimators do not incorporate the unique electrostatic effects of structural or confined interfacial water molecules [16] into the empirical environmental field [18].

7.5 Quantum Mechanical Treatment of the Protein-Water Interface

To determine the functional stimulation caused by dehydrons, we compute the pKa decrease for nucleophilic side-chain groups with dehydrons in their proximity relative to the free residue in solution. As usual, the pKa shift, ΔpKa, estimates the difference in free energy increment, $\frac{\Delta\Delta G}{RT} = \frac{[\Delta G(p) - \Delta G(w)]}{RT}$, of the proton abstraction process in the protein environment ($\Delta G(p)$) relative to the bulk aqueous environment ($\Delta G(w)$). The results are contrasted against experimental data on ΔpKa. The free energy computation follows the multiple steering ansatz [19], where the molecular dynamics (MD) trajectories are generated by treating classically all groups except for those implicated in the chemical step of proton transference. The latter are treated in the quantum-mechanics (QM) density functional theory (DFT) setting [20]. The QM treatment is thus restricted to the side chain of the catalytic residues containing the weak-acid pro-nucleophilic group and to the reactive dehydron-associated water molecule. The reactive water molecule is defined as having its oxygen within 2.5 Å of the transitional proton that is initially covalently attached to a heavy atom (O, S or N) in the pro-nucleophilic group. The latter is generically denoted AH (or AH^+) and the neighboring distance cutoff is set so that the covalent bond A–H (or $[A–H]^+$ if protonation bestows charge) turns into hydrogen bond in the deprotonated state A^----$H(H_2O)^+$ (or A---$H(H_2O)^+$) that results as the proton is transferred to the nearby dehydron-functionalized water molecule.

We denote by X the proton transference coordinate indicating the distance of the proton to the heavy atom initially covalently attached to it in the weakly acidic pro-nucleophile. Thus, $X(t=0) = X_0$ is the bond length corresponding to covalent bonding to the heavy atom in the nucleophile and $X - X_0$ measures departure from covalent bond length. We denote by $\vec{R}$ the structural-coordinate vector for protein chain and water. In accord with Jarzynski's identity [19] conveniently written as

$$\frac{\Delta G(p)}{RT} = -\log\left\langle e^{-\left[\frac{W(X,\vec{R}(X))}{RT}\right]}\right\rangle \tag{7.13}$$

where the average $\left\langle e^{-\left[\frac{W(X,\vec{R}(X))}{RT}\right]}\right\rangle$ (W = computed work) extends over all trajectories $\vec{R}(X(t))$ with structural conformations steered by the pulling $X = X_0 \rightarrow X = X_0 + v(t_f)$ at constant speed v along the harmonic proton-transference linear coordinate. The pathway ensemble $\{\vec{R}(X(t))\}$ is generated by choosing a set of initial conformations $\{\vec{R_0} = \vec{R}(X_0)\}$ within an isothermal/isobaric equilibrated ensemble (T = 298 K). This ensemble realizes the condition $X = X_0$ and is generated by a set of 20 classical thermalization trajectories, each lasting 1 ns, with the PDB-reported structure fixed at the initial condition. The trajectory multiplicity arising from X-pulling is provided by the conformational dispersion in the initial ensemble $\{\vec{R_0} = \vec{R}(X_0)\}$, with each initial conformation responding differently to the X-pulling.

The QM region is treated using flexible basis sets of linear combinations of finite atomic orbitals in a real space grid optimized to N-scaling. The basis functions enable the matching of the radial wave function to the core region described by pseudopotentials by using pseudoatomic orbitals (PAOs) [23]. Split valence bases are generated by combining numerical Gaussian orbitals with the minimal basis described. The nuclei and core electrons are represented by norm-conserving pseudopotentials to avoid the computation of core states, a procedure that smoothens out the valence charge density in accord with grid requirements. Within the nonlocal pseudopotential approximation, a Kohn-Sham Hamiltonian is adopted incorporating the Hartree and exchange-correlation potentials, and a pseudo-potential with additive contributions to account for local effects, long-range interactions and operation on valence electrons [24]. Calculations are performed on contracted Gaussian basis sets of double-zeta valence polarized (DZVP) quality (pseudoatomic orbital energy shift = 30 meV, grid cutoff = 135 Ry) [25].

The MM region is treated as detailed in Chaps. 2 and 3 [17, 26], where torsional degrees of freedom of backbone and side chains are coarse grained modulo basins of attraction in the potential energy surface in accord with Ramachandran (energetically allowed) regions in local conformation space. Interfacial water dipoles confined to dehydron cavities are subject to a torque resulting from the hindrance to alignment with the electrostatic field [17]. To equilibrate the PDB-reported structures with the solvent, we generated MD trajectories driven by the coarse-grained stochastic process, incorporating the potential energy associated with solvent orientation steering as the reversible work needed to align polarization-induced dipoles due to interfacial water confinement with the Debye electrostatic field [26]. To cover relevant timescales (~10 ns), the dynamics are entrained by the coarser "protodynamics" (see Chap. 3), where the backbone dihedral torsions are specified "modulo basins of attraction" in the potential energy surface. Coarse moves are defined as transitions between basins of attraction (R-basins) in the Ramachandran torsional map for each residue. Thus, each residue is assigned an R-basin after a coarse move, and the coarse state of the chain becomes a conformational ensemble, with each conformation generated by selecting individual backbone torsional coordinates within the assigned R-basins [17].

The hybrid Hamiltonian incorporated includes QM-MM coupling comprised of three contributions: (1) electrostatic interaction between electrons and classical charges, (2) electrostatic interactions between nuclei in the QM subsystem and the classical point charges, and (3) a Lennard-Jones 6–12 potential to account for the van der Waals interactions between the atoms in MM and QM regions constructed using the force-field parametrization of Wang et al. [27]. The forces on the QM nuclei are obtained by taking the gradient with respect to atomic positions, and include derivation of the QM-MM coupling energy.

7.6 Dehydron Chemistry

The quasi-equilibrated work plots harvested in the free-energy computation associated with the dehydron-induced proton transfer event do not portend and are not required to reproduce the actual kinetics, which occurs on much faster timescales. The work performed by the system to reach the point $X = X_0 + vt$ along the proton transference coordinate is shown in Fig. 7.5a for the imidazole >NH group in His149 of xylanase (PDB.1XNB) for ten realizations $\{\vec{R}_0 = \vec{R}(X_0)\}$ of the initial condition $X = X_0$ with harmonic force constant 48 kJ/mol. The thin lines indicate the work performed on the system at each point $X = X_0 + vt$ along the X-pulling

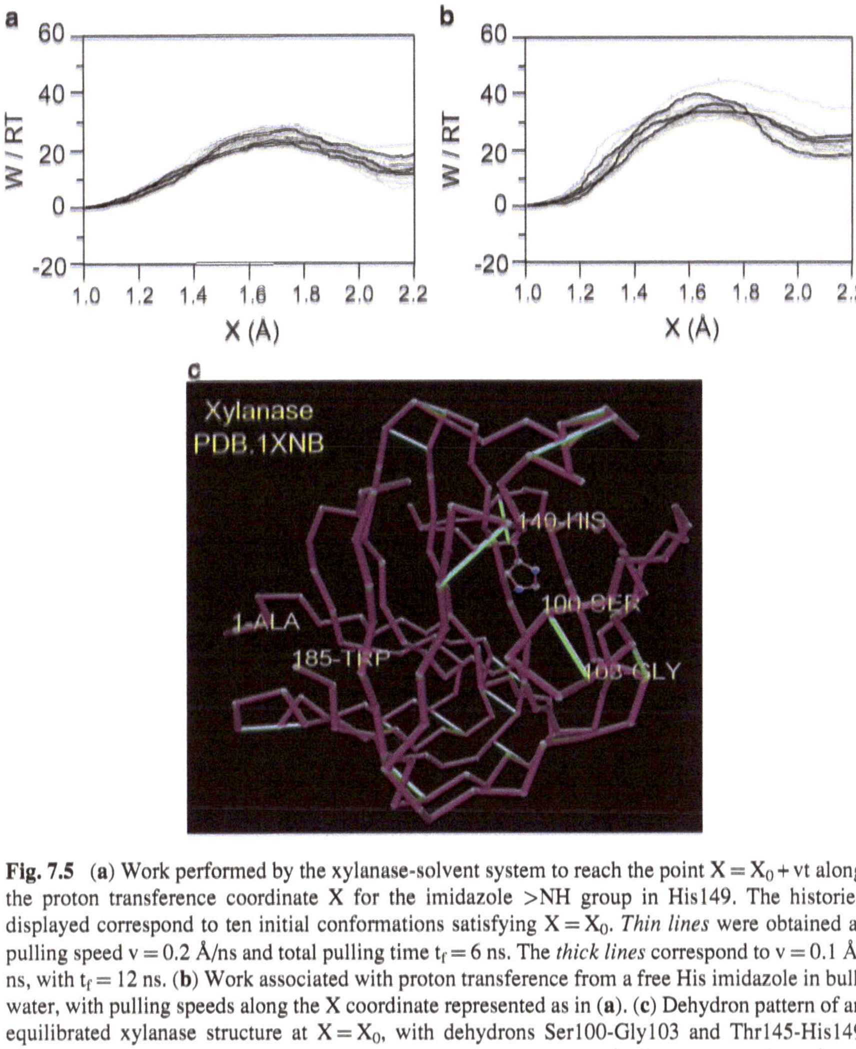

Fig. 7.5 (**a**) Work performed by the xylanase-solvent system to reach the point $X = X_0 + vt$ along the proton transference coordinate X for the imidazole >NH group in His149. The histories displayed correspond to ten initial conformations satisfying $X = X_0$. *Thin lines* were obtained at pulling speed v = 0.2 Å/ns and total pulling time $t_f = 6$ ns. The *thick lines* correspond to v = 0.1 Å/ns, with $t_f = 12$ ns. (**b**) Work associated with proton transference from a free His imidazole in bulk water, with pulling speeds along the X coordinate represented as in (**a**). (**c**) Dehydron pattern of an equilibrated xylanase structure at $X = X_0$, with dehydrons Ser100-Gly103 and Thr145-His149 flanking His 149 (side chain displayed). The backbone is schematically represented as virtual bonds joining alpha-carbons and dehydrons are displayed as sticks joining the non-adjacent alpha-carbons of the hydrogen bonded residues. (**d**) Dehydronic environment for His149 in xylanase at $X = X_0$ with the same structural representation as in (**c**). The oxygen atoms of interfacial water molecules are displayed and the interfacial water molecule with proton-accepting capability induced by dehydron Ser100-Gly103 is indicated by the *arrow*. (**e**) Chemical event along proton transference coordinate X. Reprinted from [34], copyright 2015 with permission from Wiley

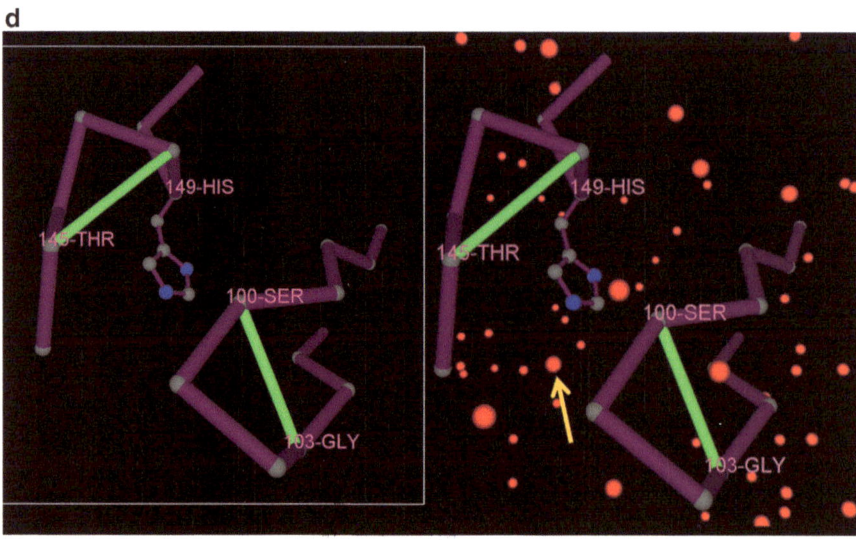

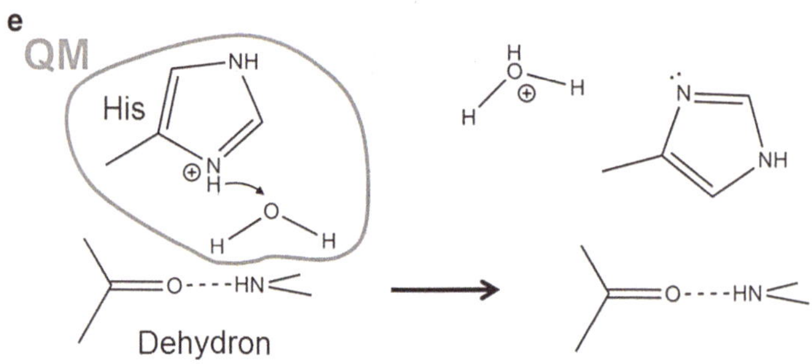

Fig. 7.5 (continued)

steering trajectory with v = 0.2 Å/ns and $t_f = 6$ ns. The thick lines correspond to slower pulling at v = 0.1 Å/ns with $t_f = 12$ ns. The work histories for proton transference from imidazole in a free His amino acid in bulk water are shown in Fig. 7.5b. The dehydronic environment of His149 in xylanase structure is displayed in Fig. 7.5c. The His149 residue is required to be deprotonated in its structure-stabilizing function exerted by hydrogen bonding Ser130, internal water and by engaging in a putative aromatic-aromatic interaction with Tyr105 [28]. The case illustrated represents a dramatic pKa shift, with $\Delta pKa < -3.8$, from the pKa value ~6.1 for free protonated imidazole in bulk water to <2.3 in the xylanase environment [18]. The dehydron-functionalized water molecule around vicinal dehydron Ser100-Gly103 in the initial conformation picked from the set $\left\{\vec{R_0} = \vec{R}(X_0)\right\}$ shown in Fig. 7.5d serves as acceptor of the imidazole proton in His149, with the associated chemical event schematized in Fig. 7.5e. Other conformations use the environment around the other vicinal dehydron Thr145-Asn148 for proton

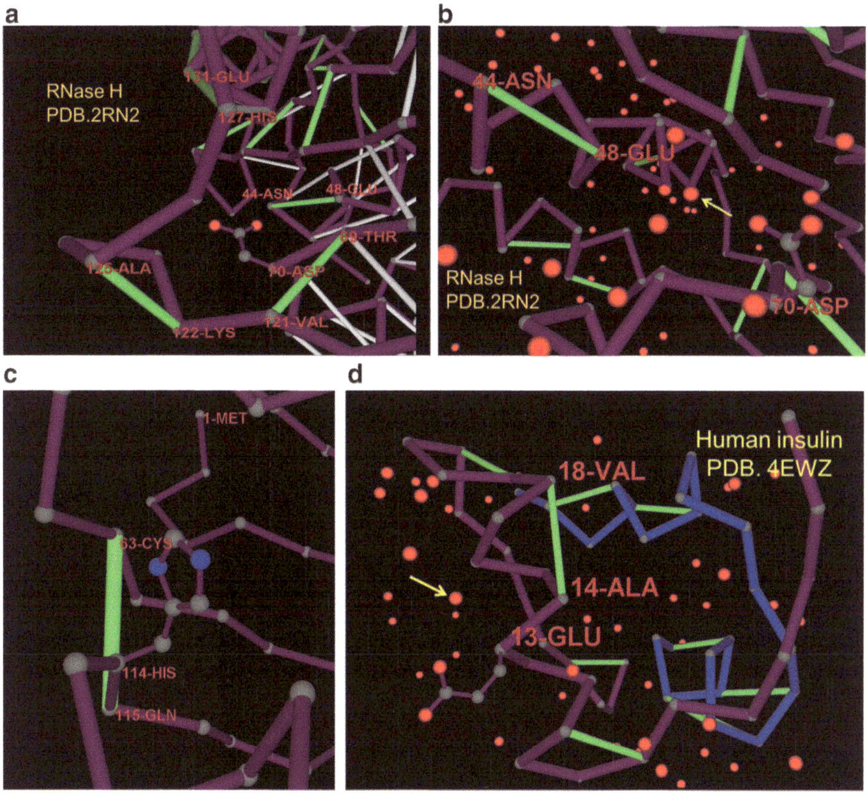

Fig. 7.6 (**a**) Dehydronic environment for Asp70 in RNase H. The surrounding dehydrons are Asn44-Glu48, Thr69-Val121, Lys122-Ala125 and His127-Glu131. The representation convention is as described for Fig. 7.5c. (**b**) Interfacial water around Asp70 in RNase H. The proton-receptive vicinal water marked by the *arrow* is associated with dehydron Asn44-Glu48 and promotes the proton transference from carboxyl group in Asp70. (**c**) Dehydron environment for His114 in RNase H with imidazole deprotonation induced by dehydron Cys63-Gln115. (**d**) Dehydronic environment for Glu13 in chain B of human insulin. The water molecule marked by the *arrow* hydrates dehydron Ala14-Val18 and behaves as a proton acceptor for the side chain of Glu13. Reprinted from [34], copyright 2015 with permission from Wiley

acceptance. A computed value of $\Delta pKa = -3.77$ is obtained for 20 trajectories generated by X-pulling at $v = 0.2$ Å/ns, while $\Delta pKa = -3.64$ obtained by collecting 20 trajectories at slower velocity $v = 0.1$ Å/ns, with both pKa shifts in good agreement with the experimental measurement [18].

A second computation of extreme pKa shifting, shown here to be due to dehydronic stimulation involves Asp70 in RNase H (PDB.2RN2), flanked by dehydrons Asn44-Glu48, Thr69-Val121, Lys122-Ala125 and His127-Glu131 (Fig. 7.6a). The carboxyl in Asp70 must remain highly nucleophilic since it is implicated in enzyme activation by ion coordination through lone electron-pair donation to Mn^{2+} [29]. In this case, the proton transference from the carboxyl group in Asp70 to proton-receptive vicinal water involves the water molecule associated with dehydron Asn44-Glu48 (Fig. 7.6b). The multiple steering QM-MM computation yields a shift $\Delta pKa = -1.3$ in coincidence with the experimental value [18] and in sharp contrast with other computations ($\Delta pKa = +0.3$ [18]) that ignored dehydronic effects.

A third study case revealing dehydronic effects in group functionalization is furnished by the pKa shift due to protein environment at His114 in RNase H. This residue in its deprotonated state is implicated in structural stabilization [30]. A shift $\Delta pKa \approx -1.6$ is calculated by defining the reaction coordinate X through the imidazole deprotonation induced by dehydron Cys63-Gln115 (Fig. 7.6c). The pKa shift obtained is identical to the experimental value and at a variance with other computations ($\Delta pKa = +1.2$ [18]) that do not incorporate dehydronic effects.

Finally, residue Glu13 on chain B of human insulin is known to play a crucial role in oligomerization [31] and association with the islet amyloid polypeptide (IAPP) through an intermolecular salt bridge [32]. These structural roles require a functionalization evidenced in the shift $\Delta pKa = -1.9$. Current pKa computations that do not incorporate the dehydron influence yield the incorrect value $\Delta pKa = -0.3$ [18]. The shift is correctly estimated by taking into account the enabling role of vicinal dehydron Ala14-Val18 on chain B (Fig. 7.6d). The chemical event of deprotonation of Glu13 is captured by implicating the interfacial water molecule (marked by an arrow in Fig. 7.6d) turned into a proton acceptor by dehydron Ala14-Val18. The multiple steering MD computation pulling along the proton transference coordinate yields $\Delta pKa = -1.81$ is satisfactory agreement with experiment [18].

The biochemical consequences of dehydron-stimulated catalytic activity are probed by examining the effects of oncogenic mutations yielding constitutively active enzymes. We focus on cancer-related kinases altered by dehydron-generating mutation. As it is known, kinases get phosphorylated by performing a nucleophilic attack on the terminal ATP phosphoester linkage. Thus, our working hypothesis is that the nucleophilicity of the active side chain is enhanced in the mutant variant provided the mutation generates stimulating dehydrons nearby. With this rationale in mind, we investigate the oncogenic mutation D816V (Asp816Val) in the c-Kit kinase, found in mastocytosis [33] and other tumors, and known to induce a transforming capacity in c-Kit by turning it constitutively active [22]. The mutation D816V occurs in a floppy region of the kinase known as activation loop. A set of five 1.1 μs MD trajectories were generated on the wild-type kinase and on the D816V substitution by equilibrating the PDB-reported conformation 1T46 following the coarse-graining protocol previously described [17, 26]. The hydrophobic replacement resulting from mutation D816V promotes water exclusion (wrapping) at backbone hydrogen bond Gly812-Arg815 that is retained as a dehydron (Fig. 7.7a). This hydrogen bond is disrupted in the wild type due to backbone hydration, as the activation loop becomes more solubilized by the presence of polar residue Asp816 (Fig. 7.7b). In turn, the presence of the Gly812-Arg815 dehydron in the vicinity of the phosphorylation site Tyr823 in the c-Kit mutant, lowers significantly the pKa of the phenolic hydroxyl in Tyr823, as shown in the structural display for the mutant (Fig. 7.7c). The enhanced nucleophilicity of Tyr823 is the resultant of the proton acceptor activity of interfacial water (Fig. 7.7d) promoted by mutation-induced dehydron Gly812-Arg815. Multiple steering MD computations yield $\Delta pKa = -2.7$ as the mutation-generated dehydron functionalizes its adjacent water molecule into a proton acceptor (Fig. 7.7c) for the phenolic hydroxyl proton

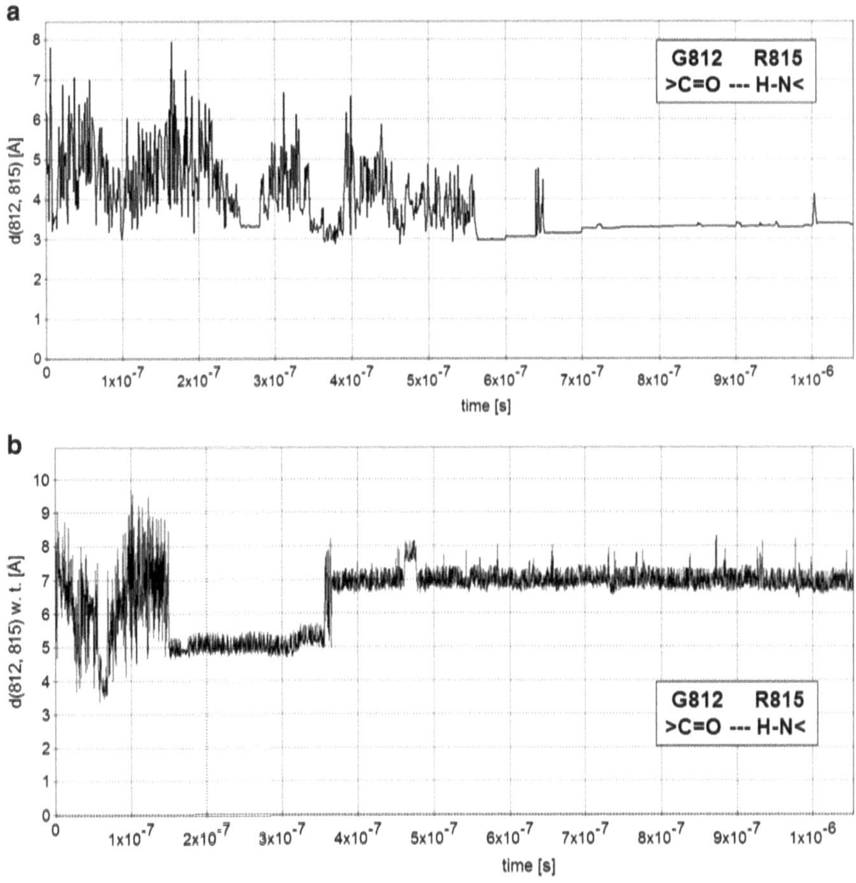

Fig. 7.7 (**a**) Time-dependent O-N distance between backbone carbonyl of Gly812 and backbone amide of Arg815 in the activation loop of c-Kit D816V mutant along a typical MD trajectory generated as described in main text. The initial torsional state is the PDB-reported in entry 1T46. A stable Gly812-Arg815 backbone hydrogen bond emerges after 0.65 ± 0.20 μs in all five trajectories generated. (**b**) Time-dependent backbone O-N distance between Gly812 and Arg815 in wild type c-Kit along a typical MD trajectory. No stable backbone hydrogen bond is observed in any trajectory. (**c**) Snapshot of the activation loop of mutant D816V after 1.1 μs equilibration (cf. (**a**)) displaying dehydron Gly812-Arg815. The interfacial water molecule implicated in proton transference and functionalization of Tyr823 is marked by the *arrow*. (**d**) Chemical event of dehydron-based functionalization of a Tyr side chain as required for transesterification reaction with ATP. Reprinted from [34], copyright 2015 with permission from Wiley

c

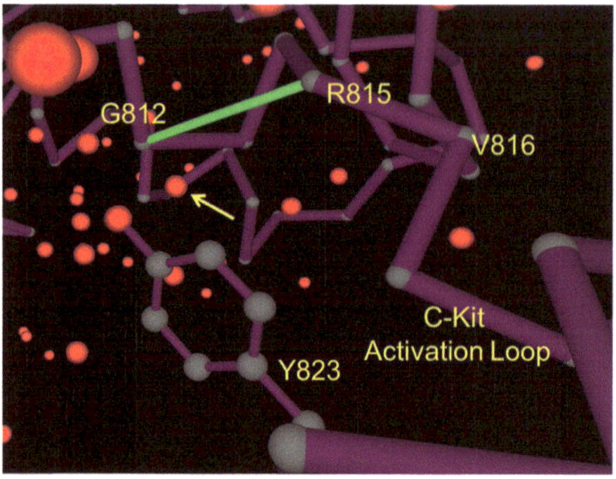

d

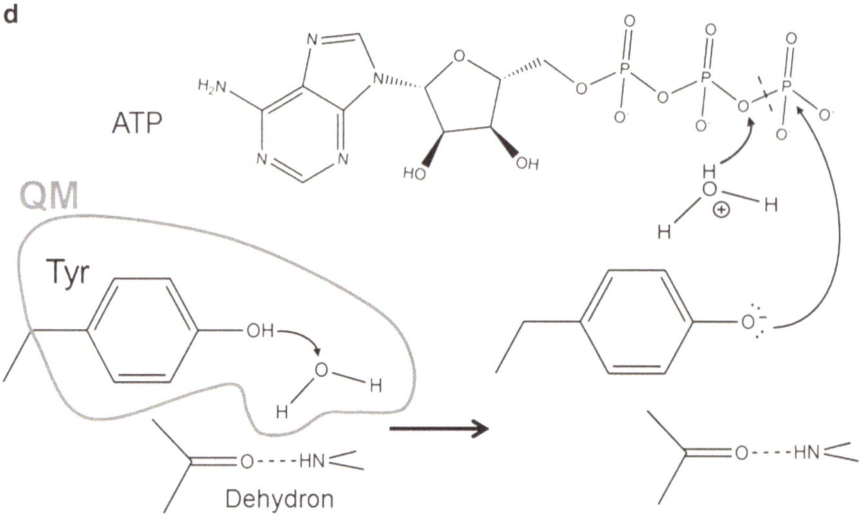

Fig. 7.7 (continued)

of Tyr823 (Fig. 7.7d). This enhancement in the nucleophilicity of the Tyr823 also enhances significantly its phosphorylation since the latter is contingent on nucleophilic attack on the ATP phosphoester linkage. Consequently, the mutation D816V turns the kinase constitutively active. It is now known that Tyr823 phosphorylation is essential to recruit anti-apoptotic pathways that bestow the transforming capability to c-Kit, leading to significantly longer tumor lifetime [22]. Thus, the oncogenic nature of the mutation may be now traced to its physico-chemical roots.

With these quantum mechanical computations we have mechanistically explored and validated the conjecture posed in Sects. 7.1 and 7.2 that dehydrons participate as quasi-reactants in biochemical events. We focused on protein functions involving nucleophiles activated through deprotonation promoted by nearby dehydrons. Through molecular dynamics steered by pulling along the proton-transference coordinate, we have unraveled a general mechanism for dehydron-based chemical functionalization.

7.7 Frustration-Related Biological Chemistry

As indicated in Chaps. 1, 2, and 4, interfacial water enveloping the structure of a soluble protein is often partially confined due to sub-nanoscale topographic detail in the protein surface [1, 3, 34]. When there is a dearth of coordination partners, this confinement brings about frustration of water molecules in regards to their hydrogen-bonding opportunities. Frustration makes water chemically reactive [34], a possibility that needs to be systematically explored. This picture is validated in this section focusing on uncompensated anticontacts or mismatches that classical electrostatics would deem untenable and classical molecular dynamics would treat as unstable. To delineate the frustration-related functionality, we need to focus on the physico-chemical properties of interfacial water that arise from the emergence of interfacial tension, known to be a consequence of frustration [3]. As described in the preparatory chapters, such properties are not enshrined in a permittivity coefficient stemming from the "classical" Debye dielectric picture [35]. Due to subnanoscale confinement, the water dipoles are often precluded from aligning with the electrostatic field (**E**), signaling a breakdown in the Debye *ansatz*. This breakdown is marked by the presence of an **E**-orthogonal polarization contribution $\mathbf{P}^{\#}$ that generates a net charge $\gamma^{\#} = -\nabla \cdot \mathbf{P}^{\#}$. The non-Debye term $\mathbf{P}^{\#}$ stores electrostatic energy that may be alternatively computed as interfacial tension [1], and is responsible for the dielectric anisotropy at the protein interface (problem 2.4, Chap. 2). In accord with the charge asymmetry with respect to hydration [36], we provide crystallographic evidence supporting the fact that hydrogen-bond frustration due to nanoscale confinement functionalizes interfacial water as a general base [6], thus determining a *frustration-related chemistry*. The consequences of this chemical role become apparent as we examine conformations featuring same-charge uncompensated anticontacts reported in crystallographic data [37] deemed "unstable" by classical molecular dynamics (MD) due the absence of nearby counterions. The stabilization of such configurations emerges in a quantum-mechanical treatment [34, 38], a suitable framework to incorporate frustration-based chemistry. As shown, the basicity of frustrated water enables titration of ammonium and guanidinium, both weaker acids than hydronium, implying that this effect cannot be accounted for in classical terms. This novel functionality of frustrated water is supported by the crystallographic record and by detailed information on water molecules in intramolecular proton transfer obtained by FTIR difference spectroscopy [9]. The results reveal a functionality of biomolecular interfaces intractable by mean-field dielectrics.

As previously described in this chapter, the Debye polarization accounts for the screening of the solute charge, the $\mathbf{P}^{\#}$-component is responsible for chemical functionality associated with induced charge, hence not accounted for by protein chemical composition. The functionality is in accord with the sign of $\gamma^{\#}$, i.e. proton donor if $\gamma^{\#} > 0$ and proton acceptor if $\gamma^{\#} < 0$. Only one alternative becomes feasible, however. The consequences of such asymmetry will be the subject of this section.

To describe the local dielectric distortion of interfacial water, we introduce a "frustration scalar field" $\phi = \phi(\mathbf{r}) = 4 - g(\mathbf{r})$ that quantifies the level of unrealized water hydrogen bonding at spatial location $\mathbf{r}$ relative to the bulk hydrogen-bond pattern [6]. As expected, the non-Debye field $\mathbf{P}^{\#}$ is proportional to the distortion gradient [6]: $\mathbf{P}^{\#} = -\xi\nabla\phi$. The net charge $\gamma^{\#}$ induced by $\mathbf{P}^{\#}$ is then $\gamma^{\#} = -\nabla\cdot\mathbf{P}^{\#} = \xi\nabla^2\phi$, and the interfacial energy ΔU_{ϕ} associated with spanning the solute-water interface becomes $\Delta U_{\phi} = (1/2)\lambda\int\|\nabla\phi\|^2\, d\mathbf{r}$, where integration is carried over a spatial domain Ω containing the interface, so that its border $\partial\Omega$ is fully contained in bulk water, that is, subject to the condition: $\{\phi(\mathbf{r}) = 0$ and $\nabla\phi(\mathbf{r}) = 0\}$ $\forall\mathbf{r} \in \partial\Omega$. As previously shown in this chapter, the mean value theorem (MV) of integral calculus yields $\gamma^{\#}_{MV} = -(\lambda\varepsilon_o)^{-1/2}\int\|\nabla\phi\|^2\, d\mathbf{r}/\int\phi d\mathbf{r} \leq 0$. Thus, frustrated interfacial water resulting from nanoscale confinement yields negatively induced charge $\gamma^{\#}_{MV} \leq 0$, consistent with a proton-acceptor role resulting from frustrated hydrogen-bond coordination to water oxygen.

This result has important methodological consequences, explaining the acid-base interfacial chemistry and may be also obtained from an intuitive spatial model based on the hydration differences between positive and negative charges [36]. Such differences arise because the water dipole center differs from the center of its van der Waals sphere, implying that the positive charges on the water protons are closer to the surface of the van-der-Waals sphere than the negative charge distributed on the lone electron pairs of the oxygen atom. Thus, a solute with negative charge may be more favorably hydrated than one with positive charge since the electrostatic interaction with water protons is facilitated relative to an electrostatic interaction involving the more internalized negative charge in the water molecule. This preferential hydration of negative over positive charges is in accord with the charge asymmetry ($\gamma^{\#}_{MV} \leq 0$, but not $\gamma^{\#}_{MV} > 0$) resulting from non-Debye polarization. A more detailed analysis requires that we consider dehydrons, water-exposed backbone hydrogen bonds that cause sub-nanoscale confinement and thereby frustration in surrounding water [1, 3, 34, 35]. Such dehydrons are primarily hydrated by water molecules interacting with the dehydron carbonyl, rather than the amide (cf. Fig. 7.8a, b). Since the water molecule is presented in the dehydron cavity as a carbonyl solvator, the question remains as to whether it becomes frustrated as proton acceptor (Fig. 7.8a), by developing a frustration-related negative charge $\gamma^{\#}_{MV} < 0$, or a proton donor ($\gamma^{\#}_{MV} > 0$), with a frustrated dangling hydroxyl (Fig. 7.8b). The option $\gamma^{\#}_{MV} < 0$ is favored since the negative charge expands due to electronic repulsion within the oxygen shell (Fig. 1a), bringing the charge closer to the water van der Waals surface. This makes it susceptible to better electrostatic interaction with vicinal water protons,

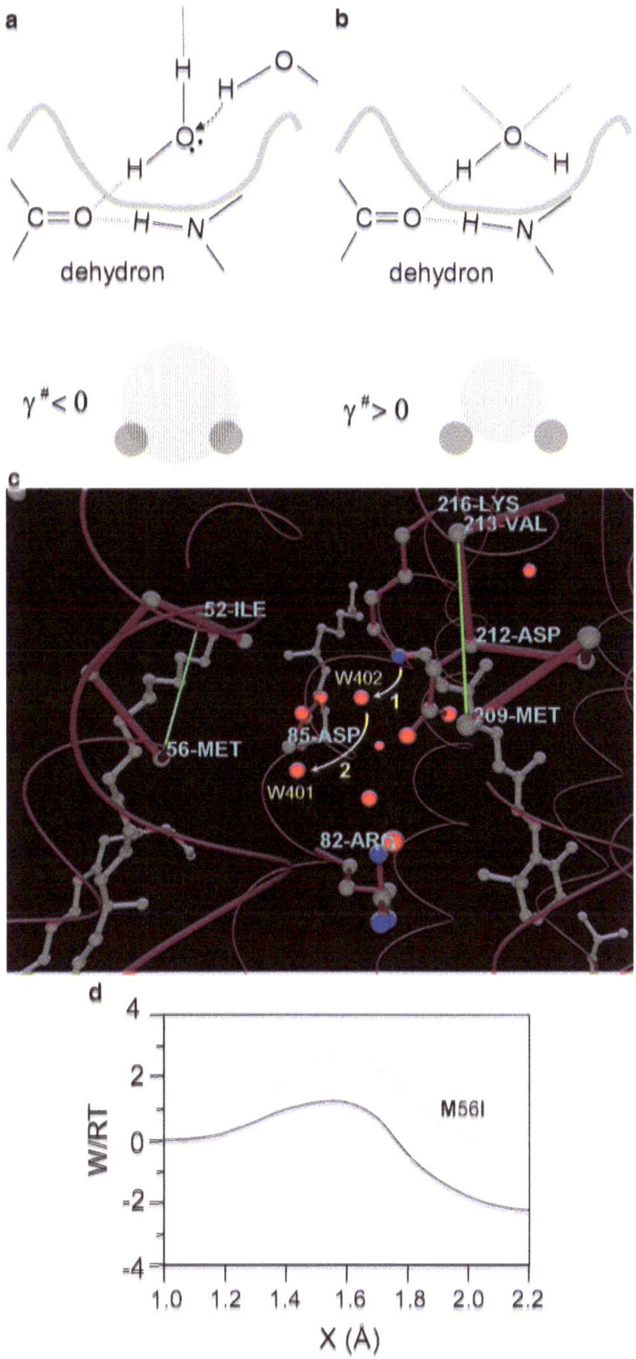

Fig. 7.8 Dehydrons functionalize interfacial water by frustrating water hydrogen-bonding opportunities. (**a**) Schematic representation of the favored presentation of an interfacial water molecule within the dehydron cavity, featuring a proton acceptor role resulting from hydrogen bond frustration due to sub-nanoscale confinement. The van-der-Waals sphere is represented in *light*

located close to the van der Waals surface [36]. This picture has been corroborated in molecular dynamics computations of water-filling dehydron cavities (Fig. 7.8b, [6]) revealing a negative induced charge with maximum magnitude 3.31×10^4 Coul/mol at curvature radius $\xi = 2.95$ Å.

The presence of nearby dehydrons inducing a proton-accepting role in vicinal interfacial water has been conjectured to enable and steer the intramolecular Grotthuss mechanism of proton transference [38, 39]. The free energy change, ΔG, associated with frustration-promoted proton transference is obtained by multiple steering molecular dynamics computation [14] as described in [34]. The proton donor and nearby proton-receptive water molecule at the dehydron interface are treated within a quantum mechanical (QM) scheme, while the rest of the molecule and TIP3P solvent model are treated using a classical molecular mechanics (MM) package, within a QM-MM hybrid approach [34]. The dehydron-functionalized water molecule involved in frustration-related chemistry (cf. Fig. 7.8a) is defined as having its oxygen within 2.5 Å of the transitional proton that is initially covalently attached to a heavy atom (O, S or N) in the proton-donating moiety, while constrained by a distance between backbone carbonyl and water oxygen atoms <3.3 Å. The proton donor is generically denoted AH (or AH^+, A = O, S, N) and the neighboring distance cutoff is set so that the covalent bond A–H ($[A\text{–}H]^+$ if protonation generates charge) turns into hydrogen bond in the deprotonated state A^----$H(H_2O)^+$ (or A---$H(H_2O)^+$) that results as the proton is transferred to the nearby dehydron-functionalized water molecule.

Fig. 7.8 (continued) grey, positive charges are shown in *dark grey*, and the net negative charge ($\gamma^\# < 0$) resulting from unsatisfied lone electron pairs within the oxygen orbitals (patterned disk, *lower panel*) is expanded due to electronic repulsion, approaching the van-der-Waals border of the water molecule. (**b**) Disfavored water presentation within the dehydron cavity, yielding $\gamma^\# > 0$. The positive polarization-induced charge ($\gamma^\# > 0$) favors weaker electrostatic interactions with another water molecule because the latter's negative charge is far from the molecule's van-der-Waals border when compared with the positive charges on the protons. (**c**) - Dehydron-based water functionalization enables the directed Grotthuss mechanism of intramolecular proton transfer in *bacteriorhodopsin* (PDB.1C3W). The hydronium that results from condensation of K216 and ligand retinal (step **1**) is identified as water molecule W402, whose proton acceptance role is induced by nearby dehydron M209-V213. For simplicity, backbone hydrogen bonds are represented as segments joining α-carbons of paired residues. The transference of this proton to W401 (step **2**) is enabled by the proton-accepting role of water molecule W401, frustrated by the confinement brought about by nearby dehydron I52-M56. Step **2** initiates the directed Grotthuss mechanism. (**d**) Free energy change $\Delta G(X)/RT = -\log\langle\exp[-W(X, \mathbf{R}(X))/RT]\rangle$ reported as work (W) in RT units along the pulling coordinate X for the proton transfer Wat402 → Wat401 for wild type (*black line*) and *in silico* mutant M56I (*grey line*). The computation is described as noted in main text. The control *in silico* mutant M56I is designed to remove the 52–56 dehydron by increasing the number of nonpolar side chains in the vicinity the backbone hydrogen bond. Since ΔG = 0 for the M56I variant, the proton transference value ΔG = −1.41 ± 0.33 kcal/mol provides a reliable estimation of the effect of dehydron-induced water frustration. Reprinted from [Fernández A. Acid–base chemistry of frustrated water at protein interfaces. FEBS Letters. 2016;590:215–223.] copyright 2016 with permission from Wiley

We denote by X the proton transference coordinate indicating the distance of the proton to the heavy atom A initially covalently attached to it. Thus, $X(t=0)=X_0$ is the bond length corresponding to covalent bond A–H, and $X–X_0$ measures departure from covalent bond length. We denote by **R** the structural-coordinate vector for protein chain and water system. We compute the free energy change as $\Delta G/RT=-\log\langle\exp[-W(X, \mathbf{R}(X))/RT]\rangle$, where the average $\langle\exp[-W(X, \mathbf{R}(X)]\rangle$ (W = computed work) extends over all trajectories $\mathbf{R}(X(t))$ spanned over the time interval $[0, t_f=t_f(v)]$ with conformations steered by the pulling $X=X_0\rightarrow X=X_0+v(t_f)$ at constant speed v along the harmonic proton-transference linear coordinate during timespan t_f [14, 34]. The pathway ensemble $\{\mathbf{R}(X(t))\}$, is generated by choosing initial conformations $\mathbf{R_0}=\mathbf{R}(X_0)$ within an isothermal/isobaric equilibrated ensemble (T = 298 K). This ensemble fulfills the condition $X=X_0$ and is generated by a set of 20 classical thermalization trajectories, each lasting 1 ns, with the PDB-reported structure fixed at the initial condition.

We also compute the pKa-shift, ΔpKa, for a proton-donating side-chain group within the protein environment relative to the free residue in solution. The quantity ΔpKa estimates the difference in free energy increment, $\Delta\Delta G/RT=[\Delta G(p)-\Delta G(w)]/RT$, of the proton abstraction process in the protein environment (ΔG(p)) relative to the chemical event taking place in the bulk aqueous environment (ΔG(w)).

The proton-accepting role of interfacial water surrounding dehydrons has been previously conjectured to promote the intramolecular Grotthuss mechanism of proton transference [38, 39]. This is clearly not a standard effect of bulk water since the basicity of dehydron-associated water enables titration of ammonium and guanidinium, which are both weaker acids than hydronium. This picture is now validated vis-a-vis detailed information on functional waters in the light-driven proton pump *bacteriorhodopsin* obtained from FTIR difference spectroscopy (PDB.1C3W) [39]. The key role of water molecule Wat401 as initiator of the proton transference cascade is actually enabled by its basicity ($\gamma^{\#}_{MV}=0.31\times10^5$ Coul/mol), in turn conferred by nearby dehydron I52-M56 (Fig. 7.8c). Hence, the proton in hydronium formed by Wat402, the byproduct of Schiff base condensation between K216 amonium and the retinal carbonyl, is accepted by Wat401 with $\Delta G=-1.41\pm0.33$ kcal/mol according to our quantum mechanics computation of dehydron-induced basicity of Wat401. The free energy changes $\Delta G(X)/RT=$ (*black line*) are evaluated every 50 ps for pulling speed v = 0.2 Å/ns (Fig. 7.8d). The average is evaluated over a pathway ensemble $\{\mathbf{R}(X(t))\}$ generated from 20 initial microstates $\mathbf{R_0}=\mathbf{R}(X_0=1\text{ Å})$ within an isothermal/isobaric equilibrated ensemble (T = 298 K) fulfilling the condition $X=X_0$, with the PDB-reported structure fixed at the initial condition [14]. The results are reported in Fig. 7.8d with the control *in silico* mutant M56I purposely designed to remove the 52–56 dehydron by increasing the number of nonpolar side chains in the vicinity of residues 52 and 56. As expected, ΔG vanishes for the proton transference from Wat402 to Wat401 in the M56I mutant as no dehydron-associated functionalization occurs (grey line, Fig. 7.8d). Thus, we can be confident that the proton transference value $\Delta G=-1.41\pm0.33$ kcal/mol for the wild type corresponds to the chemical basicity generated by the (I52-M56)-dehydron-induced frustration of Wat401.

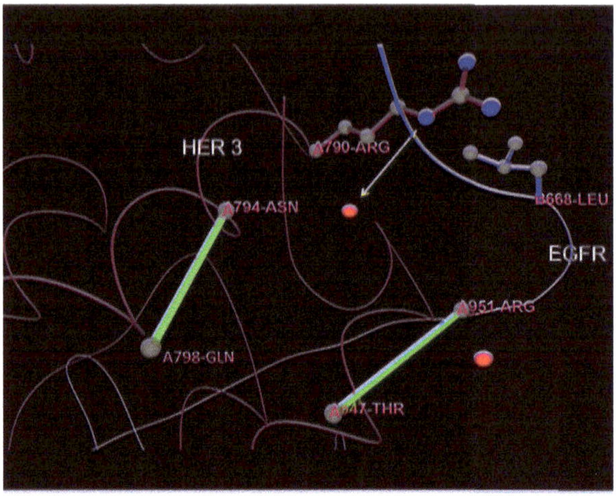

Fig. 7.9 Role of frustrated interfacial water in the recruitment of the functionally active heterodimer HER3(Q790R)-EGFR (PDB.4RIX, chain A = HER3 Q790R-mutant kinase, chain B = EGFR kinase). The HER3 mutation enables protein association as frustrated interfacial water at dehydron N794-Q798 in chain A accepts the proton from R790 (the *arrow marks* the chemical event), which is then able to pair favorably with L668 from EGFR kinase at the heterodimer interface, as there is no dehydration cost associated with (neutral-arginine)-leucine pairing. Without invoking frustration-related chemistry, the recruiting effect of mutation Q790R becomes untenable. Reprinted from [Fernández A. Acid–base chemistry of frustrated water at protein interfaces. FEBS Letters. 2016;590:215–223.] copyright 2016 with permission from Wiley

An illustration of frustration-related chemistry is provided by the effect of the oncogenic mutation Q790R in HER-3 (ERBB-3) kinase, presumed responsible for recruiting the catalytically active heterodimer HER-3-EGFR kinase complex [40]. Direct inspection of the interface between mutant HER-3 and EGFR kinase (PDB.4RIX) reveals that the side chain of R790 in HER-3 is interacting with L668 in EGFR kinase, an unfavorable charged-nonpolar interaction (Fig. 7.9), treated as anticontact by classical MD that attributes a positive charge to guanidinium in R790 and a penalty for its hydration hindrance due to proximity to the nonpolar L668. A QM computation, accounting for the frustration-related functionalization of interfacial water vicinal to dehydron N794-Q798, reveals the true recruiting role of the Q790R substitution in HER-3 that, in effect, makes the L668-R790 contact favorable. The recruiting improvement is realized through the deprotonation of R790 promoted by the dehydron-induced basicity of the vicinal water molecule (Fig. 7.9), with a transference free energy change relative to free arginine in bulk water estimated at $\Delta\Delta G = -2.34 \pm 0.25$ kcal/mol.

A more dramatic manifestation of frustration-related chemistry is provided by the L858R oncogenic mutation of the EGFR kinase [41, 42], known to promote a more ordered state than the wild type, thereby lowering the entropic cost of autophosphorylation through homo-dimerization [43]. Strikingly, the crystal structures of the monomer mutant EGFR kinase (4I20) capture the inactive state with an R836-R858 anticontact (Fig. 7.10a) which would be deemed as repulsive in

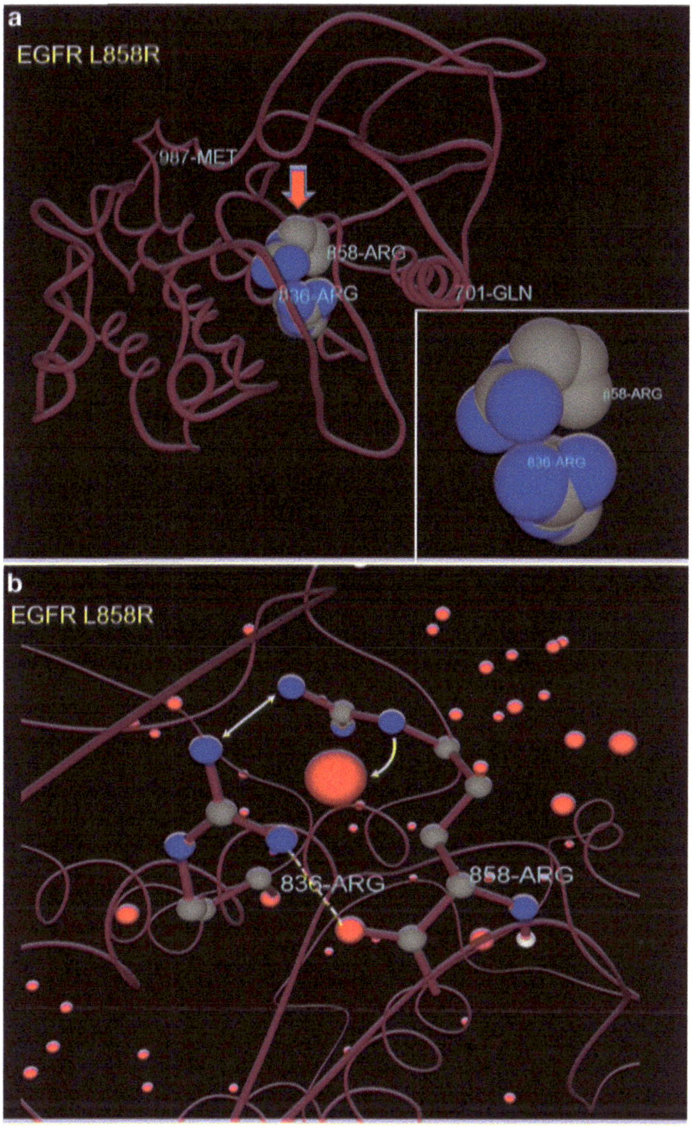

Fig. 7.10 Frustration-related chemistry of the aqueous interface and its stabilizing role on the classically untenable R836-R858 anticontact in the L858R mutated EGFR kinase (PDB.4I20 and PDB.2ITV). (**a**) Crystal structure of EGFR mutant reports the mutated residue R858 (signaled by *red arrow*) making a van der Waals contact with R836 (inset with contact detail). The paired residues are represented at atomic level, while the rest of the chain is featured schematically in tube representation. There is no counterion in the proximity of the association reported in PDB.4I20. The closest negatively charged residues D837, D855, E758 have the carboxyl center farther than 8 Å from the (R836, R858) guanidinium centers. None of these residues is engaged in salt bridges with R858, at a variance with the MD analysis in [43]. (**b**) Quantum effects accounting for frustration-related chemistry of the aqueous interface stabilize the R836-R858 interaction. The R858 guanidinium proton transfers (event marked by *arrow*) to a water molecule functionalized by

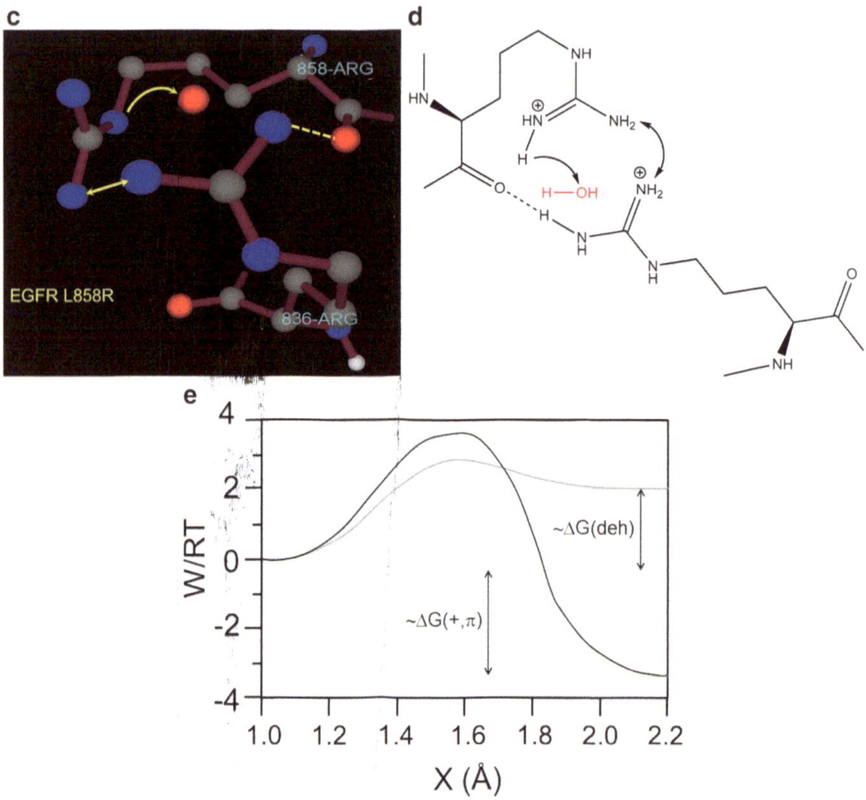

Fig. 7.10 (continued)

Fig. 7.10 (continued) the side-chain-main-chain R836-R858 dehydron (*dashed line*). The proton abstraction from R858 enables a favorable cation-π interaction between protonated R836 and the neutral R858, with its six delocalized π-electrons. (**c**) Conformation of residues R836 and R858 within the crystal structure of the inactive state (PDB.4I20) featuring the chemically competent side-chain-main-chain dehydron that induces the required basicity needed to remove the electrostatic repulsion between the two vicinal arginines. (**d**) Frustration-related chemistry enabling the interaction of two arginines through proton transference to a water molecule (water oxygen in *red*) functionalized by a side-chain-main-chain dehydron. (**e**) A QM-MM multiple steering ΔG(X)-computation of the R836-R858-frustrated-water system (*black line*) contrasted with the control proton-transference thermodynamics of guanidine to bulk water (*grey line*). Computational details are as in Fig. 7.8d. The frustrated water molecule (Fig. 7.10b,c) accepts the proton from the guanidinium in R858 with $\Delta\Delta G = -3.25 \pm 0.27$ kcal/mol ($\Delta pKa \approx -5.4 \pm 0.4$). This result yields $\Delta G(+, \pi) = -1.83 \pm 0.18$ kcal/mol for the cation-π (836–858)-interaction, since a free energy contribution -1.41 ± 0.33 kcal/mol (Fig. 7.8d) is attributed to the dehydron-induced functionalization of the vicinal water molecule turned into a chemical base. Reprinted from [Fernández A. Acid–base chemistry of frustrated water at protein interfaces. FEBS Letters. 2016;590:215–223.] copyright 2016 with permission from Wiley classical MD due to a complete absence of compensating counterions in the vicinity of the van-der-Waals contact region (cf. Fig. 7.10b) [42]. The closest negatively charged residues D837, D855 and E758 all have their carboxyl centers farther than 8 Å from the R836, R858 guanidium centers (at the ζ-carbons), in contrast with MD computational models which have them involved in salt bridges with R858 [43]. We rely on the crystal structure information which reveals that the R836-R858 "classical anticontact" prevails and is retained even in the active structure (PDB.2ITV), as shown subsequently.

As expected, classical MD computations report the R836-R858 interaction as unstable and suggest that the oncogenic mutation constitutively deregulates the kinase by destabilizing the inactive state [42]. This interpretation is at odds with the fact that the R836-R858 interaction is sufficiently stable to be captured by the X-ray diffraction patterns of the EGFR kinase mutant and present in all reported structures that are spatial renderings of both the inactive state (PDB.4I20) [41] and of the active state (PDB.2ITV) [44]. The fact that the R836-R858 interaction is found in the crystal structures of inactive and active states of the EGFR kinase mutant suggests that the mutation is not destabilizing the inactive state in the manner modeled by classical MD that computes the van der Waals contact between R836 and R858 as repulsion [42].

As it turns out, the stabilization of the R836-R858 interaction in the inactive state results from a quantum effect with significant decrease in pKa, implicating a water molecule functionalized into basicity by the dehydron sustained by the R836 guanidinium and the backbone carbonyl of R858 (Fig. 7.10c–e). A QM-MM $\Delta G(X)$-computation of the R836-R858-frustrated-water system (Fig. 7.10c–e) is contrasted with the control thermodynamics of proton-transference of guanidine to bulk water (Fig. 7.10e), revealing that the frustrated water molecule accepts the proton from the guanidinium in R858 with $\Delta\Delta G = -3.25 \pm 0.27$ kcal/mol ($\Delta pKa \approx -5.4 \pm 0.4$). The R858 side-chain deprotonation promotes a favorable cation-π interaction between protonated R836 and the neutral R858, with its six delocalized π-electrons [45]. The results shown in Fig. 7.10e yield $\Delta G(+, \pi) = -1.83 \pm 0.18$ kcal/mol for the cation-π (836–858)-interaction. To obtain this estimation, we took into account the contribution of the dehydron-induced water frustration ($\Delta G(deh)$) into $\Delta\Delta G$, estimated at $\Delta G(deh) = -1.41 \pm 0.33$ kcal/mol according to Fig. 7.8d, and use the approximate relation: $\Delta\Delta G \approx \Delta G(deh) + \Delta G(+, \pi)$.

The quantum stabilization of this R-R interaction differs significantly from what we may term the "classical stabilization" achieved through a pair of adjacent counterions. An illustration of the classical R-R stabilization is provided by the homodimer of p53 DNA-binding domain (PDB.2GEQ, Fig. 7.11a), where the two interacting R178 residues at the dimer interface are compensated by adjacent negatively charged E177 residues pairing two salt bridges (Fig. 7.11b).

The quantum stabilization of the classically repulsive R836-R858 contact by frustrated interfacial water becomes enhanced in the active state of the EGFR kinase mutant L858R (PDB.2ITV) when this state is compared with the inactive state. In the active state, a favorable interaction between both arginines is enabled

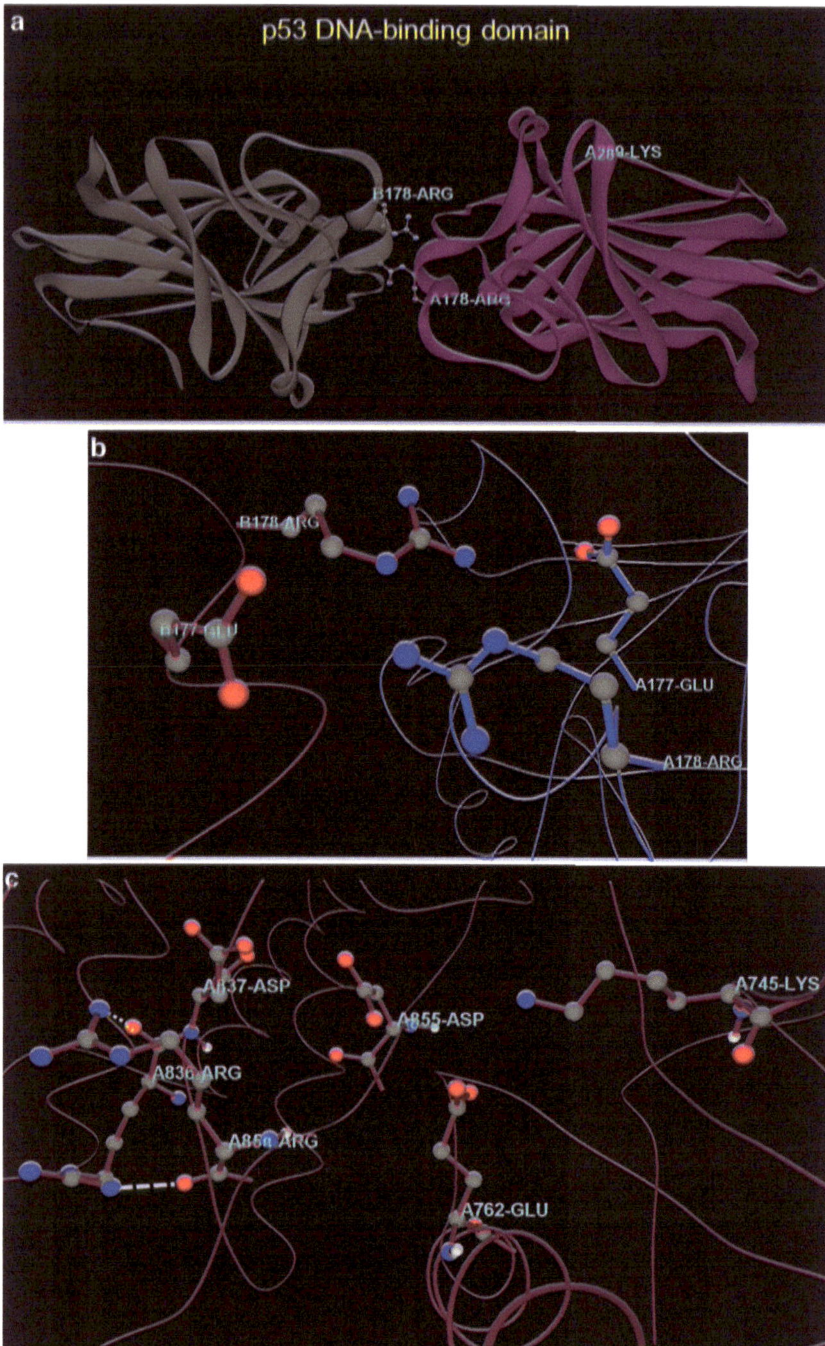

Fig. 7.11 Illustration of classical electrostatic stabilization of same-charge (R-R) pair (**a**, **b**), QM-based stabilization of R-R anticontact (**c**, **d**) and of Lys^{+}-$^{+}$AdoMet anticontact within a PKMT complex (**e**). (**a**) Functional homodimer of p53-DNA-binding domain displaying the compensated R178-R178 interaction at the protein-protein interface. (**b**) Detailed structure of R-R association classically stabilized by nearby negative counterions provided by the carboxyls from the twin E177s. (**c**) Active conformation of EGFR L858R mutant kinase featuring the van-der-Waals R836-

d

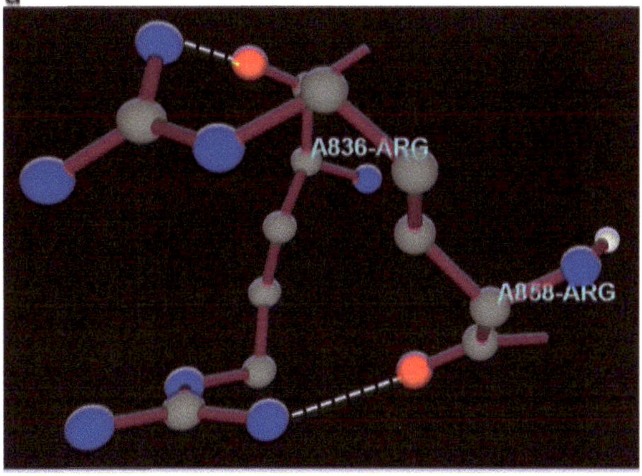

e

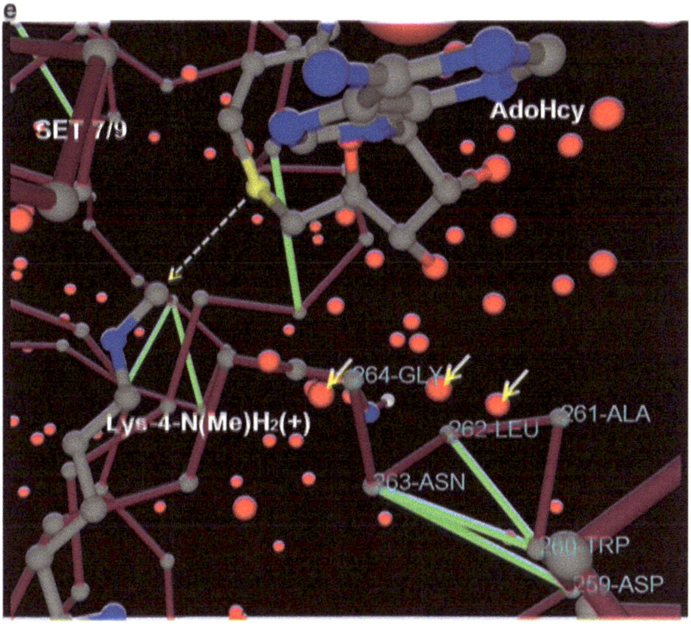

Fig. 7.11 (continued)

Fig. 7.11 (continued) R858 anticontact and the two adjacent main-chain-side-chain dehydrons involving the same interacting residues. These dehydrons confer the aqueous interface the required basicity that removes the electrostatic repulsion between the two adjacent R836-R858 guanidiniums. The closest negatively charged residues D837, D855 and E758 all have their carboxyl centers farther than 8 Å from the R836, R858 guanidium centers (at ζ-carbon atoms). (**d**) Double R836-R858 main-chain-side-chain dehydron stabilizing the R-R side-chain-side-chain interaction in the active structure of the EGFR L858R mutant kinase. (**e**) Dehydrons (*green lines*) in PKMT SET7/9 in post-methylation complex (PDB.1O9S) with ligand AdoHcy and methylated substrate Lys-$(Met)H_2^+$. Methyl transference is indicated by *dashed arrow*. The Grotthuss string of dehydron(W260-N263)-functionalized frustrated water molecules (*solid arrows*) is involved in the proton abstraction from Lys-(NH_3^+), a necessary step to enable nucleophilic attack of the neutral Lys amino group on the S-Met bond in $^+$AdoMet, yielding Lys-$N(Met)H_2^+$

by two adjacent side-chain-main-chain dehydrons pairing R858 guanidinium with the backbone carbonyl of R836, and R836 guanidinium with the backbone carbonyl of R858 (Fig. 7.11c, d). This dehydron pair further enhances the basicity of the vicinal water molecule (Fig. 7.10c–e), thereby promoting a higher propensity of proton transference ($\Delta\Delta G = -3.75 \pm 0.25$ kcal/mol, or $\Delta pKa \approx -6.2 \pm 0.4$) from the R858 guanidinium relative to the inactive state, where only one neighboring dehydron is present (Fig. 7.10c, d).

The chemical functionality of frustrated water at the protein interface is therefore responsible for the removal of the electrostatic repulsion caused by the special proximity of mutated residue R858 to R836 in the active state of the EGFR kinase. Given that no counterions are found in the vicinity within the crystal structure reported for the inactive state (PDB.4I20), a classical molecular dynamics computation leads to the erroneous conclusion that the repulsion caused by the L858R substitution destabilizes the inactive state, thereby favoring the active state [41]. The main trouble with this argument is that the R836-R858 "anticontact" (distance between ζ-carbons from R836, R858 is 4.28 Å) is present in *both* the inactive and the active state. Thus, the assumption arising from MD analysis may be challenged since the inactive conformation reported in PDB actually features a tight van der Waals R836-R858 contact (Fig. 7.10a) in absence of counterions (Fig. 7.10b), which actually persists in time, so as to be captured by X-ray diffraction. Furthermore, this contact, far from getting removed, becomes further stabilized in the active conformation (Fig. 7.11c, d).

In conclusion, at a variance with [43], the R-R interaction in EGFR kinase promoted by mutation L858R is stable in the inactive state. It is not the case that the R836-R858 contact entailing a classical electrostatic repulsion will be disrupted in the inactive state in a conformational exploration ending up in the active state [42]: the R836-R858 interaction is ***also*** present in the active state (Fig. 7.11c, d). This interaction was never unstable to begin with, even in the complete absence of counterions (Fig. 7.10b), because the dehydron-promoted interfacial basicity enabled acceptance of the guanidinium proton from Arg858, in turn, enabling a favorable cation-π interaction between protonated R836 and neutral R858. Far from vanishing, this frustration-related quantum effect is exacerbated in the active state.

The chemistry of interfacial water also plays a key role in enzyme processivity. This assertion is illustrated by the epigenetic mechanism of $AdoMet^+$ (S-adenosyl-methionine) methylation of histone(Lys^+) catalyzed by a protein lysine methyltransferase (PKMT) [46]. In this context, the histone(Lys^+)-$^+$AdoMet anticontact is removed as histone Lys^+ becomes deprotonated, which is a necessary step to enable the nucleophilic attack by the (neutral) Lys amino group on the Met-S bond of $AdoMet^+$ [46], yielding epigenetic methylation product $Lys(Met)H_2^+$. The $Lys\text{-}NH_3^+$ deprotonation occurs via a directed Grotthuss mechanism within a water channel in the PKMT/substrate/ligand complex [46]. However, *H_3O^+ is a stronger acid than Lys-$NH3^+$*, and so *the Grotthuss mechanism can only become feasible if water molecules in the channel are frustrated thereby becoming proton acceptors.*

Direct examination of the post-methyl-transfer complex of PKMT SET7/9 with de-methylated ligand AdoHcy (S-adenosyl-homocysteine) and Lys(Met)H_2^+ (PDB.1O9S, Fig. 7.10e) shows that this is indeed the case. A string of interfacial water molecules (solid arrows in Fig. 7.10e), beginning with crystal water Wat559, are frustrated due to confinement within the nanocavity of dehydron W260-N263. Thus, a QM computation of frustration-related basicity [34] yields $\Delta\Delta G = -1.65 \pm 0.25$ kcal/mol, or $\Delta pKa = 2.7 \pm 0.4$ associated with proton transference from substrate Lys-NH_3^+ to the dehydron-functionalized Wat559, in good agreement with other computations [46]. At a variance with the quoted study [46], we note that the decrease in Lys pKa is not the result of same-charge contact within the PKMT complex but of the presence of vicinal frustrated water in the complex, behaving as a general base.

To conclude this delineation of frustration-related Biological Chemistry, we may say that we have investigated crystal structures reporting uncompensated positive-charge side-chain anticontacts and point mutations generating a positive charge within a pre-existing positively charged environment or within a nonpolar environment. This picture is untenable under the premises of classical electrostatics and prompts a justification of the fact that close positioning of positively charged groups lowers the pKa of the titratable groups [46]. The conundrum is resolved by showing that a crucial chemical feature of interfacial water is missing from the standard analysis. A newly discovered chemical role of interfacial water arising from hydrogen-bond frustration is responsible for the stabilization of classical anticontacts. Thus, the basicity of frustrated water enables titration of ammonium and guanidinium, both weaker acids than hydronium, implying that this effect cannot be captured by a standard approach applicable to bulk water. The appropriate analysis is built upon a quantum mechanical treatment of the proton acceptance event that removes the electrostatic repulsion, causing significant decrease in pKa values relative to the free amino acid in bulk solvent. This frustration-based chemistry is corroborated by the crystallographic record and by detailed spectroscopic information on water molecules in directed Grotthuss mechanisms for proton transfer. Recent PDB-wide bioinformatics data [47] established the anticontact-dehydron motif, upholding the frustration-based chemistry picture put forth in this section.

7.8 Problems

(a) Draw the chemical mechanism of dehydron-based activation of histidine (His) enabling its phosphorylation.
(b) Draw the chemical mechanism for the dehydron-enabling second His phosphorylation.
(c) Draw the chemical mechanism for dehydron-enabling threonine (Thr) phosphorylation.

References

1. Fernández A. The principle of minimal episteric distortion of the water matrix and its steering role in protein folding. J Chem Phys. 2013;139:085101.
2. Giovambattista N, Lopez CF, Rossky P, Debenedetti P. Hydrophobicity of protein surfaces: separating geometry from chemistry. Proc Natl Acad Sci U S A. 2008;105:2274–9.
3. Fernández A. Epistructural tension promotes protein associations. Phys Rev Lett. 2012;108:188102.
4. Kumar P, Han S, Stanley HE. Anomalies of water and hydrogen bond dynamics in hydrophobic nanoconfinement. J Phys Condens Matter. 2009;21:504108.
5. Schutz CN, Warshel A. What are the dielectric "constants" of proteins and how to validate electrostatic models? Protein Struct Funct Genet. 2001;44:400–8.
6. Fernández A. Communication: chemical functionality of interfacial water enveloping nanoscale structural defects in proteins. J Chem Phys. 2014;140:221102.
7. Fernández A. Nanoscale thermodynamics of biological interfacial tension. Proc R Soc A. 2010;467:559–68.
8. Parai MK, Huggins DJ, Cao H, Nalam MN, Ali A, Schiffer CA, Tidor B, Rana TM. Design, synthesis, and biological and structural evaluations of novel HIV-1 protease inhibitors to combat drug resistance. J Med Chem. 2012;55:6328–41.
9. The Uniprot Consortium. Activities at the Universal Protein Resource (UniProt). Nucleic Acids Res. 2014;42:D191–8.
10. Hardie DG, editor. Protein phosphorylation: a practical approach. Oxford: Oxford University Press; 1999.
11. Zanzoni A, Carbajo D, Diella F, Gherardini PF, Tramontano A, Helmer-Citterich M, Via A. Phospho3D 2.0: an enhanced database of three-dimensional structures of phosphorylation sites. Nucleic Acids Res. 2011;39:D268–71.
12. Liebschner D, Dauter M, Brzuszkiewicz A, Dauter Z. On the reproducibility of protein crystal structures: five atomic resolution structures of trypsin. Acta Crystallogr Sect D. 2013;69:1447–62.
13. Pietrosemoli N, Crespo A, Fernández A. Dehydration propensity of order–disorder intermediate regions in soluble proteins. J Proteome Res. 2007;6:3519–26.
14. Dodson G, Wlodawer A. Catalytic triads and their relatives. Trends Biochem Sci. 1998;23:347–52.
15. Fernández A, Crespo A. Protein wrapping: a molecular marker for association, aggregation and drug design. Chem Soc Rev. 2008;37:2373–82.
16. Fernández A. Protein structural defects are enablers and stimulators of enzyme catalysis, Scientist Ariel Fernandez finds. Market Watch (The Wall Street Journal). http://www.marketwatch.com/story/protein-structural-defects-are-enablers-and-stimulators-of-enzyme-catalysis-scientist-ariel-fernandez-finds-2014-07-14. Accessed 14 July 2014.
17. Fernández A, Lynch M. Non-adaptive origins of interactome complexity. Nature. 2011;474:502–5.
18. Li H, Robertson AD, Jensen JH. Very fast empirical prediction and rationalization of protein pKa values. Proteins Struct Funct Bioinf. 2005;61:704–21.
19. Jarzynski C. Nonequilibrium equality for free energy differences. Phys Rev Lett. 1997;78:2690–3.
20. van der Kamp MW, Mulholland AJ. Combined quantum mechanics/molecular mechanics (QM/MM) methods in computational enzymology. Biochemistry. 2013;52:2708–28.
21. Senn HM, Thiel W. QM/MM methods for biomolecular systems. Angew Chem Int Ed. 2009;48:1198–229.
22. Agarwal S, Kazi JU, Ronnstrand L. Phosphorylation of the activation loop tyrosine 823 in c-Kit is crucial for cell survival and proliferation. J Biol Chem. 2013;288:22460–8.

23. Sankey OF, Niklewski DJ. Ab initio multicenter tight-binding model for molecular-dynamics simulations and other applications in covalent systems. Phys Rev B Condens Matter. 1989;40:3979–95.
24. Kleinman L, Bylander DM. Efficacious Form for Model Pseudopotentials. Phys Rev Lett. 1982;48:1425–8.
25. Chiodo S, Russo N, Sicilia E. Newly developed basis sets for density functional calculations. J Comput Chem. 2005;26:175–84.
26. Fernández A. Fast Track Communication: water promotes the sealing of nanoscale packing defects in folding proteins. J Phys Condens Matter. 2014;26:202101.
27. Wang J, Cieplak P, Kollman PA. How well does a restrained electrostatic potential (RESP) model perform in calculating conformational energies of organic and biological molecules? J Comput Chem. 2000;21:1049–74.
28. Plesniak LA, Connelly GP, Wakarchuk WW, McIntosh LP. Characterization of a buried neutral histidine residue in Bacillus circulans xylanase: NMR assignments, pH titration, and hydrogen exchange. Protein Sci. 1996;5:2319–28.
29. Goedken ER, Marqusee S. Co-crystal of Escherichia coli RNase HI with Mn2+ ions reveals two divalent metals bound in the active site. J Biol Chem. 2001;276:7266–71.
30. Kanaya S, Katayanagi K, Morikawa K, Inoue H, Ohtsuka E, Ikehara M. Effect of mutagenesis at each of five histidine residues on enzymatic activity and stability of ribonuclease H from Escherichia coli. Eur J Biochem. 1991;198:437–40.
31. Bentley GA, Brange J, Derewenda Z, Dodson EJ, Dodson GG, Markussen J, Wilkinson AJ, Wollmer A. Role of B13 Glu in insulin assembly. The hexamer structure of recombinant mutant (B13 Glu → Gln) insulin. J Mol Biol. 1992;228:1163–76.
32. Wei L, Jiang P, Yau YH, Summer H, Shocha SG, Mu Y, Pervushin K. Residual structure in islet amyloid polypeptide mediates its interactions with soluble insulin. Biochemistry. 2009;48:2368–76.
33. Piao X, Bernstein A. A point mutation in the catalytic domain of c-kit induces growth factor independence, tumorigenicity, and differentiation of mast cells. Blood. 1996;87:3117–23.
34. Fernández A. Packing defects functionalize soluble proteins. FEBS Lett. 2015;589:967–73.
35. Fernández Stigliano A. Breakdown of the Debye polarization ansatz at protein-water interfaces. J Chem Phys. 2013;138:225103.
36. Mobley DL, Barber AE, Fennell CJ, Dill KA. Charge asymmetries in hydration of polar solutes. J Phys Chem B. 2008;112:2405–14.
37. Magalhaes A, Maigret B, Hoflack J, Gomes JN, Scheraga HA. Contribution of unusual arginine-arginine short-range interactions to stabilization and recognition in proteins. J Protein Chem. 1994;13:195–215.
38. Fernández Stigliano A. Biomolecular interfaces: interactions, functions and drug design, epilogue. Heidelberg: Springer; 2015.
39. Garczarek F, Gerwert K. Functional waters in intraprotein proton transfer monitored by FTIR difference spectroscopy. Nature. 2006;439:109–12.
40. Littlefield P, Liu L, Mysore V, Shan Y, Shaw DE, Jura N. Structural analysis of the EGFR/HER3 heterodimer reveals the molecular basis for activating HER3 mutations. Sci Signal. 2014;7:ra114.
41. Gajiwala KS, Feng J, Ferre R, Ryan K, Brodsky O, Weinrich S, Kath JC, Stewart A. Insights into the aberrant activity of mutant EGFR kinase domain and drug recognition. Structure. 2013;21:209–19.
42. Sutto L, Gervasio FL. Effects of oncogenic mutations on the conformational free-energy landscape of EGFR kinase. Proc Natl Acad Sci U S A. 2013;110:10616.
43. Shan Y, Eastwood MP, Zhang X, Kim ET, Arkhipov A, Dror RO, Jumper J, Kuriyan J, Shaw DE. Oncogenic mutations counteract intrinsic disorder in the EGFR kinase and promote receptor dimerization. Cell. 2012;149:860–70.

44. Yun CH, Boggon TJ, Li Y, Woo MS, Greulich H, Meyerson M, Eck MJ. Structures of lung cancer-derived EGFR mutants and inhibitor complexes: mechanism of activation and insights into differential inhibitor sensitivity. Cancer Cell. 2007;11:217–27.
45. Dougherty DA, Ma JC. The cation-π interaction. Chem Rev. 1997;97:1303–24.
46. Zhang X, Bruice TC. Enzymatic mechanism and product specificity of SET-domain protein lysine methyltransferases. Proc Natl Acad Sci U S A. 2008;105:5728–32.
47. Scott LR, Fernández Stigliano A. Mismatched ions indicate quantum effects in proteins. The University of Chicago, Department of Computer Science Technical Report TR-2015-10; 2015.

Chapter 8
Epistructural Selectivity Filters for Molecular Targeted Therapy

Evolutionary biology tells us that 3-dimensional structure is conserved across homologous proteins. Thus, a protein target for a specific therapeutic purpose may have off-target homologs that will likely be impacted by standard structure-based drug designs. This fact introduces a major hurdle in the quest to control specificity. On the other hand, as shown in Chap. 6, the epistructural features of a protein, such as the dehydron pattern or the hot spots of epistructural tension, are not conserved across homologs. This argues for an *epistructure*-based rather than a structure-based approach to drug design. The main lesson to be drawn is that the specificity filter is enshrined in the aqueous interface and not in the protein structure. To date, rational drug design remains essentially structure-based, therefore, the impact of drug-based inhibition often spreads to off-target members of the target protein family. This uncontrollable cross-reactivity leads to undesirable side effects that may be toxic or even health-threatening. The problem becomes particularly acute when attempting to interfere with signaling pathways involved in cell fate and cell proliferation, the type of molecular intervention often exploited in molecular anticancer therapy. In this context, the therapeutically relevant targets are the kinases, signal transducers with extensive homology in humans (and in most other species) thereby sharing significant structural similarity. However, as shown in Chap. 6, *dehydrons* are typically *not* conserved across proteins of common ancestry, in particular within kinase superfamilies. This evolutionary property combined with the fact that dehydrons promote their own dehydration, thereby triggering protein-ligand associations, strongly suggest that dehydrons constitute valuable *a-priori* targets to enhance specificity. Validating this assertion and exploring its ramifications and actual implementation constitute the purview this chapter. There are nearly equivalent yet more quantitatively precise epistructural attributes that may be utilized as selectivity switches across homologous targets, as this chapter shows while it explores the paradigm of epistructure-based drug design. The main rationale for this design strategy can be summarized as follows: If the packing defect is an enabler and stimulator of catalytic function (Chap. 7), then its removal through intermolecular wrapping upon drug/target association should

A. Fernández, *Physics at the Biomolecular Interface*, Soft and Biological Matter,
DOI 10.1007/978-3-319-30852-4_8

impair the function of the targeted protein and do so with high specificity, which is precisely the goal of molecular targeted therapy. Thus, the paradigmatic concept of "drug as dehydron wrapper" is established in this chapter, notwithstanding the fact that there are other epistructural markers for highly specific protein-ligand associations, *albeit* harder to implement computationally.

8.1 Controlling Drug Specificity: A Therapeutic Imperative

Rational drug design and optimization have had only a modest impact in the pharmaceutical endeavor, where sheer serendipity and high-throughput screening have play a far larger role. As we shall argue in this chapter, this is partly due to inherent shortcomings of structure-based design and to an inadequate knowledge of the underlying scientific fundamentals. This state of affairs prompts us to advocate for a generation of epistructure-based drugs guided by systems biology annotation. Engineering potent drug-based inhibitors of protein functions with controlled specificity and without harmful side effects remains a major goal in drug design. Drug-promoted inhibition of enzymatic activity may have at least two sources of side effects: (a) the drug may interfere with pathways other than the one of therapeutic relevance due to the manifold roles of the target protein in off-target cellular contexts; (b) paralogous proteins (homologs within the same species, i.e., humans), being structurally similar, may associate with the inhibitor, causing undesired cross-reactivities. Both sources of side effects have an evolutionary root that is traceable to the fate of duplicate genes, as shown in Chap. 6. Building on the evolutionary knowledge acquired in Chap. 6 and on the understanding of biomolecular interfaces acquired in Chaps. 1–4, this chapter introduces new and paradigmatic concepts to minimize drug interactions with toxicity-related targets. Thus, this chapter and the ones that follow advocate for *epistructure*-based rather than structure-based drug design, noting that the filter for drug specificity is subsumed in the epistructural interface of the protein and not in the protein structure *per se*. In essence, the epistructure of the protein varies significantly across members of the protein family, while the structure remains topologically invariant. The lesson that distills from this argument is that *mastering epistructural targeting of the protein is an imperative to achieve drug specificity*.

In principle, the control of specificity is enabled as we exploit characteristics of protein-ligand interactions that are not preserved across paralogs, despite the similarity of their three-dimensional structures [1–3]. Thus, the insights from the previous chapters prompt us to focus on a singular feature, the dehydron, a packing defect shown to be of paramount importance in promoting protein associations [4–6] due to its role as water frustrator (cf. Chaps. 1–5). Quantifiable features of these interactions allow us to introduce a target phylogeny that we can exploit to infer drug cross-reactivities and modulate the inhibitory impact of a drug, funneling it onto the realm of clinical relevance.

Considerable research has helped clarify many aspects of protein-ligand interaction [7]. Various methods enable one to predict such interactions, such as the assessment of packing defects [8], analysis of protein interface geometry [9], docking dynamics [10], modeling free energy [11], computational molecular probing [12], sequence threading [13], and learning-theory approaches [14]. The first four of these are based on properties of protein interfaces; the latter two are based only on sequences, and attempt to 'learn' features encoded in those sequences. Combined methodologies are also being developed [14]. Some of these techniques seek features of protein surfaces that promote protein-protein interactions. In this chapter we show how features with a quantifiable metric can be used to compare paralogs in a way that provides a guidance to develop more selective drugs.

To function, soluble proteins must retain relatively stable structures, a condition often requiring protein-protein associations [15–17]. Isolated structures with packing defects arising from poorly protected backbone hydrogen bonds typically do not prevail in water [3, 18]. On the other hand, such defects are inherently sticky, promoting removal of surrounding water through protein associations [8], as indicated in Chaps. 1 and 2. Accordingly, in this chapter we show that *dehydrons may be targeted in a new generation of highly selective drug-based inhibitors*. As shown in Chap. 4, known structures of protein-inhibitor complexes [19–24] provide the blueprint to design inhibitors that in effect become wrappers or protectors of dehydrons. Yet, the latter feature belongs to a meta-category, the *epistructure*, requiring an understanding of the dynamic relation between protein and solvent structure, as described in Chaps. 1–5.

As shown in Chap. 6, while folds are typically conserved across homologous proteins, dehydron patterns are typically not conserved [3], and this property provides the main motivation for advocating a paradigmatic shift in drug design. While every targettable protein family may in principle be subject to the same analysis, we shall narrow down the scope of the discussion for the sake of clarity. Thus, we shall restrict ourselves to the goal of designing molecular therapeutic agents that selectively interfere with cell signaling. In this context, structure-based design becomes particularly daunting because the basic signal transducers in the cell, the kinases, share a common molecular ancestry and hence possess an uncanny structural similarity. This fact turns drug specificity into a major issue.

Kinase inhibitors are designed to impair the ATP-dependent signal transduction (Fig. 8.1a, b). Hence they should be capable of binding to the target competitively or noncompetitively vis-à-vis the natural kinase ligand ATP. The main structural features of the binding of ATP to a kinase are illustrated in Fig. 8.1a, b, where an active (phosphorylated) tyrosine kinase, that of the insulin receptor (IR), is shown bound to ATP and to a substrate peptide. The kinase substrate becomes phosphorylated at a tyrosine residue through a trans-esterification reaction involving transference of the γ-phosphate of ATP (Fig. 8.1b). Thus, a major challenge to design ATP-competitive inhibitors is the fact that the ATP-binding regions of most kinases are structurally very similar (with a RMSD <1.2 Å in backbone coordinates) and have a very high level of amino acid conservation.

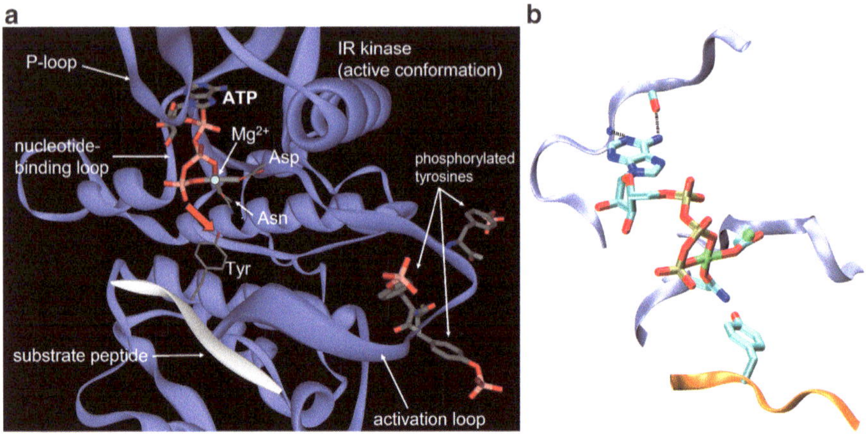

Fig. 8.1 **(a)** Generic structural features of a ternary complex involving a phosphorylated (active) kinase, its natural ligand ATP, and its phosphorylation substrate peptide (*white*). The phosphorylation event (*red arrow*) involves trans-phosphoesterification of the substrate tyrosine with transference of the γ-phosphate from ATP to the substrate, concurrent dismantling of the anchoring Mg(II)-coordination complex, and finally, release of ADP. **(b)** Detailed rendering of the ATP position with respect to the phosphorylation substrate. Atoms are conventionally colored with Mg (II) in *green*, substrate peptide in *gold* and IR kinase backbone in *light blue*

Contrasting with this troublesome aspect, the lack of conservation of the wrapping (dehydron) pattern across kinases has paramount potential to promote drug specificity, as illustrated in Fig. 8.2a, b. Thus, the structural alignment of the focal adhesion kinase (FAK, PDB.2J0L) and the insulin receptor kinase (PDB.1GAG) reveal a significant similarity, with RMSD (root mean square displacement) <1.3 Å (Fig. 8.2a). Yet, there are crucial differences in the dehydron patterns of both proteins (Fig. 8.2b), particularly in their ATP-binding pockets, where the ATP-competitive inhibitors bind [18]. These differences can be exploited with considerable therapeutic advantage: FAK is a target of major clinical relevance to treat cancer metastasis and yet the probably activity of a purported FAK inhibitor against the structurally related insulin receptor kinase is likely to have potentially devastating effects, arising mainly from the induction of a diabetic coma in the patient [18]. However, *there are dehydrons unique to FAK and not found in its paralog (arrows in Fig. 8.2b) which may be targeted to satisfy the therapeutic imperative of specificity*. This peculiar way of targeting a molecular feature materializes as inter-molecular wrapping, as illustrated in Fig. 8.2c. The example shown in Fig. 8.2a, b dramatically reveals the difficulty in telling apart targets of clinical relevance from toxicity-related targets in molecular targeted therapy and unravels a possible avenue to solve this problem.

Thus, in principle, side effects resulting from off-target ligand binding may be minimized by selectively targeting nonconserved dehydrons unique to the desired target protein with the guidance of a measure of wrapping proximity between structures. In this regard, this chapter reveals that packing defects may act not

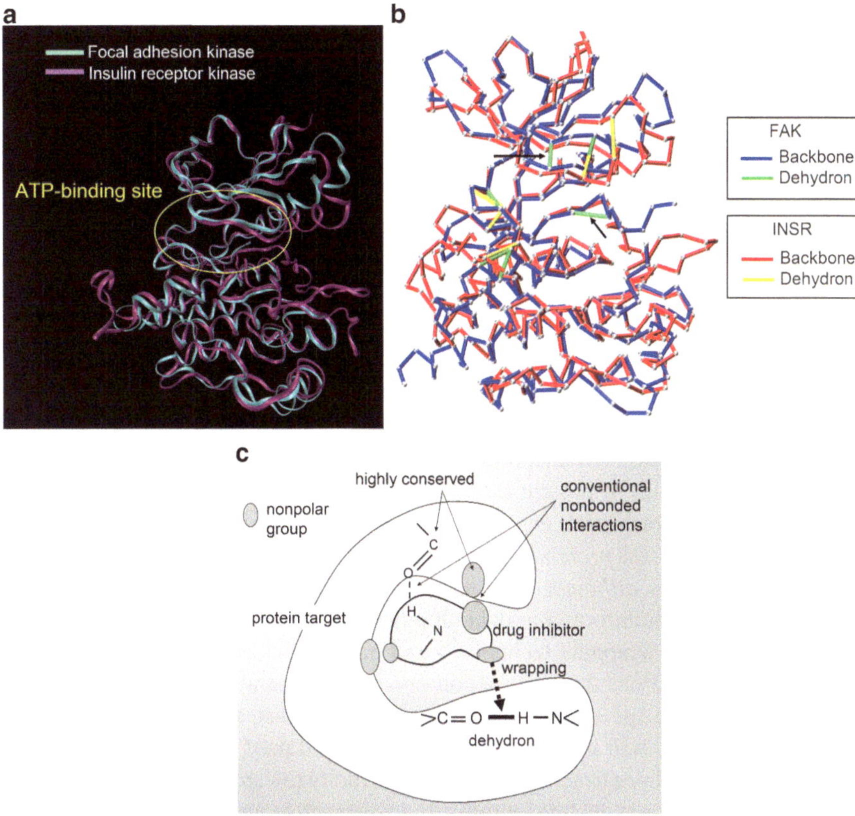

Fig. 8.2 Structural (**a**) and epistructural (**b**) comparison between the focal adhesion kinase (FAK, PDB.2J0L) and the insulin receptor kinase (INRK, PDB.1GAG). The dehydrons in the ATP-binding region that are unique to FAK are marked by *arrows*. (**c**) Drug paradigmatically represented as a wrapper of pre-formed dehydrons in the target protein. Conventional nonbonded interactions between drug and target such as hydrophobic contacts and intermolecular hydrogen bonds typically foster drug promiscuity as they involve conserved features in the target protein

only as selectivity filters operative across homolog proteins, but also as selectivity switches, to re-engineer a protein inhibitor to redirect its impact towards another target. This latter role requires taking advantage of the spatial displacement of packing defects across aligned homologous structures.

Functional differences across homolog proteins sharing a common fold are very frequent in biology. Previous studies revealed that the wrapping constitutes a molecular dimension explored when evolution is constrained to preserve the fold for functional reasons. Thus, it is likely that wrapping differences across paralogs are indicative of the functional fine tuning needed to avoid dosage imbalances (see Chap. 6). It is precisely this fine tuning that needs to be discriminated in drug-based molecular therapy. *As this chapter reveals, such wrapping differences offer a new control in constructing the selectivity filters and selectivity switches that may endow the wrapping technology with a high therapeutic value.*

Structural descriptors of protein binding sites, such as hydrophobicity [25], curvature [26] and accessibility [27] are routinely used to guide inhibitor design. However, examination of the 814 non-redundant protein-inhibitor PDB complexes reveals that 488 of them have binding cavities with only average hydrophobicity [25], not significantly higher than the rest of the surface. In such cases, ligand affinity is attributed to the intermolecular hydrogen-bonding propensities of the inhibitor, inferred from protein-substrate transition-state mimetics [21–24]. However charge screening in water renders putative intermolecular hydrogen bonds unlikely promoters of protein-ligand association, unless water removal is favored at the interface by some pre-existing feature on the surface of the protein [8].

One obvious such feature is the solvent-exposed nonpolar moiety, except that, as stated above, such features occur infrequently on binding sites and if they do, they are highly conserved across homologs. A more common and less troublesome water-excluding feature has been recently identified. We have reported [3, 8, 18, 28, 29] that dehydrons constitute sticky sites with a propensity to become dehydrated as a result of protein associations.

To explore the potential of dehydrons as promoters of drug selectivity, we start by noting that *in most PDB protein-inhibitor complexes, the ligand is in effect a wrapper of protein dehydrons, although it was not purposely designed to fulfill this role*. In this way, the available bioinformatics data support the novel design concept. Ultimately, we shall develop a wrapping technology and provide evidence supporting the idea that targeting dehydrons that are not conserved across paralogs constitutes a useful strategy to enhance the selectivity of the inhibitor. Thus, the notion of wrapping (packing) similarity will be rigorously introduced and used selectively to modify a multiple-target inhibitor to achieve a higher specificity towards a particular target and subsequently to redirect its impact towards an alternative target.

8.2 Epistructure-Based Drug Design

Given that the dehydron has a catalytic role (Chap. 7) and frustrates nearby water, its removal through intermolecular wrapping upon drug/target association becomes possible and should impair the function of the targeted protein. This line of thought marks the genesis of the paradigmatic concept of "drugs as dehydron wrappers". This leads us to pose a simple question: Is intermolecular wrapping a hitherto undetected factor contributing to ligand affinity and selectivity? To answer this question, the interfaces of the 814 protein-inhibitor PDB complexes were re-examined to determine whether inhibitors were "dehydron wrappers", that is, whether nonpolar groups of inhibitors penetrated the desolvation domain of dehydrons. This feature was found in 631 complexes, and *invariably* found in the 488 complexes whose the binding cavities presented average or no surface hydrophobicity [25]. Figure 8.3 illustrates such wrapping for the HIV-1 protease [21, 30]. The inhibitor contribution to improve the protein packing is not fortuitous since the substrate must be anchored and water must be expelled from the enzymatic site. *Strikingly, the wrapping of dehydronsis not purposely targeted in current drug design.*

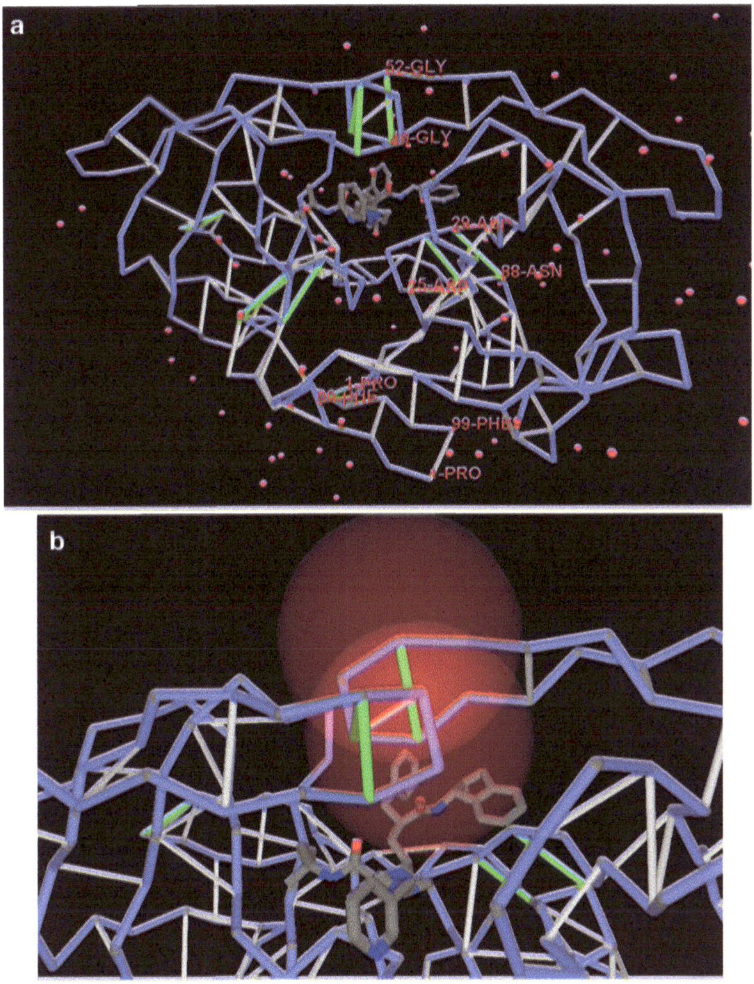

Fig. 8.3 HIV-1 protease in complex with inhibitor that acts as dehydron wrapper. (**a**) Indinavir (Crixivan) inhibitor crystallized in complex with HIV-1 protease (PDB.2BPX). The packing defects in the dimeric HIV-1 protease are displayed jointly with their spatial relation to the inhibitor position. The chain backbone is represented by *blue* virtual bonds joining α-carbons, well wrapped backbone hydrogen bonds are shown as *light grey* segments joining the α-carbons of the paired residues, and dehydrons are shown as *green* segments. (**b**) Desolvation spheres for flap Gly49-Gly52 dehydron containing nonpolar groups of the wrapping inhibitor. Reprinted from [1], copyright 2005 with permission from Elsevier

The Merck inhibitor *Indinavir* (Crixivan) bound to the functionally dimeric HIV-1 protease (PDB.2BPX) is shown in Fig. 8.3 [21, 30]. The dehydrons in the protease are marked in green. On each monomer, these are backbone hydrogen bonds between these residue pairs: Ala28-Arg87; Asp29-Asn88; Gly49-Gly52 and Gly16-Gln18. The cavity associated with substrate binding contains the first three dehydrons, with dehydron 49–52 located in the flap and dehydrons 28–87 and 29–88 adjacent to the catalytic site (Asp25), to anchor the substrate. This "sticky track" determined by dehydrons 28–87 and 29–88 is required to align the substrate peptide across the cavity as needed for nucleophilic attack by the Asp25. The flap, on the other hand, needs an exposed and hence labile hydrogen bond to confer the flexibility associated with the gating mechanism. The lack of protection on the flap (49–52) hydrogen bond becomes the reason for its stickiness, as the bond can be strengthened by the exogenous removal of surrounding water (Fig. 8.3b). The positioning of all three dehydrons in the cavity (six in the dimer) promotes inhibitor association (Fig. 8.3a).

Indinavir is a wrapper of packing defects in the enzymatic cavity: it contributes 12 desolvating groups to the 49–52 hydrogen bond (Fig. 8.3b), 10 to the 28–87 hydrogen bond and 8 to the 29–88 hydrogen bond. All functionally relevant residues are either polar or expose the polarity of the peptide backbone (Asp25; Thr26; Gly27; Ala28; Asp29; Arg87; Asn88; Gly49; Gly52), and thus are not *themselves* promoters of association to hydrophobic groups of the ligand. The strategic position of dehydrons involving these residues in their microenvironments becomes a decisive factor to drive water removal or charge de-screening required in facilitating the enzymatic nucleophilic attack.

Central to drug design is the minimization of toxic side effects and the enhancement of safety. Achieving such goals is proving to be extremely challenging and yet a more modest yet crucially related aim, the control of specificity, seems clearly within reach, as advocated in this section and proved in the subsequent chapters. Because paralog proteins are likely to share a common structure [31], the possibility of multiple binding partners for a given protein inhibitor arises, unless the inhibitor specifically targets nonconserved features. Thus, the problem of controlling specificity may in principle be circumvented by targeting dehydrons because, in contrast to the fold, the wrapping is generally *not* conserved across homologs [3]. This evolutionary fact makes the dehydron pattern a possible selectivity filter for drug design. The usefulness of this filter of course depends on our ability to design drugs that may contribute intermolecularly to a better wrapping of the target.

To determine whether dehydron targeting is likely enable a control of specificity, we first investigated the extent of conservation of dehydrons across human paralogs in PDB. The paralogs for every crystallized protein-inhibitor complex were identified and dehydron patterns at binding cavities were compared. A 30 % minimal sequence alignment was required for paralog identification. Packing defects were found to be a differentiating marker in paralogs of 527 of the investigated 631 proteins crystallized in complex with inhibitors. Protein chains are often reported in complexes with different inhibitors.

The PDB contains 440 redundancy-free pairs of human paralogs. Of these, 308 involve some of the 527 proteins containing binding dehydrons. In 269 pairs, the intramolecular wrapping at the binding cavity differs in the location or presence of at least one dehydron, and in 203 pairs, the difference extends to two or more dehydrons. Thus, *the probability of avoiding side effects by selectively wrapping packing defects is estimated at 88 %* (269 pairs out of 308).

8.3 Poor Dehydron Wrappers Make Poor Drugs

The natural product *staurosporine* has been singled out as the ligand inhibitor most promiscuously interactive with the human kinome [32]. Its complexes are extensively reported in the PDB and high-throughput screening reveals nanomolar inhibitory impact over ~80 % of the screened kinases. In spite of this extremely high affinities, staurosporine has shown very limited therapeutic possibilities, and that is—we anticipate—because it is a poor wrapper of preformed dehydrons in the proteins to which it binds. We have examined all 96 staurosporine-protein PDB complexes and found no instance of the inhibitor acting as dehydron wrapper. All interactions were hydrophobic in nature, involving nonpolar groups on the protein surface. An illustration of such type of binding is given in Fig. 8.4, where the Syk tyrosine kinase-staurosporine complex (PDB.1XBC) is shown to hinge on three pivotal hydrophobic interactions engaging Phe382, Val385 and Leu377. It should be noticed that staurosporine does not wrap the single dehydron Gly380-Gly382 closest to the binding site. Thus, since surface hydrophobic residues tend to be conserved as markers for interactivity and staurosporine does not discriminate among packing differences in the kinome, its promiscuity becomes entirely expected, and so is its uselessness as a therapeutic agent.

8.4 The Biomolecular Interface as a Selectivity Filter

A major paradox arises as we attempt to exploit the subtle structural differences across paralog proteins: *Structure-based drug design proves inadequate to tell appart paralog structures*. It is as if structural biology requires a meta-category that has not been created so far, and we are forced to resort to epistructural biology, as developed in Chaps. 1–7. The subtle structural differences among paralogs of a kinase target may be quantified by introducing a "wrapping homology". Assessing wrapping similarity across paralogs requires comparing the microenvironments of backbone hydrogen bonds [8] that can be effectively carried out following the structural alignment [33]. This "wrapping homology" becomes a useful tool to control specificity by targeting a packing defect. For a protein chain of length N, a matrix of dehydrons D_{ij}, $i, j = 1, 2, \ldots, N$ may be constructed by choosing $D_{ij} = 1$ if residues *i* and *j* are paired by a dehydron and $D_{ij} = 0$, otherwise. This representation is illustrated in Fig. 8.5a. When the dehydron matrices for two proteins *m* and *n* are aligned, they are trimmed by restricting them to those residues that have been structurally aligned in a one-to-one correspondence. Then a Hamming distance $M_H(m, n) = \sum_{i<j} |D_{ij}(m) - D_{ij}(n)|$ may be computed to serve as indicator of the wrapping similarity between proteins *m, n*. Given a collection of paralogs m, $n = 1, \ldots, F$, and its corresponding packing-similarity matrix $\mathbf{M_H} = [M_H(m, n)]$, a wrapping-similarity tree (WST) may be easily constructed by adopting the premise that the minimal pathway between protein-nodes *m* and *n* crosses a number of nodes equal to $M_H(m, n)$.

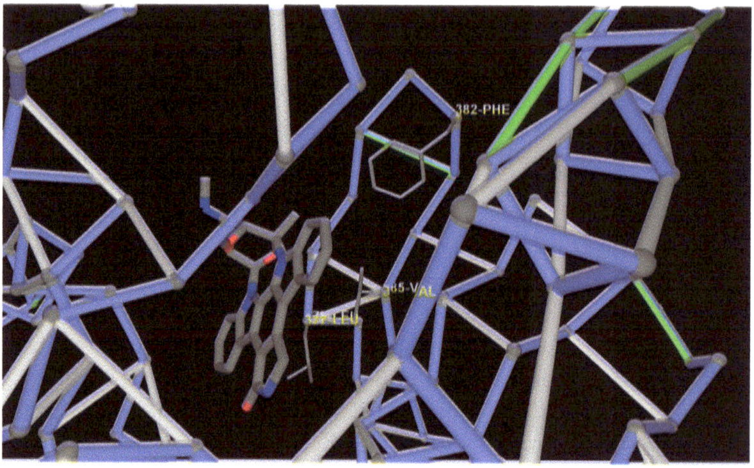

Fig. 8.4 Staurosporine does not wrap dehydrons and thus, it does not discriminate proteins with different packing, which makes it highly promiscuous within the kinome. The staurosporine-Sykcomplex (PDB.1XBC) reveals a binding mode determined exclusively by hydrophobic interactions engaging residues Leu377, Phe382, Val385. The closest Gly380-Gly383 dehydron is not wrapped by staurosporine

A comparison of dehydron matrices for proteins with aligned structure enables an automated digitalization of the drug selectivity problem. The alignment of dehydron matrices reveals the pattern of *a-priori* cross reactivities: Thus, targeting a specific dehydron in one particular protein is likely to promote cross reactivities with other targets where this dehydron is found at the aligned position and preclude inhibitory activity whenever the dehydron aligns with a well-wrapped hydrogen bond (Fig. 8.5b).

8.5 Implementing an Epistructure-Based Drug Design

We now present an exercise on controlling selectivity based on the wrapping design concept. Because of the evolutionary proximity of kinases, side effects arising from off-target ligand binding often arise with kinase inhibitors, especially in cancer therapy [32]. Thus, a need arises to sharpen the binding affinity within the pharmacokinome associated with a specific drug. For example, the selective inhibition of the constitutively active or deregulated Brc-Abl (Abl: Abelson tyrosine kinase), the fusion product of an aberrant chromosomal translocation, has proven decisive to treatchronic myeloid leukemia (CML) [34]. Brc-Abl has been proven to be a primary target for the potent inhibitor imatinib (Gleevec) [34], but is not its only target [32]. Of the alternative targets with reported structure, the C-Kit tyrosine kinase has been recognized as binding partner with clinical relevance, making Gleevec a therapeutic agent to treat certain solid tumors like gastrointestinal

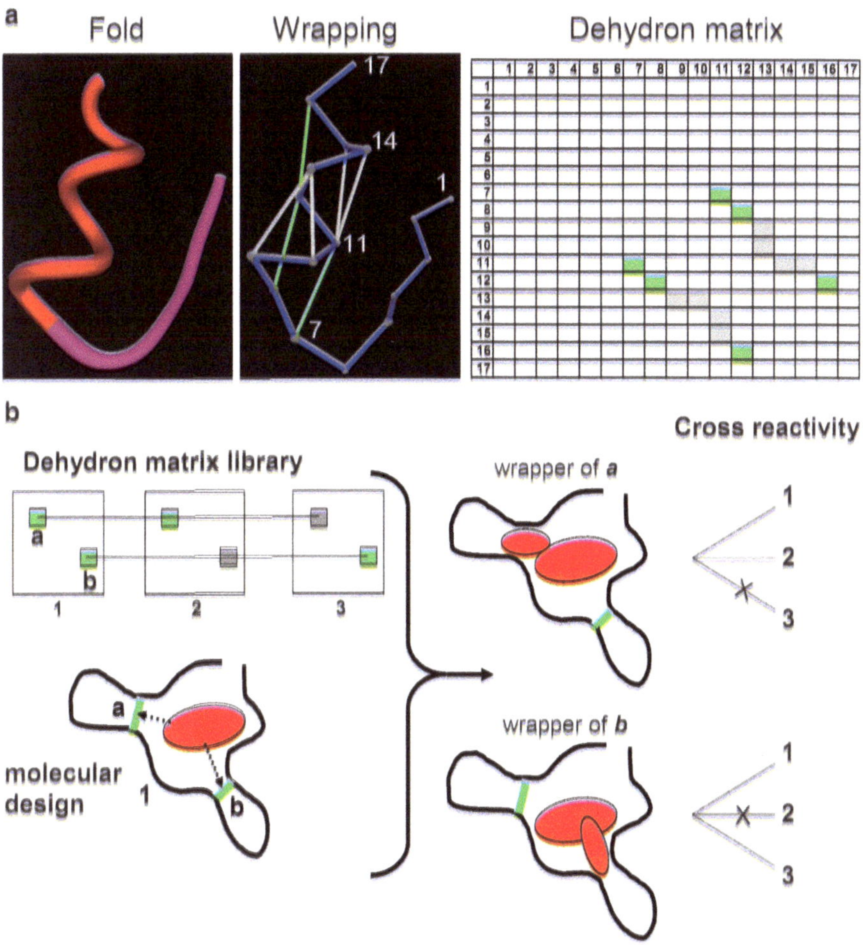

Fig. 8.5 Dehydron matrices and *a-priori* cross reactivities derived from their alignment. (**a**) Three different representations of a folded peptide chain of length N = 17: fold, wrapping and dehydron matrix. The *grey* entries in the matrix indicate well-wrapped hydrogen bonds and the *green* entries represent dehydrons. (**b**) A-priori cross reactivities across three proteins (**1**, **2**, **3**) of drugs/ligands designed to wrap specific dehydrons (**a**, **b**) in protein **1**. The cross-reactivities are inferred from the alignment of the dehydron matrices

stromal tumor [35, 36]. In addition Gleevec binds tightly to the lymphocyte-associated kinase (Lck) [37], an immunosuppression related target probably undesirable and to the platelet-derived growth factor receptor (PDGFR) kinase, a close homolog of Kit. Thus, we sought to modify Gleevec to improve its selectivity for Brc-Abl by targeting dehydrons not conserved across paralogs.

The protein-inhibitor complex PDB.1FPU (Fig. 8.6a), reveals three electrostatic interactions in Brc-Abl, the dehydrons Gly249-Gln252, Gln300-Glu316 and the salt bridge Lys271-Glu286, that can be better wrapped by methylatingGleevec at the positions indicated. Thus, methylation at positions I, II, and III would presumably

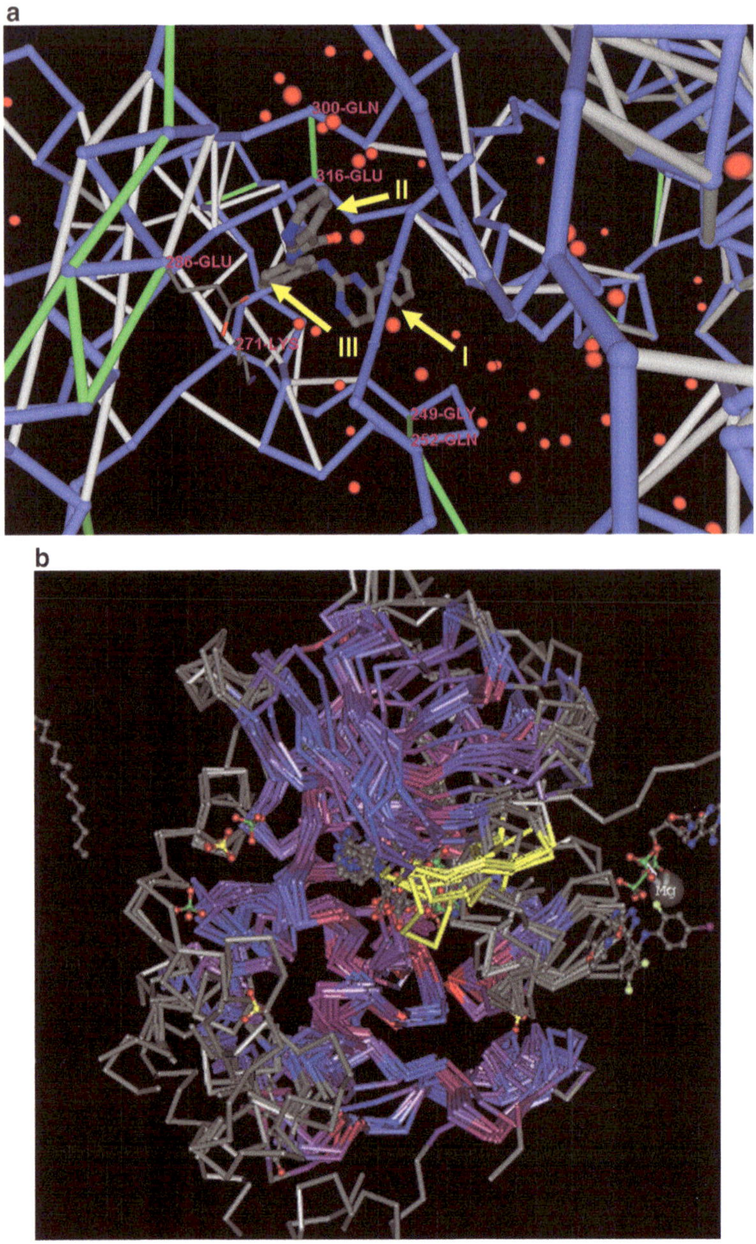

Fig. 8.6 Wrapping as selectivity filter in inhibitor drug design: Modifications of Gleevec geared at improving selectivity and affinity for Brc-Abl. **(a)** Three possible sites for Gleevec methylation (I–III) aimed at selectively improving the wrapping of packing defects of Brc-Abl (PDB.1FPU). **(b)** Structural alignment of Brc-Abl and its six paralogs using the program Cn3D. The *yellow region* corresponds to residues 247–257 in Abl

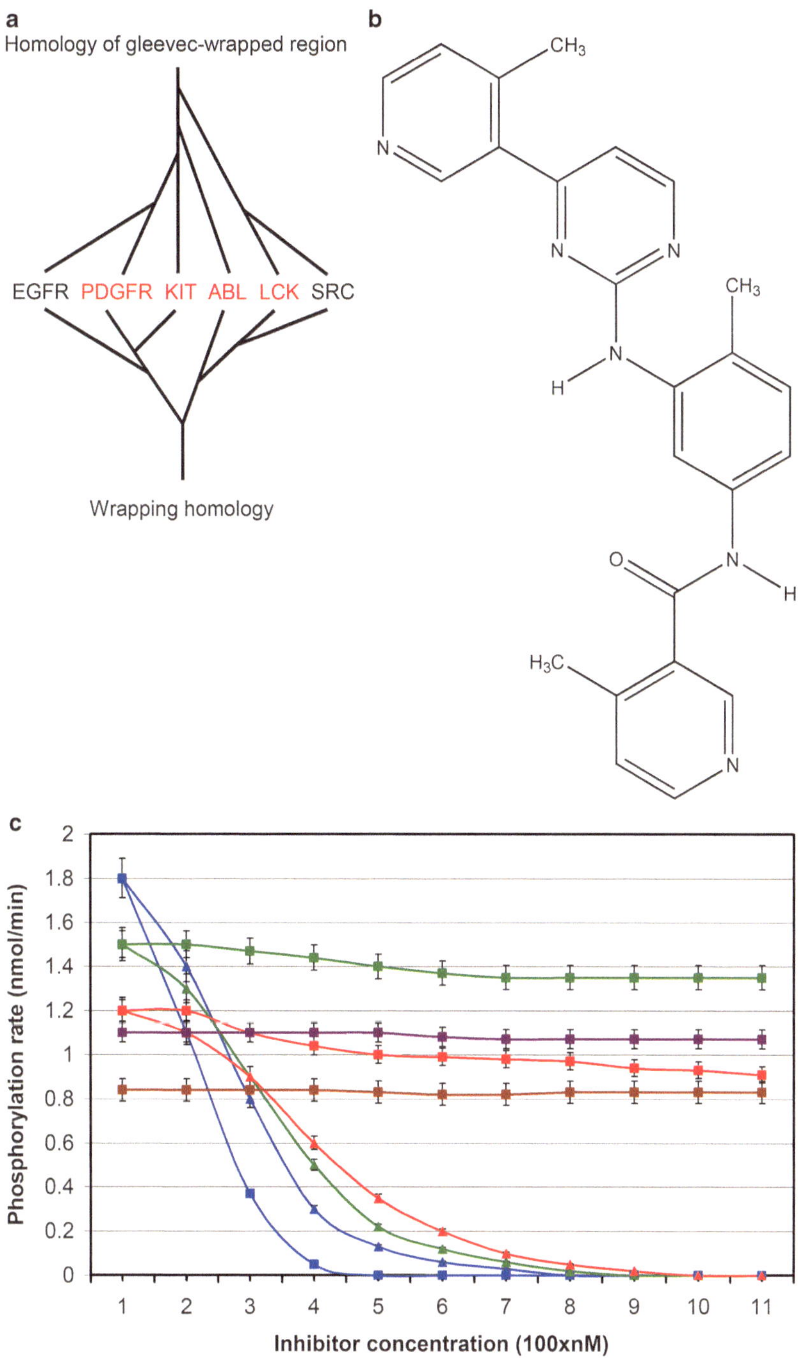

Fig. 8.7 (**a**) Wrapping similarity tree (WST) for the six structurally aligned paralogs including Brc-Abl. The WST restricted to the alignments of the Gleevec-wrapped region in Brc-Abl is shown above. The paralogs in *red* have the most similar wrapping in the region that aligns with the Gleevec-wrapped region in Brc-Abl. (**b**) The modified Gleevec-based molecule methylated at sites

improve the wrapping of dehydrons 249–252, 300–316 and salt bridge 271–286, respectively. A structural alignment of the paralogs of Brc-Abl was performed using the program Cn3D [33], to investigate the microenvironment conservation for these intramolecular interactions. The alignment of six paralog proteins is shown in Fig. 8.6b. The dehydron 249–252 (crankshaft-like kink marked in yellow in Fig. 8.6b) is not conserved in *any* of the six paralogs of Brc-Abl, while the dehydron 300–316 becomes well wrapped in the paralogs. On the other hand, the microenvironment of the salt bridge 271–286 is conserved.

The WST for Abl and its paralogs is shown in Fig. 8.7*a.* We have also restricted the WST computation to the hydrogen bonds/dehydrons in Brc-Abl which become wrapped by Gleevec upon association. *Adopting this restricted measure, the closest proteins to Brc-Abl become Lck, C-Kit and PDGFR, precisely the alternative Gleevec targets with reported structure.*

In order to enhance affinity and selectivity for Brc-Abl, we modified the inhibitor methylating at positions I and II (Fig. 8.7b). The synthesis of the wrapping prototype recapitulates imatinib synthesis [39], as described in [38]. To test whether the specificity and affinity for Brc-Abl improved, we conducted a spectrophotometric kinetic assay to measure the phosphorylation rate of peptide substrates in the presence of the kinase inhibitor at different concentrations. This assay couples production of adenosine diphosphate (ADP), the byproduct of downstream phosphorylation, with the concurrent detectable oxidation of reduced nicotinamide adenosine dinucleotide (NADH). The oxidation results upon transfer of phosphate from PEP (phosphoenolpyruvate) to ADP followed by the NADH-mediated reduction of PEP to lactate. Thus, phosphorylation activity is monitored by the decrease in 340 nm-absorbance due to the oxidative conversion NADH $\rightarrow$ NAD^+ [34, 38].

As indicated in Fig. 8.7c, the spectrophotometric assay revealed that the inhibition of the unphosphorylated Brc-Abl by the wrapper of the 249–252 and 300–316 dehydrons (the I, II-methylation product) improved over Gleevec levels. Furthermore, the inhibitory impact of the dehydron wrapper became selective for Brc-Abl*vis-à-vis* C-Kit and Lck. Dehydrons 249–252 and 300–316 are absent in the latter kinases, and consistently, the drug designed to better wrap them has very low inhibitory impact against C-Kit and Lck.

Fig. 8.7 (continued) I and II and assayed *in vitro*. (**c**) Phosphorylation rates of Brc-Abl (*blue*), C-Kit (*green*), Lck (*red*), PDGFR (*purple*) and Src (*brown*) in the presence of Gleevec (*triangles*) and in the presence of the I, II-methylated modified Gleevec (*squares*). The latter compound was designed to better wrap the nonconserved dehydrons in Brc-Abl. To measure the rate of phosphorylation due to kinase activity in the presence of inhibitors, a standard spectrophotometric assay has been adopted in which the adenosine diphosphate (ADP) production, a marker of phosphorylation activity, is coupled to NADH (reduced nicotinamide adenosine dinucleotide) oxidation and determined by absorbance reduction at 340 nm, as described in [34, 38]. Basically, ADP production is coupled to the final steps of glycolysis, so that the enzymatic reaction of ADP with reactant phosphoenol pyruvate generates pyruvate whose production can be measured by reducing it to lactate with NADH. Reprinted from [1], copyright 2005 with permission from Elsevier

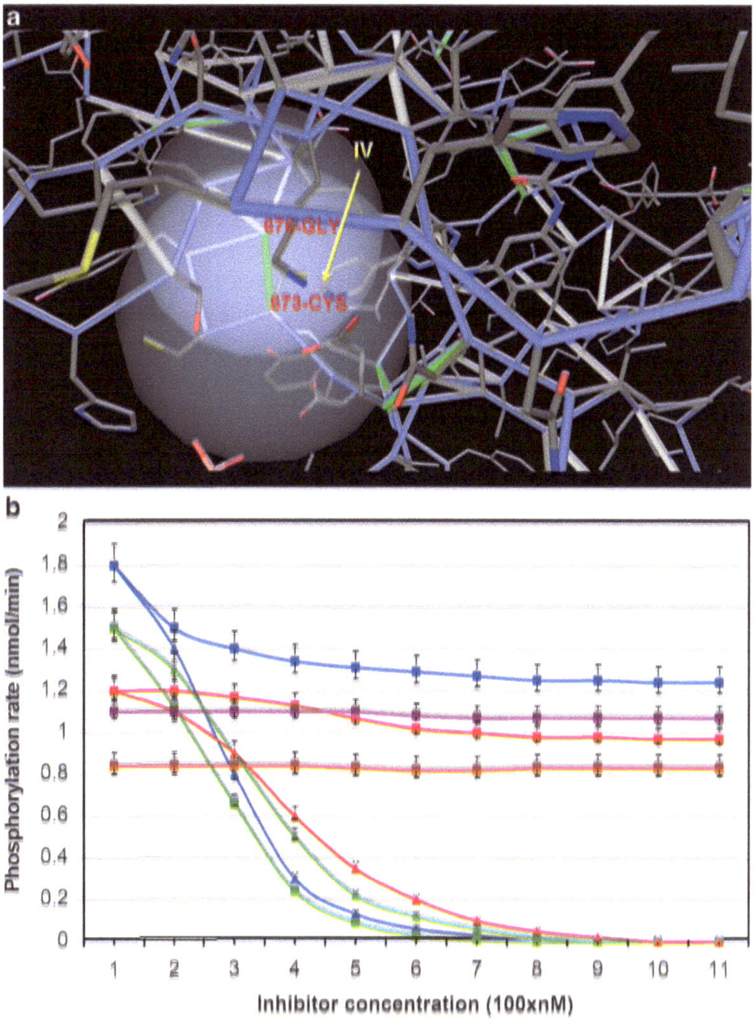

Fig. 8.8 Selectivity switch determined by packing differences across paralog proteins. The structural alignment (Fig. 8.6b), reveals that dehydron Gly249-Gln252 in Brc-Abl maps onto the spatially displaced dehydron Cys673-Gly676 in C-Kit. (**a**) Relative position of 673–676 dehydron of C-Kit with respect to Gleevec in protein-inhibitor complex (PDB.1T46). The desolvation domain of the dehydron is displayed by two intersecting spheres. By methylating Gleevec at the *para* position (IV) rather than the *ortho* position (I, Fig. 8.6a), the inhibitor becomes a selective wrapper of the packing defect in C-Kit. (**b**) Phosphorylation rates from spectrophotometric assay on the kinases indicated in Fig. 8.7c inhibited by Gleevec methylated at positions II and IV. The *squares* correspond to the Gleevec variant methylated at positions II, IV. Notice the selective and enhanced inhibition of C-Kit. Reprinted from [1], copyright 2005 with permission from Elsevier

8.6 Selectivity of Epistructure-Based Drug Designs

Can we really control imatinib specificity through a reworking of the parent compound as suggested in Fig. 8.5b? If this is indeed the case, we should be capable of re-funneling the inhibitory impact towards any of the alternative primary targets of imatinib. A structural alignment of the Abl and Kit kinases, reveals that dehydron Gly249-Gln252 in Abl (Fig. 8.6a) maps onto dehydron Cys673-Gly676 in Kit (Fig. 8.8a), spatially displaced with respect to the former. Given the relative position of the 673–676 dehydron of Kit with respect to Gleevec in the protein-inhibitor complex (PDB.1T46), we may determine which modification of Gleevec would selectively direct its impact towards C-Kit. By methylating Gleevec at the *para* position (IV) in the terminal ring, rather than at the *ortho* position (I, Fig. 8.6a), the inhibitor becomes a selective wrapper of the packing defect in C-Kit, while it no longer wraps the 249–252 dehydron in Brc-Abl. The inhibitory impact of Gleevec methylated at positions II, IV becomes apparent in the phosphorylation rates obtained from the spectrophotometric assay on the same kinases studied previously (Figs. 8.7c and 8.8b). *The inhibitory impact of the II, IV Gleevec variant is now concentrated on C-Kit, rather than Brc-Abl. Thus, by redesigning Gleevec according to the wrapping differences between Brc-Abl and C-Kit, we redirected the inhibitory impact from one kinase to the other.* Taken together, both design exercises reveal that wrapping technology may enable us to introduce a selectivity switch in inhibitor design.

8.7 Targeting Dehydrons to Selectively Block Protein Functions

This chapter already revealed how and why dehydrons may be exploited as selectivity filters in drug design. Yet, a more fundamental question has not been dealt with, at least explicitly: Why would targeting dehydrons by wrapping them with small-molecule drugs be effective at blocking specific protein functions, as required for therapeutic applications? One would argue that the dehydron wrappers discussed so far displace the natural ligand ATP of the target protein in a competitive manner, but this limits the discussion to dehydrons in the ATP-binding pocket, and we are interested in a more general answer. The answer may be found in Chap. 7, where the chemical role of dehydrons as participants in biochemical reactions has been highlighted and their function as enablers and stimulators of catalytic activity (in this case phosphorylation) has been unraveled. For the sake of the argument, let us say that we intend to block a signaling cascade that promotes cancer-related cellular activity. Then, our focus is to block the phosphorylation of a signal transducer, a kinase. Phosphorylation is an enzymatic activity whereby a kinase residue with a pro-nucleophilic side chain (tyrosine, serine, threonine and sometimes even histidine) performs a nucleophilic attack on the terminal phosphor-

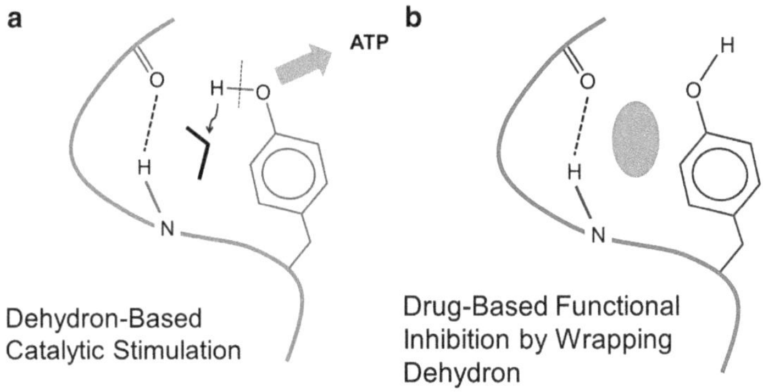

Fig. 8.9 Drug-based dehydron wrapping impairs phosphorylation of a tyrosine kinase. The latter is schematically represented by the side chain susceptible to phosphorylation (Tyr) and the nearby dehydron that stimulates the enzymatic activity. (**a**) The dehydron-based enablement of phosphorylation depends on the capacity of the dehydron to induce basicity in nearby interfacial water (represented by *thick sticks*), thereby triggering proton transference (indicated by the *arrow*) from the nucleophilic group susceptible to phosphorylation. The latter group is now enabled to perform the nucleophilic attack on the ATP terminal ester linkage. (**b**) A drug that wraps the dehydron displaces the interfacial water that stimulates kinase catalytic activity, and thus the drug impairs phosphorylation upon binding

ester linkage of ATP triggering an intermolecular trans-esterification reaction which in effect amounts to phosphorylating the nucleophilic residue through covalent attachment of the terminal (gamma) phosphate group from ATP. If our intent is to block this chemical process, one way of doing it is to wrap and thereby annihilate the dehydron involved in stimulating the kinase catalytic activity. The dehydron-based enablement and stimulation of phosphorylation hinges upon the capacity of the dehydron to induce basicity in interfacial water, thereby triggering proton transference from the nucleophilic group assumed to undergo phosphorylation (Fig. 8.9a). Once deprotonated, the latter group is habilitated to perform the nucleophilic attack on the ATP terminal ester linkage. A drug that wraps the dehydron displaces the basic interfacial water molecule that enables and promotes the kinase catalytic activity (Fig. 8.9b), and thus a dehydron-wrapping drug impairs phosphorylation, which is the goal of the molecular targeted treatment.

8.8 Caveats in Targeting Dehydrons in Floppy Regions

The activation loop in a kinase is in some ways very well suited as a drug target. Its high amino acid variability relative to other regions makes it a good selectivity feature, since it is fairly unique to each kinase and its floppy nature implies that it contains high dehydron density. In addition, dehydrons created by oncogenic mutation in the activation loop (such as D816V in c-Kit, Chap. 7) are particularly dangerous, especially if they are close to side chains susceptible of phosphorylation

(Ser, Thr, Tyr, His), as they functionalize such side chains for transesterification reactions. One such dehydron-creating mutation may in principle deregulate the kinase into a constitutively active signal transducer, with devastating consequences. As said, this is due to the catalysis-stimulating role of the dehydron (Chap. 7). On the other hand, activation loops tend to be structurally disordered in their phosphorylated state, and highly fluctuating in their unphosphorylated state. Thus, targeting dehydrons in the activation loop may require a mastery of drug-induced folding that we currently do not fully possess. Be as it may, Chap. 11 describes a number of situations where such control of induced folding is possible, in particular when the latter is restricted to inducing dehydrons.

8.9 Epistructural Thermodynamics Filters for Drug Specificity

The structure-based identification of dehydrons exploited so far utilizes a geometric criterion to determine the wrapping of the backbone hydrogen bond. Dehydrons thus identified constitute a proxy for hot spots of protein-water interfacial tension, as shown in Chap. 1. The latter represent more accurate indicators of the singularities of the aqueous interface that may be exploited as selectivity filters in ligand/ drug design. This strategy in fact provides a foundational scientific basis for other efforts in the same direction, such as those under the label Watermap®, a Schroedinger product. Counting nonpolar groups surrounding the hydrogen bond may not yield the optimum indicator of solvent exposure: the spatial distribution of wrapping groups needs to be also factored in to assess the extent of water exclusion from the bond microenvironment. The latter criterion is ultimately captured by an interfacial thermodynamics filter, whereupon hot spots in interfacial tension not conserved across homologous proteins become the true targets for selectivity control. As shown in Fig. 8.10, the thermodynamic filter operates by evaluating the interfacial free energy difference $\Delta\Delta G_{if}$ associated with mutating a poor wrapper residue (Gly in Fig. 8.10) paired by a backbone hydrogen bond (BHB) into a better wrapper residue (Val in Fig. 8.10). Thus, if the BHB happens to be a "thermodynamic dehydron" (a dehydron based on the thermodynamic criterion) that becomes a well-wrapped BHB in the mutated version, then $\Delta\Delta G_{if} < 0$, since dehydrons generate interfacial tension and this tension is removed by the wrapping-enriching mutation. In fact, the thermodynamic criterion for identifying interface hot spots is more precisely formulated by the inequality $\Delta\Delta G_{if} \leq -2k_BT$, since the free energy decrease associated with removing a dehydron by improving its wrapping is of the order of $-2k_BT$, as shown in Chap. 1. On the other hand, if the dehydron aligns with a well-wrapped hydrogen bond in a homologous protein, then in the latter protein the effect of the mutation replacing a poor wrapper for a good wrapper becomes negligible: $\Delta\Delta G_{if} \approx 0$. Thus, if two homologous proteins differ in a hot spot for interfacial tension so that $\Delta\Delta G_{if} \leq -2k_BT$ in one protein and $\Delta\Delta G_{if} \approx 0$ in the other, then the thermodynamic dehydron thus identified

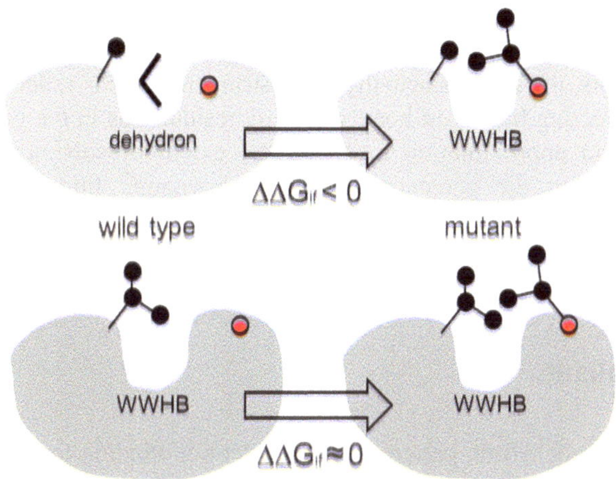

Fig. 8.10 Interfacial thermodynamics filter telling apart two homolog proteins that do not share a specific "thermodynamic dehydron" ($\Delta\Delta G_{if} \leq -2k_BT$). A "thermodynamic dehydron" in one protein is identified by quantifying the interfacial free energy change associated with a mutation that improves the wrapping of the hydrogen bond (G → Val in the figure, with α-carbon of the mutation site in *red*). The thermodynamic dehydron aligns with a well-wrapped backbone hydrogen bond (BHB) in the homologous protein. This is evidenced by quantifying the negligible thermodynamic effect ($\Delta\Delta G_{if} \approx 0$) of the mutation replacing a poor hydrogen-bond wrapper (G, *red spot*) for a good wrapper (Val) in the homolog protein

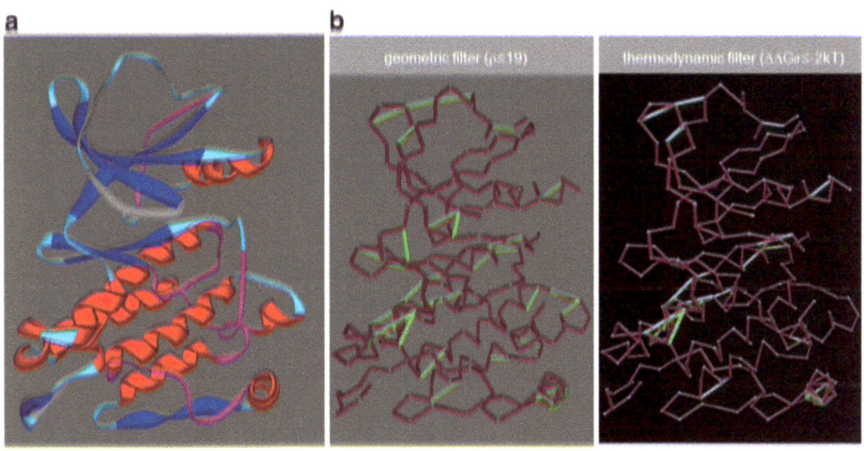

Fig. 8.11 (**a**) Ribbon representation of Src kinase (PDB.3G5D) included as an aid to the eye. (**b**) Dehydron pattern of Src kinase identified by adopting a geometric (*left panel*) or a thermodynamic criterion (*right panel*)

constitutes a filter for selectivity, meaning that if a ligand is designed to wrap it upon association, then the affinity of the ligand for the protein containing the dehydron will be significantly higher than that for the protein that does not contain the dehydron upon alignment.

As it can be easily surmised, because the geometric criterion ($\rho \leq 19$) is only a proxy for the thermodynamic criterion ($\Delta\Delta G_{if} \leq -2k_BT$) that identifies interfacial tension hot spots, the latter is actually more stringent. This is reflective of the fact that merely counting hydrogen-bond wrapping residues, as in the geometric criterion, is only an approximation to gauge the extent of solvent exposure. The difference between the geometric and thermodynamic filter is illustrated in Fig. 8.11a, b for the Src kinase, where the set of thermodynamic dehydrons is shown to be strictly contained in the set of geometric dehydrons.

8.10 Problems

8.1. Analyze the dehydron patterns at the nucleotide-binding loops of the homologous kinases Lck and Lyn, belonging to the Src tyrosine kinase family. Identify wrapping differences that may be exploited to tell them apart. This distinction is particularly important if the design effort is geared at removing immunosuppressive effects arising from Lck inhibition (Chaps. 9 and 14) while maintaining activity against Lyn.

References

1. Fernández A. Incomplete protein packing as a selectivity filter in drug design. Structure. 2005;13:1829–36.
2. Fernández A, Scott R, Berry RS. The nonconserved wrapping of conserved folds reveals a trend towards increasing connectivity in proteomic networks. Proc Natl Acad Sci U S A. 2004;101:2823–7.
3. Fernández A, Berry RS. Molecular dimension explored in evolution to promote proteomic complexity. Proc Natl Acad Sci U S A. 2004;101:13460–5.
4. Fernández A, Berry RS. Extent of hydrogen-bond protection in folded proteins: a constraint on packing architectures. Biophys J. 2002;83:2475–81.
5. Fernández A, Berry RS. Proteins with hydrogen-bond packing defects are highly interactive with lipid bilayers: Implications for amyloidogenesis. Proc Natl Acad Sci U S A. 2003;100:2391–6.
6. Fernández A, Rogale K, Scott LR, Scheraga HA. Inhibitor design by wrapping packing defects in HIV-1 proteins. Proc Natl Acad Sci U S A. 2004;101:11640–5.
7. Deremble C, Lavery R. Macromolecular recognition. Curr Opin Struct Biol. 2005;15:171–5.
8. Fernández A, Scheraga HA. Insufficiently dehydrated hydrogen bonds as determinants of protein interactions. Proc Natl Acad Sci U S A. 2003;100:113–8.
9. Ban Y-E, Edelsbrunner H, Rudolph J. interface surfaces for protein-protein complexes. J Assoc Comput Mach. 2006;53:361–78.
10. Adesokan AA, Roberts VA, Lee KW, Linz RD, Briggs JM. Prediction of HIV-1 integrase/viral DNA interactions in the catalytic domain by fast molecular docking. J Med Chem. 2004;47:821–8.
11. Kortemme T, Baker D. Computational design of protein-protein interactions. Curr Opin Chem Biol. 2004;8:91097.
12. Kortvelyesi T, Silberstein M, Dennis S, Vajda S. Improved mapping of protein binding sites. J Comput Aided Mol Des. 2003;17:173–86.

13. Lu L, Lu H, Skolnick J. Multiprospector: an algorithm for the prediction of protein-protein interactions by multimeric threading. Proteins Struct Funct Genet. 2002;49:350–64.
14. Sen TZ, Kloczkowski A, Jernigan RL, et al. Predicting binding sites of hydrolase-inhibitor complexes by combining several methods. BMC Bioinf. 2004;5:205–15.
15. Huber R. Conformational flexibility in protein molecules. Nature. 1979;280:538–9.
16. Dunker KA, Brown C, Obradovic Z. Identification and function of usefully disordered proteins. Adv Protein Chem. 2002;62:25–49.
17. Verkhivker G, Bouzida D, Gehehaar D, Rejto P, Freer ST, Rose P. Simulating disorder–order transition in molecular recognition of unstructured proteins: where folding meets binding. Proc Natl Acad Sci U S A. 2003;100:5148–53.
18. Fernández A, Crespo A. Protein wrapping: a molecular marker for association, aggregation and drug design. Chem Soc Rev. 2008;37:2372–82.
19. Stevens RC. Long live structural biology. Nat Struct Mol Biol. 2004;11:293–5.
20. Fauman EB, Hopkins A, Groom C. Structural bioinformatics in drug discovery. In: Bourne P, Weissig H, editors. Structural bioinformatics. New York: Wiley-Liss; 2003.
21. Wlodawer A, Vondrasek J. Inhibitors of HIV-1 protease: a major success of structure-assisted drug design. Annu Rev Biophys Biomol Struct. 1998;27:249–84.
22. Arkin MR, Wells JA. Small-molecule inhibitors of protein-protein interactions: progressing towards the dream. Nat Rev Drug Discov. 2004;3:301–17.
23. Katz BA. Structural basis for selectivity of a small molecule, S1-binding, submicromolar inhibitor of urokinase-type plasminogen activator. Chem Biol. 2000;7:299–307.
24. Steinmetzer T, Hauptmann J, Sturzebecher J. Expert Opin Invest Drugs. 2001;10:845–64.
25. Nicholls A, Sharp KA, Honig B. Protein folding and association: insights from the interfacial and thermodynamic properties of hydrocarbons. Proteins Struct Funct Genet. 1991;11:196–281.
26. Liang J, Edelsbrunner H, Woodward C. Anatomy of protein pockets and cavities: measurement of binding site geometry and implications for ligand design. Protein Sci. 1998;7:1884–97.
27. Lee B, Richards FM. The interpretation of protein structures: estimation of static accessibility. J Mol Biol. 1971;55:379–400.
28. Fernández A, Scott LR. Adherence of packing defects in soluble proteins. Phys Rev Lett. 2003;91:18102.
29. Fernández A, Scott LR. Dehydron: a structure encoded signal for protein interactions. Biophys J. 2003;85:1914–28.
30. Munshi S, Chen Z, Li Y, et al. Rapid X-ray diffraction analysis of HIV-1 protease-inhibitor complexes: inhibitor exchange in single crystals of the bound enzyme. Acta Crystallogr D. 1998;54:1053–60.
31. Mount DW. Bioinformatics. New York: Cold Spring Harbor Laboratory Press; 2001.
32. Karaman MW, Herrgard S, Treiber DK, et al. A quantitative analysis of kinase inhibitor selectivity. Nat Biotechnol. 2008;26:127–32.
33. NIH/NCBI website for the structural alignment program Cn3D. http://www.ncbi.nlm.nih.gov/structure/CN3D/cn3d.shtml
34. Schindler T, Bornmann W, Pellicena P, et al. Structural mechanism for STI-571 inhibition of Abelson tyrosine kinase. Science. 2000;289:1938–42.
35. Attoub S, Rivat C, Rodrigues S, et al. The c-Kit tyrosine kinase inhibitor STI-571 for colorectal cancer therapy. Cancer Res. 2002;62:4879–83.
36. Skene RJ, Scheibe DN, Kraus ML. Structural basis for autoinhibition and STI-571 inhibition of C-Kit Tyrosine kinase. J Biol Chem. 2004;279:31655–63.
37. Perlmutter RM, Marth JD, Lewis DB, et al. Structure and expression of lck transcripts in human lymphoid cells. J Cell Biochem. 1988;38:117–26.
38. Fernández A, Sanguino A, Peng Z, Ozturk E, Chen J, Crespo A, Wulf S, Shavrin A, Qin C, Ma J, Trent J, Lin Y, Han HD, Mangala LS, Bankson JA, Gelovani J, Samarel A, Bornmann W, Sood AK, Lopez-Berestein G. An anticancer c-Kit kinase inhibitor is reengineered to make it more active and less cardiotoxic. J Clin Invest. 2007;117:4044–54.
39. Li J, Johnson D, Sliskovic D, Roth B. Contemporary drug synthesis. Hoboken: Wiley Interscience; 2004.

Chapter 9
Epistructural Re-engineering of *Imatinib* to Eliminate Adverse Side Effects

Molecular cancer treatments are often geared at interfering with signaling pathways that govern cell fate or proliferation. These therapeutic strategies target kinases, the signal transducers in the cell. This approach remains challenging because kinases are evolutionarily and therefore structurally related and thus kinase inhibitors often lack the required specificity, which may lead to toxic or health-threatening side effects. This chapter illustrates the power of the wrapping concept in engineering safer kinase inhibitors, thus unraveling a rational approach to fulfill this therapeutic imperative. The focus of this chapter is the redesign of the anticancer drug imatinib (Gleevec) used to treat chronic myeloid leukemia, where its primary target is the chimeric Bcr-Abl kinase, as well as certain solid tumors based on its impact on the C-Kit kinase. Imatinib also has potentially cardiotoxic effects traceable to its impact on the Abl kinase in off-target cells, and immunosuppressive effects through its impact on Lck. This chapter describes a wrapping modification of imatinib purposely designed to prevent inhibitory impact against Abl (and Bcr-Abl) and Lck, to re-focus the impact towards C-Kit and to promote inhibition of an additional target, JNK, required to reinforce the prevention of cardiotoxicity and recognized as a target of significant therapeutic interest in its own right. The molecular blueprint for target discrimination, described in Chaps. 4 and 6, is now validated through assays that span a vast testing ground, from *in silico* assessment and test-tube probes to animal models. The findings surveyed in this chapter identify the wrapping-based re-engineered imatinib as an agent to treat gastrointestinal stromal tumors and possibly ovarian cancer with reduced side effects, thereby illustrating the therapeutic impact of the wrapping technology.

9.1 Imatinib Re-engineering with a Higher Safety Bar

The possibility of molecular interference with signaling pathways that control cell fate and cell proliferation opened up many therapeutic opportunities for the treatment of malignancy. Thus, protein kinases, the signal transducers of the cell, have

A. Fernández, *Physics at the Biomolecular Interface*, Soft and Biological Matter,
DOI 10.1007/978-3-319-30852-4_9

become paradigmatic targets for drug-based cancer therapy [1–13]. However, their evolutionary—and hence structural—relatedness often leads to unforeseen and uncontrollable cross-reactivities [13–16]. Although the relationship between specificity, anti-cancer activity and therapeutic efficacy remains opaque to our understanding, a lack of controlled specificity towards targets of therapeutic relevance typically underlies toxic side effects [5, 14]. Undesired and even health-threatening side effects can even be traced to inhibitory impact on a primary molecular target in off-target (i.e. normal) cells. This scenario is superbly illustrated by the reported potential cardiotoxicity of imatinib (STI571, Gleevec) attributed to its impact on Bcr-Abl [15], the chimeric Abelson (Abl) kinase target in the treatment of CML (chronic myeloid leukemia) already introduced in the preceding chapter.

In this chapter we survey a rational drug-redesign strategy to control and re-focus the inhibitory impact but adopt a more sophisticated and physically rigorous blueprint than the dehydron pattern to guide molecular design [14]. Furthermore, for validation of the design strategy, we present a broader testing ground, from *in silico* assay to animal model, than that presented in the preceding chapter.

While wrapping remains a structure-based target discriminator, we now exploit its link to the physically more tangible and unique hydration structure of the target protein (Chap. 4, [16]) to guide our molecular design. Thus, exploiting the target dehydration or de-wetting hot spots as a blueprint for drug design, we modified imatinib to significantly reduce its impact on Bcr-Abl, *prima facie* curbing its cardiotoxicity, while retaining the anticancer activity associated with its other primary targets.

In concrete terms, by sculpting into the ligand the de-wetting differences across imatinib targets, we are able to focus the inhibitory impact specifically on the C-Kit kinase, a target for GIST (gastrointestinal stromal tumor) [9–11] while suppressing Bcr-Abl inhibition (Fig. 9.1a). In addition, our wrapping-based redesign is expected to avoid binding LCK, because this kinase is implicated in antigen-triggered T-cell activation (Fig. 9.1b), and therefore LCK inhibition is likely to bring about immunosuppressive effects in a therapeutic context where they would be most unwelcomed (Chap. 14). Finally, we encourage JNK inhibition, as needed to reinforce the prevention of cardiotoxicity [15] and because of the importance of JNK as a target in the treatment of ovarian cancer [17]. The discriminating molecular design is reflected in selective anti-cancer activity on GIST cell lines and a GIST animal model, and a *prima facie* reduction of Abl-related cardiotoxicity. The program for imatinib redesign is sketched in Fig. 9.1a.

9.2 Epistructural Blueprint for Imatinib Redesign

As noted in Chap. 4, dehydrons generate de-wetting hot spots, that is, defects in the hydration layer of the protein. This physical property clearly suggests a molecular design strategy where, upon association with the target, the drug/ligand contributes to the removal of the hydrating molecules at the de-wetting hot spot. This design

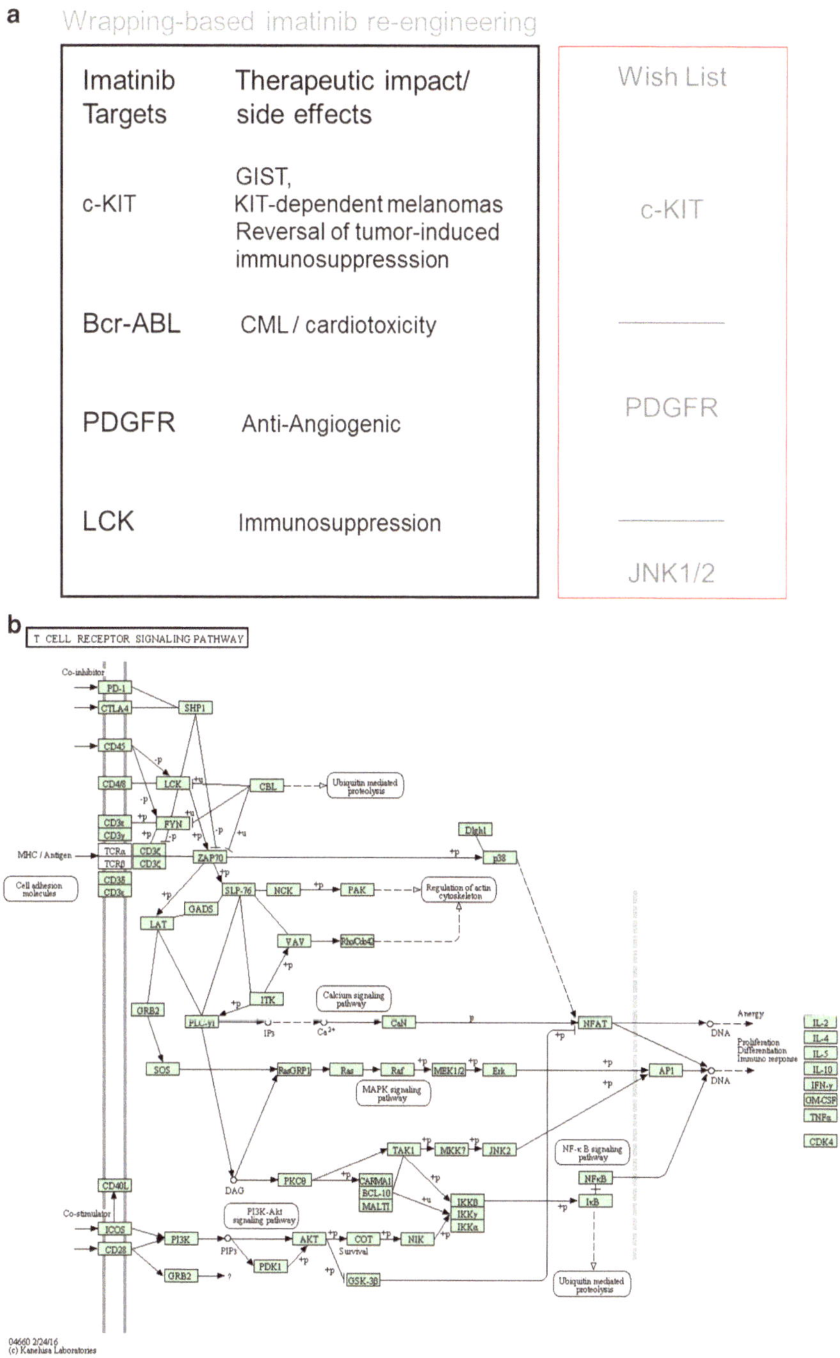

Fig. 9.1 (**a**) Imatinib targets and specific plans for its epistructure-based re-engineering. (**b**) Role of undesired target LCK as signal transducer in antigen-triggered T-cell maturation. The signaling pathways for this immune response were reported in *KEGG*, the Kyoto Encyclopedia of Genes and Genomes

strategy is often but not always equivalent to the dehydron-wrapping approach. The latter is based on an arithmetic counting of side-chain nonpolar groups in protein microenvironments disregarding their relative positions, while the former is rooted in the physical properties of interfacial biological water, providing a more rigorous approach.

To engineer an affinity discriminator for C-Kit following the de-wetting blueprint, we compare the patterns of residence times of water molecules that solvate the aligned interfacial regions of PDB-reported imatinib targets Bcr-Abl (PDB.1FPU), C-Kit (PDB.1T46) and Lck (lymphocyte specific tyrosine kinase, PDB.3LCK) (Fig. 9.2a–c, generated as indicated in Chap. 4). Thus, we introduce a descriptor of hydration tightness for targeted soluble proteins defined as the mean residence time of hydrating molecules within a domain around each residue on the protein surface. Hydrating molecules with short residence times constitute our blueprint for ligand re-engineering since they signal a local propensity for water removal. *The crux of the redesign strategy is then the sculpting in the ligand of nonconserved local de-wetting propensities in the aligned targets: the ligand is engineered to remove interfacial water upon association according to weaknesses in the target hydration shell (see Chap. 4).* This blueprint is typically not conserved across drug targets (cf. Fig. 9.2a), enabling us to modulate the inhibitory impact of the drug and funnel it on the realm of therapeutic relevance. The key de-wetting hot spots correspond to amide-carbonyl backbone hydrogen bonds pairing backbone-exposed residues [13]. As noted in Chap. 4, such bonds become energetically enhanced and stabilized upon removal of surrounding water and thus constitute de-wetting sites.

A de-wetting hot spot in C-Kit not conserved in the alternative targets corresponds to the residue pair C673-G676, which is mapped into the M318-G321 pair in Bcr-Abl and the M319-G322 pair in Lck (Fig. 9.2a–c). Thus, this local difference in de-wetting propensity, prompted us to conceive and synthesize a methylated variant of imatinib, named WBZ_4 [14] (Fig. 9.2d, e). To achieve a higher level of favorable and selective interactivity required an exogenous modulation of the target microenvironment, in turn, requiring a modification of the parental compound. As dictated by the de-wetting blueprint, the added methyl group would promote the favored dehydration of the pair C673-G676 upon binding to C-Kit, while it would hamper the association with Bcr-Abl and Lck, since the latter kinases favor hydration of the nucleotide-binding loop residues aligned with the targeted C-Kit residues C673-G676.

A comparison of the aligned dehydron matrices (see also Chap. 8) of the three imatinib-targeted kinases restricted to the imatinib-binding region single out the C673-C676 dehydron in C-Kit as the selectivity-promoting feature (red circles in Fig. 9.2f–h), in good agreement with the de-wetting analysis. The dehydron matrix is precisely the representational tool that enables the simplest and fastest *in silico* assessment of drug cross reactivity across the entire human kinome, as described in Chap. 10.

Reported assays on imatinib-promoted cardiomyocyte damage suggest that JNK (JNK1) inhibition may be beneficial to reduce the induced collapse of

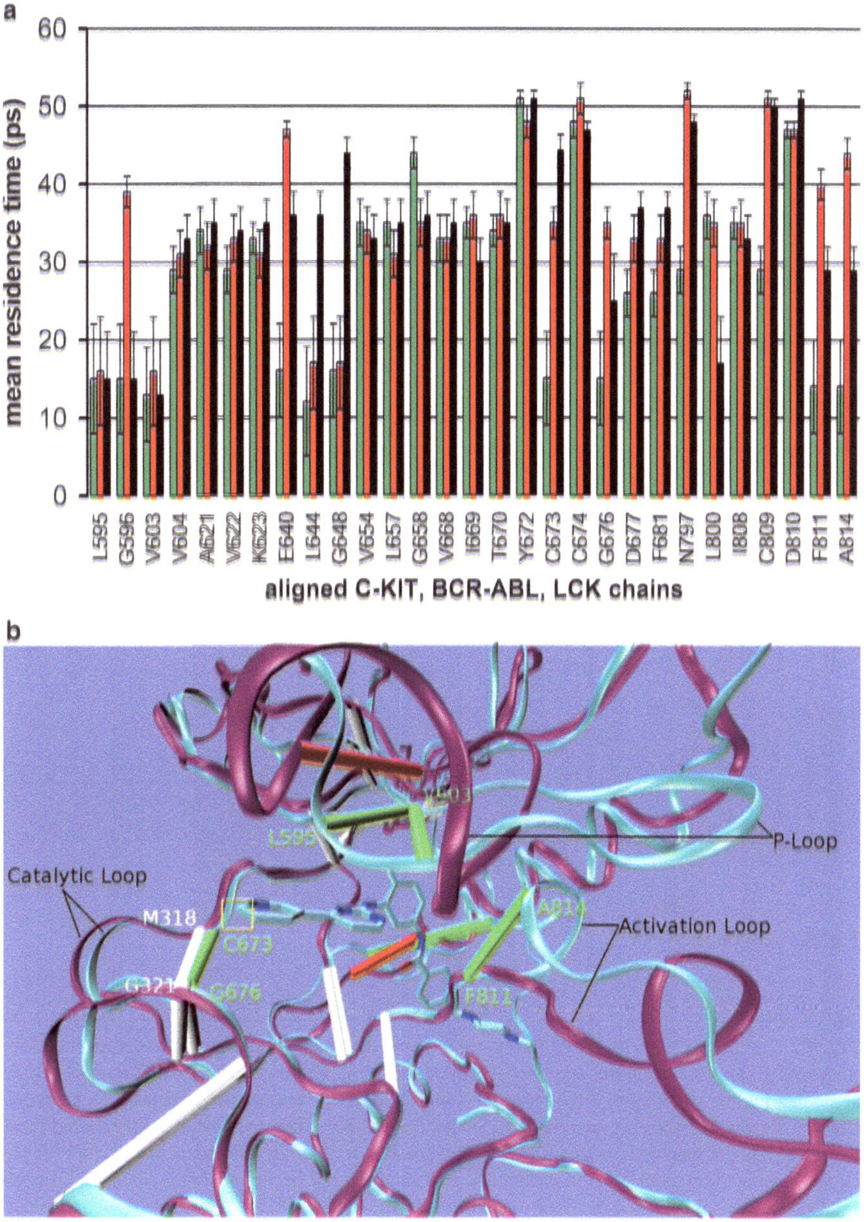

Fig. 9.2 (**a**) Dewetting propensities of C-Kit residues in contact with imatinib (PDB.1T46, *green bars*) and of aligned residues in Bcr-Abl kinase (PDB.1FPU, *red bars*) and Lck (PDB.3LCK, *black bars*). Residue i is in contact with the ligand if an atom of the latter lies within its domain $D(i)$. The de-wetting propensity is quantified by the mean lifetime of solvating water molecules. Error bars denote Gaussian dispersion over 5 MD (molecular dynamics) runs. (**b**) Pattern of de-wetting hot spots that arise as backbone hydrogen-bonded residues in C-Kit kinase (ribbon backbone representation in *light blue*, de-wetting hot spots in *green*), aligned with Bcr-Abl kinase (*magenta* backbone, de-wetting hot spots in *red*). The imatinib methylation site leading to the expulsion of water from C-Kit C673-G676 de-wetting hot spot is highlighted (*yellow rectangle*). Imatinib is

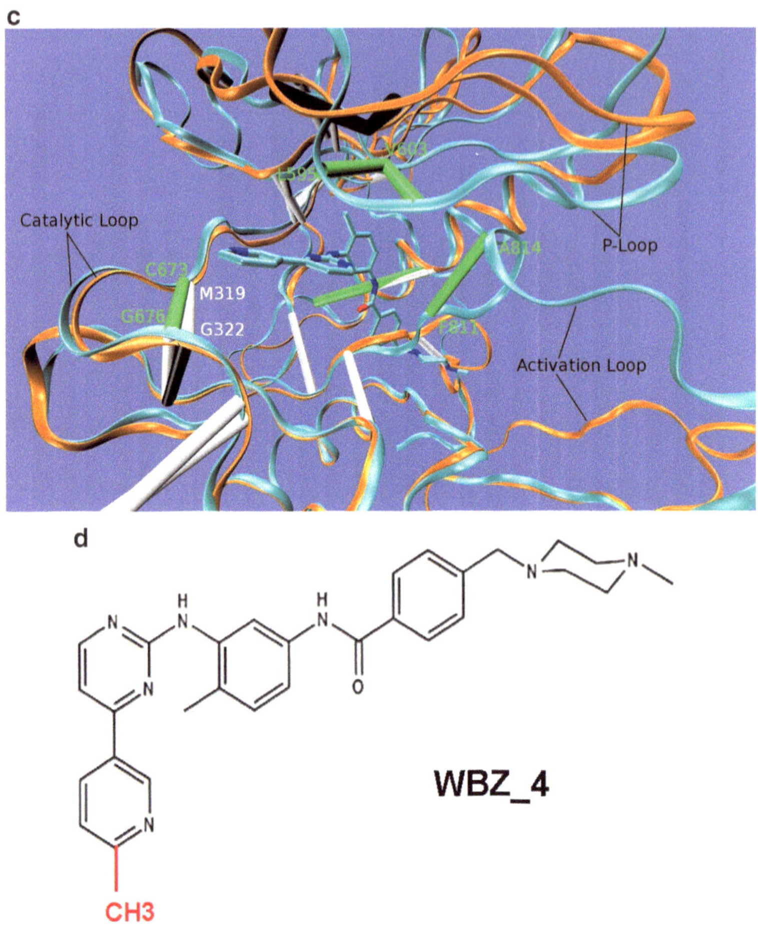

Fig. 9.2 (continued)

Fig. 9.2 (continued) thus modified to favorably expel interfacial water molecules from the C-Kit microenvironment, a feature not conserved in Bcr-Abl. Hydrogen bonds are represented as segments joining α-carbons of the paired residues, and those lacking the propensity for dehydration are shown as *light grey* segments. Residues from C-Kit chain are labeled in green and those from Bcr-Abl are labeled in *white*. (**c**) Differences in de-wetting hot spots upon alignment of C-Kit (*light blue* backbone, hot spots in *green*) and Lck kinase (*gold* backbone, *black* hot spots). Labels for Lck residues are shown in *white*. (**d**) Prototype molecule WBZ_4 (N-{5-[4-(4-methyl piperazine methyl)-benzoylamido]-2-methylphenyl}-4-[3-(4-methyl)-pyridyl]-2-pyrimidine amine). The added methyl group is indicated in *red*. (**e**) Total synthesis of WBZ_4 [14]. The wrapping modification (methylation) of imatinib is indicated by the *thick arrow*. (**f–h**) Dehydron matrices of Kit, Abl and Lck kinases respectively, derived from PDB entries 1T46, 1FPU and 3LCK and restricted to the residues in contact with imatinib. Rows and columns are labeled by residue numbers, well-wrapped bonds are displayed in *grey* and dehydrons are shown in *green*, *red* and *black* for Kit, Abl and Lck, respectively. The entry denoting a selectivity-promoting feature for the wrapping design is circled in *red*. Figures a–e reprinted from [14], copyright 2007 with permission from the American Society for Clinical Investigation

e

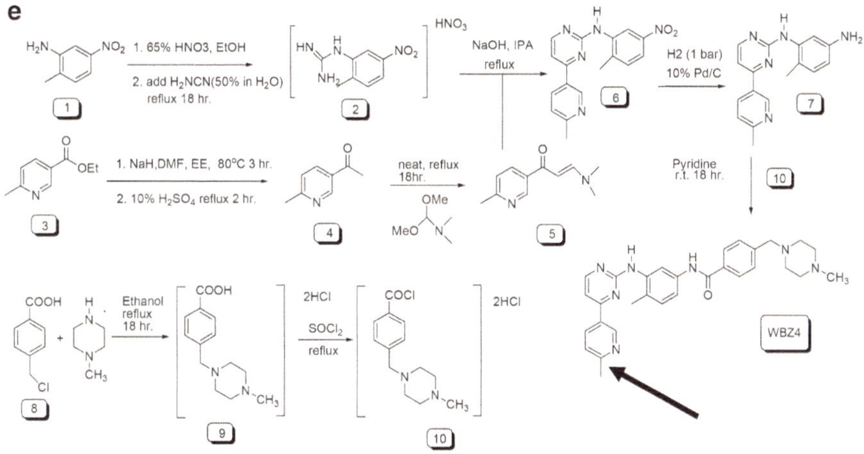

f

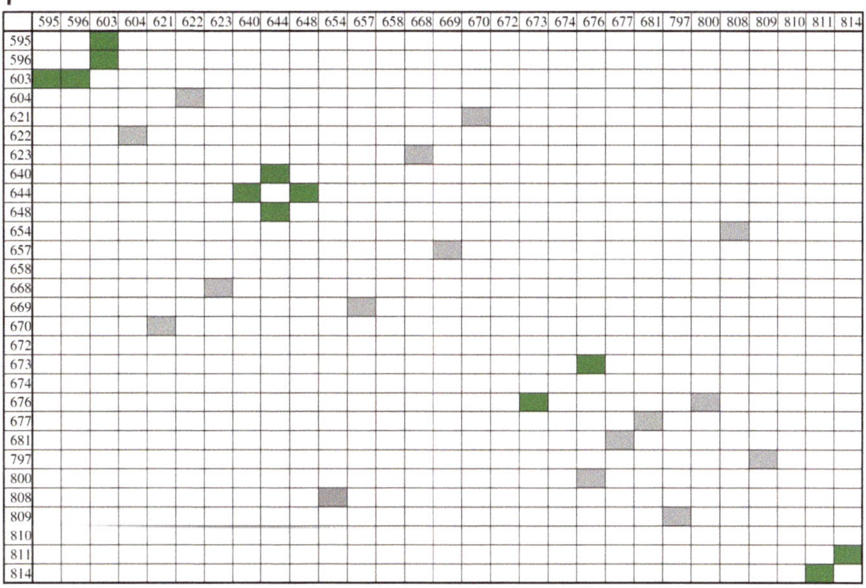

Fig. 9.2 (continued)

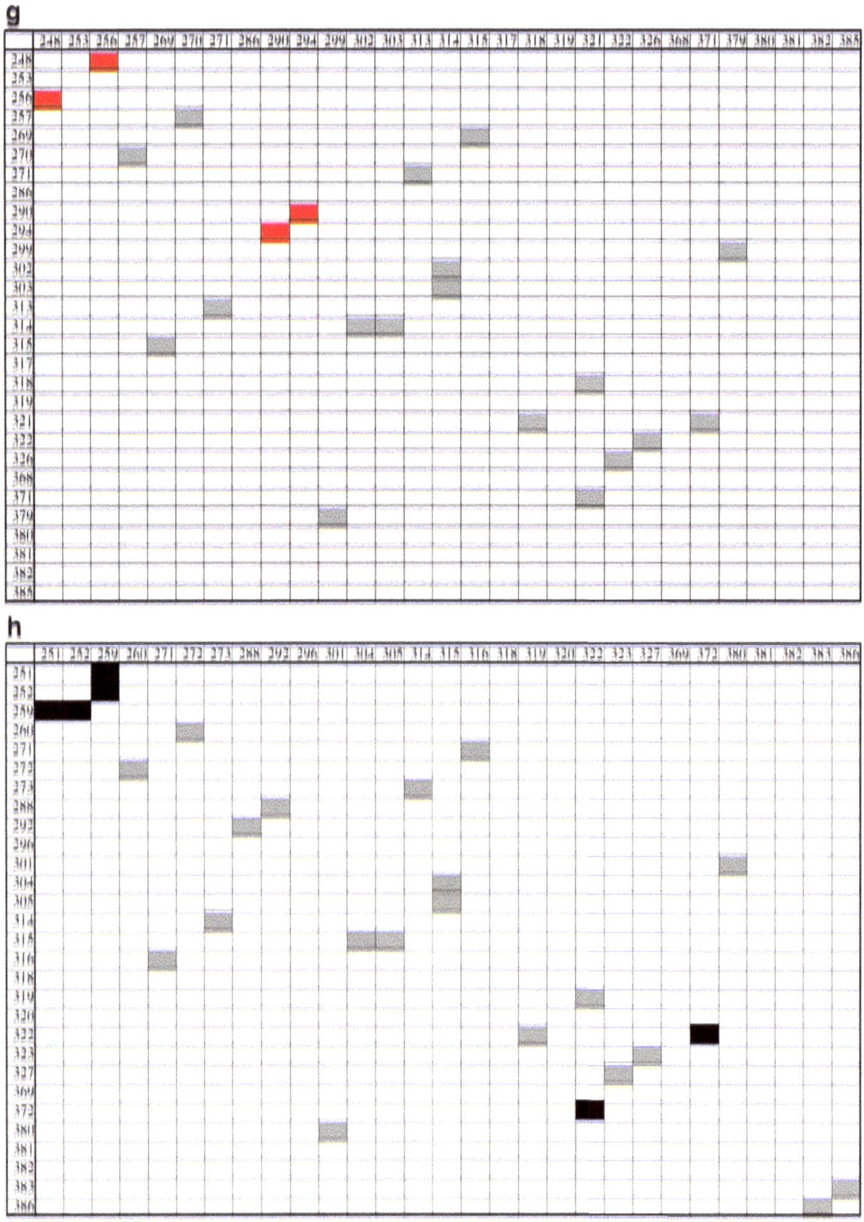

Fig. 9.2 (continued)

mitochondrial membrane potential and preserve sarcolemmal integrity [15]. For this reason, we tested the binding of WBZ_4 as it anchors on the de-wetting pattern of JNK1 (PDB.2G01). This analysis required alignment of JNK1 with the imatinib-C-Kit complex (PDB.1T46) (Fig. 9.3a, b). The high de-wetting propensity at JNK1 residue M111, the residue that aligns with C673 in C-Kit, instills confidence in the affinity of WBZ_4 for JNK.

9.3 Epistructural Thermodynamics for In Silico Assessment of the Efficacy of Imatinib Wrapping Variant WBZ_4

As described in Chap. 4, molecular dynamics of the target-dewetting patterns has proven to be a useful tool to assess the efficacy of drugs or ligands in displacing interfacial water from the target. In this regard, we have tested WBZ_4 *in silico* by monitoring the spatio-temporal effect of the specific methylation on the kinase-inhibitor association through molecular dynamics (Fig. 9.4a, b). As C-Kit associates with imatinib, the loop facing the para-position on the terminal ring (Figs. 9.2b and 9.4a) was found to be unstable. In all five simulation trajectories, the backbone hydrogen bond between Cys673 and Gly676 was irreversibly replaced by water-mediated interactions within 1 ns (Fig. 9.4a). When WBZ_4 replaced imatinib within the C-Kit complex, the same loop was found to be stabilized due to the improved dehydration of the hydrogen bond by the added methyl group on the ligand (Fig. 9.4b). Thus, for C-Kit, WBZ_4 introduces favorable interactions engaging the solvent-exposed pre-formed hydrogen bond and consequently stabilizes the inhibitor binding. The effect of WBZ_4 is the opposite on Bcr-Abl. The simulations demonstrate that in this case, the imatinib complex has a better stability in the same loop region, while WBZ_4 significantly destabilizes the loop. This is due to the fact that the Met318-Gly321 backbone hydrogen bond on the Bcr-Abl loop is well dehydrated intramolecularly and positioned closer to imatinib, thus this bond becomes well protected [18] by imatinib and the addition of the methyl group clashes sterically with the loop. Thus, molecular dynamics provides a rationale for the discriminatory power of WBZ_4 relative to the parent drug.

The solvent-centric physical understanding of the mechanism of action of the wrapping drug WBZ_4 requires evaluation of the change in epistructural free energy, ΔG_{if}, associated with interfacial water dynamics in the C-Kit/imatinib versus the C-Kit/WBZ_4 complex. The C673-G676 dehydron in the C-Kit/imatinib complex becomes irreversibly disrupted due to backbone hydration in ~5 ns. The hydrogen-bond disruption is thermodynamically favored as the means of releasing the interfacial tension created by the C673-G676 dehydron. Based on the decrease in interfacial water frustration, from $\phi = 1$ ($g = 3$) around the dehydron-sustaining conformation to $\phi = 0$ ($g = 4$) in the conformation with the dehydron disrupted, and applying the thermodynamic functional methods described in Chap. 1, we estimate the decrease in interfacial free energy associated with dehydron disruption in the

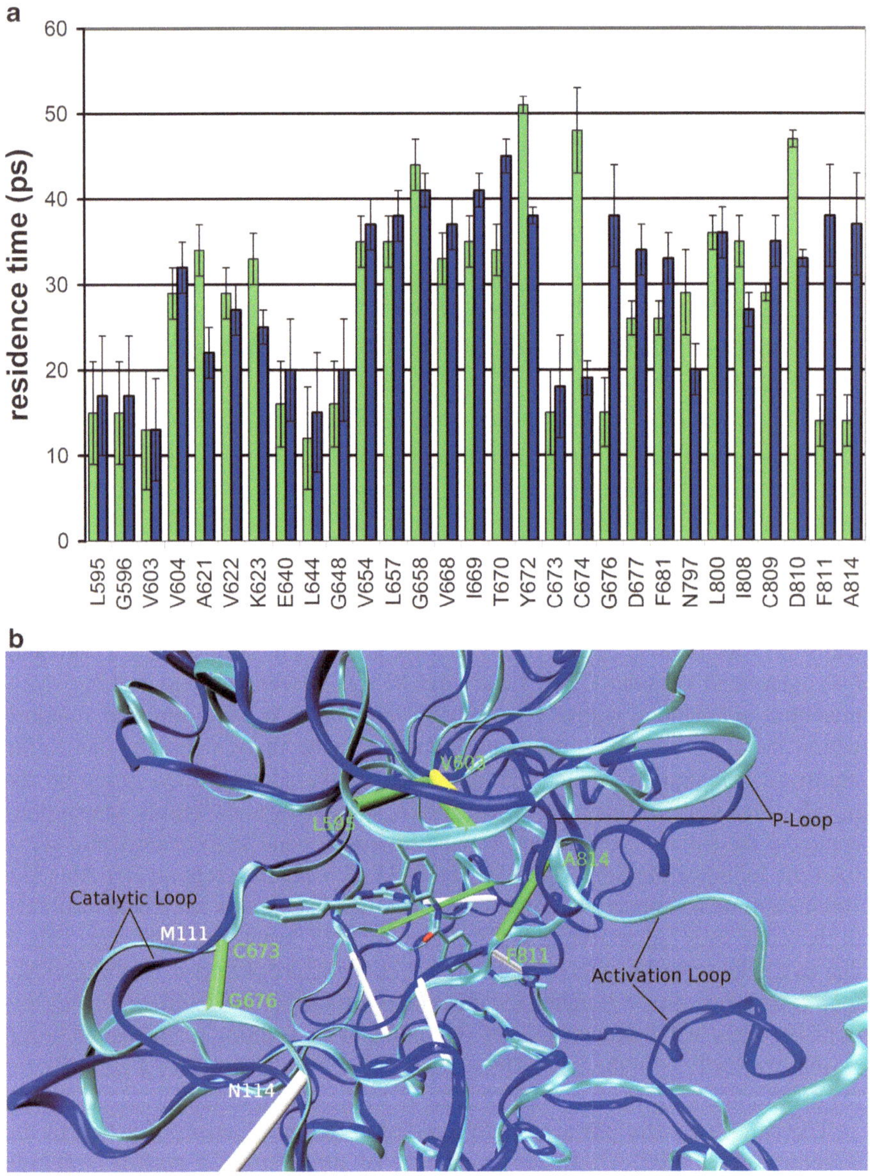

Fig. 9.3 (**a**) Aligned de-wetting patterns for C-Kit kinase (*green*) and JNK1 (*blue*) restricted to the C-Kit residues in contact with imatinib. (**b**) De-wetting hot spots arising as backbone hydrogen-bonded residues in C-Kit kinase (backbone in *light blue*, de-wetting hot spots in *green*), aligned with JNK (*blue* backbone, de-wetting hot spots in *yellow*). The JNK residues M111 and N114, aligned with the de-wetting hot spot C673-G676 in C-Kit are not paired by a hydrogen bond. Yet, M111 is a de-wetting hot spot for JNK (**a**), hence a harnessing spot for the designed imatininb modification. Reprinted from [14], copyright 2007 with permission from the American Society for Clinical Investigation

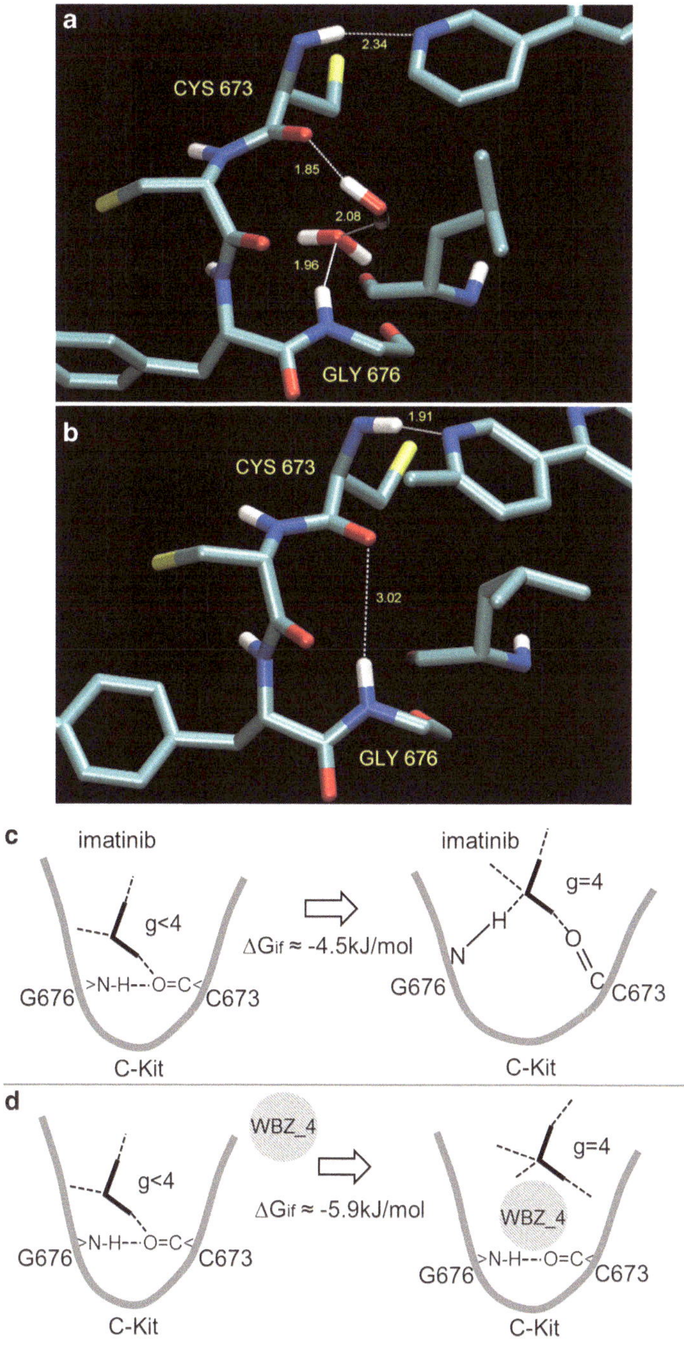

Fig. 9.4 (**a**) Snapshot of the molecular dynamics simulation for C-Kit kinase bound to imatinib at 1 ns. The main-chain hydrogen bond between Cys673 and Gly676 is competitively and irreversibly replaced by hydrogen bonding to a water molecule, revealing the instability of the intramolecular interaction. (**b**) Snapshot of the simulation of C-Kit kinase in complex with WBZ_4 at 1 ns. The main-chain hydrogen bond between Cys673 and Gly676 is stabilized by the water expulsion

C-Kit/imatinib complex at $\Delta G_{if} \approx -4.5$ kJ/mol. On the other hand, the mitigation of interfacial tension generated by the C673-G676 dehydron upon formation of the C-Kit/WBZ_4 complex follows a completely different mechanism, as it befits the wrapping effect of WBZ_4 upon the C673-G676 dehydron. In this case the frustrated water molecule hydrating the C673-G676 dehydron becomes displaced to a bulk-like locus with $\phi = 0$ ($g = 4$) upon C-Kit binding to WBZ_4. The reduction in water frustration in this case yields $\Delta G_{if} \approx -5.9$ kJ/mol, a figure obtained as the difference in association free energy between WBZ_4 and the parental drug imatinib, which only evaluates the extra dehydron wrapping.

Thus, as shown in Fig. 9.4c, d, we observe a qualitative difference in the mechanism of mitigation of the interfacial tension generated by the unique nonconserved C-Kit dehydron C673-G676. If the drug is unable to wrap the dehydron upon associating with the target, the dehydron simply breaks apart to reduce water frustration (Fig. 9.4c), whereas, a dehydron-wrapping drug stabilizes the dehydron upon binding, as it displaces surrounding frustrated water (Fig. 9.4d). This dynamic aspect of epistructure-based drug design will be further developed in Chap. 11, but it is important to highlight its importance at this early stage of conceptual maturation of the wrapping technology.

9.4 Test-Tube Validation of the Engineered Specificity in Wrapping Imatinib Variant WBZ_4

Screening methodologies for kinase inhibitors have been revolutionized in their breadth with the advent of bacteriophage libraries of kinase display [6], enabling a huge amplification of the affinity signal of the test compound. With panels now covering 442 of the 519 kinases that comprise the human kinome, this screening methodology has become the benchmark of high-throughput assays in the test tube. Thus, kinases are expressed on the capside of bacteriophages and this display construct is subject to a competitive binding assay. In the assay, the test compound and the kinase-display construct are in solution and the high-affinity and promiscuous ATP-mimicking ligand is in a fixed phase within a chromatographic column. Thus, the affinity of the test compound is measured relative to that of the fixed-phase ligand through its ability to displace the latter and hence enable the kinase-display bacteriophage to elute from the column.

Fig. 9.4 (continued) promoted by the added methyl on the inhibitor. Reprinted from [14], copyright 2007 with permission from the American Society for Clinical Investigation. (**c**) Mechanism of mitigation of interfacial tension generated by the unique nonconserved C-Kit dehydron C673-G676 within the nucleotide-binding loop of the kinase. If the drug is unable to wrap the dehydron upon associating with the C-Kit target (as is the case with imatinib), the dehydron simply breaks apart to reduce water frustration ($g < 4 \rightarrow g = 4$) and thereby reduce interfacial tension. (**d**) Dehydron-wrapping drug WBZ_4 stabilizes the dehydron C673-G676, as it displaces interfacial frustrated water surrounding the dehydron upon associating with C-Kit

To test whether wrapping-based design truly enables the control of specificity, WBZ_4 was screened for affinity against the T7-bacteriophage kinase-display library (Fig. 9.5), using imatinib screening as control [14]. The selective affinity of WBZ_4 for C-Kit kinase and JNK1 is noteworthy. Predictably, the affinity of WBZ_4 for ABL1 is reduced by 75 % and by 95 % or more on all other Abl variants, while, in contrast with imatinib, WBZ_4 shows no detectable affinity for Lck. The impact of WBZ_4 on all additional imatinib targets was comparable, while its controlled specificity is apparent.

9.5 *In vitro* Assays for Imatinib Wrapping Variant WBZ_4

As shown in [14], the specificity of WBZ_4 towards C-Kit was established first through *in vitro* assays that probes the phosphorylation kinetics of competitive inhibition. These spectrophotometric assays couple ADP production, the byproduct of phosphorylation activity, with oxidative steps following the glycolysis cycle [19, 20], as described in Chap. 8. The inhibitory impact of imatinib (triangles) and WBZ_4 (squares) on the rate of phosphorylation was determined by spectrophotometry, assaying for the activity of C-Kit and Bcr-Abl (blue and red plots, respectively, Fig. 9.6). These kinetic assays revealed a high specificity of WBZ_4 for C-Kit, in contrast with imatinib. WBZ_4 enhances the inhibition of C-Kit activity even beyond imatinib levels, revealing a higher competitive affinity of the prototype compound for the adenosine tri phosphate (ATP)-binding pocket (K_I(imatinib) $\approx 55 \pm 7$ nM; K_I(WBZ_4) $\approx 43 \pm 5$ nM). On the other hand, the pattern of inhibition for Bcr-Abl is dependent on the phosphorylation state of this kinase [20, 21]. At 1 μM concentration, imatinib reduces 66 % of the activity of Tyr412-phosphorylated kinase [8] ($K_I \approx 5 \pm 1$ μM) and ~100 % of the activity of the unphosphorylated state ($K_I \approx 50 \pm 5$ nM). By contrast, 1 μM WBZ_4 reduces by less than 20 % the activity of both states of Bcr-Abl. The prototype compound has reciprocal affinity constants $K_I \approx 18 \pm 3$ μM and $K_I \approx 11 \pm 2$ μM, for phosphorylated and unphosphorylated Bcr-Abl, respectively.

In vitro colorimetric assays were performed over a 1 pM–100 μM range in ligand concentration to assess the inhibition of phosphorylating activity by antibody recognition of phosphorylated peptide substrates [14]. The IC50 (50 %-inhibition concentration) for imatinib/Abl is ~1 μM, while the WBZ_4/Abl value is above 100 μM (Fig. 9.7a). The active recombinant Abl kinase and its substrate (Abl-tide) were incubated in the presence of various WBZ_4 or imatinib concentrations and ATP (100 nM). Phosphorylation of Abl-tide peptide was detected by spectrophotometry following incubation with phospho-Abl-tide antibodies.

The specificity of the inhibitory impact on C-Kit is significantly enhanced as WBZ_4 substitutes imatinib (Fig. 9.7a, b), while C-Kit inhibition is 23 ± 12 % greater for WBZ_4 (Fig. 9.7b). The impact of the wrapping prototype ligand on alternative imatinib targets, such as the platelet dependent growth factor receptor (PDGFR) kinase, could not be modulated since such proteins are not reported in

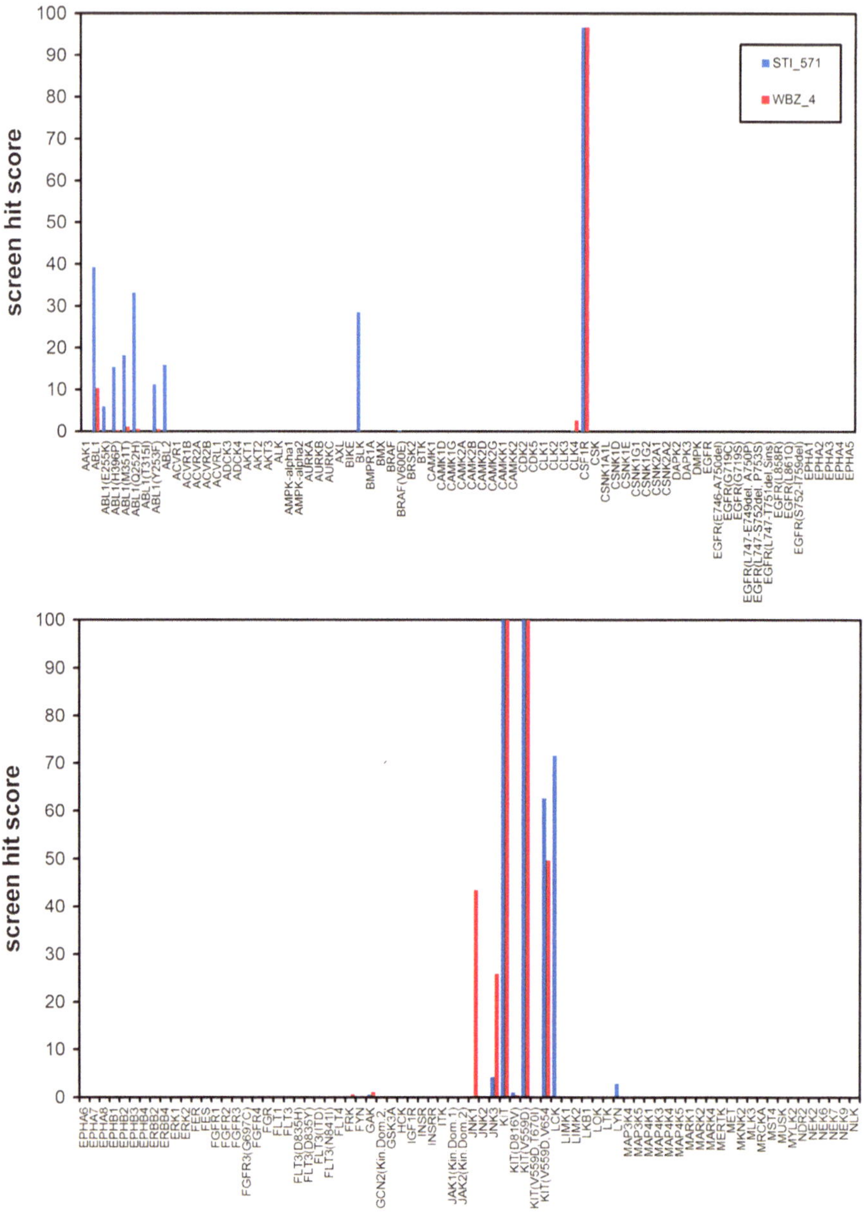

Fig. 9.5 High-throughput screening at 10 mM for WBZ_4 (*red*) and imatinib (STI_571, *blue*, control) over a battery of human kinases displayed in a T7-bacteriophage library (Ambit Biosciences, San Diego, CA). Hit values are reported as percentage bound kinase. Reprinted from [14], copyright 2007 with permission from the American Society for Clinical Investigation

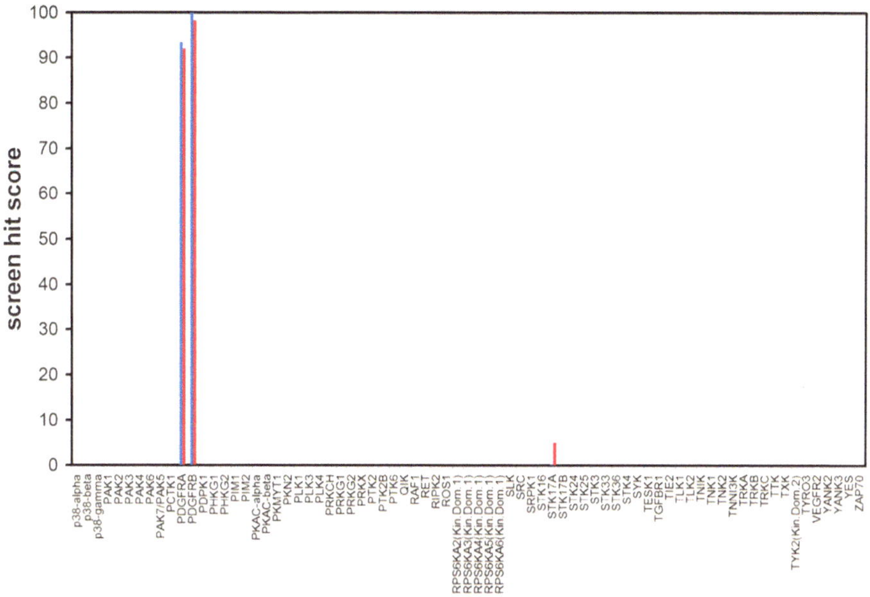

Fig. 9.5 (continued)

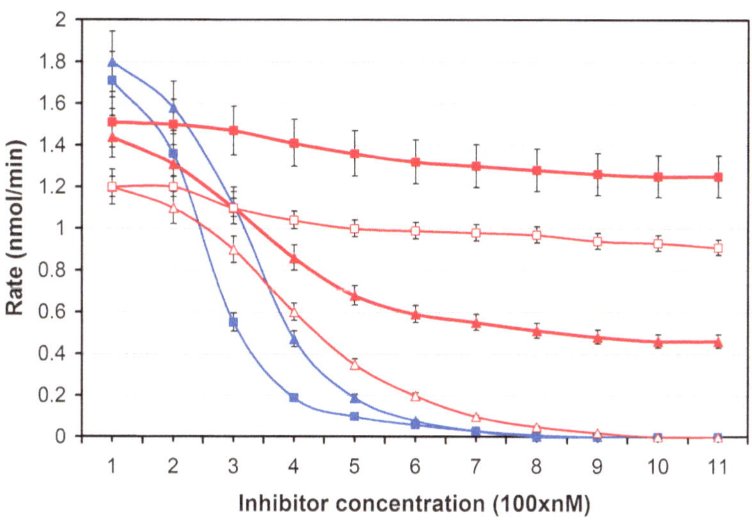

Fig. 9.6 Kinetic inhibitory impact of compounds WBZ_4 and imatinib determined by measuring phosphorylation rates through spectrophotometric assays of C-Kit and Bcr-Abl kinase activity. The kinases are inhibited by WBZ_4 (*squares*), and by the parental compound (*triangles*). Phosphorylation rate plots are given for Bcr-Abl (*red*) and C-kit (*blue*). The open *red* symbols correspond to inhibition of unphosphorylated Bcr-Abl, while the full *red* symbols correspond to the Tyr412-phosphorylated form. Error bars represent dispersion over five runs for each kinetic assay. Reprinted from [14], copyright 2007 with permission from the American Society for Clinical Investigation

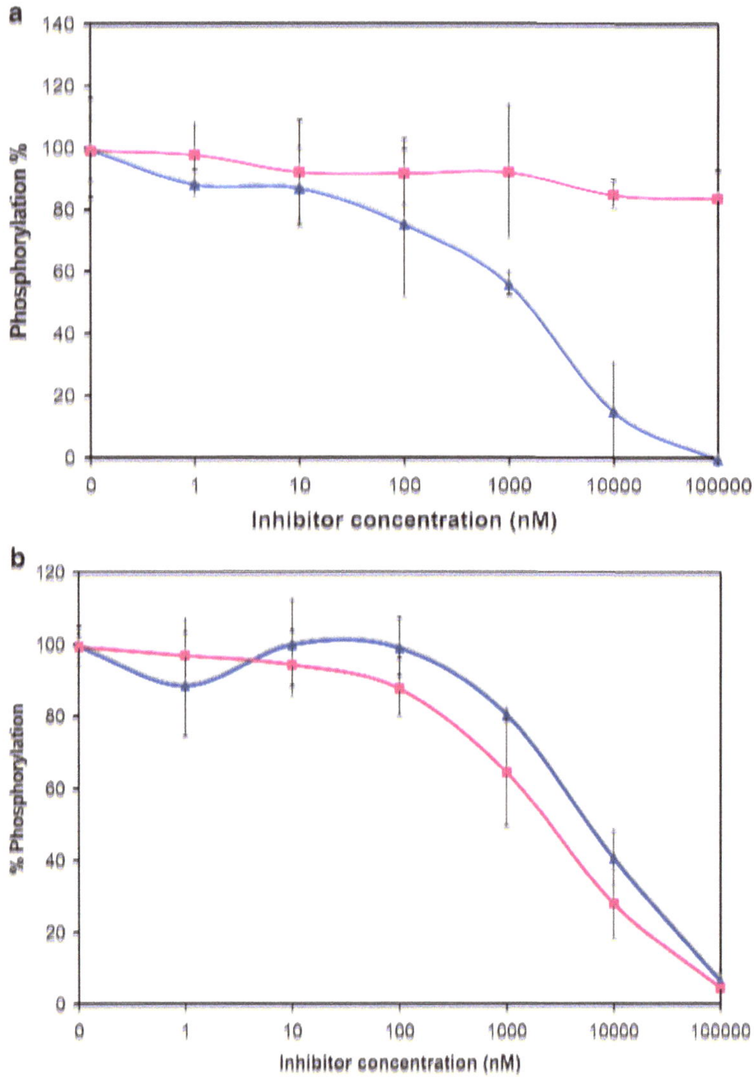

Fig. 9.7 (**a**) *In vitro* phosphorylation inhibition assay for Abl enzyme in the presence of WBZ_4 (*pink*) or imatinib (*blue*). Active recombinant Abl enzyme (1 μg/mL) and its substrate (Abl-tide, 1 μg/mL) were incubated for 1 h at 37 °C in the presence of various WBZ_4 or imatinib concentrations. ATP (100 nM) was added to the reaction mixture. Phosphorylation of Abl-tide peptide was detected by incubation in consecutive order with anti-rabbit phospho-Abl-tide antibody and anti-rabbit horseradish peroxidase (HRP) antibody. Phosphorylation of the substrate was quantified as absorbance units (AU) by spectrophotometry at 450 nm. Values obtained with the enzyme without the inhibitors (WBZ_4 or imatinib) were assumed to be 100 % phosphorylation and were compared to the values obtained with the addition of the inhibitors. (**b**) *In vitro* phosphorylation inhibition assay for C-Kit in the presence of WBZ_4 (*pink*) or imatinib (*blue*). Active recombinant C-Kit kinase (25 ng/mL) and its substrate Poly (Glu4-Tyr, 150 nM) were incubated for 1 h at 37 °C in the presence of various WBZ_4 or imatinib concentrations. ATP (100 nM) was added to the reaction mixture. Phosphorylation of Poly (Glu4-Tyr) peptide was detected by incubation in consecutive order with anti-phosphotyrosine antibody and anti-rabbit horseradish peroxidase (HRP) antibody. Reprinted from [14], copyright 2007 with permission from the American Society for Clinical Investigation

PDB and thus, no de-wetting pattern can be reliably identified (cf. high-throughput screening of WBZ_4 versus imatinib in Fig. 9.5). On the other hand, as shown in Chap. 8, the PDGFR kinase is a close paralog of the Kit kinase and hence it is largely expected that any Kit kinase inhibitor would be cross reactive with the PDGFR kinase.

9.6 *In Vitro* Assay of Selective Anticancer Activity of the Wrapping Design

The controlled inhibitory impact of prototype WBZ_4 was further tested through *in vitro* assays on cancer cell lines representative of CML and GIST [14]. The initial tests entailed treating the GIST882 [12] cell line and the CML K562 [7] cell line with WBZ_4 and contrasting its inhibitory impact on cell proliferation and in-cell specificity with those associated with imatinib. Because of its relative insolubility, WBZ_4 was incorporated into liposomes to promote cellular delivery. The proliferation of C-Kit-expressing GIST882 cells treated with WBZ_4 was significantly inhibited in a quantitative dose-dependent manner similar to imatinib, with maximum impact at 1 μM bulk concentration (Fig. 9.8a). By contrast, the inhibitory impact of WBZ_4 on CML K562 cells at the same bulk physiological dose of 1 μM is almost negligible (cell proliferation less than 10 % lower than the proliferation of untreated cells), while 1 μM imatinib promotes a decrease in cell proliferation of approximately 66 % (Fig. 9.8b). These results demonstrate the higher specificity in anti-cancer activity of WBZ_4.

A Western blot assay on treated cancer-derived cell lines (Fig. 9.9a, b) was performed to determine the *in-cell* specificity of WBZ_4. The immunoblots revealed specificity towards C-Kit consistent with the selective anticancer activity on the GIST882 cell line that expresses C-Kit. Thus, the activating phosphorylation of C-Kit at sites Tyr703 and Tyr721 in ST882 cells is inhibited by WBZ_4 in a dose-sensitive manner similar to imatinib (Fig. 9.9a). By contrast, phosphorylation of Bcr-Abl at Tyr 412 [22] in K562 cells was not significantly inhibited (<15 %) by WBZ_4, while densitometry revealed an imatinib-induced inhibition of ~85 % (Fig. 9.9b).

Since imatinib is a micromolar inhibitor of phosphorylated Bcr-Abl (Fig. 9.6), the attack on CML cells at 1 μM bulk concentration (growth decrease 63 %, Fig. 9.8b) must be attributed to the inhibition of phosphorylated Bcr-Abl (Fig. 9.9b) combined with the effective nanomolar inhibition of the unphosphorylated form (Fig. 9.6). By contrast, WBZ_4 hinders the phosphorylation of Bcr-Abl only partially (Fig. 9.9b) and in any case, it is an ineffective (micromolar) inhibitor of *both* forms of Bcr-Abl (Fig. 9.6). Hence, its anti-tumor activity is predictably minimal on CML cells (Fig. 9.8b). In the case of GIST cells, comparable anti-tumor activity of both compounds (Fig. 9.8a) is likely to arise from comparable inhibitory impact (~85 %, Fig. 9.9a) on C-Kit phosphorylation, and the fact that both compounds are nanomolar-affinity inhibitors of C-Kit (Fig. 9.6).

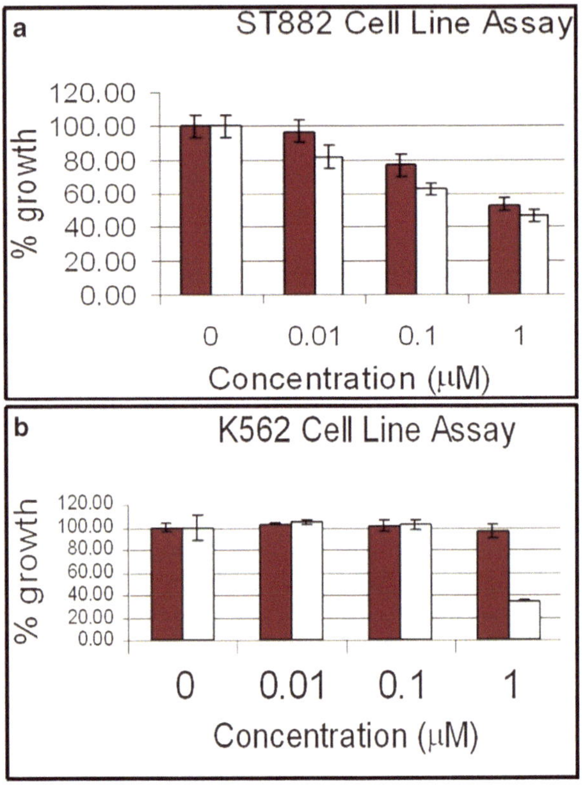

Fig. 9.8 (**a**) Cell-proliferation assay for GIST 882 cells. WBZ_4 inhibits cell proliferation of C-Kit positive ST-882 cells. GIST cancer cells ST882 (GIST 882) were seeded in 96 wells plates at a density of 8×10^3 cells per well. The cells were treated with various concentrations of WBZ_4 (*crimson*) and imatinib (*light blue*) for an additional 48 h. Cell proliferation is expressed as the percentage of proliferating cells relative to untreated cells. (**b**) Cell-proliferation assay for K562 cells. WBZ_4 does not significantly inhibit cell proliferation of Bcr-Abl-expressing K562 cells. K562 cells were seeded in 96 wells plates at a density of 1×10^4 cells per well in 50 μL of medium. Two hours later, 50 μL medium containing different concentrations 0.01, 0.1, 1 μM of liposome-encapsulated WBZ_4 (*crimson*) or soluble imatinib (*light blue*) were added to the wells to reach a final volume of 100 μL per well. Following 48 h of exposure, the Alamar blue assay was performed. Plates were read at dual wavelength (570–595 nm) in an Elisa plate reader. Reprinted from [14], copyright 2007 with permission from the American Society for Clinical Investigation

9.7 Enhanced Safety Tested in Animal Models of Gastrointestinal Stromal Tumor

To assay WBZ_4 for anticancer activity *in vivo*, an animal model for GIST growth based on female C.B-17/IcrHsd-*Prkdc*^{SCID} mice was developed *de novo* [14]. The model involved the subcutaneous injection of GIST882 cells [23]. The efficacy of WBZ_4 was found to be comparable to that of imatinib, as determined by the

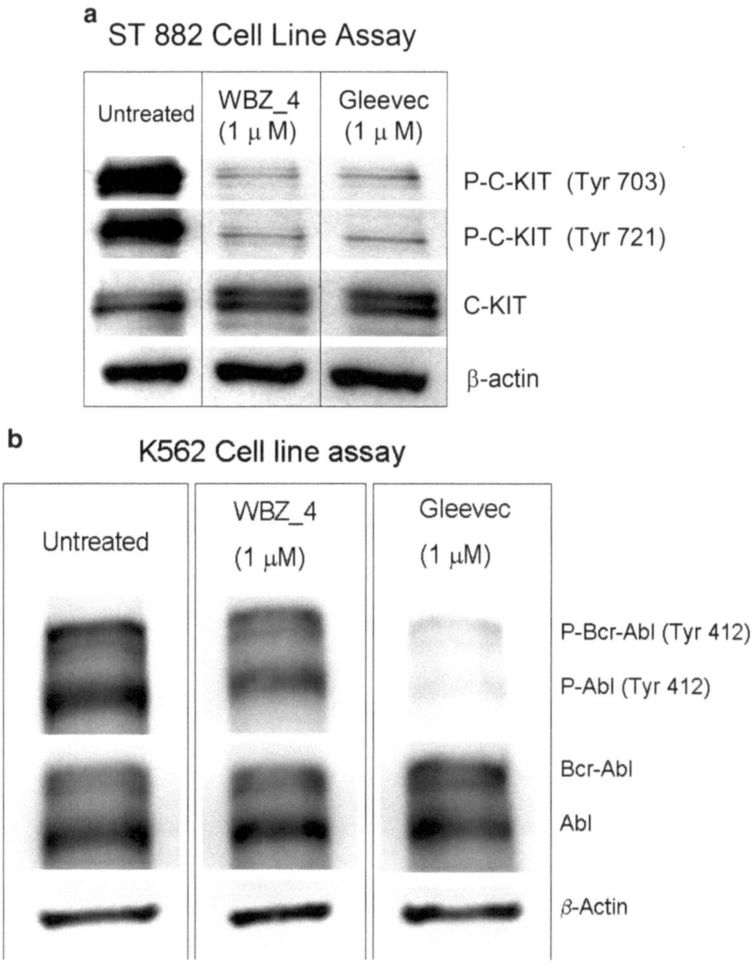

Fig. 9.9 (**a**) Western blot of C-Kit inhibition. WBZ_4 inhibits phosphorylation of C-Kit kinase in ST-882 GIST cells. Gel bands from the Western blot assays of C-Kit and its phosphorylated (P-) form in GIST cells treated with WBZ_4 and imatinib. The β-actin assay was adopted as control. (**b**) Western blot of Bcr-Abl inhibition. Phosphorylation of Bcr-Abl kinase is not significantly inhibited by WBZ_4 in K562 CML cells. Electrophoretic gel bands for Western blots for Bcr-Abl kinase and its phosphorylated (P-) form in CML cells treated with WBZ_4 and imatinib. Reprinted from [14], copyright 2007 with permission from the American Society for Clinical Investigation

decrease in tumor volume and weight (Fig. 9.10a, b). No obvious toxicities were observed in the animals during treatment. Selectivity of WBZ_4 in the animal model was corroborated by assaying on the xenograft induced by CML cells K562 (Fig. 9.10c). While imatinib is shown to significantly impair tumor growth ($p < 0.01$), the prototype WBZ_4 has virtually no effect, in accord with its engineered specificity and hence lack of inhibitory impact on Bcr-Abl kinase.

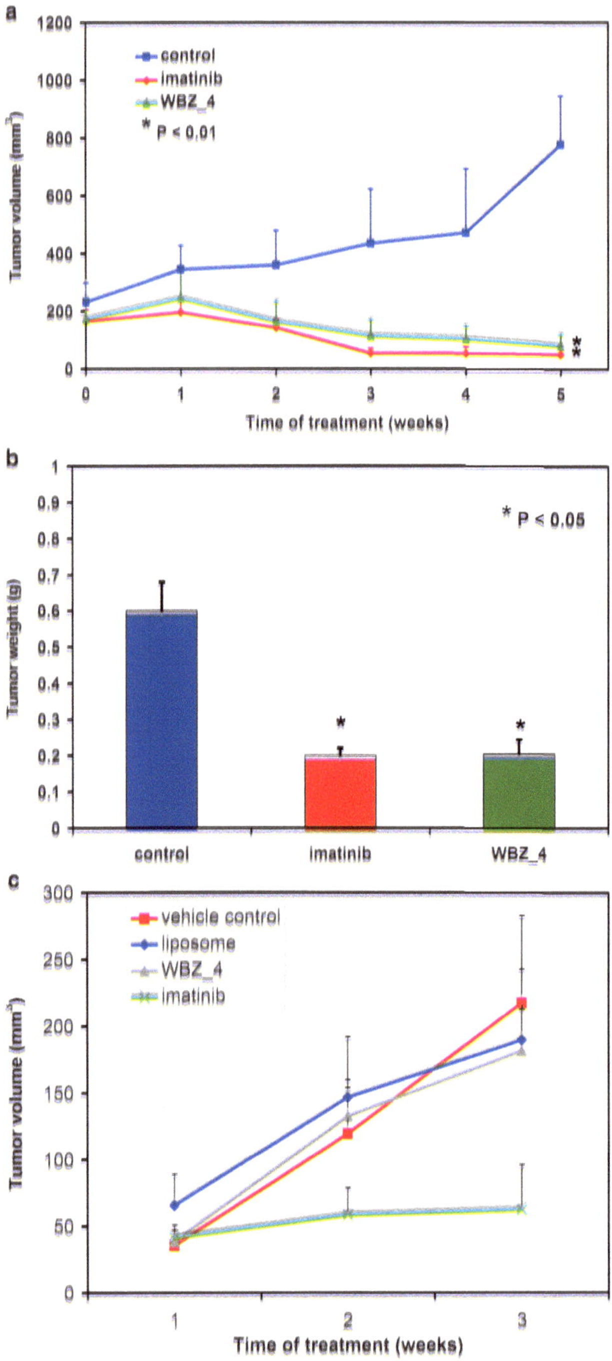

Fig. 9.10 Xenograft models of anticancer activity. (**a**) Effect of WBZ_4 or imatibib therapy on *in vivo* GIST growth determined by longitudinal tumor volume measurements. Mice were randomized to treatment with either control (normal PBS and empty liposomes give indistinguishable results within experimental uncertainty), imatinib or liposome-formulated WBZ_4. (**b**) Effect of

Previous studies of imatinib-induced cardiotoxicity identified a protective JNK (JNK1) inhibition as a means to reduce the collapse of mitochondrial membrane potential [15]. Accordingly, WBZ_4 was engineered to also inhibit JNK, a feature preliminarily corroborated *in vitro* (Fig. 9.5). This protective role of WBZ_4 is also validated *in vivo*. Western blotting of cell extracts derived from the same imatinib- and WBZ_4-treated cardiomyocytes (Fig. 9.11a) revealed that both drugs increased the phosphorylation state of ERK1/2 and $p38^{MAPK}$ as compared to untreated cultures. In contrast, cells treated with WBZ_4 showed reduced levels of JNK activation at each concentration tested, as compared to untreated cells, or cells treated with imatinib. These results confirm other observations [15] demonstrating that imatinib induces the endoplasmic reticulum (ER) stress response and activates JNKs in neonatal rat ventricular myocytes (NRVM). JNK activation has been clearly associated with cardiomyocyte apoptosis via collapse of the mitochondrial membrane potential and release of cytochrome C [23]. It was further confirmed that imatinib (and WBZ_4) had little effect on $p38^{MAPK}$ activation at concentrations below 50 μM, and actually increased ERK activation in these cultured cells. In contrast, WBZ_4 appeared to reduce levels of pJNK at each dose tested, suggesting that upstream activation of JNKs was inhibited by WBZ_4. Thus, a direct inhibitory effect on JNK activity would predict that WBZ_4 should have reduced cardiotoxicity *in vivo*.

This expectation is supported by other observations using the surrogate marker brain natriuretic peptide (*BNP)*, a most sensitive indicator of myocardial hypertrophy and cardiac impairment [24]. The expected curbing of cardiotoxity in WBZ_4 anticancer therapy was confirmed by examining the mRNA levels of BNP in the left ventricle of mice from the same groups assayed for anticancer activity. The *BNP* mRNA levels were about 58 % higher in the ventricles from imatinib-treated animals (Fig. 9.11b), while no significant difference was detected in the WBZ_4-treated animals when compared with untreated mice. Finally, the reduced cardiotoxicity of WBZ_4 was directly tested in mice by determination of the percentage ejection fraction (EF) in the left ventricle by magnetic resonance (MR) imaging of treated mice subject to imatinib or WBZ_4 therapy (Fig. 9.11c, d). Following 6 weeks of therapy, the cardiac EF was significantly lower in the imatinib group compared to controls ($p = 0.02$), which is consistent with previous findings [15]. Remarkably, WBZ_4 treatment had no effect on cardiac EF despite prolonged therapy (Fig. 9.11c, d).

Fig. 9.10 (continued) WBZ_4 or imatibib therapy on *in vivo* GIST growth determined by weight measurements. Animals from all groups were sacrificed after 6 weeks of therapy, tumors were excised, and the weight was recorded. (**c**) Effect of WBZ_4 or imatibib therapy on *in vivo* CML growth induced through a xenograft of K562 tumor cells, determined by longitudinal tumor volume measurement. The WBZ_4 selectivity is hereby corroborated *in vivo*. Reprinted from [14], copyright 2007 with permission from the American Society for Clinical Investigation

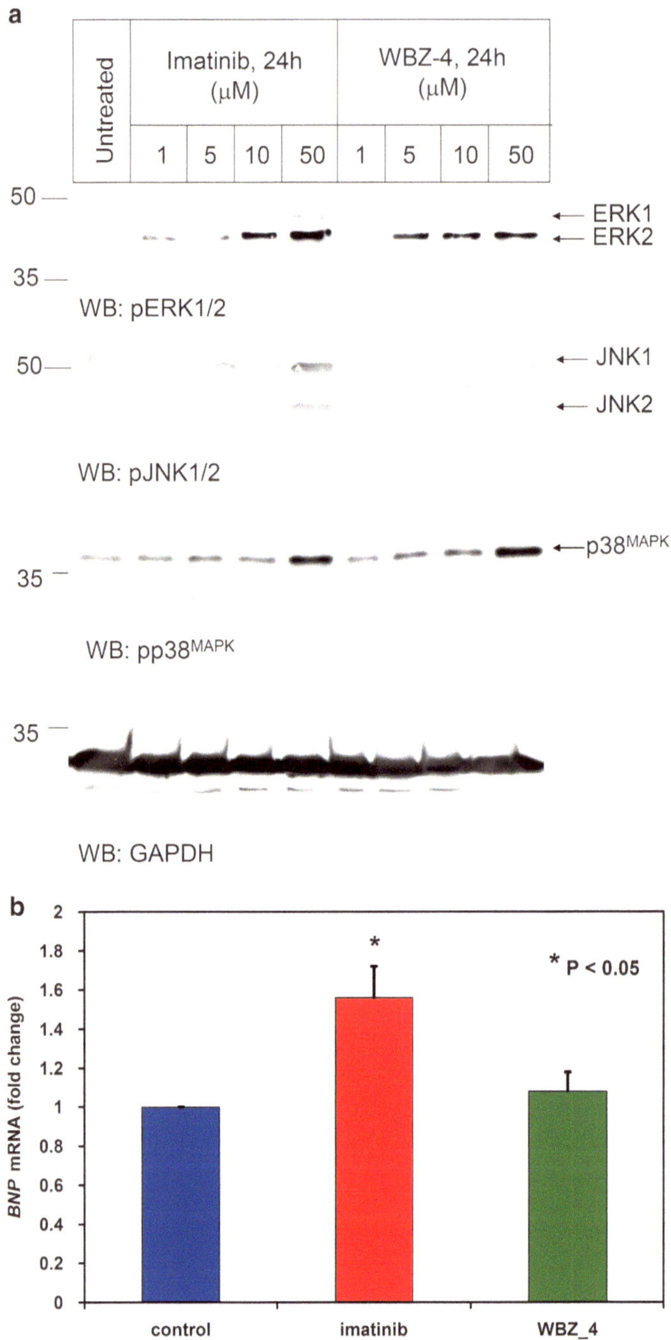

Fig. 9.11 Cardiotoxicity assays. (**a**) Western blot of JNK inhibition in cardiomyocites. Western blots were probed with primary antibodies specific for the phosphorylated forms of ERKs, JNKs, and $p38^{MAPK}$. The position of molecular weight standards is indicated to the left of each blot. (**b**) Effect of WBZ_4 or imatinib therapy on mouse heart brain natriuretic peptide (*BNP*). The mRNA levels of *BNP* (a sensitive marker of myocardial hypertrophy and cardiac impairment) were examined in the left ventricle of mice from the groups in Fig. 9.10a. The *BNP* mRNA levels were about 58 % higher in the ventricles from imatinib-treated animals ($p = 0.02$), but no significant difference was noted in the WBZ_4 treated animals. (**c**) Comparison of left ventricular

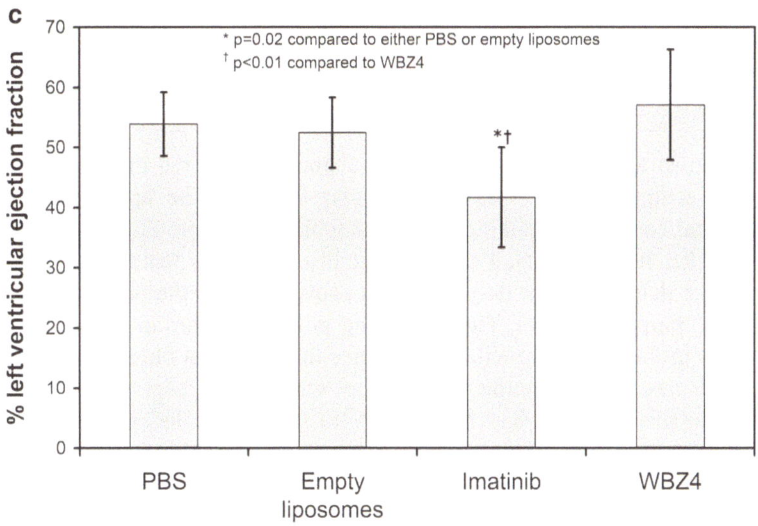

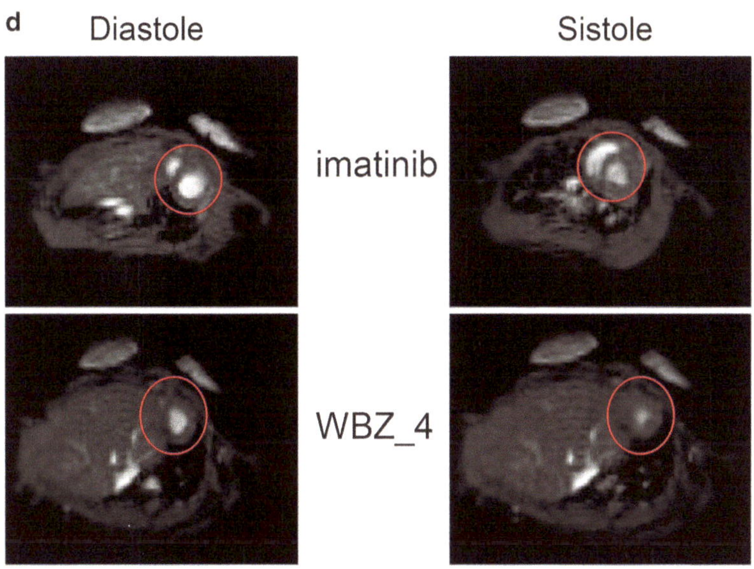

Fig. 9.11 (continued)

Fig. 9.11 (continued) ejection fraction after 6 weeks of control (groups treated with either PBS or empty liposomes), imatinib, or WBZ_4 therapy in mice. (**d**) Magnetic resonance imaging of cardiac function. Assessment of left ventricular function (*red circle*, lower cavity) was performed as follows: representative *in vivo* axial images of the left ventricle in diastole and systole of control and test mice were acquired in a 4.7 T MR scanner cardiac gating using a magnetization-prepared spoiled gradient echo sequence. To assess cardiac function, short axis cardiac cine images were acquired using a magnetization-prepared, cardiac-gated spoiled gradient echo sequence (TE/TR 2.1 ms/~23.5 ms; 16 phases covering 1.5 R-R cycles; in-plane resolution 312 μm × 312 μm; 1.25 mm slice thickness). For assessing EF, regions of interest encompassing the cavity of the left ventricle were drawn. For each animal, the average region of interest of four central slices in systole was divided by the average region of interest of four central slices in diastole taken at the same location in the left ventricle to derive the EF. Figures (**a**)–(**c**) reprinted from [14], copyright 2007 with permission from the American Society for Clinical Investigation

9.8 Controlled Specificity Through Epistructural Design

The top-down approach presented in this chapter is centered in a rational ligand re-design sculpting target-specific molecular features. The approach harnesses fundamental concepts pertaining to the discrimination of proteins with a common ancestry, sketched in Chaps. 4 and 6. These discriminatory features are associated with unique dehydration or de-wetting hot spots found on the protein/water interface of the purported target. The de-wetting propensities are in turn promoted by dehydrons in the protein structure, and hence the hydration blueprint of the target becomes a design tool complementing the wrapping concept. In this particular illustration, the re-design had four goals: (a) to refocus the primary impact on C-Kit kinase; (b) to reduce the inhibitory impact on Bcr-Abl kinase; (c) to reduce the inhibitory impact against LCK in order to prevent immunosuppression (cf. Fig. 9.1b); and (d) to promote JNK inhibition in order to reduce cardiotoxic effects and because JNK is a recognized target in the treatment of ovarian cancer [17]. These three goals have been achieved [14], as attested by the *in vitro* and *in vivo* results described in this chapter, clearly revealing that *drug specificity can be rationally controlled.* The imatinib re-design turned the wrapping ligand into a selective proto-therapeutic agent to treat GIST with a *prima facie* reduction of cardiotoxicity and removal of drug-induced immunosuppressive effects.

9.9 Problems

9.1. *WBZ_4 as a chemo-sensitizer to treat Pancreatic Ductal Adenocarcinoma.* Pancreatic Ductal Adenocarcinoma (PDA) is probably the most lethal cancer (median survival <6 months, 5-year survival rate <5 %), with no cure in sight. Part of the therapeutic challenge arises from the inflammatory context concurrent with cancer development and progression [25]. Thus, probably in response to the stem cell factor (SCF) produced by the tumor, mast cells infiltrate the tumor stroma and secrete pro-inflammation cytokines, promoting the angiogenic and invasive phenotype [26]. The epithelial tumor cells apparently secrete mediators, especially SCF, that induce degranulation and migration of mast cells from the bone marrow to the tumor stroma, where they induce the rapid progression of the disease within an inflammatory setting. The inhibition of mast cell activation has been shown to be sufficient to result in tumor death in animal models with activated Myc oncogene [1]. This analysis suggests a promising adjuvant therapeutic opportunity based on an antimastocyte agent.

The stem cell factor (SCF) is the ligand of the c-Kit receptor, which is expressed by mast cells. Thus, a kinase inhibitor (KI) that targets the c-Kit kinase with high affinity and specificity is expected to provide the type of therapeutic agent required to treat PDA by acting as chemosensitizer, used in synergy with a chemotherapeutic agent like gemcitabine (Gemzar®, Eli Lilly). The rationale for this synergic modality is that the removal of mast cells from the tumor stromal microenvironment is likely to sensitize the tumor to chemotherapy. The affinity profile of the KI must fulfill a number of additional constraints arising from important therapeutic imperatives:

(a) A massive dose, far higher than the prescribed or tolerated dose of other KIs like imatinib or dasatinib on their anticancer applications, will be required in this context to dismount or circumvent the desmoplastic mass and reduce mast-cell infiltration. Thus, c-Kit-affinity should be subnanomolar, significantly higher than that of the first-generation KIs. A superior safety standard should be met by the proposed KI.
(b) Side effects, especially cardiotoxicity, pose a significant risk especially at such high doses (much higher for pancreatic ductal carcinoma than for CML or GIST) and this risk can only be avoided if the c-Kit inhibitor does not exhibit cross reactivity with the Abelson kinase, a cancer target whose inhibition causes mitochondrial depolarization in cardiomyocytes as described in this chapter. This again rules out first-generation KIs like imatinib.
(c) Mast cells are key players in the development of the adaptive immune response, serving as effectors, antigen presenters and co-stimulators of T-cells. Thus, the c-Kit KI intended to target mast cells should not introduce an additional compromise to the immune system. This is especially important since complete therapeutic success hinges also on the possibility of recruiting the immune response by harnessing the antigenic activity generated by chemotherapy. Thus, as we target mast cells to diffuse inflammation, we should be keenly aware that we are losing a trigger of antigen-based T-cell receptor signaling conducive to T-cell activation/development. An additional hindrance of T-cell activation due to an undesired cross reactivity of the KI (anti-Lck activity) would probably completely impair the success of treatment. These considerations rule out c-Kit inhibitors like imatinib that are also inhibitors of the lymphocyte-associated kinase (Lck), as shown in this chapter. This kinase is known to be a crucial signal transducer in the development and activation of T-cells as well as in NK-mediated cytotoxicity, and its inhibition in these contexts would compromise the immune response beyond therapeutically advisable levels.
(d) The c-Kit kinase is also on-target to create additional synergies with the immune response. These cooperative effects are likely to be generated by our chemosensitizing agent. Thus, c-Kit is not only a target for the removal of mast cell infiltration but also a target to reverse the tumor-induced immunosuppression [27]. The regulatory T-cells (Tregs) are major modulators of the antigen-triggered T-cell response and their proliferation and development is crucially dependent on the SCF generated by the tumor. Hence, since as stated previously, SCF is the ligand of

the c-Kit receptor, c-Kit becomes also on-target to reverse tumor-induced immunosuppression (see Chap. 14 for a comprehensive discussion).

Based on the preceding considerations, rationalize the use of WBZ_4 as a precursor or lead compound in the engineering of antimastocyte-chemosensitizer agents to synergize with *Gemzar* in the cure of pancreatic ductal adenocarcinoma. Consider the fact that WBZ_4 has already been proven to synergize favorably with chemotherapy in the context of ovarian cancers with highly angiogenic phenotype [17].

9.2 *WBZ_4 for the Treatment of Ovarian Cancer*. Through a literature search, identify functional roles of JNK that could justify the exploitation of anti-JNK inhibitor WBZ_4 for the treatment of ovarian cancer (cf. [17]).

9.3 Dehydron C673-G676 in C-Kit kinase has been exploited in this chapter as a selectivity filter to engineer a safer version of *imatinib*, named WBZ_4. The molecular prototype WBZ_4 has a superior affinity for cancer targets C-Kit and JNK relative to *imatinib*, while the immunosuppressive activity of WBZ_4 attributable to Lck inhibition is significantly reduced relative to its parental compound. Upon structural alignment, C-Kit dehydron C673-G676 (PDB.1T46) becomes the well-wrapped hydrogen bond M319-G322 in Lck (PDB.3LCK) and the severely underwrapped (structurally unstable) dehydron M111-N114 in JNK (PDB.2G01). By adopting the interfacial free energy expression

$$\Delta G_{if}(\phi) = \int \left[\frac{1}{2}\omega \| \vec{\nabla}\phi \|^2 - k_B T(\rho_0 - \rho) \ln\left(1 - \frac{\phi}{4}\right) \right] d\vec{r}$$

(1.7), compute the change $\Delta\Delta G_{if}$ of interfacial free energy associated with the mutation G676V in C-Kit, G322V in Lck and N114I in JNK. Prove that the trend associated with $\Delta\Delta G_{if}$ is indicative of the selectivity filter described.

References

1. Dancey J, Sausville EA. Issues and progress with protein kinase inhibitors for cancer treatment. Nat Rev Drug Discov. 2003;2:296–313.
2. Levitski A, Gazit A. Tyrosine kinase inhibition: an approach to drug development. Science. 1995;267:1782–8.
3. Tibes R, Trent J, Kurzrock R. Tyrosine kinase inhibitors and the dawn of molecular cancer therapeutics. Annu Rev Pharmacol Toxicol. 2005;45:357–84.
4. Gibbs J, Oliff A. Pharmaceutical research in molecular oncology. Cell. 1994;79:193–8.
5. Donato NJ, Talpaz M. Clinical use of tyrosine kinase inhibitors: therapy for chronic myelogenous leukemia and other cancers. Clin Cancer Res. 2000;6:2965–6.

6. Fabian MA, Biggs WH, Treiber DK, et al. A small molecule kinase interaction map for clinical kinase inhibitors. Nat Biotechnol. 2005;23:329–36.
7. Gambacorti-Passerini C, le Coutre P, Mologni L, et al. Inhibition of the ABL kinase activity blocks the proliferation of BCR/ABL+ leukemic cells and induces apoptosis. Blood Cells Mol Dis. 1997;23:380–94.
8. Schindler T, Bornmann W, Pellicena P, et al. Structural mechanism for STI-571 inhibition of Abelson tyrosine kinase. Science. 2000;289:1938–42.
9. Attoub S, Rivat C, Rodrigues S, et al. The c-kit tyrosine kinase inhibitor STI-571 for colorectal cancer therapy. Cancer Res. 2002;62:4879–83.
10. DeMatteo RP. The GIST of targeted cancer therapy: a tumor (gastrointestinal stromal tumor), a mutated gene (c-kit), and a molecular inhibitor (STI571). Ann Surg Oncol. 2002;9:831–9.
11. Skene RJ, Kraus ML, Scheibe DN, et al. Structural basis for autoinhibition and STI-571 inhibition of C-kit Tyrosine kinase. J Biol Chem. 2004;279:31655–63.
12. Tuveson DA, Willis NA, Jacks T, et al. STI571 inactivation of the gastrointestinal stromal tumor c-Kit oncoprotein: biological and clinical implications. Oncogene. 2001;20:5054–8.
13. Chen JP, Zhang X, Fernández A. Molecular basis for specificity in the druggable kinome: sequence-based analysis. Bioinformatics. 2007;23:563–72.
14. Fernández A, Sanguino A, Peng Z, Ozturk E, Chen J, Crespo A, Wulf S, Shavrin A, Qin C, Ma J, Trent J, Lin Y, Han HD, Mangala LS, Bankson JA, Gelovani J, Samarel A, Bornmann W, Sood AK, Lopez-Berestein G. An anticancer c-Kit kinase inhibitor is reengineered to make it more active and less cardiotoxic. J Clin Invest. 2007;117:4044–54.
15. Kerkela R, Grazette L, Yacobi R, et al. Cardiotoxicity of the cancer therapeutic agent imatinib mesylate. Nat Med. 2006;12:908–16.
16. Druker BJ. Molecularly targeted therapy: have the floodgates opened? Oncologist. 2004;9:357–60.
17. Vivas-Mejia P, Benito JM, Fernández A, Han HD, Mangala L, Rodriguez-Aguayo C, Chavez-Reyes A, Lin YG, Carey MS, Nick AM, Stone RL, Kim HS, Claret FX, Bornmann W, Hennessy BT, Sanguino A, Peng Z, Sood AK, Lopez-Berestein G. c-Jun-NH2-kinase-1 inhibition leads to antitumor activity in ovarian cancer. Clin Cancer Res. 2010;16:184–94.
18. Fernández A, Scheraga HA. Insufficiently dehydrated hydrogen bonds as determinants of protein interactions. Proc Natl Acad Sci U S A. 2003;100:113–8.
19. Barker S, Kassel D, Weigl D, et al. Characterization of $pp^{60c\text{-}src}$ tyrosine kinase activities using a continuous assay: autoactivation of the enzyme is an intermolecular autophosphorylation process. Biochemistry. 1995;34:14843–51.
20. Songyang Z, Carraway KL, Eck M, et al. Catalytic specificity of protein-tyrosine kinases is critical for selective signalling. Nature. 1995;373:536–9.
21. Clarkson B, Strife A, Wisniewski D, Lambek CL, Liu C. Chronic myelogenous leukemia as a paradigm of early cancer and possible curative strategy. Leukemia. 2003;17:1211–62.
22. Baines CP, Molkentin JD. Stress signaling pathways that modulate cardiac myocyte apoptosis. J Mol Cell Cardiol. 2005;38:47–62.
23. Prenen H, Deroose C, Vermaelen P, Sciot R, Debiec-Rychter M. Establishment of a mouse gastrointestinal stromal tumour model and evaluation of response to Imatinib by small animal positron emission tomography. Anticancer Res. 2006;26:1247–52.
24. Scheuermann-Freestone M, Simon Freestone N, Langenickel T, et al. A new model of congestive heart failure in the mouse due to chronic volume overload. Eur J Heart Fail. 2001;3:535–43.
25. Theoharides T. Mast cells and pancreatic cancer. N Engl J Med. 2008;358:1860–1.
26. Huang B, Lei Z, Zhang GM, Li D, Song C, Li B, Liu Y, Yuan Y, Unkeless J, Xiong H, Feng ZH. SCF-mediated mast cell infiltration and activation exacerbate the inflammation and immunosuppression in tumor microenvironment. Blood. 2008;112:1269–79.
27. Kao J, Ko EC, Eisenstein S, Sikora AG, Fu S, Chen SH. Targeting immune suppressing myeloid-derived suppressor cells in oncology. Crit Rev Oncol Hemat. 2011;77:12–9.

Chapter 10
Epistructural Informatics for the Drug Designer

The dehydron-based re-engineering of kinase inhibitor Gleevec to make it a safer drug (Chap. 9) illustrated the power of epistructural physics as enabler of a novel platform for drug design. As shown in this bioinformatics chapter, epistructural analysis provides a universal selectivity filter, applicable to the entire human kinome and its idiosyncratic variations. A kinome-wide bioinformatics analysis reveals that epistructure-based design heralds a new generation of drugs that enable a tighter control of specificity and a personalization of the treatment. The universality of this selectivity filter in the field of therapeutic interference with cell signaling is thus revealed.

Furthermore, the impact of epistructure-based design for personalized molecular therapy is established, thereby fulfilling an imperative of the post-genomic era. This perception is reinforced as promising therapeutic agents are so often recalled because of idiosyncratic side effects detected in small subpopulations of patients. This chapter addresses this issue and describes how epistructure-based design can be brought to fruition in the personalization of targeted medicine. The chapter invites the exploitation of proteome-wide epistructural bioinformatics for personalized medicine, creating an enabling platform to tailor drugs to patient idiosyncrasies.

Within this epistructural bioinformatics pursuit, an *in-silico* affinity profiler of kinase inhibitors is developed and tested. In the initial stages of drug discovery, a thorough experimental screening of lead compound libraries becomes critically limited by the size of the kinase sample. This chapter introduces a computational approach for extensive screening entrained by small-scale affinity profiles as input data. The profiler then makes affinity predictions for a larger kinase test set. These predictions result from a propagation of the reduced profile, exploiting an epistructural comparison of kinases based on a feature-similarity matrix. The comparison focuses on molecular markers for specificity and promiscuity. Our approach enables the computational high-throughput screening of entire libraries of compounds to search for suitable leads, mapping their inhibitory impact on a sizable sample of the human kinome. The *in silico* tool is validated by contrasting predictions against reported high-throughput screening experiments.

A. Fernández, *Physics at the Biomolecular Interface*, Soft and Biological Matter,
DOI 10.1007/978-3-319-30852-4_10

10.1 Epistructural Selectivity Filter for Drug Design

Kinase targeting is a central theme molecular cancer therapy [1–5]. Be as it may, a scientific basis for the rational design of safe and efficacious kinase inhibitors with controlled specificity appears to be sorely absent [1, 3]. In practice, most leads and scaffolds for drug development are actually discovered through screening techniques, and are generally transformed into therapeutic tools through a painstaking trial-and-error process that often renders a final product plagued with clinical uncertainties.

While the paradigm of target specificity may be shifting to controlled multi-target drugs [3], the structural factors determining cross-reactivities are not yet fully understood, in spite of some progress. For instance, the solvent-accessible nonpolar surface on the surface of a protein, frequently invoked as a major determinant of protein associations [6, 7], actually fosters promiscuity in the context of kinase inhibition [3, 8]. This is so because nonpolar groups on the surface of proteins tend to be highly conserved across paralog proteins and even within the entire kinase superfamily [9, 10]. Progress in controlling specificity is also hampered by the lack of structural information. For instance, only about 22 % of the human kinome is represented in the PDB.

In this chapter we focus on assessing the universality of wrapping as the molecular feature controlling nonpromiscuous drug targeting. To fulfill this objective an extensive bioinformatics exploration of drug profiles is required. Recent high-throughput screenings of commercially available kinase inhibitors lead us to an operational definition of promiscuity as a significant cross reactivity (dissociation constant $K_d < 100\,\text{nM}$) extended over >30 % of the sampled kinome. The problem is complicated due to the scarcity of affinity-profiled kinases with reported structure: To date, only about 24–25 % of the human kinome (123 out of the ~519 discovered protein kinases) is reported in the PDB. Furthermore, kinase homology models are quite unreliable to make inferences on such detailed features as wrapping patterns since the level of pairwise sequence identity across the kinase superfamily is typically low (<30 %) [11].

To mitigate these problems, we take advantage of the high degree of topological conservation of the kinase folds, a known consequence of their common ancestry. Thus, reliable sequence-based attributes such as disorder propensity [12] are here mapped onto structurally threaded models to make inferences about dehydron patterns and ultimately, about drug specificity/promiscuity patterns. As described in Chap. 5, these cross-validations of dehydron inferences in proteins with unreported structure take advantage of the crucial link between disorder propensity and the degree of exposure of the protein backbone in the native fold [13]. The latter is itself a measure of the propensity of backbone hydration, the main factor competing with the hydrophobic collapse of the polypeptide chain. To make specific sequence-based predictions of 3D dehydron locations, a learned resolution enhancement (LRE) algorithm like *Twilighter* needs to be deployed in conjunction with the predictor of structural disorder propensity, as described in Chap. 5.

Thus we seek a sequence-based attribute that enables a classification of kinase space in a way that accurately reproduces similarities/differences in affinity profiling against a background of drugs [14, 15]. The classification introduces a partition into disjoint sets of kinases within which pharmacological differences are accurately predicted from molecular differences. This methodology could be implemented because the relevant classification of kinases was performed by comparing the regions deprived of adequate packing or intramolecular dehydration, i.e. the flexible regions, which are precisely the markers differentiating kinases at the structural level [4].

The chapter is organized as follows: First, we determine the type of kinase molecular similarity that promotes promiscuity whenever targeted. Thus, we demonstrate that solvent-exposed nonpolar regions promoting ligand associations foster promiscuity. To draw statistics at a kinomic scale, we formally define a region, the nonpolar hull, enabling comparison of targeted exposed nonpolar regions across different kinases. Then, we examine other targeted features in molecular design aimed at fostering pair-wise interactions between ligand and kinase and show that the high degree of conservation of the partner groups on the protein surface does not enable a suitable control of specificity. Subsequently, we introduce a means of calculating dehydration propensities on polar-paired regions on the protein surface and formally define a region, the so-called environmental (wrapping) hull, enabling a comparison of polar dehydration propensities across targets. We demonstrate that this molecular attribute is responsible for controlling specificity. To carry out the analysis at a sequence level, a technique named *environmental alignment* is implemented. This technique enables the identification of residues whose microenvironments are likely to be perturbed by ligand association. Such residues are identified by aligning the kinase sequence against a background of sequences of homologous kinase-ligand complexes reported in the PDB. Finally, these universal filters are applied to the challenges of personalized medicine, where idiosyncratic mutations that generate or suppress dehydrons in an individual kinase may be targeted by wrapping designs.

10.2 Epistructural Comparative Analysis of the Human Kinome

10.2.1 Sequence-Based Dehydron Inference

To establish the universal validity of wrapping as a selectivity filter, the comparative analysis of molecular attributes in target paralogs cannot be constrained to PDB-reported kinases (32 out of the 119 assayed in [14]). Homology-based predictions of native structures [16, 17] become plausible given the extent of PDB representation (~24 %) of paralogs within the kinase superfamily and also because dehydrons may be directly inferred from sequence (Chap. 5, [13]), and ultimately contrasted with the homology-based predictions for mutual validation. Thus, the disorder-based prediction of dehydrons hinges on the fact that native disorder arises essentially from the impossibility to dehydrate intramolecular hydrogen bonds, as noted in chapter 5 and further developed in Chap. 16. The choice of disorder score to benchmark homology predictions is an obvious one given the particularly low accuracy of the latter on loopy or flexible regions [10].

The disorder propensity is given by a score determined by the program PONDR® (predictor of native disorder), a neural-network predictor of native disorder. Only 0.4 % of more than 900 non-homologous PDB proteins give false positive predictions in regions with 40 or more consecutive sites of predicted disorder. Even this 0.4 % of false positives is an overestimation, as many disordered regions in monomeric chains become ordered upon ligand binding or in crystal contacts [12]. The false negatives error rate (~11 % for regions of 40 or more consecutive predicted ordered residues) is also compelling in regards to the predictor quality.

The correlation between solvent exposure of hydrogen bonds and disorder propensity implies that it is possible to predict dehydrons directly from sequence [13]: It suffices to determine the PONDR-generated pattern associated with the desired feature. The correlation implies that the propensity to adopt a natively disordered state becomes pronounced for proteins which, due to a chain composition reflecting high hydrophilicity, cannot protect the backbone hydrogen bonds even minimally. Thus, we can infer the existence of dehydrons from the PONDR score (f_d) with 92 % accuracy in regions with $f_d > 0.35$ provided such regions are flanked by well-protected regions ($f_d < 0.35$), to ensure the existence of structure. The accuracy of this sequence-based dehydron predictor was established by inferring the location of dehydrons in proteins with reported structure, for which the microenvironment of each hydrogen bond can be determined unambiguously [13]. The false negatives constitute 368 of the 8215 dehydrons in a PDB database of 1466 proteins free from structural redundancy and less than 25 % sequence identity in pairwise alignment. The false positives correspond to 2721 of the 133,623 backbone hydrogen bonds examined.

10.2.2 Epistructural Alignment of Targetable Regions Across the Human Kinome

In order to assess differences in the exposed nonpolar regions of kinase targets that interact with different ligands we define a common region, named the *nonpolar hull*. A residue *a* is defined as making contact with a ligand L within a PDB-reported complex if a side-chain heavy atom (H excluded) is found to be within 3.6 Å (upper bound for any bond length) of a heavy atom in the ligand. The nonpolar hull for protein chain *i*, $H_{np}(i)$, is dependent on a "structural background set" of chains, S(i), which includes all homolog chains that align with chain *i* [18] for which there are protein-ligand complexes with reported structure. Thus, the nonpolar hull is defined as $H_{np}(i) = \mathbf{U}_{j\in S(i)}\Phi_i(R_{np}(j))$, where $\Phi_i(a)$, with $a\in$chain j, is the residue in chain i that aligns with residue a in chain j, and $R_{np}(j)$ is the set of nonpolar residues in chain j in contact with its respective ligand L_j. For any pair i, j, the following relation holds: $\Phi_i(H_{np}(j)) = H_{np}(i)$, enabling a comparison of kinases by examining differences in nonpolar hulls.

To assess differences in the dehydration propensities of polar regions for purported kinase targets, we introduce a common region for comparison named *environmental hull*. First, the set $R_{env}(j)$ is defined for protein chain j as the set of residues paired by dehydrons in chain j within a protein-ligand complex with reported structure subject to the following condition: ligand L_j contributes to the dehydration of the dehydron, that is, it has some carbonaceous nonpolar group within the dehydration domain of the dehydron. Then, the set $E_{env}(j)$ is defined as the set of residues from chain j that contribute to the dehydration microenvironment of a dehydron contained in $R_{env}(j)$, that is they are either paired by the dehydron or they contain a side-chain nonpolar group within the microenvironment of the dehydron. Then, the environmental hull for chain j, $H_{env}(j)$, is defined as the union of the residues in chain j that align with residues framing the environments of dehydrons that in turn are environmentally affected by ligands in PDB-reported complexes: $H_{env}(j) = \mathbf{U}_{i\in S(j)}\Phi_j(E_{env}(i))$ (notation has been followed consistently, the structural background S(j) is defined above). As with nonpolar hulls, for any pair i, j, the following relation also holds: $\Phi_i(H_{env}(j)) = H_{env}(i)$. This property is needed to actually compare environments of different proteins.

10.3 Pharmacological Relevance of the Epistructural Analysis

In this section we report on previous attempts to correlate different molecular attributes of the protein targets with available drug-screening data for a sizable set of kinase targets. Essentially, the scope of this section is to contrast differences in the pharmacological profiles of kinases with structure-based dissimilarities that take into account specific molecular attributes. The large drug/kinase panel experimentally assayed [14] and exploited in our analysis is highly underreported in the PDB. Consequently, a reliable sequence-based predictor of the relevant molecular attributes of kinases had to be implemented as shown in the previous sections.

We first define a pharmacological distance, d_{phar}, that quantifies differences in the affinity profiling of kinases against a background of available drugs. This metric is the Euclidean distance between affinity vectors with entries given in negative logarithmic or $\Delta G/RT$ units (ΔG = free energy change for protein-ligand association, R = universal gas constant, T = absolute temperature). A positive cutoff value $\Delta G/RT = \ln 10 \approx 2.3$ is adopted for affinities reported as "no hit" in the screening ($K_d > 10\,\mu M$). Figure 10.1a displays the matrix $\mathbf{D}_{phar} = [d_{phar}(i, j)]$ for all pairs (i, j)

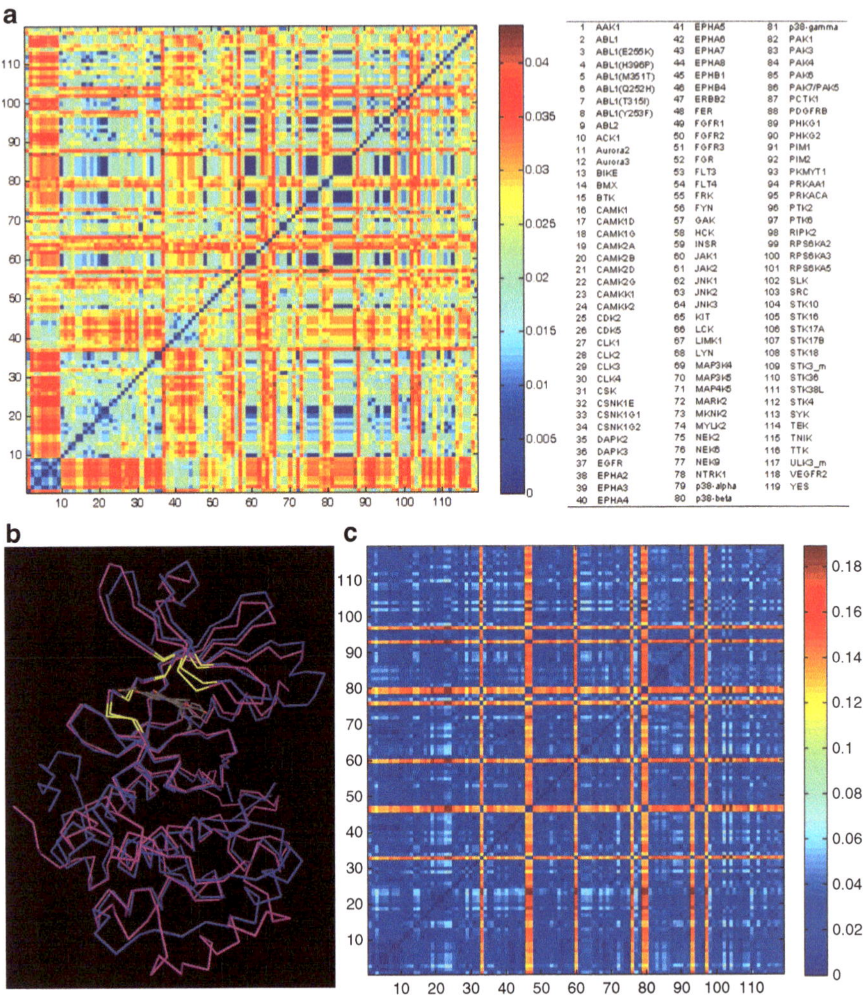

Fig. 10.1 (**a**) Pharmacological distance matrix $\mathbf{D}_{\text{phar}} = \left[d_{\text{phar}}(i,j)\right]$ for all pairs (i, j) from the 119 kinases assayed through affinity profiling against a background of 19 drugs [14]: SB202190; SB203580; sp600125; imatinib (Gleevec); VX-745; BIRB 796; BAY-43-9006; GW-2016; gefitinib; erlotinib; CI-1033; EKB-569; ZD-6474; Vatalanib; SU11248; MLN-518; LY-333531; roscovitine/CYC202 and flavopiridol. The distance is given by $d_{\text{phar}}(i,j) = \left[\sum_{m\in\text{inhibitors}} (K(i,m) - K(j,m))^2\right]^{1/2}$, where K(i, m), K(j, m) represent respectively the negative logarithm of equilibrium constants for complexation of kinase *i* and kinase *j* with drug inhibitor *m*. (**b**) Aligned backbones (RMSD ≈ 0.33Å) for paralog kinases PDK1 (*blue*) and CHK1 (*lilac*) in their active folds. The structures were reported in complex with ligands BIM8 (PDB.1UVR) and 3A3 (PDB.2GCU), respectively. The nonpolar hulls are depicted in *yellow*. (**c**) Nonpolar distance matrix $\mathbf{D}_{\text{np}} = \left[d_{\text{np}}(i,\ j)\right]$ over the 119 assayed kinases. (**d**) Plot of nonpolar distance versus pharmacological distance. Each *circle* represents a kinase pair. No correlation is observed, while there is some bimodality in each dimension. (**e**) Correlation between pseudopharmacological distance (including staurosporine in the drug screening background) and nonpolar distance between kinases. The sole outliers are pairs involving the EGFR kinase, the kinase whose affinity vector is **not** dominated by staurosporine [14]. Reprinted from [10], copyright 2007 with permission from Oxford University Press

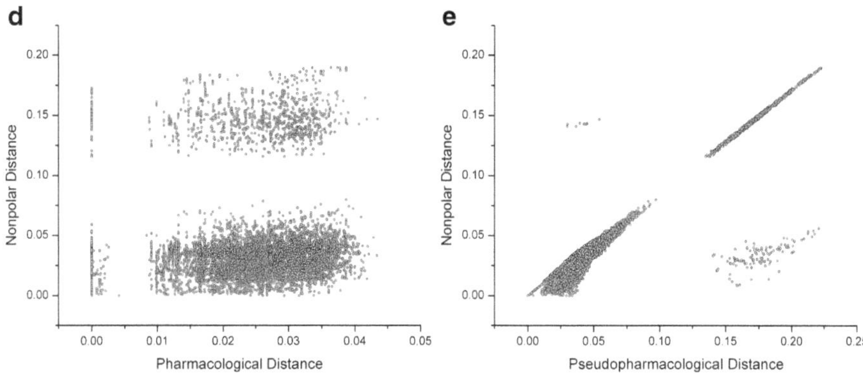

Fig. 10.1 (continued)

from the 119 assayed kinases. The affinity profiling adopted included 19 of the 20 drugs originally screened [14]: only the promiscuous staurosporine was excluded since it does not belong to the pharmacology realm.

To determine whether pharmacological differences are dictated by differences in nonpolar accessible surfaces of the targets within ligand-binding sites, a nonpolar distance, $d_{np}(i, j)$, between the affinity-assayed kinases i, j is introduced. The $d_{np}(i, j)$ is determined by differences in accessible nonpolar surface areas of the respective nonpolar hulls, $H_{np}(i)$, $H_{np}(j)$. As rigorously defined in the previous section, the nonpolar hull of a kinase is comprised of the nonpolar residues of the kinase in contact with its drug ligands, whenever the complexes are reported in PDB, and residues in the kinase (not necessarily nonpolar) that align with nonpolar residues in contact with ligands in homolog PDB-reported ligand-kinase complexes (Fig. 10.1b). The latter constitute the structural background for the comparison. Introducing the hull becomes necessary to compare kinases at this level. Thus, the nonpolar distance is defined as

$$d_{np}(i, j) = \left[A\left(H_{np}(i)\right)\right]^{-1} \sum\nolimits_{a \in Hnp(i)} \left|A(a) - A\left(\Phi_j(a)\right)\right| \qquad (10.1)$$

with A = nonpolar accessible area [19, 20]; a = generic residue in the nonpolar hull $H_{np}(i)$ of chain i; $A(a) = 0$ if a is polar; and $\Phi_j(a)$ = residue in chain j that aligns with a.

Only 32 of the 119 assayed kinases are reported in PDB complexes, yet, structural inferences can be made with confidence given the significant extent of structural alignment (RMSD < 1 Å) across reported structures [10]. The prediction accuracy decreases somewhat but in a quantifiable manner (~12 %) on loopy regions [10]. The matrix $\mathbf{D}_{np} = \left[d_{np}(i, j)\right]$ (Fig. 10.1c) reveals remarkable nonpolar-similarity across kinases: $\langle d_{np} \rangle / \text{maximum}(d_{np}) \approx 11\ \%$, where the angular brackets represent average over kinase pairs); $\left\langle \left[d_{np} - \langle d_{np} \rangle\right]^2 \right\rangle^{1/2} / \langle d_{np} \rangle \approx 16\%$. *This similarity implies a high conservation of accessible nonpolar surface, an indication that ligands whose affinity is dominated by hydrophobic interactions should be highly promiscuous.*

The plot d_{np} vs. d_{phar} for all (119 × 118)/2 kinase pairs (i, j) shows no correlation between the two metrics (Fig. 10.1d). However, when the highly promiscuous affinity-dominant staurosporine is incorporated to the affinity profile [14] and the affinity-based distance matrix is recalculated ($d_{phar} \rightarrow d_{ps}$ = pseudopharmacological distance), a good correlation ($R^2 = 0.875$) between d_{ps} and d_{np} is obtained (Fig. 10.1e). *This correlation reveals that promiscuity, the dominant affinity trait when staurosporine is incorporated, is fostered by targeting accessible nonpolar moieties, in turn, a highly conserved feature of protein interfaces* [21]. The strong correlation shown in Fig. 10.1e implies that staurosporine should bind mainly through hydrophobic contacts, as it is indeed the case in its PDB complexes [22].

With typical ligand-dissociation constants in the nanomolar range, the coordinate for staurosporine dominates the affinity vector for each kinase to such an extent that most affinity profiles look similar when this compound is included in the drug background. This similarity is reflected in the low rank (~7) of the pseudopharmacological distance matrix. The nonpolar matrix possess a comparable rank (~6) because the basis for comparison, i.e. exposed nonpolar residues, is highly conserved, leading to high levels of target similarity.

Sequence alignment including the 32 assayed kinases with PDB-reported complexes reveals that residues in ATP-binding sites engaged in hydrogen bonding with ligands are highly conserved, with $0 \leq \sigma(n) \leq 0.21$ (n = chain position for hydrogen-bonding residue; σ = information entropy reflecting aminoacid variability after sequence alignment; average σ in $H_{np} = 0.87$; maximum ($\sigma = ln20 \approx 4.2$). As expected, differences in hydrogen-bonding capabilities do not appreciably correlate with d_{ps} ($R^2 \approx 0.19$), and there is no correlation with d_{phar}.

These observations lead to the question: Is wrapping the *universal* feature to be targeted to promote specificity? *In other words, is the dehydron pattern a molecular feature with sufficient variability across homologs and capable of significantly influencing ligand affinity*? To answer this question, we focus on "environmental residues", i.e. those framing the dehydron microenvironments. As we recall from Chaps. 1, 3 and 5, dehydrons promote protein associations because the enhancement and stabilization of the underlying electrostatic interactions overcomes the thermodynamic cost associated with removing the surrounding water molecules that hydrate amide or carbonyl groups [23, 24]. Environmental residues include the hydrogen-bonded residues themselves. To compare environments of different kinases, we define the environmental hull of a kinase, H_{env}, as the reunion of all environmental residues in the chain and residues aligning with environmental residues from other chains (Fig. 10.2a, b). Thus, an environmental distance $d_{env}(i, j)$ between kinases *i* and *j* is obtained by comparing the aligned hydrogen-bond microenvironments within $H_{env}(i)$ and $H_{env}(j)$: $d_{env}(i, j) = M(i, j)^{-1}\left[\sum_{n=1,\dots,M(i,j)} \Delta_n(i, j)\right]$, where M(i, j) = number of residue pairs in $H_{env}(i)$ corresponding to dehydrons in kinase *i* or to hydrogen bonds or nonbonded residue pairs that align with dehydrons in $H_{env}(j)$, n = dummy index denoting residue pair, and $\Delta_n(i,j) = 0$ if residue pair n corresponds to a SAHB in $H_{env}(i)$ that aligns with a dehydron in $H_{env}(j)$, and $\Delta_n(i,j) = 1$, otherwise. Thus defined, the environmental distance compares local dehydration propensities associated with SAHB patterns in kinases. The validity of the relation: $\Phi_i(H_{env}(j)) = H_{env}(i)$ enables a rigorous comparison between environments of two different proteins.

The matrix $\mathbf{D}_{env} = [d_{env}(i,j)]$ for the 119 kinases assayed (Fig. 10.2c) is obtained after inference of the dehydrons for the 87 kinases unreported in PDB from direct structure prediction. The predictions are validated through a correlation with an independent and accurate sequence-based prediction of the disorder propensity. This attribute was chosen because loopy regions, in the twilight between order and disorder (Chap. 5), compromise somewhat the accuracy of a structure prediction.

The strong correlation between d_{env} and d_{phar} ($R^2 \approx 0.917$) (Fig. 9.2d) reveals that the impact of drugs on the human kinome is dictated by differences in hydration microenvironments or dehydron patterns across the ligand-binding regions of the kinases. The diversity in hydration microenvironments needed to yield specificity across paralog kinases results from the variability ($\langle\sigma\rangle \approx 1.38$) of environmental residues ($\langle\sigma\rangle \approx 0.21$ when the average is restricted to residues paired by hydrogen bonds that are environmentally affected by the ligands).

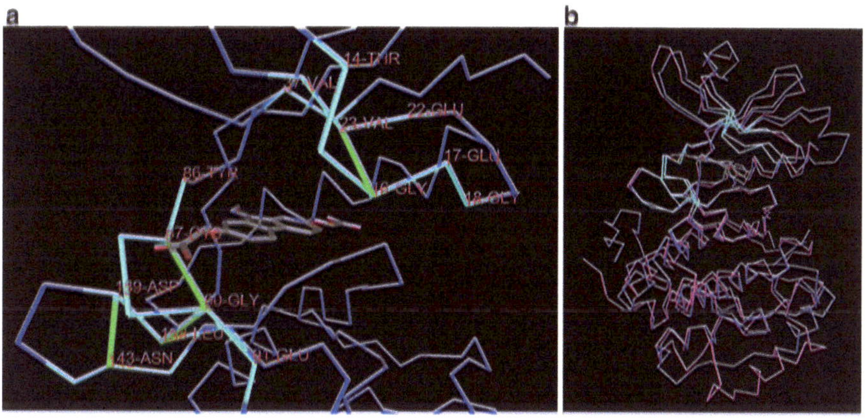

Fig. 10.2 (**a**) Environmental hull (*light blue*) for CHK1 (obtained from alignment with PDK1). Dehydrons are indicated as *green* segments joining the α-carbons of the paired residues. The virtual bonds are shown as *blue* segments. The three dehydrons perturbed by the ligand (named 3A3) are C87-G90; G90-L138; G16-V23. (**b**) Aligned backbonesfor PDK1 (*blue*) in complex with BIM8 (PDB.1UVR) and CHK1 (*lilac*) in complex with 3A3 (PDB.2GCU), with the environmental hulls depicted in *light blue*. (**c**) Environmental distance matrix $\mathbf{D}_{env} = [d_{env}(i,j)]$ for the 119 kinases assayed. (**d**) Correlation of environmental versus pharmacological distance. The *line* indicates the optimal linear fit obtained from linear regression. The *red diamonds* correspond to the six pairs including ABL1, the primary target for imatinib, and its six mutants (listed in Fig. 10.1a) that confer drug resistance. Reprinted from [10], copyright 2007 with permission from Oxford University Press

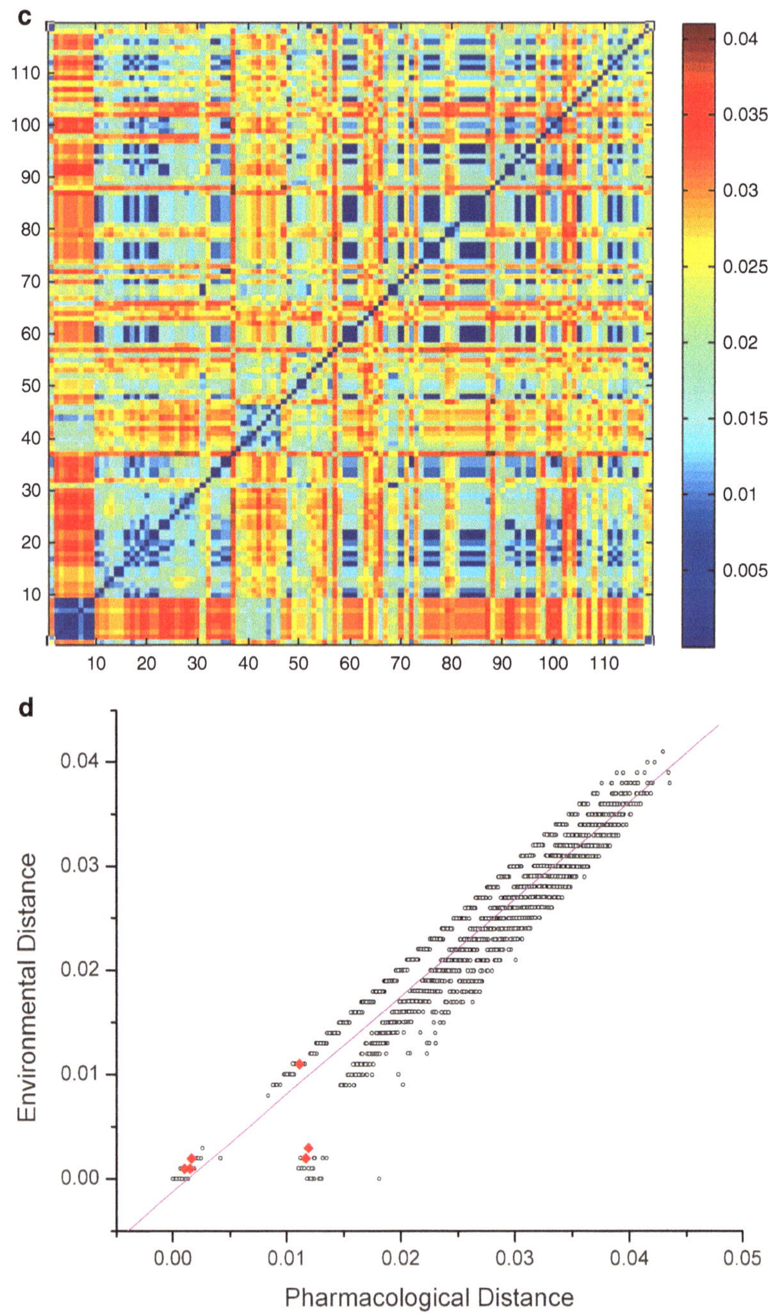

Fig. 10.2 (continued)

The strong correlation shown in Fig. 10.2d *provides compelling evidence in support of the universality of protein wrapping as the selectivity filter for molecular therapeutic approaches targeting the human kinome.*

10.4 Epistructural Target Library for the Human Kinome

As shown in the previous section, the wrapping (dehydron) pattern of a target kinase provides a universal selectivity filter for the drug designer. This fact begets the obvious question: How do we take advantage of this selectivity filter to actually broaden the technological basis of drug discovery? To address this question we envision the construction of a dehydron library in drug-targeted proteins and then exploit this bioinformatics library to generate a map of controlled cross-reactivity that result when such specialized patterns are targeted. The controlled cross-reactivity map will reflect the superior level of specificity that may be achieved by targeting packing patterns. This resource will exploit the matrix representation of the wrapping pattern of a protein that enables comparative alignment across an entire target superfamily and will digitalize information on subtle packing differences across protein targets. The long-term application of this resource is the control of side effects in molecular-targeted therapy adopting the new drug-design paradigm, *the drug as dehydron wrapper*.

As shown in Chaps. 8 and 9, specific drugs may be chemically modified to turn them into wrappers of dehydrons, protecting them from water attack. This type of ligand optimization has been shown in specific cases to retain anticancer activity while reducing potential life-threatening side effects. Thus, the molecular basis for target discrimination was delineated *in vitro* and through *in vivo* assays for antitumor activity, *albeit* over limited sets of purported targets.

Against this background, it becomes essential to introduce bioinformatics results of wide applicability that precisely map out the optimizing chemical modifications for all drug chemotypes and all kinase targets with the goal of controlling and enhancing inhibitor specificity across the entire human kinome. Thus, in principle at least, the technological basis of drug design can indeed be broadened contingent on our ability to construct a comparative and annotated dehydron library. This library should compile the cross-reactivity map that will contain "raw" information on wrapping defects as selectivity-promoting features and contain the annotated information to guide the drug designer to develop superior drugs with controlled specificity. The raw data should include:

- Identification of all packing defects—*dehydrons*—targetable by ATP-competing inhibitors for each human kinase.
- Comparative alignment of the dehydron pattern of one kinase against the respective patterns in all the others to obtain a map of *a-priori* cross reactivities for dehydron-wrapping drugs.

- Identification of dehydrons with the lowest levels of conservation for each kinase together with the kinases where the dehydron is replaced by a well-wrapped hydrogen bond upon alignment. Thus, the optimal specificity-controlling dehydrons will be determined through a "conservation score".
- Determination of accessibility of optimal dehydrons vis-à-vis wrapping/targeting within a given drug chemotype and the set of kinases that are discriminated as the optimal dehydrons are targeted. Obviously, a specificity-promoting dehydron would be of no practical use if its exogenous wrapping is not accessible from the particular lead chemotype considered.

Thus, the results expounded in this chapter pave the way for the construction of the matrix-based library of dehydrons for human kinases enabling direct comparative alignment across the structure-reported kinome. This library is destined to be used to decompose PDB-reported target space into clusters of cross-reactive kinases, enabling direct assessment of the full potential of the novel selectivity filter on targets with reported structure. The results can be validated by examining compounds independently screened using a phage-display kinase library and contrasting their affinity (K_D)-profile against predicted cross-reactivity clusters. The results should also be cast in usable format to guide the engineering of safer inhibitors with controlled specificity. Dehydrons should be prioritized according to their decreasing levels of conservation across homologs in the human kinome, hence serving as the most reliable molecular features to be targeted in order to control and reduce cross reactivity.

10.5 Specificity-Promoting Target Features

In the first stages of the resource building, we envision a comparative analysis confined to kinases with PDB-reported structures, thus guaranteeing unambiguous dehydron identification. The structural alignment required for comparative analysis will be largely facilitated by the kinase fold conservation resulting from common ancestry. Targetable dehydrons should be reported for each kinase jointly with several annotations instrumental to the drug-optimization strategy: (a) extent of dehydron conservation across the superfamily and expected cross reactivity arising if targeted by a wrapping ligand; (b) accessibility of the dehydron vis-à-vis a specific drug chemotype; (c) activation state of the kinases under comparison; (d) variations in dehydron patterns due to different induced fits of the same target; and (e) indication on whether a dehydron identified as targetable belongs to an induced fit region within a kinase-ligand complex (in which case the lead optimization should preserve the ligand chemotype) or is an inherent feature of the kinase.

Such a digital library of packing defects enabling a comparative kinome-wide alignment will represent a powerful tool to guide lead optimization geared at controlling specificity and reducing side effects. Thus, the results of this chapter call for a superfamily-wide bioinformatics platform to safely and reliably control cross reactivity of kinase inhibitors based on the dehydron concept.

We also envision extensions of the library to targets with unreported structure. Specifically, we envision a homology-based extension of the dehydron-matrix library to the entire human kinome and concurrent extension of the map of controlled cross-reactivity enabling a comprehensive assessment and validation of the novel packing-based design paradigm. As described in the preceding sections, only about 25 % of the human kinome has PDB-representation and hence many kinase targets of clinical interest become intractable when adopting structure-based strategies for drug design/optimization. Thus, homology-based approaches become essential to turn the dehydron library into a useful tool to guide the optimization of therapeutic agents. Our adopted homology-based predictors can generate kinase 3D-structures by satisfying spatial restraints imposed by threading with the template kinase structure and applying a force field with refinement protocols based on simulated annealing. The predicted dehydron patterns should be corroborated by two independent routes: (I) through comparison with sequence-based inferences of disorder propensity, harnessing on a known correlation between poor packing and disorder propensity (see Chap. 5); and (II) by contrasting inferred dehydron-based cross-reactivities with experimental high-throughput screening of kinase inhibitors. The annotations for homology-based dehydron inference will include those indicated previously (a–e) and should incorporate additional annotations pertaining to inference reliability: (f) extent of template-probe sequence identity (s.i.); (g) quality of template-probe alignment defined by number of sequence insertions and deletions.

The construction of the dehydron library and its validation as resource of kinome-wide applicability for the drug developer involves nine interrelated milestones described in Fig. 10.3. As indicated above, such a comparative dehydron library should provide "raw" information on packing defects as selectivity-promoting features in drug design. This information ought to be supplemented with annotations that should aid the drug designer in the interpretation and exploitation of the raw data and in addressing the problems pertaining to the concrete application of the concept of "drug as wrapper of packing defects". These annotations are now described in detail:

(A) *Extent of conservation of each dehydron within the kinase superfamily and expected cross reactivity arising when targeted by a wrapping ligand.* In general, the dehydron pattern of a kinase may not enable a *complete* discrimination against all other ~518 kinases. For example, the wrapping distances across pairs from the set of 119 kinases originally screened for drug affinity by AMBIT [14, 15], reveal that the focal adhesion kinase FAK (synonym: PTK2) has the same dehydron pattern as members of the CAMK, JAK, JNK, NEK and PAK families (Figs. 10.1a and 10.2c). This example illustrates the fact that *certain degree of cross reactivity may be unavoidable when adopting the wrapping technology, pointing at the very limitations of this approach. The whole issue then becomes whether the particular cross reactivity is clinically tolerable or whether it provides the right compromise between activity and toxicity.*

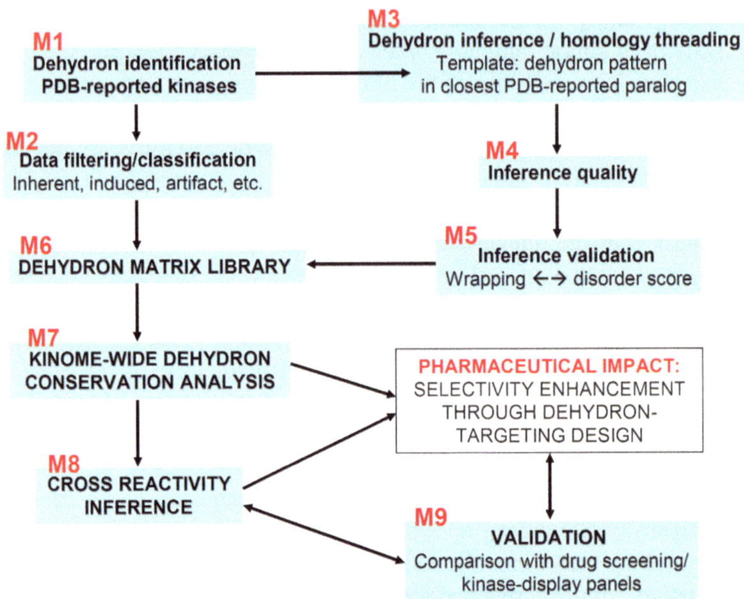

Fig. 10.3 Nine milestones (M1-M9) for the construction and validation of the kinome-wide dehydron library as a resource to guide drug design

Besides giving the level of conservation of each packing defect, the library should provide an assessment of the expected cross reactivity for a drug that wraps the defect. As a concrete example, let us determine the level of control of cross reactivity that may be achieved when wrapping the KIT dehydron C673-G676, as described in Chap. 9. Table 10.1 provides a comparative analysis of the conservation of the KIT dehydron across KIT paralogs and other members of the kinase superfamily.

Thus, a drug inhibitor that wraps the KIT dehydron C673-G676 is *a-priori* likely to also hit the following kinases: CSF1R, FLT3, TIE2, MET, MUSK, IGFR1 and INSR, where the KIT dehydron aligns with a dehydron. However, the actual levels of cross reactivity may be smaller depending on the drug chemotype and its associated steric clashes with the target. For instance, of this group of kinases, the imatinib modification that wraps KIT dehydron C673-G676 binds only to CSF1R, as shown in Chap. 9.

(B) *Accessibility of a dehydron as a targettable feature vis-à-vis a specific drug chemotype*. The mere identification of a specificity-promoting feature in a kinase is of limited practical use unless it can be targeted through a chemical modification within a specific chemotype. For example, the selectivity-promoting E282-E286 dehydron in the Abelson (Abl) kinase may be targeted through a wrapping chemical modification (methylation) of imatinib (Fig.10.4a). However, the same dehydron is not accessible through wrapping

Table 10.1 Conservation of KIT dehydron C673-G676 within an assortment of PDB-reported kinases

Group	Protein	PDB accession code	Sequence identity	Wrapping analysis of HB aligning with KIT dehydron C673-G676
TK	KIT	1T46	100	**DH**
		1PKG	99	DH
		1T45	95	DH
	CSF1R	2I1M	70	DH
	FLT3	1RJB	65	DH
	VEGFR2/ KDR	2YWN	59	–
	FGFR2	1GJO	54	HB
	FGFR1	1FGI	52	HB
	RET	2IVS	49	HB
	TIE2	2OO8	44	DH
	HSK	1QCF	44	HB
	MET	1R0P	42	DH
	MUSK	1LUF	42	DH
	ABL	2HZ0	42	HB
		2HYY	41	HB
	LCK	2OG8	42	HB
		2PL0	41	HB
		2OFV	41	HB
	INSR	1IRK	39	DH
	IGF1R	1M7N	36	DH
TKL	BRAF	1UWH	36	–
CMCG	CDK6	1BLX	31	–
CAMK	MARK1	2HAK	31	HB
	TTN	1TKI	29	–
STE	PAK1	1F3M	28	–
AGC	AKT2	1MRV	28	–
	MSK1	1VZO	27	–

DH dehydron, *HB* well-wrapped hydrogen bond, "–" no bond in alignment

methylation of the compound PD173955 (Fig. 10.4b) since the incorporated nonpolar group lies outside the microenvironment of the E282-E286 dehydron.

(C) *Variations in dehydron patterns due to different induced fits*. The identification of a targettable feature to promote drug specificity is not independent of the chemotype of the drug to be redesigned. This is not only due to the fact that the chemotype determines the accessibility of the dehydron (point B), but also because a kinase may present different induced fits depending on the ligand to which it is associated, and hence different ligand-dependent dehydron patterns. Thus, if a dehydron in a kinase is identified as targettable but only occurs in a

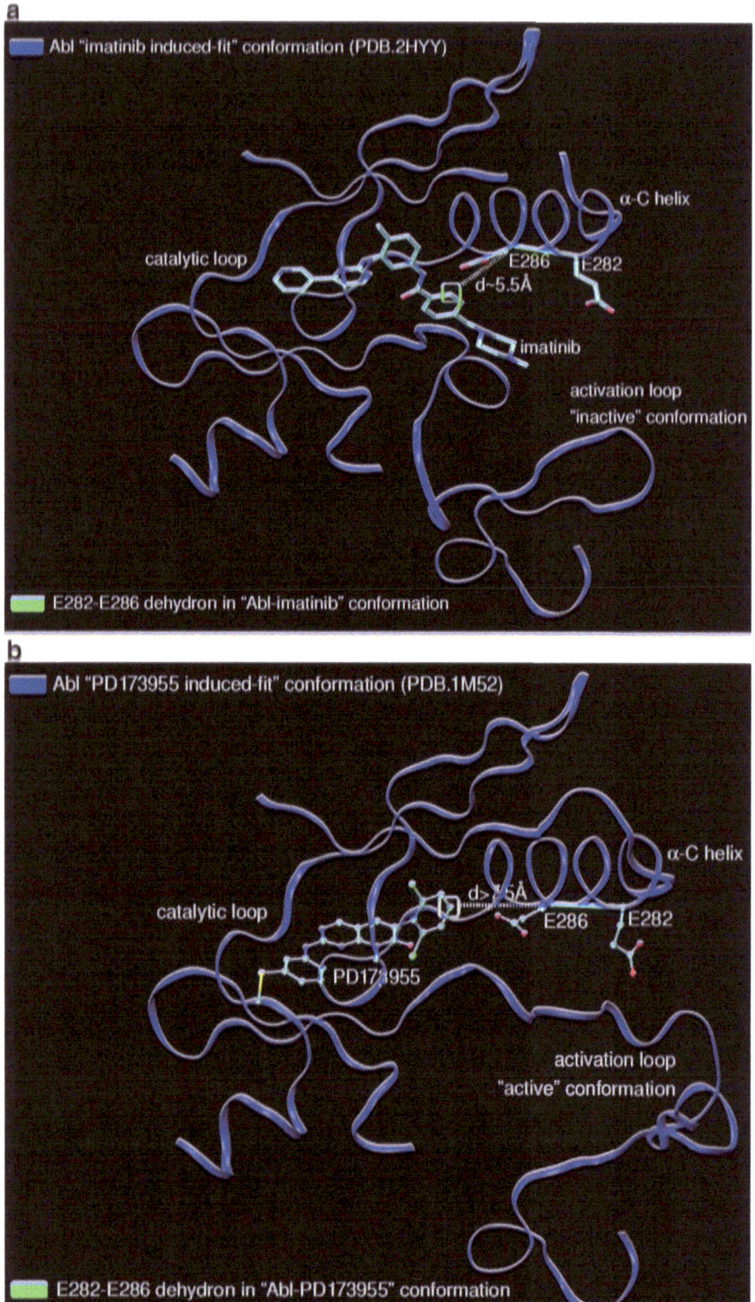

Fig. 10.4 Differences in accessibility of ABL dehydron E282-E286 according to drug chemotype. (**a**) Dehydron-wrapping methylation of imatinib. (**b**) inaccessibility of dehydron to wrapping methylation in compound PD173955

particular induced fit with a particular drug ligand, it may be most advantageous to respect the chemotype of the ligand when targeting/wrapping the dehydron in order to enhance specificity. To address this problem, when identifying a particular selectivity-promoting dehydron to be targeted, a dehydron library should always include the ligand and PDB accession of the complex where the dehydron has been identified.

For example, the Q300-E316 dehydron in Abl kinase may be a good target to enhance the specificity of *dasatinib* and drugs within its chemotype since this dehydron is present in the induced fit of the Abl upon complexation with dasatinib (Fig. 10.5a). However, that dehydron becomes a well-wrapped hydrogen bond in the induced fit resulting upon Abl complexation with imatinib (Fig.10.5b). Thus, the library should contain annotation guarding *against* adopting the imatinib chemotype when targeting the Q300-E316 dehydron in Abl, since this dehydron is not present upon imatinib association.

10.6 Epistructural Target Library as a Biotechnological Resource

The dehydron library will generate chemotype-dependent decompositions of target space (human kinome) into cross-reactivity clusters. These results are crucial to assess the full potential of the packing-based drugs as therapeutic agents of superior controllable selectivity. Cross-reactivity clusters include only kinases that are inhibited by the same wrapping modification within a fixed lead chemotype. The wrapping modification, in turn, targets a dehydron in a particular kinase, and all other kinases presenting a dehydron at the aligned position will belong to the same chemotype-dependent cross-reactivity cluster. The cartoon in Fig. 10.6 represents the construction/validation of the cross-reactivity map combining the dehydron library and structural information on a single kinase/inhibitor complex for the particular drug chemotype.

The cross-reactivity map is thus obtained by a straightforward procedure that combines structural input from a kinase/inhibitor complex with a kinome-wide comparative alignment of packing defects provided by the bioinformatics library. The results will be validated by examining compounds independently screened using the T7-phage display kinase library (~360 kinases in latest panels) [15] and contrasting their affinity (K_D)-profile with their predicted cross-reactivity cluster: if a compound is identified to wrap a particular dehydron in a kinase, then this compound should be cross-reactive with all other kinases for which the dehydron is present at the aligned position except for kinases where structural alignment reveals steric clashes. This target-space decomposition will assess the full potential of the wrapping-based design motif.

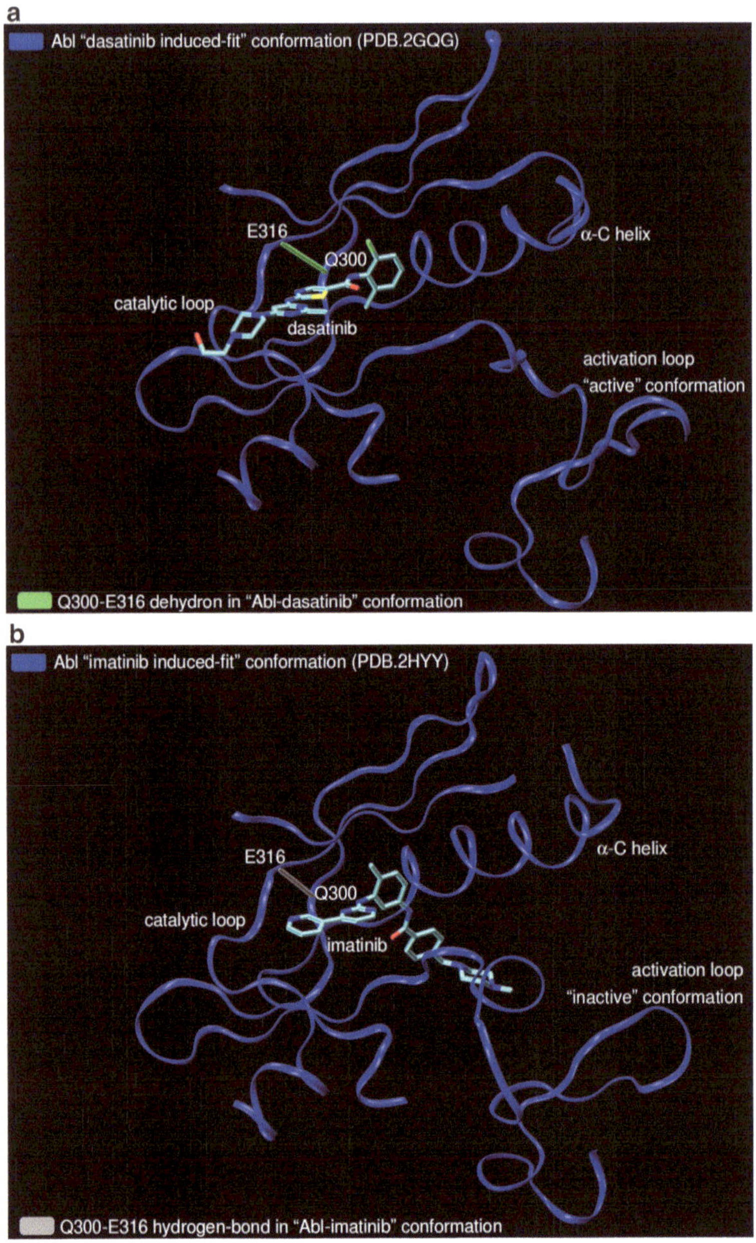

Fig. 10.5 Variations in dehydron patterns for the ABL kinase due to different induced fits. (**a**) ABL/*dasatinib* complex and (**b**) ABL/*imatinib* complex

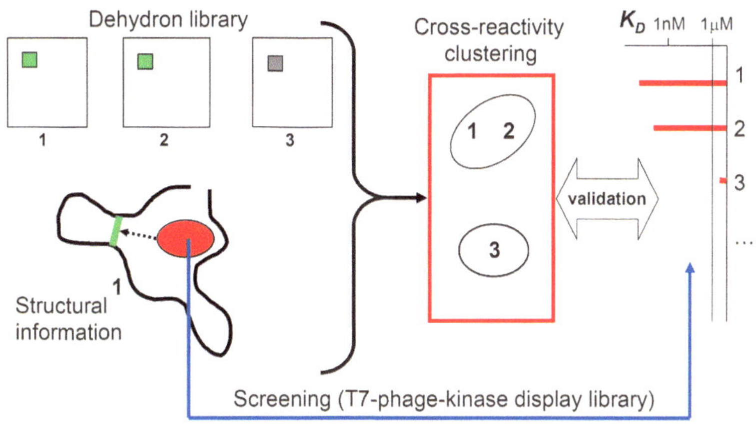

Fig. 10.6 Schematic representation of the generation of a chemotype-dependent cross-reactivity map. A dehydron (*green*) in a particular protein (**1**) that may be wrapped (*dashed-line arrow*) by a chemical substitution of a lead within a specific chemotype (*red*) will be singled out for comparison across the kinome and to define the cross-reactivity cluster to which this protein belongs. The extent of conservation of the dehydron will be then established by comparative alignment of dehydron matrices within the bioinformatics library. Conservation of the dehydron (across proteins **1** and **2** in the scheme) promotes cross reactivity of the wrapping compound, defining the cluster for protein **1** within the chemotype-dependent decomposition of target space (*red rectangle*). Proteins that lack drug-accessible dehydrons or those that contain unique (not conserved) drug-accessible dehydrons constitute singleton clusters. Validation is readily obtained by examining compounds independently screened using AMBIT's high-throughput kinase panels [14, 15] and contrasting the predicted cross-reactive clusters with the reported affinity (K_D)-values

10.7 Epistructural Analysis for Personalized Molecular Medicine

In a context of rational design, personalized molecular therapy is expected to exploit idiosyncratic differences to tailor drugs to the patient's individual needs. In practice, one of the stumbling blocks in this effort arises because individual idiosyncrasies do not usually reveal themselves at the structural level: deleterious genetic variation is unlikely to prevail if the structure of gene products is compromised or even altered. Rather, such individual variations surface in a more subtle way at the epistructural level, where structure is conserved but the wrapping pattern of the protein is altered introducing functional distortions. Such distortions typically arise from the dehydron-based functionalization or deregulation of vicinal side-chain groups (see Chap. 7) as idiosyncratic dehydrons are incorporated, as illustrated in Fig. 10.7a. Since the dehydron is generally targettable by wrapping drugs, we envision personalized therapy arising from the epistructural signature of the patient idiosyncrasies. Personalized molecular therapy is a major imperative of post-genomic medicine. This perception is reinforced with the frequent news that promising therapeutic agents are recalled because of idiosyncratic side effects detected in patient subpopulations. Rational approaches to personalized

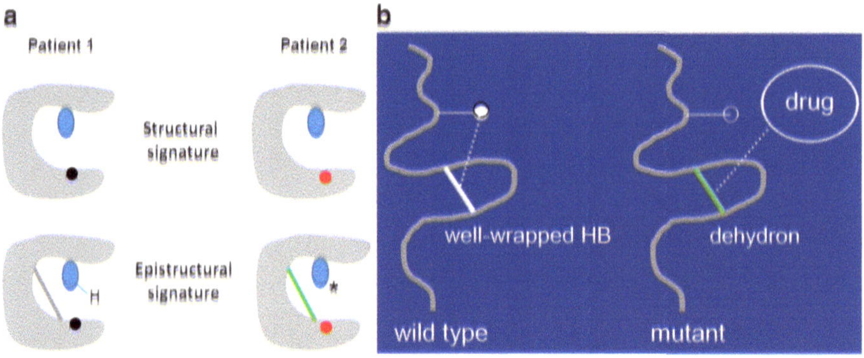

Fig. 10.7 Exploitable idiosyncratic variability at the epistructural level. (**a**) An aminoacid variation (*black to red disk*) between two patients in a specific protein does not typically manifest at the structural level. Thus, the structure of the gene product is retained. However, the epistructural pattern and associated epi-function are altered as the relation between the structure and the solvent is altered: a well-wrapped HB (grey in patient 1) becomes a dehydron (marked in green in patient 2). In this illustration, patient 2 has a dehydron enrichment due to deprivation of backbone shielding as a consequence of the mutation. As a result, a nearby proto-nucleophilic group (blue ellipse) becomes functionalized in patient 2 (indicated by the asterisk) thorugh dehydron-promoted deprotonation. (**b**) An idiosyncratic variation substituting a good hydrogen-bond wrapper (*solid circle*) for a bad wrapper (*empty circle*) can produce a new dehydron in the protein structure. This "idiosyncratic dehydron" may in turn be targeted or wrapped intermolecularly by a purposely designed drug

drug therapy will ultimately depend on our ability to translate genomic idiosyncratic variations into molecular biomarkers that can guide a patient-tailored design. This aspect is particularly challenging when the target proteins have no reported structure, and the molecular features promoting specificity need to be predicted from sequence (Chaps. 5 and 16). This section addresses this issue and describes how the wrapping design concept can be brought to fruition in this area. This section introduces plausible scenarios in which oncogenic idiosyncrasies may promote targettable differences in the wrapping patterns of the gene products. Ultimately an invitation is extended to adopt and exploit protein wrapping as a molecular biomarker for personalized medicine within an enabling platform to tailor drugs to patient idiosyncrasies.

The therapeutic imperative to make more efficacious and safer drugs by exploiting the enormous output of genomic forays has proven to be a daunting challenge, far more difficult than originally expected. No reliable rational approach has been identified so far to meet such a challenge. In addressing this imperative, the dearth of ideas of general applicability with the potential to become transformative and broaden the technological base of the pharmaceutical industry is quite evident [25, 26]. From the perspective of a rational drug designer, one can only speculate that a personalized approach would require an effective translation of genomic idiosyncrasies and disease-related variations into structural differences that may guide the molecular engineering of the therapeutic agent. While this conclusion may be inevitable, it has not been pursued beyond a limited number of anecdotal instances [27].

It is seldom the case that idiosyncratic or disease-related genomic differences would translate as significant structural changes in the gene products. The functional demands of the wild-type locus cannot generally be compromised without a serious fitness cost and misfolded proteins are generally degraded unless they can aberrantly aggregate (as in poly-glutamine or poly-asparagine in-frame insertions). On the other hand, subtler variations, leading for instance to an oncogenic deregulation of structural elements, may still be exploitable from the vantage point of the drug designer. Thus, while such accidents are unlikely to alter the functionally-competent topology of the protein, they may affect its wrapping, as shown below. We may thus speculate that wrapping patterns may well be on their way to become molecular biomarkers for drug personalization in the particular context where a genetic accident or idiosyncrasy impacts a coding region and produces a change in the microenvironment of a preformed hydrogen bond (Fig. 10.7b).

In other words, a single amino acid substitution can turn a wrapping residue into a bad wrapper (that is, one of the following amino acids: G, A, N, D, S, T or P) and hence generate a dehydron in *lieu* of the well-wrapped hydrogen bond in the wild type. Thus, a mutation can produce a new dehydron, actually an "idiosyncratic dehydron", which in turn may be wrapped intermolecularly by a purposely designed drug, as shown schematically in Fig. 10.7b.

Conversely, an amino acid substitution that generates a well-wrapped hydrogen bond by increasing the number of nonpolar groups in the bond microenvironment may not yield a productive opportunity from the perspective of the wrapping designer: if the bond is well wrapped, the wrapping drug would sterically clash with the intramolecular wrappers and has nothing to add in terms of hydrogen-bond stabilization (dehydration). Yet, there is a conspicuous exception arising in the case where the augmentation in the number of wrapping side-chain groups triggered by the mutation *induces* a dehydron that was absent in the wild type due to a local scarcity of wrapping residues (the $\rho < 7 \rightarrow \rho \geq 7$ transition upon drug binding).

To illustrate this case, we focus on the Asp816 → Val substitution in the activation loop that confers resistance to the C-Kit kinase against the powerful anticancer drug imatinib [27]. This substitution renders imatinib ineffective to treat the gastrointestinal stromal tumor (GIST), which has C-Kit as the primary target. Yet, Val is a good wrapper residue while Asp is not. Hence, the Val816-Kit mutant is able to sustain the dehydron Phe811-Ala814, which is absent or unstable in the wild-type Kit kinase due to insufficient wrapping (see Chap. 7). In turn, the induced dehydron may be exploited to design a wrapping variant of imatinib with an extra methyl group, as needed to target the idiosyncratic dehydron Phe811-Ala814 by contributing to its dehydration upon association (Fig. 10.8). This wrapping modification has been named WBZ_7 and has been shown in cellular assays to be highly effective against imatinib-resistant GISTs [27]. Thus, the subpopulation of GIST patients with the Asp816 → Val substitution in the C-Kit kinase domain may potentially be treated with a drug that targets the induced dehydron arising from the single-site mutation.

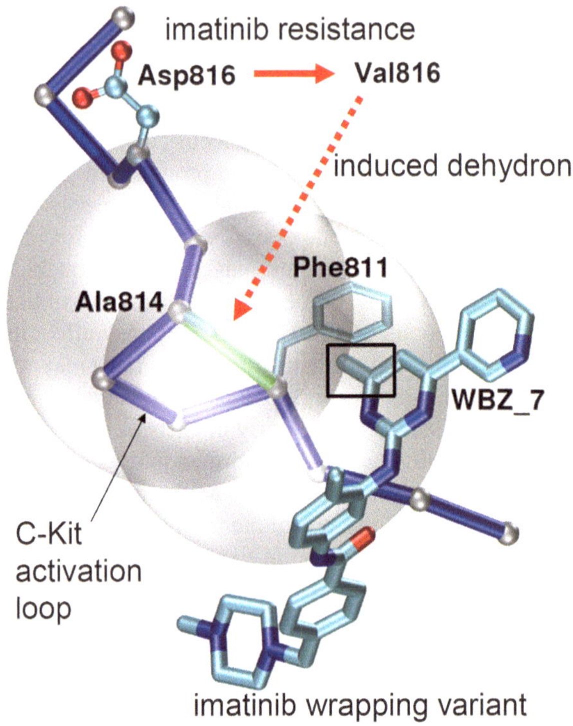

Fig. 10.8 Imatinib wrapping modification engineered to overcome drug resistance in the C-Kit kinase. This resistance is promoted by the somatic mutation Asp816 → Val. The mutation induces the dehydron Phe811-Ala814, in the activation loop which in turn can be targeted or wrapped by suitably modifying imatinib at the position highlighted by the *rectangle* [3]

The impact on protein structure of amino acid substitutions stemming from oncogenic mutations has not been systematically studied. In the particular context of cell signaling related to mitosis or to the onset of anti-apoptotic pathways, it is likely that many such mutations will promote deregulation, yielding constitutively active kinases. Under normal conditions, these enzymes typically become activated through phosphorylation of specific residues in the activation loop. This is a reversible event that fundamentally alters the conformation of the loop, making it more soluble and less obstructive, thereby promoting ATP binding. Thus, a single aminoacid substitution within the activation loop turning a nonpolar residue into a negatively charged one can promote the deregulation of the kinase, turning it into a constitutively active signal transducer. This accident can of course promote tumor development and cancer progression [28], depending on whether the transduced signal recruits pathways leading to mitosis, cell immortalization, focal adhesions, etc.

On the other hand, since the oncogenic substitution solubilizes the activation loop, it should promote a loss of structure. This process is likely to yield a loop conformation far worse wrapped than that of the autoinhibited state and hence, far richer in dehydrons (Fig. 10.9). Ergo, a constitutively deregulated kinase should be more susceptible to be targeted by wrapping/crating drugs than the wild type. This operational hypothesis begs a validation in the light of the ideas described in this book.

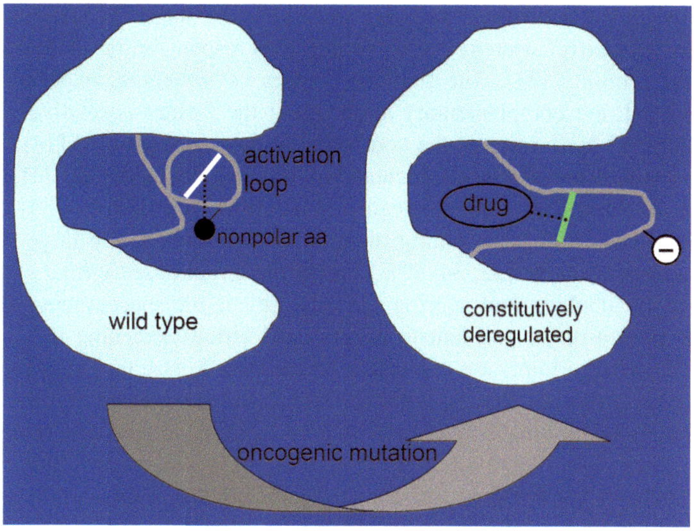

Fig. 10.9 An oncogenic mutation in a kinase domain produces a constitutively active signal transducer by permanently solubilizing the activation loop. Being less structured, this deregulated kinase is richer in dehydrons than the wild type in its autoinhibited state and hence may be targeted by a wrapping drug

By interfering with cell signaling pathways, single-molecule kinase inhibitors hold great promise as anticancer agents and probably hold potential for the molecular targeted treatment of other diseases. Yet, their therapeutic efficacy and clinical benefit is typically limited to a fraction of the treated patients. Genomic variations across patients and idiosyncratic patterns of drug resistance are likely to contribute significantly to such differences in clinical outcome. The acknowledgment of these facts has triggered the need for a personalized drug design. Perhaps this challenge is not so daunting in so far as we can translate idiosyncratic or disease-related variations into molecular differences that can be systematically targeted by molecular engineers. The wrapping pattern may prove to be a transformative concept that may be brought to fruition in this key area of molecular medicine.

10.8 Virtual Screening of Chemical Libraries Based on Epistructural Bioinformatics

The identification of compounds with controlled specificity against clinically relevant targets, a primary problem in drug discovery, is usually handled in early stages of development by high-throughput screening, experimentally and computationally (*in silico*) [29, 30]. Due to the increasing cost resulting from high clinical failure rates in a downstream stage of development, drug designers are prone to terminate the efforts on those compounds likely to fail in late stages as early as possible [30]. This "fail early" strategy places a significant responsibility on the early-stage compound profiling [31]. The two basic types of screening, experimental and computational, are complementary approaches: the former is relatively accurate but cost-limited by the size of the compound library to be profiled [14], while the latter is more time- and cost-efficient but less reliable [30, 32, 33]. *In silico* screening methods [34] can be categorized as either ligand-based [35] or target-based [36]. The latter is becoming mainstream for cases where target structural information is available [32, 34, 37]. Most of the target-based virtual screening is performed by docking and scoring [36]. However, the inaccuracies of scoring functions pose a major problem in target-based virtual screening [32, 33]. Such docking-based algorithms are inadequate to examine kinase targeting [1, 3–5, 10, 38–40] because most of them do not take into account the induced fits upon binding, which are crucial for kinase-ligand associations that typically involve loopy regions [4] (see Chap. 11). There exist some docking-based algorithms that take into account induced fits, but they still cannot handle extensive induced fit adaptation including large movements of the backbone [38, 39], which is just the case for the loopy regions in kinases. As discussed in Chap. 11, kinase ATP-pockets are partly framed by floppy regions, including the activation loop, catalytic loop and P-loop [4]. These parts of the structure undergo an order-upon-binding transition upon association with the ligands (natural or otherwise) which cannot be captured with a docking algorithm, no matter how well it is able to handle flexibility. The induced fit problem as it stands today is almost as hard as the protein folding problem and no algorithm deals with it effectively from first principles.

This problem notwithstanding, the recent discovery of novel molecular markers for specificity and promiscuity [10] may herald the advent of novel predictive tools for drug affinity profiling by providing a new vantage point for kinase comparisons, as described in this contribution. These comparisons will enable us to predict cross reactivity. Unexpected cross reactivity became recently apparent with the advent of high-throughput experimental screening techniques [14] based on bacteriophage kinase display. Thus, the affinity profiles of 20 inhibitors against a battery of 119 kinases have been reported [14]. However, the operational value of these assays to identify leads from within a compound library (~1000 compounds) is limited by cost, justifying our development of *in silico* profiling tools.

In this section we introduce a predictive profiler based on the assumption that a structure-based feature-similarity comparison of molecular targets can be used as a surrogate for the differences in their pharmacological behavior [10]. Our approach exploits the recent discovery of a structural marker governing specificity, a result holding even for kinases lacking PDB representation [10] and, based on this feature, it introduces a comparison of kinases including purported targets and experimentally confirmed targets. The core of our affinity-profile predictor involves determining a linear propagator of profiling data. This propagator consists of the structure-based estimation of pharmacological distances across kinases. Once the propagator is determined, the inference of affinity profiles for test drugs becomes a problem in distance geometry.

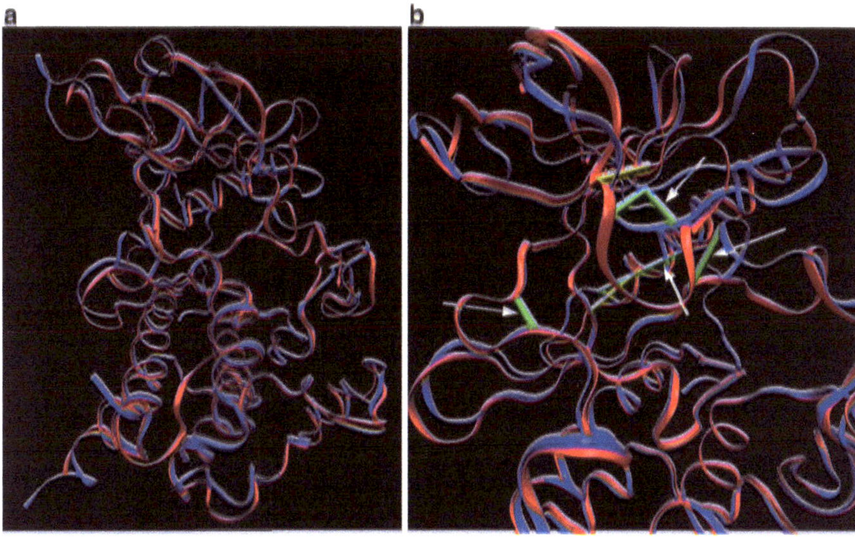

Fig. 10.10 Structural alignment of C-Kit kinase and Bcr-Abl kinase and their SAHB patterns in ATP-binding pockets. The figure highlights the relevance of SAHBs as selectivity filters distinguishing drug targets. (**a**) Structural alignment of C-Kit kinase (pdb code 1T46, ribbon representation in *blue*) and Bcr-Abl kinase (pdb code 1FPU, ribbon representation in *red*). The two structures are aligned by DaliLite (http://www.ebi.ac.uk/DaliLite/) and rendered with VMD (http://www.ks.uiuc.edu/Research/vmd/). Only backbones are indicated for clarity. The structure similarity (RMSD~1.4 Å) between the two kinases makes it difficult to specifically target one without touching the other. (**b**) SAHB distributions in the ATP-binding pockets of C-Kit and Bcr-Abl shown in the same alignment of (**a**). SAHBs of C-Kit are represented by *green* virtual bonds joining α-carbons while SAHBs of Bcr-Abl by *yellow*. While the backbone structures are highly alignable, pointing to high levels of structure conservation as shown in (**a**), SAHBs are not conserved. The *white arrows* indicate the nonconserved SAHBs

The identified specificity marker is the solvent-accessible hydrogen bond (SAHB) or dehydron in the target, inferred from protein structure by quantifying the extent of intramolecular dehydration of the amide-carbonyl pair (Fig. 1.10). Thus, as shown in Chaps. 1–4, dehydrons are insufficiently dehydrated and become molecular markers for specificity due to the following properties: (1) dehydrons are sticky since they promote the removal of surrounding water upon associations (Fig. 8.1e, [24]) and (2) dehydrons are *not conserved across paralog kinases, regardless of their degree of structural similarity* (Chap. 8 and Fig. 10.10), thus promoting selectivity when and if appropriately targeted [10]. Based on these properties, we introduced the environmental or wrapping distance matrix ($\mathbf{D}_{env}$) for all kinase pairs obtained from the respective differences in the SAHB patterns within aligned structures. On the other hand, the differences in the affinity profiles of kinases against a background of drugs are quantified and arrayed in the so-called pharmacological distance matrix $\mathbf{D}_{phar}$. Our previous work revealed a tight linear correlation between the environmental and the pharmacological distances, thus delineating the molecular basis for cross reactivity. The profiling method presented

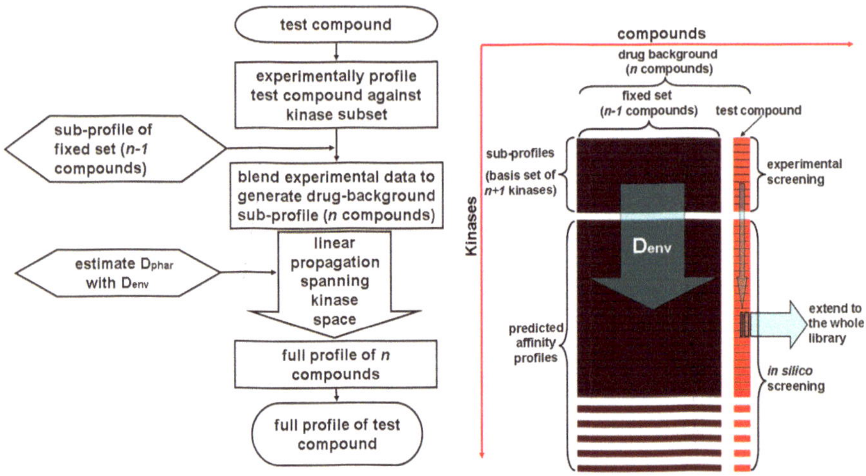

Fig. 10.11 Flow chart (*left*) and process diagram (*right*) of the *in silico* profiling method. In the diagram, each column corresponds to one compound and each row to one kinase. The *brown* columns correspond to the compounds with profiles already known and the *red* column corresponds to the test compound. The *upper rows* represent the sub-profiles that are obtained from experiments and the *lower rows* represent the profiles predicted by the profiler

in this work is an application of this correlation to estimate $\mathbf{D}_{\text{phar}}$ from $\mathbf{D}_{\text{env}}$ and further, to construct a full affinity profile for a test compound from its sub-profile obtained from experimental affinity assays against a small kinase subset. This problem becomes a linear algebra problem, where the full affinity vector for a given drug may be determined from suitably defined distances between kinases and the experimentally obtained affinity vector.

The implementation of the profiling method is thus carried out according to the following steps (Fig. 10.11):

1. Estimate the pharmacological distance matrix ($\mathbf{D}_{\text{phar}}$) defined over all kinase pairs by appropriately rescaling the environmental distance matrix ($\mathbf{D}_{\text{env}}$). This estimation will increase in accuracy as a more comprehensive background of drugs, i.e. one covering all kinase targets, is used to determine pharmacological behavior.
2. For every test compound to be fully profiled *in silico*, choose a subset of kinases to obtain a small-scale experimental affinity profile, heretofore noted the "*sub-profile*". The required size of the kinase subset sampled to produce the sub-profile depends on the fixed number of inhibitors that define the pharmacological matrix $\mathbf{D}_{\text{phar}}$.
3. Determine the full affinity profile for a test compound from its sub-profile and our estimated $\mathbf{D}_{\text{phar}}$.
4. Repeat the steps above for each test compound in the library.

The sub-profile is obtained from small-scale experiments, most advantageous in the case when a battery of a large number of kinases needs to be screened.

To infer $\mathbf{D}_{\text{phar}}$, it becomes necessary to identify the structural feature that governs drug cross reactivity across paralog kinases. Recent research [10] revealed that SAHBs constitute such structural markers. Thus, we estimate $\mathbf{D}_{\text{phar}}$ from $\mathbf{D}_{\text{env}}$ obtained by comparing the SAHB patterns of purported targets. The matrix $\mathbf{D}_{\text{env}}$ quantifies differences in the SAHB patterns within the ATP pockets (i.e. the drug binding site) across all kinase pairs. Thus, the environmental distance is based on structural alignment followed by comparison of poorly conserved features.

To determine SAHBs, we assess the hydrogen-bond microenvironment by calculating the extent of intramolecular dehydration (wrapping) of the backbone hydrogen bond, as described in Chap. 1. To compare the SAHB patterns for different kinases, we define the environmental hull of a kinase, H_{env}, as the set of all residues contributing to the microenvironment of hydrogen bonds, i.e. the environmental residues, as well as residues aligning with environmental residues from other paralog chains. The i-j matrix element of $\mathbf{D}_{\text{env}}$, associated to kinases i and j is then defined by comparing the aligned hydrogen-bond microenvironments within the environmental hulls of kinases i and j:

$$\mathrm{D}_{env}(i,j) = M(i,j)^{-1}\left[\sum\nolimits_{\mathrm{n}=1,\ldots,M(i,j)} \Delta_n(i,j)\right], \tag{10.2}$$

where $M(i, j)$ = number of residue pairs in $H_{\text{env}}(i)$ corresponding to SAHBs in kinase i or to hydrogen bonds or nonbonded residue pairs that align with SAHBs in $H_{\text{env}}(j)$; n = dummy index denoting residue pair; and $\Delta_{\text{n}}(i,j) = 1$ if residue pair n corresponds to a SAHB in $H_{\text{env}}(i)$ that aligns with a non-SAHB in $H_{\text{env}}(j)$ or *vice versa*, and $\Delta_{\text{n}}(i, j) = 0$, if n corresponds to SAHBs or non-SAHBs in both $H_{\text{env}}(i)$ and $H_{\text{env}}(j)$. Thus defined, the environmental distance compares local dehydration propensities associated with SAHB patterns in kinases.

On the other hand, $\mathbf{D}_{\text{phar}}$ is used to quantify differences in the affinity profiling of kinases against a background of inhibitors [22]. $\mathbf{D}_{\text{phar}}$ is defined as the matrix of Euclidean distances between affinity vectors for kinase pairs with entries given in negative logarithmic or, equivalently, $\Delta G/RT$-units:

$$\mathrm{D}_{phar}(i,j) = \left\|A_i - A_j\right\| = \left[\sum\nolimits_n \left(A_{i,n} - A_{j,n}\right)^2\right]^{1/2} \tag{10.3}$$

where A_i and A_j are the affinity profiles of kinases i and j, with components $A_{i,n}$ and $A_{j,n}$ being the values of the negative logarithms of binding constants for complexation of inhibitor n with kinases i and j, respectively. A "no hit" entry in the screening table of Fabian et al. [14] is assigned the value -23.026 $\Delta G/RT$-units, corresponding to a large dissociation constant $K_d = 10^{10}\,\mu\text{M}$. Notice that our "no hit" value is well above the cutoff ($K_d = 10\,\mu\text{M}$) adopted previously [10].

Figure 10.12 shows the correlation between $\mathbf{D}_{\text{phar}}$, and $\mathbf{D}_{\text{env}}$ when the former is obtained from the T7-bacteriophage kinase display screening against a background

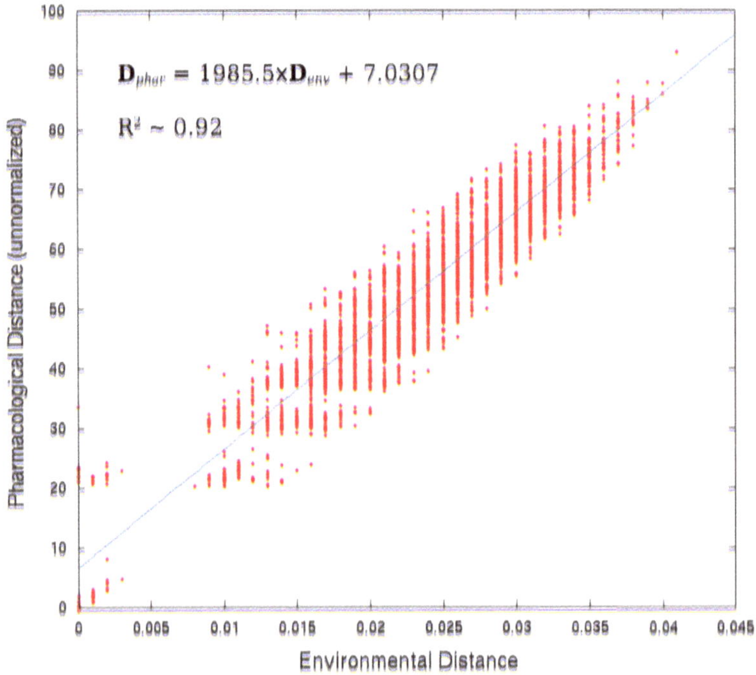

Fig. 10.12 Correlation between environmental and pharmacological distances. Each diamond represents a pair of kinases with horizontal coordinate being the environmental distance and vertical coordinate being the pharmacological distance between them. Unlike Fig. 2d in [10], the environmental distances in this work are not normalized. The *straight line* indicates the linear fit by least-squares method: $D_{phar}(i, j) = 1985.5D_{env}(i, j) + 7.0307$. $R^2 \sim 0.92$

of 17 drugs [14]. For each kinase pair (i, j), we can then infer the pharmacological distance using the linearly fitted parameters:

$$D_{phar}(i, j) = 1985.5D_{env}(i, j) + 7.0307 \quad (10.4)$$

Note that the correlation is very tight ($R^2 \sim 0.92$).

While inferring $\mathbf{D}_{phar}$ from $\mathbf{D}_{env}$, some errors are introduced in the estimated pharmacological distances. There are at least two possible sources of errors leading to dispersion in the correlation: (a) the background of drugs used to define the pharmacological distance is limited, with uneven target coverage, and thus only approximately indicative of pharmacological behavior; or (b) the SAHBs are not the only selectivity determinant for nonpromiscuous drugs. In regards to error source (a), we are limited in our analysis by the availability of drugs chosen to define pharmacological profiles in high-throughput experiments [14]. We may need to revise $\mathbf{D}_{phar}$ as new screening data becomes available. In regards to (b), we can only claim that SAHBs are a determinant but not necessarily the only factor governing ligand specificity [10]. Even though these errors will be inherited in the following steps, we made the method less sensitive to systematic errors through adequate parametrization.

We now show how to determine the affinity profiles of kinases from structure-based estimations of pharmacological distances between kinase pairs. Just like vector coordinates cannot be uniquely determined from vector distances, affinity profiles cannot be uniquely determined solely from the pharmacological distances: additional constraints are required. To solve the profile prediction problem, we cast it in terms of linear algebra, through a (vector) ↔ (kinase profile) correspondence. The procedure boils down to determining vector coordinates given distances between vectors. To guarantee the uniqueness of the solution, a subset of vectors should be given in addition to vector distances. Thus, for given pharmacological distances between pairs of kinases, the number of independent degrees of freedom of the solution vectors (kinase profiles) is n, which corresponds to the number of inhibitors, i.e. the dimension of the affinity-profile space. Therefore, we need to know at least n independent vectors to determine all affinity profiles. These conditions narrow down solutions to two possibilities, due to symmetry. Reflection relative to the hyperplane determined by the fixed n independent vectors produces two conjugate solutions to the problem. Thus, to unambiguously determine the solution, we need the coordinates of an additional vector. Accordingly, the minimal number of fixed vectors should be $n + 1$. A constraint on these $n + 1$ vectors is that n of them should be linearly independent.

To summarize, we may re-cast the profile prediction problem in linear-algebra terms through the following correspondences:

Space dimension ↔ *Number of sampled inhibitors*
Distance ↔ *Pharmacological distance*
Vector ↔ *Affinity profile of kinase (against a background of inhibitors)*
Vector $\boldsymbol{x}_i$ ↔ *Experimentally determined profile of the* ith *kinase in the subset*
Vector $\boldsymbol{y}$ ↔ *Affinity profile of the test kinase*

Thus, the *in silico* profiling problem may be cast as a linear-algebra problem as follows: In an n-dimension space, we have $n + 1$ given vectors (the subset) with n of them being linearly independent:

$$\boldsymbol{x}_i, \ i = 0, 1, 2, \ldots, n. \tag{10.5}$$

Note that $\boldsymbol{x}_i$ is the profile vector of kinase i against the n inhibitors. For a generic vector $\boldsymbol{y}$ (affinity profile of a test kinase) that is not in the subset, we have estimated the distances d_i from $\boldsymbol{y}$ to all $\boldsymbol{x}_i$'s:

$$d_i = |\boldsymbol{y} - \boldsymbol{x}_i|, \quad i = 0, \ldots, n. \tag{10.6}$$

Our purpose is to determine the coordinates of vector $\boldsymbol{y}$ based on the conditions given above. Note that $\boldsymbol{y}$ is the profile vector of a kinase that is not in the subset to be screened by experiment, i.e., $\boldsymbol{y}$ is one of the profile vectors to be determined *in silico*. We now show that this can be achieved by linear algebra calculation.

Since n of the $n+1$ vectors are linearly independent, we can assume without loss of generality that $\boldsymbol{x}_1, \ldots, \boldsymbol{x}_n$ are the independent vectors. We first define $\boldsymbol{x}_i^{'} = \mathbf{x}_i - \mathbf{x}_0$, for $i = 1, 2, \ldots, n$. Note that all $\boldsymbol{x}_i^{'}$ are linearly independent. Similarly, we define $\boldsymbol{y}^{'} = \boldsymbol{y} - \boldsymbol{x}_0$. Note that now we only need to determine $\boldsymbol{y}'$ to obtain $\boldsymbol{y}$. We can determine the scalar product of $\boldsymbol{y}'$ with each $\boldsymbol{x}_i'$ using the information on the vector distances and the well-known relation $|\boldsymbol{y}^{'} - \boldsymbol{x}_i^{'}|^2 = |\boldsymbol{y}^{'}|^2 + |\boldsymbol{x}_i^{'}|^2 - 2\boldsymbol{y}^{'} \cdot \boldsymbol{x}_i^{'}$, in turn obtained from the definition of scalar product:

$$\begin{aligned} \alpha_i &= \boldsymbol{y}^{'} \cdot \boldsymbol{x}_i^{'} \\ &= \left(|\boldsymbol{y}^{'}|^2 + |\boldsymbol{x}_i^{'}|^2 - |\boldsymbol{y}^{'} - \boldsymbol{x}^{'}|^2\right)/2 \\ &= \left(|\boldsymbol{y}^{'}|^2 + |\boldsymbol{x}^{'}|^2 - d_i^{\,2}\right)/2, \end{aligned} \tag{10.7}$$

where the last expression indicates how α_i can be evaluated from the estimated distances. We then have a system of n-variable linear equations by rewriting the relationships above

$$\boldsymbol{x}_1^{'} \cdot \boldsymbol{y}^{'} = \alpha_1 \tag{10.8}$$

$$\begin{aligned} \boldsymbol{x}_2^{'} \cdot \boldsymbol{y}^{'} &= \alpha_2 \\ &\cdots \\ \boldsymbol{x}_n^{'} \cdot \boldsymbol{y}^{'} &= \alpha_\mathrm{n} \end{aligned} \tag{10.9}$$

Thus, the pharmacological problem of determining the profiles of test kinases boils down to solving the system of linear equations (10.9), to obtain $\boldsymbol{y}'$ and thereby $\boldsymbol{y} = \boldsymbol{y}^{'} + \boldsymbol{x}_0$.

Note that each test-kinase profile $\boldsymbol{y}$ is represented as a row in the "predicted affinity profiles" in the scheme shown in Fig. 10.11. This calculation can be repeated for each test kinase not included in the subset of experimentally screened kinases. By calculating the "predicted affinity profiles" row-by-row, we then extend the sub-profile to an entire affinity profile over all kinases for which structural information is available (and hence pharmacological distances may be estimated).

In order to find the affinity profile of a new inhibitor, we first obtain its affinity sub-profile against the kinases in the subset. We treat the inhibitor as one of the linear dimensions of the pharmacological distance space. As shown above, we calculate the affinities of all the inhibitors, including the new one, towards the kinases that are not within the subset. In this way, the entire profile of the new inhibitor is obtained. In an ideal case as a mathematical model, if the estimated $\mathbf{D}_{\mathrm{phar}}$ were highly close to the real values, we could obtain the quantitatively exact profile (i.e. exact values of K_d's). Unfortunately, the correlation of R^2~0.92 is still not tight enough for a quantitative profile prediction. Alternatively, we predict the profile in a qualitative level, i.e. "hit" or "no hit". At this stage, we use thresholds of the ligand-target dissociation constant ($K_{d,threshold}$) to determine whether an inhibitor hits a specific kinase or not. We used three thresholds: $K_{d,threshold} = 1$, 10, and 100 μM.

Repeating these steps on all test compounds, a large library profiling can be generated from a small-scale (i.e. sub-profile size) experiment. The process is schematically represented in Fig. 10.11.

The choice of kinases in the basis set (the subset) is crucial for the predictor's performance. A required condition for the subset is that the affinity vectors corresponding to the kinases span all dimensions of the affinity space. Furthermore, some of the kinases discriminate the inhibitory compounds less effectively than the others, i.e. compounds' affinities to them are more uniform than that to others. If such kinases are included in the subset, the prediction would be more sensitive to errors in the estimated pharmacological distances. This would yield larger errors in the predicted affinity vectors. We then designed a simple algorithm to examine and optimize a series of subset choices. First, the predicted results are benchmarked against the experimental results [14] and the accuracies (percentages of the correct predictions) are calculated. We compared our predicted results on 17 inhibitors out of the 20 in the screening experiment [14] excluding three promiscuous inhibitors (Staurosporine, EKB-569 and SU11248), against 119 kinases. For the 17 × 119 entries, we counted the number of entries correctly predicted, in the "hit or no hit" level, and calculated the percentage of correct predictions. Using this percentage as scoring function of the subset-choice, we performed optimizations for the basis set and found several sets with correct prediction percentages around 93 %. Our algorithm to optimize the basis set is shown in the pseudo code below:

FOR each kinase in the basis set,
replace this kinase with each kinase not in the set;
predict the profile;
benchmark the predictive results and calculate the accuracy;
IF the current percentage is higher than the existing best one;
THEN use this kinase in the place of the original kinase in the set;
END LOOP

The affinity prediction is validated by benchmarking the result for all the nonpromiscuous inhibitory compounds in the phage-display kinase assay against the experimental data. The compounds used in the benchmark are: SB202190, SB203580, VX-745, BIRB-796, SP600125, Gleevec, Iressa, Tarceva, ZD-6474, CI-1033, GW-2016, Vatalanib, MLN-518, LY-333531, BAY-43-9006, Roscovitine and Flavopiridol. For randomly chosen kinase subsets prior to any optimization, mostly the accuracies of the predictions are around 80~90 %. For instance, consider the randomly chosen subset of kinases:

AAK1, ABL1, CDK2, EGFR, ERBB2, FLT3, GAK, JNK1, KIT, LCK, p38-alpha, PDGFRB, PHKG1, SLK, SRC, STK10, VEGFR2, YES.

Using this subset, the profiler predicts affinities for all 119 kinases independently assayed [14] with 252 false positives and false negatives out of 17 × 119 predictions, for the affinity threshold 10 μM. The corresponding accuracy is 88 %. Notice that the 17 × 119 predictions cover the sub-profiles for the subset and the accuracy is calculated based on all 17 × 119 predictions. This does not weaken the validation since when predicting for a kinase within the subset, the input distances are obtained independently, i.e. estimated from environmental distances. In other words, the predictor does not discriminate kinases within or outside of the subset. In a real work, it is useless to predict the affinities of an inhibitor for kinases within the subset, but here we include these results for the sake of validation.

The next step is to optimize the subset of kinases to improve prediction accuracy. For the purpose of validation, we perform the optimization with a "leave-one-out" algorithm. That is, to predict the profile of one inhibitor, we apply the optimization algorithm discussed above to the other 16 inhibitors and thereby get an optimized subset of 17 (=16 + 1) kinases. These 17 kinases plus one more randomly chosen kinase constitute the subset used for the prediction of the inhibitor. In this way, the inhibitor profile to be predicted is left out of the subset optimization process, and thus the optimized choice of the subset is completely independent of the inhibitor left out, though it might not be the best choice since one kinase in the subset is chosen randomly. It is possible that the randomly chosen kinase plus the 17 optimized ones do not constitute a complete subset (dimension < 17), rendering equation 3 not solvable. In this case, we simply replace the randomly chosen kinase with another random one until the corresponding equation 3 becomes solvable. As an example, the optimized subset used to predict the profile of SB202190 is (the first 16 kinases result from optimization):

INSR, PTK6, CSNK1G1, TEK, Aurora2, EPHA3, PHKG2, BIKE, EPHB4, FYN, TNIK, DAPK2, FGFR1, SLK, PRKAA1, ERBB2, YES, p38-beta (randomly chosen).

We predict the complete affinity profiles of the 17 inhibitors one by one, using the profiling information of the 17 inhibitors against the 18-kinase subset. Before predicting for each inhibitor, a subset optimization based on the other 16 inhibitors is carried out. That means the subset used in the profile prediction for each one of the 17 inhibitors is not necessarily the same in each case.

The accuracy of our predictor, as revealed by Fig. 10.13, is quantitatively summarized in Table 10.2. In accord with Fig. 10.12, we use three thresholds for kinase-inhibitor hit/no-hit results: $K_{d,threshold} = 1$, 10, and 100 μM. The accuracy of the predictions are 91 %, 93 % and 93 %, respectively. When using $K_{d,threshold} = 1$ μM, the errors are mainly due to false positives, while for $K_{d,threshold} = 10$ and 100 μM they arise mostly from false negatives. Thus we can choose different $K_{d,threshold}$'s for different purposes. If we are concerned with drug specificity and promiscuity in a clinical context, it is reasonable to use $K_{d,threshold} = 1$ μM since it imposes a more stringent criterion for affinity, suitable for clinical purposes. If binding is the sole concern, a better choice would be $K_{d,threshold} = 10$ μM or 100 μM, since these filters entail higher sensitivity.

One docking-based method was recently reported and claimed to be the best for computing the affinities of inhibitors for homologous receptors [41]. The authors compared their predictions with the experimental results published by Fabian et al. [14] and found "a reasonable but not perfect correspondence" [41]. Both this work and our profiler predict the affinity profile of imatinib (Gleevec) against extended lists of kinases, and thus the results can been compared. Benchmarked by

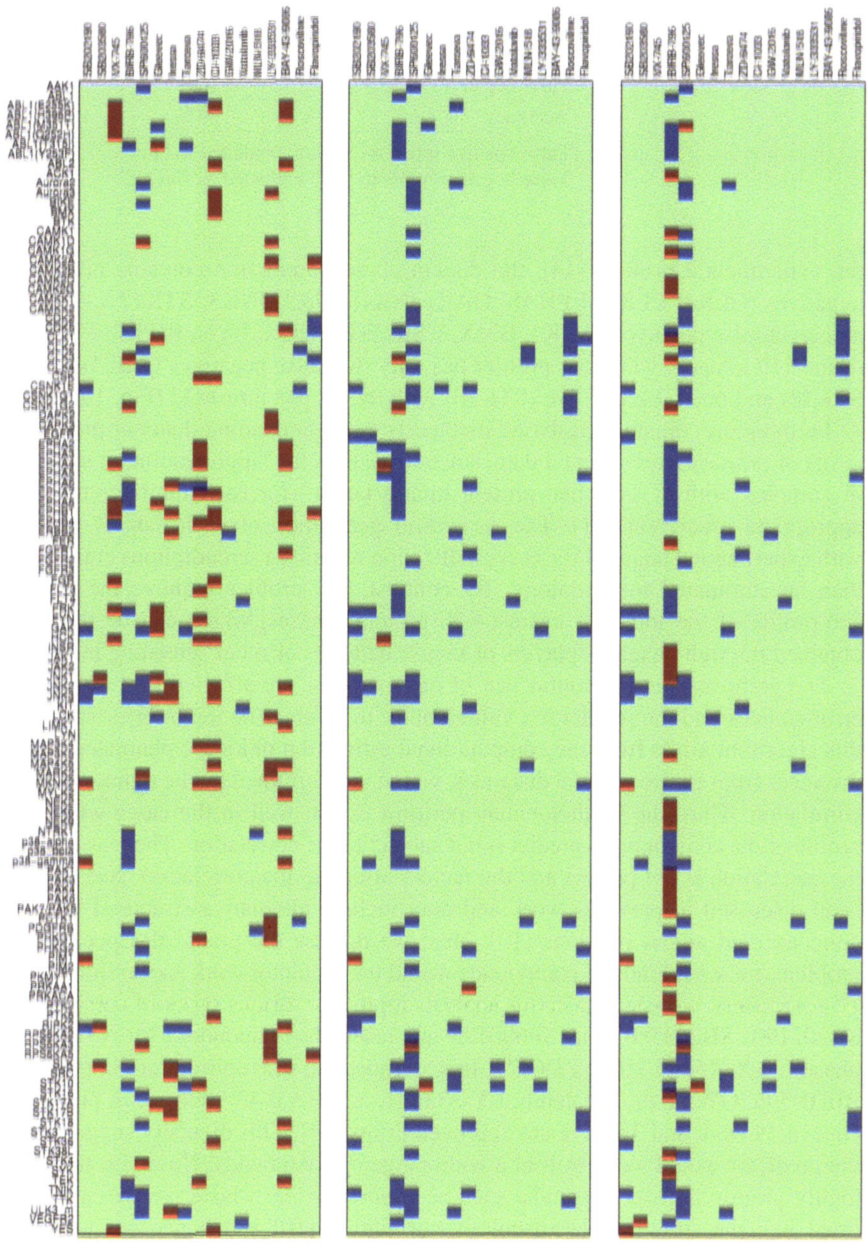

Fig. 10.13 Matrices of prediction performance, corresponding from left to right to affinity thresholds $K_d = 1\,\mu M$, 10 μM, and $100\,\mu M$, respectively. The complete affinity profiles of 17 inhibitors independently screened [14] were predicted one by one, using the experimental profiling information on the 17 inhibitors against an 18-kinase subset (different for each inhibitor). *Green* cells indicate correct predictions; *blue*, false negatives ("hit" predicted as "no hit"); *red*, false positives ("no hit" predicted as "hit"). The accuracy percentages shown in the three matrices are 91, 93 and 93 %. Detailed quantitative summary of the accuracy is in Table 10.2

Table 10.2 Prediction accuracy with different filters for hit/no-hit

$K_{d,threshold}$	1 μM	10 μM	100 μM
False positive[a]	116	9	36
False negative[b]	69	133	114
Accuracy	91 %	93 %	93 %

[a]False positive refers to "no hit" predicted as "hit"
[b]False negative refers to "hit" predicted as "no hit"

the experimental profiles [14], the docking-based prediction contains nine false negatives (CLK1, CLK4, EPHA8, GAK, JNK1, JNK2, JNK3, STK17A, STK18) and eight false positives (ACK1, BMX, CSK, FGR, HCK, LYN, RIPK2, YES) out of the 119 kinases, while our profiler has only two false negatives (ABL1(T315I), STK18) and one false positive (STK10) with the 10 μM threshold (Fig. 10.13).

Furthermore, the docking-based method is more demanding than our profiler in terms of primary experimental data, since it requires the "high-resolution structure in complex with at least one protein kinase target" for each inhibitor to make appropriate prediction [41]. The successful generation of such data is plagued with experimental uncertainty as crystallization remains a serendipitous craft rather than an automated methodology. By contrast, our profiler requires the affinity sub-profile of the inhibitor against ~20 protein kinases, which can be routinely obtained through phage displayers of kinase batteries or other screening methods.

The performance and confidence of our predictor are affected by two factors: error sources and the predictor's vulnerability to errors. The major error source in this algorithm arises from the computational estimation of kinase pharmacological distances from environmental distances, which is determined by the tightness of the correlation. Thus, the predictor may perform not so well in the cases where the profile of the compound to predict is not subject to the correlation. The reasons why the correlation is not perfect and the factors affecting the correlation have already been discussed in previous work and may include alternative structural features nonconserved across paralogs [10]. Here we discuss the practical aspect of the problem: for what kinds of compounds would the predictor work successfully? The 17 compounds we adopted in our analysis represent various types of compounds: SB202190, SB203580, and SP600125 are research compounds; MLN518 is in phase I; VX745, CI-1033, ZD6474, Roscovitine, and Flavopiridol are in phase II; BIRB-796, GW-2016, Vatalanib, LY-333531, and BAY-43-9006 are in phase III; Gleevec, Iressa, and Tarceva are approved drugs [14]. This diversity suggests that the predictor would work well in a wide range of compounds. However, there are highly promiscuous compounds, such as *staurosporine*, whose profile cannot be fitted into the structure-pharmacology correlation [10] and hence our predictor would fail to yield a reliable result.

Another important factor influencing predictor confidence is its robustness or vulnerability to errors introduced in the distance estimation. This is mainly determined by the choice of kinases in the small-scale sample, the affinities for which constitute the *sub-profile*. Some of the kinases differentiate the inhibitory compounds better than others, i.e. compounds' affinities for them are more rigorous. It is better to include such kinases in the sub-profile, since such basis subsets are less sensitive to the errors in the estimated pharmacological distances. This problem is handled by the optimization process, in which the subset of kinases that performs best within the training set is chosen.

To summarize, we introduced a method to predict affinity profiles of inhibitor compound libraries against entire batteries of human kinases based on a structural descriptor of the kinase targets [42]. The method is rooted in a molecular marker governing drug specificity and promiscuity. A feature-similarity matrix constructed based on the molecular marker is defined across kinase targets and used as an information propagator of a sub-profile where drugs are screened against a small group of kinases. The method reported makes use of distance-geometry techniques, and boils down to determining vectors (kinase profiles) from distances between vectors (feature-similarity distances regarded as surrogates for pharmacological distances between kinases). To guarantee the uniqueness of the solution, some vector coordinates need to be fixed and adopted as constraints. These constraints represent the linear algebra counterpart of the sub-profile. Our *in silico* method enables us to screen large libraries of compounds predicting their profile. To provide the input data, only a limited experimental screening against a reduced subset of kinases needs to be performed in advance. Thus, our predictor becomes a valuable tool for lead discovery.

The linear-algebra operations subsumed in our predictor are based on the construction of the informational propagator and hence do not entail any source of errors. The only systematic source of uncertainty arises from the estimation of pharmacological parameters from structure-based attributes (environmental distances). Given the high accuracy (for a qualitative prediction) of the pharmacological distance estimation, our profiling method should be deemed highly reliable: the accuracy is up to 93 % with an optimized choice of kinase subset as starting point, as shown above (Table 10.2).

Alternative virtual screening methods rooted in docking algorithms are unlikely to match this level of accuracy, not only because of their inherent parametric uncertainty and time expense, but also because kinase binding entails extensive induced fit of the loopy regions within the ATP-pocket [4]. Even those docking algorithms that incorporate induced fits into affinity calculations cannot handle the lengthy loopy regions of kinases, which undergo extensive structural adaptation [38, 39]. The induced fit problem as it stands today remains intractable from first-principle approaches (Chap. 11). This is the main reason for adopting an information-based predictor.

10.9 Problems

10.1 *Shift in the dehydron-wrapping paradigm for drug design.* The discovery of the chemical functionality of dehydrons (Chap. 7) forces us to contemplate a shift in paradigm for drug design, as we are not only targeting dehydrons as

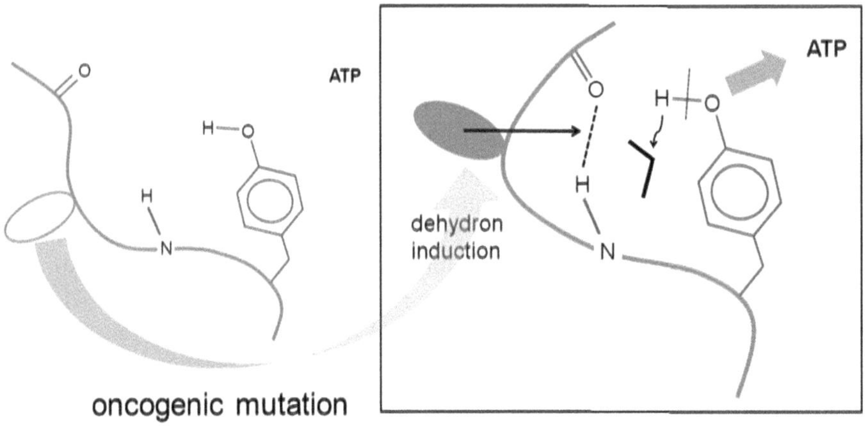

Fig. 10.14 General scheme of oncogenic mutation generating a catalytically stimulating dehydron, yielding a constitutively active signal transducer. The dehydron occurs in a disordered region in the wild type as a result of a germline or somatic mutation that improves the wrapping quality of a residue (enhancing its hydrophobicity as represented by the *unfilled-to-filled ellipse*). The proximity of the dehydron to a phosphorylation-susceptible residue (Tyr) enables and stimulates the catalytic activity of the latter, with dramatic physiological consequencesKindly check the order of the figure citationsDone. Only figure 10.10 is misplaced. It should go with problem 10.1, where it is quoted, as commented.

sticky non-conserved sites but also as catalytic enablers/stimulators. Many oncogenic mutations affect regulation of the signal transducers, often creating constitutively deregulated or constitutively active kinases. For instance, a mutation that substitutes a phosphorylation-susceptible residue (Ser, Thr, Tyr, or His) for a negatively charged one at physiological pH (Glu, Asp) often creates a constitutively active kinase whose lack of regulation constitutes the molecular ethiology of the oncogenic activity. A far less studied type of mutation that, in light of the results in Chap. 7, may promote deregulated activity is the one that would lead to the creation of a dehydron in the vicinity of a phosphorylation site (Fig. 10.14). Such mutation-induced dehydrons become enablers and stimulators of an inordinate level of catalytic activity, with dramatic physiological consequences. Discuss the therapeutic implications and advantages of targeting such a transformative mutation.

References

1. Bain J, McLauchlan H, Eliott M, Cohen P. The specificities of protein kinase inhibitors: an update. Biochem J. 2003;371:199–204.
2. Druker BJ. Molecularly targeted therapy: have the floodgates opened? Oncologist. 2004;9:357–60.

3. Hopkins AL, Mason JS, Overington JP. Can we rationally design promiscuous drugs? Curr Opin Struct Biol. 2006;16:127–36.
4. Huse M, Kuriyan J. The conformational plasticity of protein kinases. Cell. 2002;109:275–82.
5. Knight ZA, Shokat KM. Features of selective kinase inhibitors. Chem Biol. 2005;12:621–37.
6. Chothia C. Hydrophobic bonding and accessible surface area in proteins. Nature. 1974;248:338–9.
7. Whittle PJ, Blundell TL. Protein structure-based drug design. Annu Rev Biophys Biomol Struct. 1994;23:349–75.
8. Feng BY, Shelat A, Doman TN, Guy RK, Shoichet BK. High throughput assays for promiscuous inhibitors. Nat Chem Biol. 2005;1:146–8.
9. Zhang X, Crespo A, Fernández A. Trung promiscuous kinase inhibitors into safer drugs. Trends Biotechnol. 2008;26:295–300.
10. Chen JP, Zhang X, Fernández A. Molecular basis for specificity in the druggablekinome: sequence-based analysis. Bioinformatics. 2007;23:563–72.
11. Manning G, Whyte DB, Martinez R, Hunter T, Sudarsanam S. The protein kinase complement of the human genome. Science. 2002;298:1912–34.
12. Braken C, Iakoucheva LM, Romero PR, Dunker AK. Combining prediction, computation and experiment for the characterization of protein disorder. Curr Opin Struct Biol. 2004;14:570–6.
13. Fernández A, Berry RS. Molecular dimension explored in evolution to promote proteomic complexity. Proc Natl Acad Sci U S A. 2004;101:13460–5.
14. Fabian MA, Biggs WH, Treiber DK, et al. A small molecule kinase interaction map for clinical kinase inhibitors. Nat Biotechnol. 2005;23:329–36.
15. Karaman MW, Herrgard S, Treiber DK, et al. A quantitative analysis of kinase inhibitor selectivity. Nat Biotechnol. 2008;26:127–32.
16. Bonneau R, Straus CE, Rohl CA, et al. De novo prediction of three-dimensional structures for major protein families. J Mol Biol. 2002;322:65–78.
17. Chivian D, Kim DE, Malmstrom L, et al. Prediction of CASP6 structures using automated Robetta protocols. Proteins. 2005;61 Suppl 7:157–66.
18. Higgins DG, Thompson JD, Gibson TJ. Using CLUSTAL for multiple sequence alignments. Methods Enzymol. 1996;266:383–402.
19. Fraczkiewicz R, Braun W. Exact and efficient analytical calculation of the accessible surface areas and their gradient for macromolecules. J Comput Chem. 1998;19:319–33.
20. Ooi T, Oobatake M, Nemethy G, Scheraga HA. Accessible surface area as a measure of the thermodynamic parameters of hydration of peptides. Proc Natl Acad Sci U S A. 1987;84:3086–90.
21. Ma B, Elkayam T, Wolfson T, Nussinov R. Protein-protein interactions structurally conserved residues distinguish between binding sites and exposed protein surfaces. Proc Natl Acad Sci U S A. 2003;100:5772–7.
22. Fernández A, Maddipati S. A priori inference of cross reactivity for drug-targeted kinases. J Med Chem. 2006;49:3092–100.
23. Fernández A, Sosnick TR, Colubri A. Dynamics of hydrogen bond desolvation in protein folding. J Mol Biol. 2002;321:659–75.
24. Fernández A. Keeping dry and crossing membranes. Nat Biotechnol. 2004;22:1081–4.
25. Gazdar A. Personalized medicine and inhibition of EGFR signaling in lung cancer. N Engl J Med. 2009;361:1018–20.
26. Janne PA, Gray N, Settleman J. Factors underlying sensitivity of cancers to small molecule kinase inhibitors. Nat Rev Drug Discov. 2009;8:709–23.
27. Fernández A, Sanguino A, Peng Z, Crespo A, Ozturk E, Zhang X, Wang S, Bornmann W, Lopez-Berestein G. Rational drug redesign to overcome drug resistance in cancer therapy: imatinib moving target. Cancer Res. 2007;67:4028–33.
28. Hubbard SR. Oncogenic mutations in B-Raf: some losses yield gains. Cell. 2004;116:764–6.
29. Drews J. Drug discovery: a historical perspective. Science. 2000;287:1960–4.

30. Bleicher KH, Böhm HJ, Muller K, Alanine AI. Hit and lead generation: beyond high-throughput screening. Nat Rev. 2003;2:369–78.
31. Liszewski K. Drug discovery: successful lead optimization strategies. Genet Eng Biotech News. 2006;26:14.
32. Kitchen DB, Decornez H, Furr JR, Bajorath J. Docking and scoring in virtual screening for drug discovery: methods and applications. Nat Rev. 2004;3:935–49.
33. Shoichet BB. Virtual screening of chemical libraries. Nature. 2004;432:862–5.
34. Oprea TI, Matter H. Integrating virtual screening in lead discovery. Curr Opin Chem Biol. 2004;8:349–58.
35. Lengauer T, Lemmen C, Rarey M, Zimmermann M. Novel technologies for virtual screening. Drug Discov Today. 2004;9:27–34.
36. Lyne PD. Structure-based virtual screening: an overview. Drug Discov Today. 2002;7:1047–55.
37. Kuhn P, Wilson K, Patch MG, Stevens RC. The genesis of high-throughput structure-based drug discovery using protein crystallography. Curr Opin Chem Biol. 2002;6:704–10.
38. Mizutani MY, Itai A. Efficient method for high-throughput virtual screening based on flexible docking: discovery of novel acetylcholinesterase inhibitors. J Med Chem. 2004;47:4818–28.
39. Mizutani MY, Takamatsu Y, Ichinose T, Nakamura K, Itai A. Effective handling of induced-fit motion in flexible docking. Proteins Struct Funct Bioinf. 2006;63:878–91.
40. Vieth M, Higgs RE, Robertson DH, Shapiro M, Gragg EA, Hemmerle H. Kinomics-structural biology and chemogenomics of kinase inhibitors and targets. Biochim Biophys Acta. 2004;1697:243–57.
41. Rockey WM, Elcock AH. Rapid computational identification of the targets of protein kinase inhibitors. J Med Chem. 2005;48:4138–52.
42. Crespo A, Fernández A. Kinase packing defects as drug targets. Drug Discov Today. 2007;12:917–23.

Chapter 11
Drug-Target Associations Inducing Protein Folding

Structure-based drug design has met only modest success due mainly to two reasons: a) as shown in the previous chapter, pharmacologically relevant features are often enshrined in the epistructure, not in the structure itself; and b) the targettable features are often found in structurally floppy regions. In this chapter we shall focus on the latter aspect, first highlighting the inadequacy of standard rational design to deal with it. This is because structural adaptation upon ligand binding resulting in induced folding is usually very difficult to predict. Thus, dynamic information must be incorporated into rational drug design. The conformation of the protein chain in complex with the drug ligand often differs significantly from the conformation of the apo form of the protein, and the structural difference is often unpredictable. In fact, the induced folding problem is every bit as difficult as the protein folding problem, whose arduous solution required a combination of structural and epistructural approaches, as described in Chap. 3. Local conformational plasticity of the protein target is probably the main reason for the modest interest in rational drug design. Thus, floppy regions such as the activation loop or nucleotide-binding loop in a kinase are seldom targeted with small molecule inhibitors in spite of their value as selectivity filters. For example, the activation loop is the structural region that presents the largest amino acid variability within kinase families and thus its compositional uniqueness makes it an attractive target to control specificity. In this chapter we advocate for a strategy to target flexible regions, offering a way to control the induced folding and turn it into a selectivity-promoting feature. Drugs designed to wrap disordered regions in the target may be used to steer induced folding in specific controllable ways, i.e. by inducing the formation of specific dehydrons. The results surveyed in this chapter herald the paradigmatic concept of "wrapping drugs for structurally adaptable targets". The wrapping-induced folding concept is illustrated by redesigning the anticancer drug imatinib in order to redirect its affinity towards a floppy region in JNK1, which constitutes an important target in the treatment of ovarian cancer. Thus, this chapter reveals that insights from epistructural physics become essential to incorporate conformational dynamics into the technological base of the drug design platforms.

A. Fernández, *Physics at the Biomolecular Interface*, Soft and Biological Matter,
DOI 10.1007/978-3-319-30852-4_11

11.1 Induced Folding: The *Bête Noire* of Structure-Based Drug Design

As described in Chaps 8, 9 and 10, kinases offer desirable therapeutic opportunities because they transduce cell signals implicated in cancer progression and development and because they possess a natural binding pocket susceptible of being targeted by a man-made ligand [1–5]. These binding regions are framed by floppy regions (nucleotide-binding loop, activation loop, catalytic loop) whose composition determines the activation state of the kinase and its substrate specificity, hence conferring them their functional identity [5–10].

Precisely because of the conformational plasticity of the purported targets, structure-based design of kinase inhibitors has met with modest success. As an inhibitor binds to a kinase, it often promotes a largely unpredictable structural adaptation of the floppy regions. This adaptation is referred to as the induced folding and entails a considerable loss in conformational entropy that adversely affects the affinity of the drug for its target. This phenomenon poses design problems since the stability of the drug-target complex, and hence the drug affinity, cannot be meaningfully determined from the structures of the isolated binding partners [6–11]. The *ab-initio* prediction of the induced fold remains forbiddingly difficult, often precluding a rational control of cross reactivity for drugs that interact with loopy regions [6, 7]. In fact, the induced folding problem compares with the protein folding problem in terms of the level of difficulty associated with a first-principle solution rooted in the tenets of epistructural physics (Chap. 3) [7–9].

However intractable, we are compelled to deal with this problem for a very simple reason: the highest variability in aminoacid composition across kinases occurs precisely in the floppy regions, more specifically in the activation loop [12]. Thus, the most significant specificity-promoting feature becomes in practice the most difficult to target. This poses a real challenge to the drug designer.

As illustrated in this chapter, the induced folding problem may be tractable in special contexts that are especially relevant to drug design: there are restrictive instances where structural adaptation may be steered in a controllable and predictable way, and selectivity control may be achieved through the identification of targets that undergo a similar induced fit upon association.

11.2 Wrapping the Floppy Protein Target in a Controllable Way

To deal with a tractable problem, we shall focus on small regions (4–8 residues) of the protein chain which are unstructured in the free "apo" state and represent extreme cases of wrapping deficiency, that is, they are unable to hinder backbone hydration to any significant extent [11]. As discussed in Chap. 5, such regions are unequivocally characterized as being unable to contribute with at least seven nonpolar wrapping groups to the microenvironment of putative backbone hydrogen bonds. Since $\rho = 7$ is the threshold for existence of a backbone hydrogen bond, such regions are disordered with 100 % certainty in the uncomplexed state of the protein. To distinguish the solvent-shielding of induced hydrogen bonds from that of pre-formed hydrogen bonds, we shall introduce the term "crating" when referring to the former while retaining the usual term wrapping for the latter.

We shall focus on the problem of targeting regions that visit conformations where the backbone hydrogen bonds are too exposed to the solvent to be retained in soluble structure. The backbone amides and carbonyls in these regions are actually prone to become completely hydrated with concurrent loss of structure (Fig. 11.1a). Thus, drug ligands may be specifically designed to "correct" the severe yet localized packing deficiency by removing surrounding water upon association, thereby inducing a hydrogen bond which is missing in the uncomplexed state of the protein. The design principle is illustrated in Fig. 11.1a and is based on ligand designs which are protective of specific conformations *visited but not retained* by the target protein in the unbound state.

As emphasized in previous chapters, in regions where kinase structure is retained upon binding, drug selectivity may be achieved or at least controlled by targeting dehydrons [12]. This special type of targeting is realized by engineering a drug that wraps the pre-formed intramolecular hydrogen bond [13–15], effectively contributing with nonpolar groups to the exclusion of water from the hydrogen-bond microenvironment.

Thus, a backbone hydrogen bond inferred from homologous structure alignment may be absent in a reported protein structure due to severe under-protection, i.e. when $\rho < 7$ [11]. Yet, it may be selectively induced if a drug is appropriately constructed to crate it, much like the drugs designed to "correct" wrapping defects in target proteins [12].

The scheme presented in Fig. 11.1a may be translated into the thermodynamic terms of epistructural physics, as described in Fig. 11.1b. In the unfolded state, the backbone is hydrated with no frustrated water ($g = 4$). In the absence of the ligand, the spontaneous formation of solvent-exposed structure (overexposed dehydron) generates interfacial tension as dehydron hydration leads to water frustration ($g < 4$), so that the ΔG_{if} associated with the unfolded → folded transition is positive (unfavorable transition). In the presence of the ligand, the interfacial tension in the folded state is removed as the solvent is displaced from the dehydron neighborhood to the bulk, where water is not frustrated ($g = 4$). This implies that the unfolded → folded transition takes place without water frustration upon ligand binding, and hence $\Delta G_{if} \leq 0$, so the conformational transition becomes thermodynamically favorable.

The crating design concept on a structurally adaptable target is illustrated in Fig. 11.2, which displays a complex between a protective drug and a kinase with a drug-induced folded region arising from the intermolecular crating of a hydrogen bond. The F811-A814 dehydron within the activation loop (in red) of the C-kit kinase is being crated and hence induced by a purposely designed modification of the anticancer drug imatinib. The latter is known to target the C-kit kinase, while

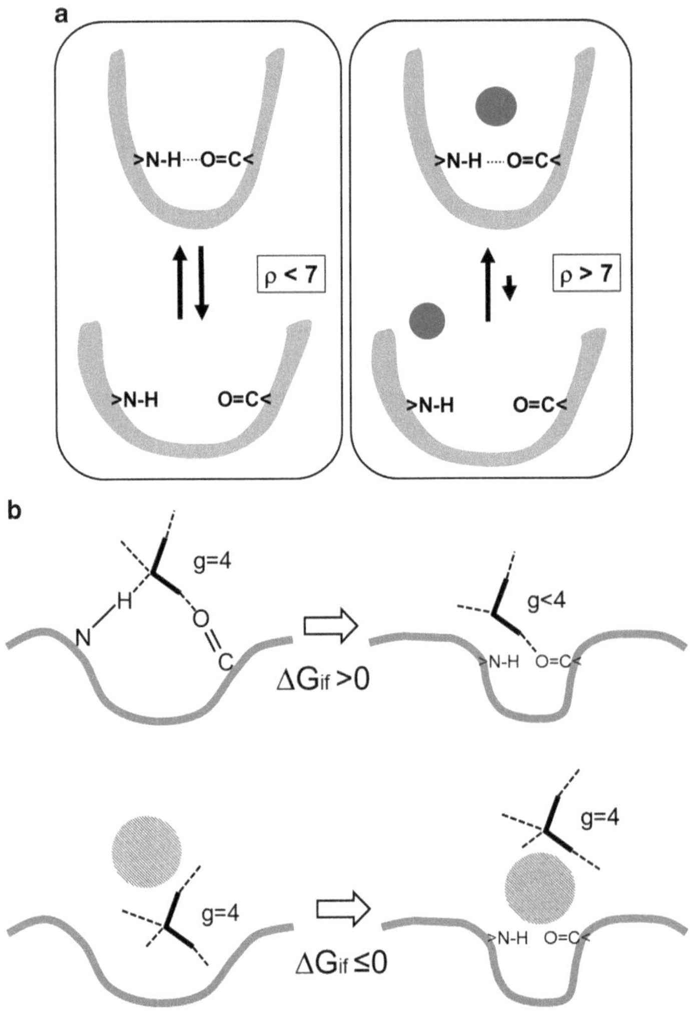

Fig. 11.1 *Induced folding upon drug-target association.* (**a**) Basic scheme of the folding-inducing drug. Due to insufficient crating in the unbound state ($\rho < 7$), the target protein *visits but does not retain* a hydrogen-bonded conformation that is in dynamic equilibrium with the fully hydrated amide and carbonyl. Since the drug (*grey sphere*) is able to provide sufficient crating ($\rho > 7$), the target protein in the drug-bound state is able to retain the hydrogen-bonded conformation. Since the proximity of a nonpolar moiety hinders hydration, the drug has no affinity for the alternative conformation of the protein target, with its solvent-exposed amide and carbonyl. (**b**) Epistructural thermodynamics of induced folding upon ligand binding. The presence of the ligand removes the interfacial tension that arises from water frustration ($g < 4$) upon formation of the overexposed dehydron in the unfolded → folded transition of the protein target

the modified version, known as WBZ_7 [16], provides additional crating to the activation loop, enhancing affinity and selectivity beyond the levels achieved by the parental compound [16]. In other words, in the unbound state of the protein, the

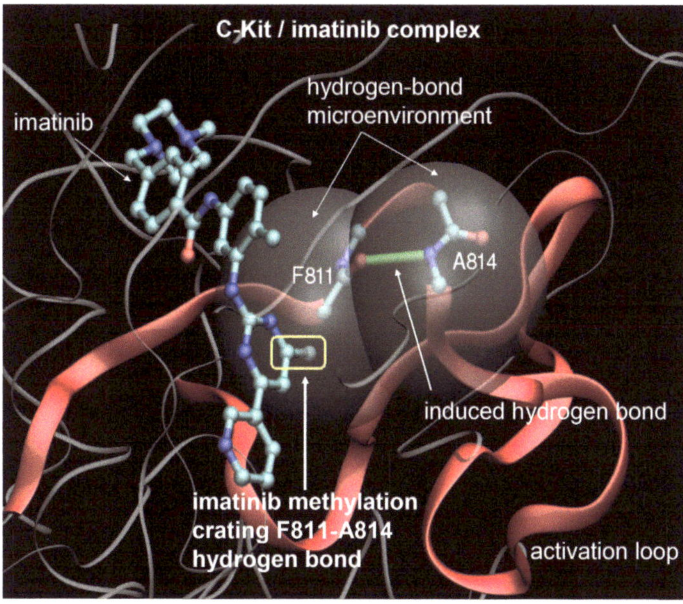

Fig. 11.2 *Design concept of protective drug illustrated by an imatinib modification.* A primary imatinib target, the C-kit kinase (PDB.1T46), displayed in ribbon representation with activation loop in red, is shown in complex with a methylated version of imatinib (methylation highlighted by yellow box). The parental drug has been modified to crate the deficiently packed F811-A814 backbone hydrogen bond within the flexible activation loop. The protection materializes because the appended methyl in the drug penetrates the desolvation domain of the backbone hydrogen bond upon association. In consistence with Chap. 1, the bond microenvironment is defined by two intersecting balls centered at the α-carbons of residues F811, A814. Thus, a specific local structure of the activation loop has been induced upon association with a drug purposely designed to crate a hydrogen bond. Reprinted from [22], copyright 2009 with permission from Elsevier

activation loop is likely to visit the conformation that contains the F811-A814 backbone hydrogen bond, but it is unlikely to retain it due to the local absence of tertiary structure. Yet, when the ligand is able to provide appropriate crating, the bond becomes inducible as the loop is able to adopt a specific conformation. Hence there is a crucial difference between wrapping and crating: while wrapping refers to an extra level of shielding or protection of a preformed hydrogen bond retained in the uncomplexed protein, crating provides exogenous shielding to a hydrogen bond that would otherwise not survive in the uncomplexed state.

The design concept of drug as crater of a specific local conformation adopted by a floppy region in the target has in fact been exploited in re-workings of imatinib [14]. As described in Chap. 9, this wrapping-based redesign targets a unique dehydron in the intended target C-Kit but also crates an otherwise floppy region in JNK, thus promoting an intended cross reactivity needed to curb the side effects of the parent compound [14]. However, only a dynamic analysis of the target/drug complex can attest that the induced folding was indeed steered by the drug design. Thus, to fully assess this possibility, we revisit the clinically relevant imatinib redesign this time in light of a dynamic analysis of structural adaptation in the drug-target complex. The focus on JNK is absolutely intended given the relevance of this kinase as a target to treat ovarian cancer and deter its progression [17].

11.3 Crating Floppy Regions in Drug Targets

The crating drug concept is evaluated by revisiting a recent redesign of imatinib [15], the anticancer drug adopted to treat chronic myeloid leukemia (CML) [3, 4]. As described in Chaps 8 and 9, this inhibitor was originally designed to target the deregulated kinase chimera from Bcr-Abl fusion, a result of chromosomal translocation causative of CML [4]. The redesign, named WBZ_4 [14, 15], was engineered to funnel the drug specificity towards some clinically relevant targets (C-Kit, JNK1 and PDGFR) [17, 18] while reducing the cardiac risk attributed to Bcr-Abl kinase targeting [19, 20]. One reinforcing way of averting this side effect is by ensuring that the redesigned version of imatinib also inhibits JNK [12, 21], a robust downstream signal transducer in the ATP-depletion and pro-apoptotic pathways triggered by Bcr-Abl inhibition in cardiomyocytes [18, 20, 21].

Another strong motivation to incorporate JNK1 in the realm of desired targets for the imatinib redesign was found *a-posteriori*, as JNK1 proved to be a target of the utmost clinical relevance in treating ovarian cancer and deter its progression [17]. *To avoid introducing new cross reactivities, the inhibition of JNK must involve the same crating modification that enabled WBZ_4 to discriminate C-Kit (and its paralog kinase PDGFRK) from the primary imatinib target Bcr-Abl. Thus, WBZ_4 functions as a wrapper of the unique dehydron in C-Kit and as a crater of an inducible hydrogen bond in JNK.*

The JNK-inhibitory activity of WBZ_4 ($K_D \approx 51$ nM) has being experimentally confirmed and is *not* found in imatinib [14]. In fact, as our dynamic analysis shows, this affinity results from a structure-inducing role of WBZ_4 that, in contrast with imatinib, acts as a protector and hence a stabilizer of a specific JNK conformation. As shown below, this conformation introduces an additional hydrogen bond in the nucleotide-binding loop. This revision of the inhibitory impact of WBZ_4 on the floppier JNK supports our proposed strategy to control the target induced folding in a selective manner.

As shown in Chap. 9, there is a dehydron present in C-Kit, the C673–G676 backbone hydrogen bond in the nucleotide-binding loop, that aligns with the well-protected M318-G321 hydrogen bond in Bcr-Abl. Targeting this unique packing defect by crating it with an added methylation in imatinib enhances drug selectivity (Fig. 11.3), telling apart C-Kit, where the wrapping defect is present, from Bcr-Abl kinase [14]. On the other hand, *the C-Kit C673-G676 packing defect does not align with any backbone hydrogen bond in JNK* (Fig. 11.3). This is so since the purported JNK M111-N114 hydrogen bond that would align with C673-G676 in C-Kit, is too severely under-protected to prevail (even as a packing defect) in the soluble structure of JNK: a homology threading of the 109–118 nucleotide-binding loop

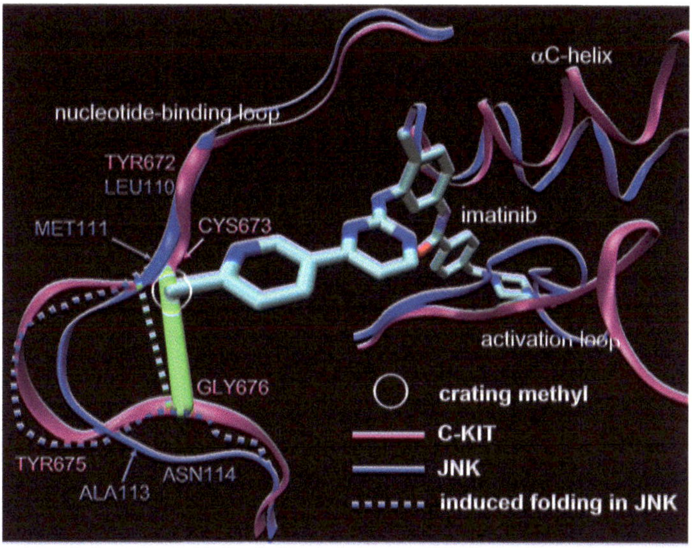

Fig. 11.3 *Protective drug as folding inducer. Structural alignment of targets JNK (isoform 1) and C-Kit for the imatinib redesigned prototype drug WBZ_4* [14]. The JNK backbone from PDB.1UKH is displayed as a thicker line than that of C-Kit (PDB.1T46). The C-Kit residues Y672 and Y675 contribute to crate the "crankshaft" C673-G676 hydrogen bond (dashed line) in C-Kit. This bond gets exogenous protection from the drug/ligand which has been purposely engineered to contribute to its crating. This crating capability represents the major difference between WBZ_4 and the parental drug imatinib. Protective interactions contributing to the dehydration of the C673-G676 hydrogen bond are indicated as thin green lines. The JNK residues aligning with C673 and G676 in C-Kit are M111 and N114. Lacking enough intramolecular protection (the protective Y675 in C-Kit is substituted for the poor protector A113 in JNK), they can only become paired by an -induced- backbone hydrogen bond (arrow) upon association with WBZ_4. This induced local folding does not prevail upon binding to the parental drug imatinib since the latter lacks the crating group at the terminal ring. Reprinted from [22], copyright 2009 with permission from Elsevier

of JNK into the backbone conformation of C-Kit produces the backbone hydrogen bond M111-N114 but with maximum possible extent of protection $\rho = 6$. This value, obtained by maximizing over side-chain rotamers in JNK 109–118 region, is below the threshold protection needed to maintain the bond. In other words, JNK is locally more flexible and less constrained than the primary imatinib targets and may be inhibited precisely by targeting its unique region of structural plasticity. This is precisely what the WBZ_4 compound accomplishes.

A molecular dynamics (MD) analysis enables a clear dissection of the protective and folding-inducing role of WBZ_4 targeting the flexible region of JNK. This analysis provides a new framework to interpret the affinity of this ligand for the kinase and supports the design concept of protective drug steering structural adaptation.

As MD computations corroborate [22], the M111-N114 hydrogen bond (N–O distance < 3.4 Å) is unsustainable in the uncomplexed JNK or even in the

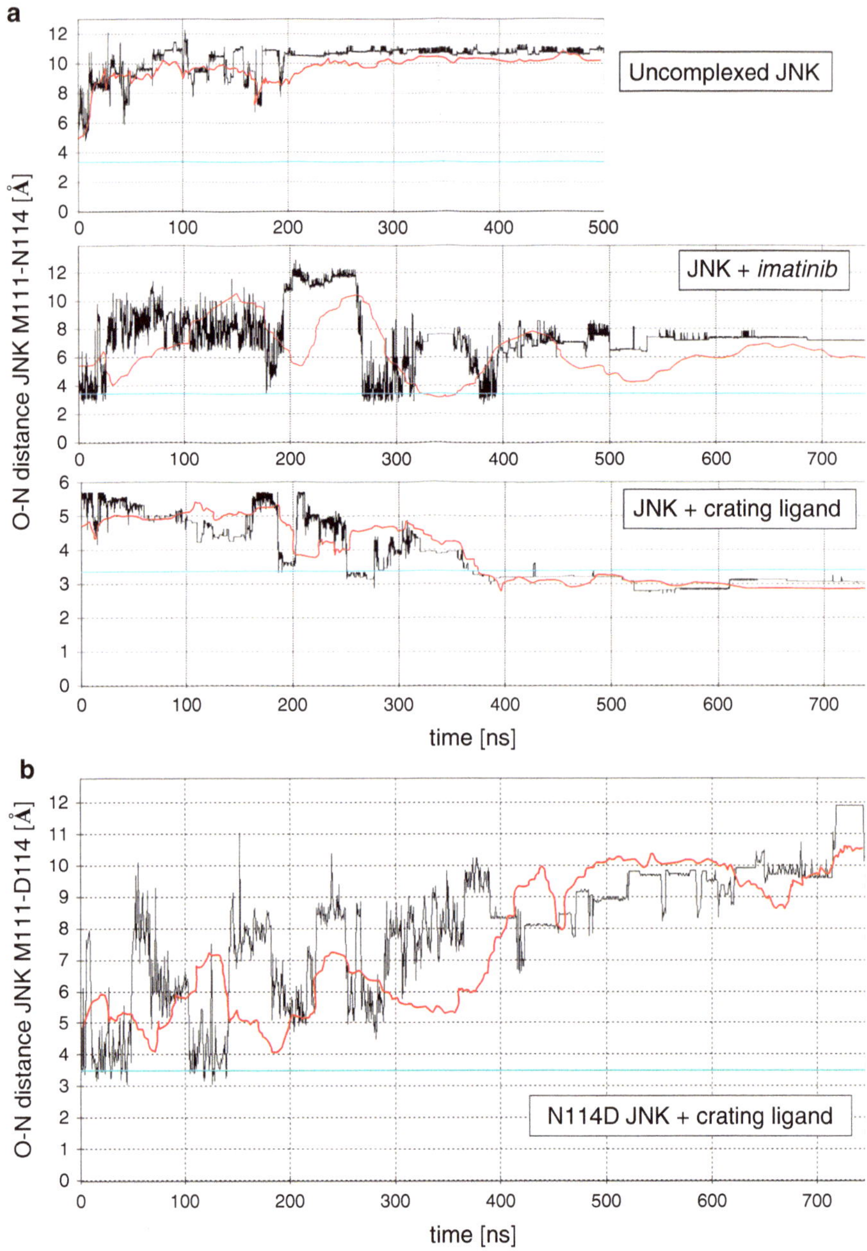

Fig. 11.4 *Folding-upon-binding dynamics of JNK in complex with a protective drug*. (**a**) Dynamics of the M111-N114 backbone hydrogen bond formation in JNK (isoform 1) in the uncomplexed form, within the complex with imatinib and within the complex with the crating ligand WBZ_4. The induced folding is monitored by the M111-N114 carbonyl-amide O–N distance and materializes when this distance is below a hydrogen-bond threshold (~3.4 Å, *blue* line). Twenty runs were generated for each case and the dynamics averaged over the trajectory ensemble is indicated by the red-line plot. The black-line plot corresponds to a single representative run. For the uncomplexed kinase, the runs covered an equilibration span of 500 ns (time extensions were

Table 11.1 Across-paralog alignment of residues that either form or crate the inducible backbone hydrogen bond M111-N114 in JNK

Kinase	Kit	Jnk1	Jnk3	MAPK6	MAPK1	p38α	p38γ
PDB code	1T46	1UKH	1PMN	2I6L	2OJI	1A9U	1CM8
Gatekeeper	THR670	MET108	MET146	GLN108	GLN103	THR106	MET109
HB crating	TYR672	LEU110	LEU148	TYR110	LEU105	LEU108	PHE111
Crankshaft HB	CYS673	MET111	MET149	MET111	MET106	MET109	MET112
HB crating	TYR675	ALA113	ALA151	THR113	THR108	ALA111	THR114
Crankshaft HB	GLY676	ASN114	ASN152	ASP114	ASP109	ASP112	ASP115
Crankshaft HB inducible?	YES	YES	YES	NO	NO	NO	NO
Crating-drug binding?	YES	YES	YES	NO	NO	NO	NO

The alignment extends over all paralogs of JNK (isoforms 1 and 3) with PDB-reported structure and enables an inference of the feasibility of inducing the hydrogen bond. This feasibility depends on the purported extent of crating of the putative hydrogen bond and on the hydration propensity of the residues paired by the bond and becomes a necessary and sufficient condition for detectable affinity of the protective drug WBZ_4

JNK/imatinib complex. The latter forms the bond intermittently and ephemerally but its average lifetime, $\tau = 12$ ps over 20 runs spanning 730 ns, indicates metastability (Fig. 11.4a). This metastability results primarily from the lack of enough intramolecular protection: the protective Y675 in C-Kit is substituted in JNK for the poor protector A113 (Fig. 11.2, Table 11.1). By contrast, the hydrogen bond is induced upon association with WBZ_4 (Fig. 11.4a): the added methyl in WBZ_4 brings extra protection upon binding.

The computational analysis supports and rationalizes the experimentally established affinity of WBZ_4 for JNK as being mainly due to its ability to crate and thus stabilize a particular JNK conformation. This conformation contains the

Fig. 11.4 (continued) deemed unnecessary due to equilibration) and represented the control results. The folding dynamics for the complexes covered 730 ns. (**b**) Induced hydrogen-bond dynamics for the N114D JNK mutant in complex with the crating drug WBZ_4. The stochastic boundary molecular dynamics (SBMD) method [22] was employed to reduce simulation times while capturing the localized binding-folding dynamics at the ATP-binding site of the JNK kinase. The system has a reaction zone of 42 residues and a reservoir region. The MD simulations were performed using the structural coordinates of free JNK (control), chimeric JNK/imatinib and JNK/WBZ_4 complexes that resulted from alignment of the JNK1 crystal structure PDB.1UKH with that of the C-Kit/imatinib complex PDB.1T46, followed by in silico removal of the Kit domain. Hence, the simulations described the equilibration process of the JNK/imatinib and JNK/WBZ_4 complexes originally obtained from alignment and combination of the PDB files 1UKH and 1T46. Reprinted from [22], copyright 2009 with permission from Elsevier

M111-N114 backbone hydrogen bond that aligns with the packing defect C673-G676 in C-Kit and becomes an induced fit in JNK upon binding to WBZ_4.

To corroborate that the induction of the M111-N114 hydrogen bond is the dominant factor dictating the drug affinity for JNK, we examined the residues aligning with the C-Kit C673 and G676 across paralogs of JNK(1/3) with PDB-reported structure: MAPK1, MAPK6, p38α and p38γ (Table 11.1). As shown elsewhere [14], WBZ_4 does not have any detectable affinity for such paralogs. Indeed, we expect that none of the paralogs is capable of maintaining the crankshaft hydrogen bond because one of the putatively paired residues is charged (Asp) rather than the polar N114 in JNK (Table 11.1). Furthermore, the extent of crating of the JNK paralogs is comparable to that of JNK, and less than that of C-Kit, mainly because of the replacement of the C-Kit Y675 for poorer protectors (A, T). The charged Asp aligning with position N114 in JNK has higher hydrophilicity and is therefore disruptive of the backbone hydrogen bond. This is corroborated not only by the absence of a crankshaft hydrogen bond in all reported structures of JNK(1/3) paralogs (Table 11.1), but also dynamically (Fig. 11.4b). In contrast with wild-type JNK, the crating drug WBZ_4 fails to induce the crankshaft hydrogen bond in the N114D in-silico mutated JNK (Fig. 11.4b). Thus, the substitution of polar N114 for charged D114 introduces disruptive negative design, preventing side-chain nonpolar groups from protectively clustering around the putative M111-D114 bond.

11.4 Target Conformational Dynamics as Selectivity Filter for Drug Design

Their therapeutic appeal notwithstanding, from a structural biology perspective, kinase targets pose daunting problems to the drug designer: their floppy regions can undergo considerable structural adaptation upon binding to the drug. Thus, due to the often unpredictable nature of induced folding, structure-based drug design has met limited success and is often relegated in favor of high-throughput screening methods and serendipitous engineering as the means to find or optimize molecular leads. This leaves us with a conceptual conundrum since floppy regions, in particular the activation loop, are the most unique in the kinase targets, i.e. the ones that confer the kinase its functional and contextual identity.

A rational control of induced folding is nevertheless possible and of course highly desirable, as shown in this chapter. This control may be achieved by focusing on structural adaptations of deficiently wrapped regions of the protein chain. Thus, we propose a targeting approach suitable for floppy regions in kinases based on ligand/drug designs which crate specific conformations of the protein that would be otherwise unsustainable in the uncomplexed state. Such drugs contribute to improve the wrapping of the target upon association to the point where the latter is now capable of retaining a specific conformation. Thus, the target affinity of folding-inducing drugs is due in part to the fact that they stabilize the drug/target complex. Support for this concept is found through re-examination of a recently re-engineered version of imatinib in light of a dynamic analysis. Our dynamic assessment shows that the inhibitory activity of the compound results precisely from its ability to steer induced folding by protecting a conformation in a floppy kinase. On the other hand, since packing deficiencies are fairly unique to kinase targets and typically not conserved across proteins of common ancestry, the specificity of such protective drugs is basically guaranteed.

We envision a new generation of protective kinase inhibitors stemming from known chemical scaffolds that may induce specific local structuring of the activation loop, the most extensive flexible region. Such drugs may exquisitely compete with ATP as they possess an extra anchoring site so far unexploited. This possibility may represent one of the few approaches to the problem of rationally exploiting induced folding in floppy targets to steer drug cross reactivity in ways that provide clinical advantage.

11.5 Kinase Inhibitor Design Based on Dynamic Information

In part, the motivation to redesign imatinib into a ligand capable of crating the floppy nucleotide-binding loop of JNK and even optimize this ligand beyond WBZ_4 levels of selective affinity comes from the *a-posteriori* realization that JNK is a highly relevant target to treat ovarian cancer, as it appears to be indispensable to promote progression through the cell cycle [17]. In fact, imatinib has been shown not to be effective against ovarian cancer, while JNK1/2 is differentially inhibited by WBZ_4. Originally regarded as a controversial cancer target, the efficacy of a JNK inhibitor is now known to be highly context-dependent [17]. In the ovarian cancer-related context, JNK transduces signals for apoptosis, differentiation, survival and carcinogenesis, and activated JNK1 has been shown to be present in a significant proportion of ovarian cancers and predictive of low progression-free survival (PFS). Thus, the crating inhibitor WBZ_4 provides a great opportunity to probe the functional, clinical, and biological significance of c-Jun-NH2-kinase (JNK)-1 in ovarian carcinoma.

A preliminary analysis of the impact of JNK on 116 epithelial ovarian cancers has already been conducted, and the role of JNK in vitro and in experimental models of ovarian cancer was assessed [17]. The role of WBZ_4 was established in cell lines and in experimental models of ovarian cancer. A significant association of pJNK with progression-free survival was first verified in the 116 epithelial ovarian cancers obtained at primary debulking therapy. WBZ_4 led to cell growth inhibition and increased apoptosis in a dose-dependent fashion in four ovarian cancer cell lines. *In vivo*, whereas imatinib had no effect on tumor growth, WBZ_4 inhibited tumor growth in orthotopic murine models of ovarian cancer.

The antitumor effect was further increased in combination with docetaxel. Silencing of JNK-1 with systemically administered siRNA led to significantly reduced tumor weights compared with nonsilencing siRNA controls, indicating that indeed the antitumor effects observed were due to JNK-1 inhibition.

All in all, the studies described in [17] identify JNK-1 as an important therapeutic target in ovarian carcinoma. Further optimizations of WBZ_4 ought to be considered for clinical development after an engineered enhancement of its JNK-crating power is carried out along the lines described in this chapter.

11.6 Induced Structural Disruption: Designing Drugs by Boosting the Conformational Entropy of the Target Protein

Structure-based drug design has met a modest degree of success so far. As discussed in this chapter, part of the problem relates to the difficulty in incorporating induced folding, the unpredictable structural adaptation of the target protein to the drug upon association. The mere existence of induced folding implies that structure-based drug design cannot be based on the current static picture but must necessarily incorporate dynamic features in most cases [23]. This makes the design problem quite daunting, especially when we consider that the induced folding problem is probably as hard as the protein folding problem, if not harder due to the chemical inhomogeneity of the ligand relative to the target.

Even in the absence of dynamic guidance to drug design, there are a few "hard facts" that have been incorporated into standard thinking about drug design. One of them is the usually accepted decrease in conformational entropy upon formation of the drug-target complex. The decrease in entropy is associated with the induced folding that takes place upon drug binding. This entropic cost obviously decreases the drug affinity since it introduces a destabilization of the protein/drug complex.

On the other hand, upon examination of drug-target complexes reported in the Protein Data Bank, we found significant local mismatches across the drug-target interface, implying that *induced unfolding* is also possible upon association [23], and almost as likely to occur as induced folding. Invariably, the drugs that generated such induced unfolding were not the result of rational design but arose from the optimization of leads discovered via high-throughput screening. This is understandable since most designers have little or no command of dynamic-based molecular engineering.

The partial induced unfolding of the target protein may appear quite striking and even counterintuitive to the unprepared mind. Yet, further reflection on the problem suggests that an entropy boost upon association can enhance affinity, as it decreases the over-all free energy content of the complex. However, we are not yet prepared to incorporate purposely engineered drug-target mismatches into the designs. This is mainly because we are as incapable of controlling induced unfolding as we are of controlling induced folding, save for the cases previously described in this chapter. Entropy is the true *bête noir* of rational design.

Rather than ignoring these counterintuitive mismatches across the drug-target interface, we need to bring them to broad daylight if we are to ever fully succeed at rational drug design.

11.7 Problems

11.1. By examining and comparing the wrapping of the protein structure of the

A. Apo form,
B. drug/target complex, and
C. drug/target complex with *in-silico* removal of the drug, determine the following:

(a) Drug-wrapping pattern, identifying the carbonaceous groups in the drug (Figs. 11.5 and 11.6) that contribute to the wrapping of pre-formed or induced backbone hydrogen bonds and dehydrons in the target protein.
(b) Drug-induced backbone hydrogen bonds.
(c) Drug-induced dehydrons.

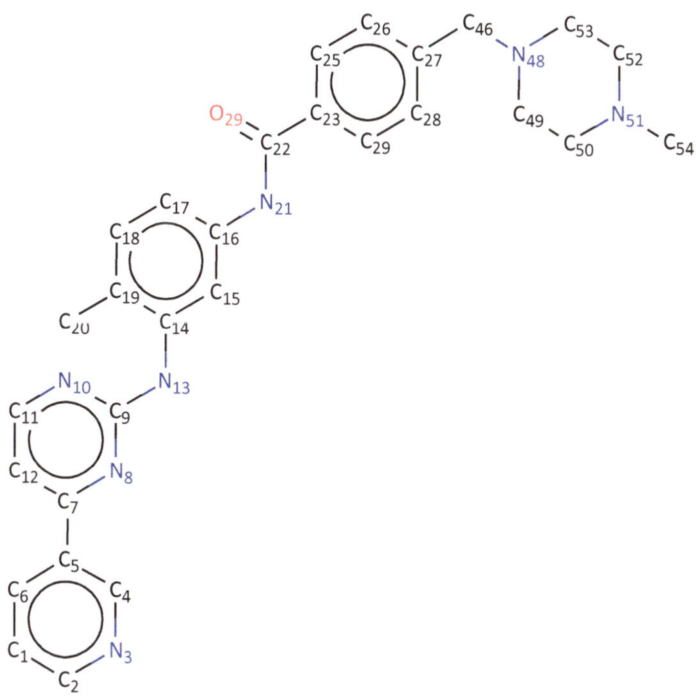

Fig. 11.5 Chemical structure of imatinib with labelled atoms

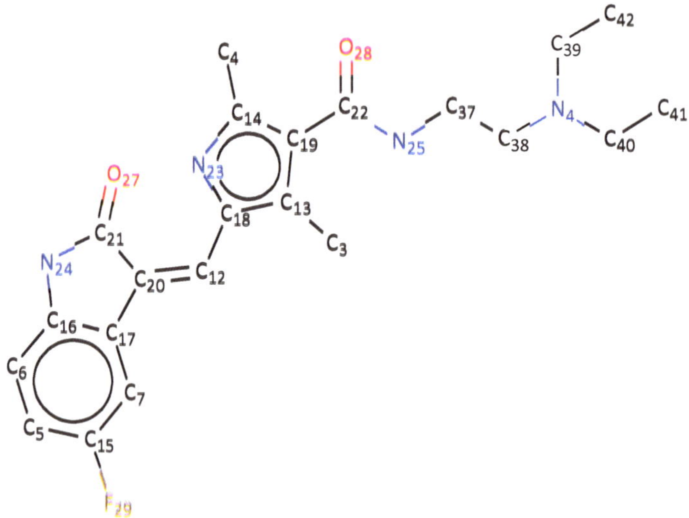

Fig. 11.6 Chemical structure of sunitinib with labelled atoms

(d) Preformed backbone hydrogen bonds that are "de-wrapped" or destabilized upon binding so they can get better wrapped exogenously by the drug. Structural adaptation to form the complex turns these hydrogen bonds into less wrapped than in the apo form, inviting exogenous wrapping upon drug binding. This phenomenon is termed productively induced metastability (PIM) [24].

Consider the following cases:

I. c-Kit kinase + imatinib. Apo form: PDB.1PKG; Drug/Target complex: PDB.1T46
II. VEGFR2 kinase + sunitinib. Apo form: PDB.1VR2; Drug/Target complex: PDB.4AGD

11.2. *Target un-wrappers*: *Next-generation drugs that induce dynamic states in target proteins*. Discuss the design possibilities emerging from hydrophobic-polar mismatches across the drug-target interface (Figs. 11.7 and 11.8), acting as likely promoters of induced dynamic states in the target protein. Discuss the illustrations of entropy-optimized drugs that are likely to enhance exploration of conformation space in the target protein through purposely engineered hydrophobic-polar mismatches at the drug-target interface.

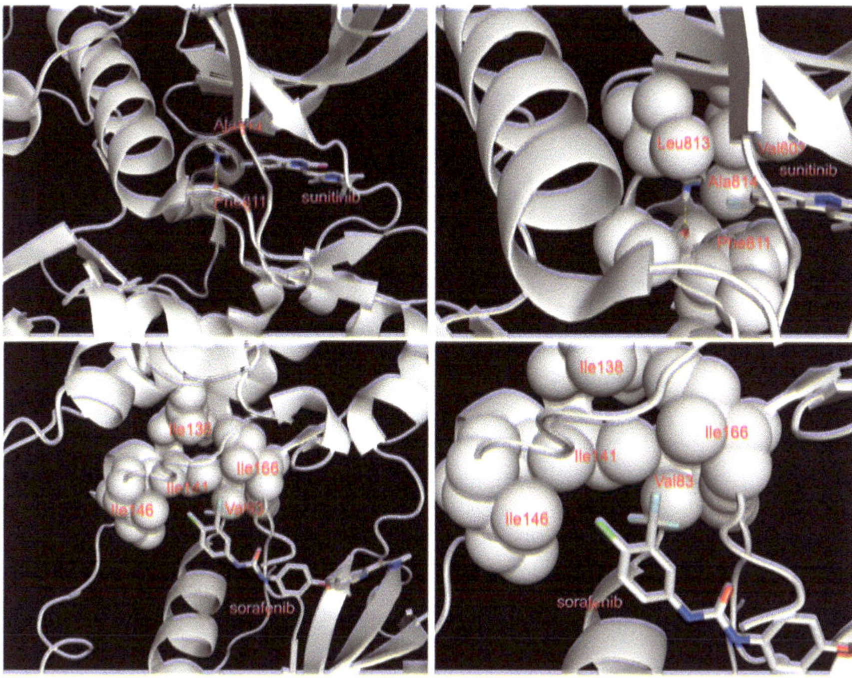

Fig. 11.7 Hydrophobic-polar mismatches across the drug-target interface are likely promoters of induced dynamic states in the target protein. Illustrations of entropy-optimized drugs that are likely to enhance exploration of conformation space in the target protein through purposely engineered hydrophobic-polar mismatches at the drug-target interface. The upper panels correspond to sunitinib in complex with c-Kit kinase (PDB.3G0E), while the lower panels correspond to sorafenib in complex with the VEGRF kinase (PDB.3GCS). Atoms are depicted following color convention, with fluorine in light blue and chlorine in green. The hydrogen bond Phe 811-Ala814 in C-Kit is likely to be a *disrupton* (a hydrogen bond disrupted upon ligand binding). Its unwrapping (and ultimate disruption) is promoted by the proximity of the sunitinib fluorine that is mismatched in the complexed state against the wrapping nonpolar residues Phe811, Leu813, Ala814, Val603. In both cases, the probable partial disruption of structure boosts conformational entropy, thereby enabling hydration of the otherwise occluded polar groups (halogen atoms) in the drug. This binding mode promoting an induced dynamic state shall be tested/corroborated through molecular dynamics, NMR-relaxation experiments and by examining the effect of halogen-to-hydrogen substitution in the thermodynamics of the drug-target association for both kinase inhibitors [23]

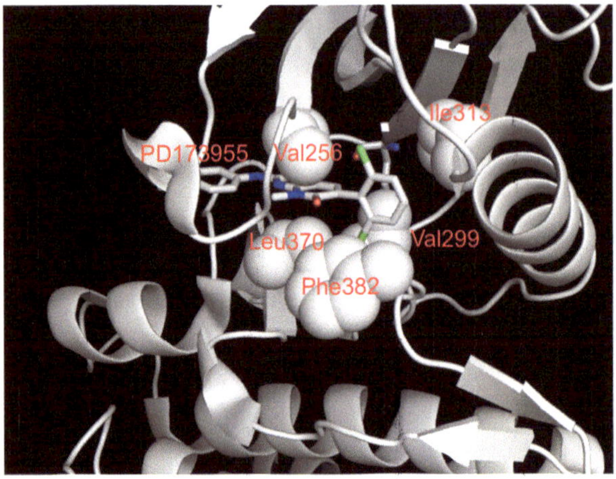

Fig. 11.8 Glaring hydrophobic-polar mismatches between high-affinity inhibitor PD173955 and the Abelson (ABL) kinase in PDB entry 1M52. The chlorines (*green*) in PD173955 are occluded upon drug-target association by nonpolar moieties from the labeled residues. This desolvation penalty is severe and likely to create an expansion in accessible conformation space of the loopy regions in order to enable proper drug hydration. This scenario may be corroborated by the thermodynamic computation of the impact on affinity of the chlorine-to-hydrogen substitution in PD173955 [23]

References

1. Dancey J, Sausville EA. Issues and progress with protein kinase inhibitors for cancer treatment. Nat Rev Drug Discov. 2003;2:296–313.
2. Tibes R, Trent J, Kurzrock R. Tyrosine kinase inhibitors and the dawn of molecular cancer therapeutics. Annu Rev Pharmacol Toxicol. 2005;45:357–84.
3. Schindler T, Bornmann W, Pellicena P, et al. Structural mechanism for STI-571 inhibition of Abelson tyrosine kinase. Science. 2000;289:1938–42.
4. Schiffer CA. BCR-ABL tyrosine kinase inhibitors for chronic myelogenous leukemia. N Engl J Med. 2007;357:258–65.
5. Crespo A, Fernández A. Kinase packing defects as drug targets. Drug Discov Today. 2007;12:917–23.
6. Teague S. Implications of protein flexibility for drug discovery. Nat Rev Drug Discov. 2003;2:527–41.
7. Damm KL, Carlson HA. Exploring experimental sources of multiple protein conformations in structure-based drug design. J Am Chem Soc. 2007;129:8225–35.
8. Hornak V, Simmerling C. Targeting structural flexibility in HIV-1 protease inhibitor binding. Drug Discov Today. 2007;12:132–8.
9. Erickson J. Lessons in molecular recognition: the effects of ligand and protein flexibility on molecular docking accuracy. J Med Chem. 2004;47:45–55.
10. Noble ME, Endicott JA, Johnson LN. Protein kinase inhibitors: insights into drug design from structure. Science. 2004;303:1800–5.
11. Pietrosemoli N, Crespo A, Fernández A. Dehydration propensity of order-disorder intermediate regions in soluble proteins. J Proteome Res. 2007;6:3519–26.
12. Chen J, Zhang X, Fernández A. Molecular basis for specificity in the druggable kinome: sequence-based analysis. Bioinformatics. 2007;23:563–72.

13. Fernández A. Keeping dry and crossing membranes. Nat Biotechnol. 2004;22:1081–4.
14. Fernández A, Sanguino A, Peng Z, Ozturk E, Chen J, Crespo A, Wulf S, Shavrin A, Qin C, Ma J, Trent J, Lin Y, Han HD, Mangala LS, Bankson JA, Gelovani J, Samarel A, Bornmann W, Sood AK, Lopez-Berestein G. An anticancer c-Kit kinase inhibitor is reengineered to make it more active and less cardiotoxic. J Clin Invest. 2007;117:4044–54.
15. Crunkhorn S. Anticancer drugs: redesigning kinase inhibitors. Nat Rev Drug Discov. 2008;7:120–1.
16. Fernández A, Sanguino A, Peng Z, Crespo A, Ozturk E, Zhang X, Wang S, Bornmann W, Lopez-Berestein G. Rational drug redesign to overcome drug resistance in cancer therapy: imatinib moving target. Cancer Res. 2007;67:4028–33.
17. Vivas-Mejia P, Benito JM, Fernández A, Han HD, Mangala L, Rodriguez-Aguayo C, Chavez-Reyes A, Lin YG, Carey MS, Nick AM, Stone RL, Kim HS, Claret FX, Bornmann W, Hennessy BT, Sanguino A, Peng Z, Sood AK, Lopez-Berestein G. c-Jun-NH2-kinase-1 inhibition leads to antitumor activity in ovarian cancer. Clin Cancer Res. 2010;16:184–94.
18. Demetri GD, von Mehren M, Blanke CD, et al. Efficacy and safety of imatinib mesylate in advanced gastrointestinal stromal tumors. N Engl J Med. 2002;347:472–80.
19. Kerkela R, Grazette L, Yacobi R, et al. Cardiotoxicity of the cancer therapeutic agent imatinib mesylate. Nat Med. 2006;12:908–16.
20. Force T, Krause D, van Etten RA. Molecular mechanisms of cardiotoxicity of tyrosine kinase inhibition. Nat Rev Cancer. 2007;7:332–44.
21. Demetri GD. Structural reengineering of imatinib to decrease cardiac risk in cancer therapy. J Clin Invest. 2007;117:3650–3.
22. Fernández A, Bazan S, Chen J. Taming the induced folding of drug-targeted kinases. Trends Pharmacol Sci. 2009;30:66–71.
23. Fernández A, Fraser C, Scott R. Purposely engineered drug-target mismatches for entropy-based drug optimization. Trends Biotechnol. 2012;30:1–7.
24. Montes de Oca J, Rodriguez Fris A, Appignanesi G, Fernández A. Productive induced metastability in allosteric modulation of kinase function. FEBS J. 2014;281:3079–91.

Chapter 12
Drug Combinations to Enhance Therapeutic Efficacy and Edit Out Side Effects and Resistance to Inhibition of Drug Resistance

This chapter exploits systems biology annotation and combines it with the wrapping technology. The chapter focuses on combination drug treatments developed based on epistructural design to enhance therapeutic power and safety beyond the levels achievable through drug monotherapy. The drugs in the combination treatments exert a therapeutic effect and also counteract/antagonize each other in contexts where one of them elicits undesirable side effects or resistance to inhibition of drug resistance.

Epistructure-based drug designs have limitations in their ability to exclude off-target proteins from the inhibitory impact of the drug. Such undesired but inevitable cross reactivity arises from similarities in the epistructure of on-target and off-target proteins. While dehydron wrapping enables the control of specificity, it may not be able to exclude every single toxicity-related target if the latter shares a similar dehydron pattern in the drug-binding region with the targets of therapeutic value. In such cases, as demonstrated in this chapter, we may resort to a combination therapy where one drug acts synergistically with the other while selectively antagonizing it in the specific cellular context where the action of the first drug promotes toxicity or resistance to the inhibition of drug resistance.

Due to their ability to block signal transduction events controlling cell proliferation and cell fate, kinase inhibitors are perceived as important anticancer agents. Nevertheless, the functional role of a kinase depends on the cellular context and hence kinase inhibition in an off-target cell may lead to undesired side effects. Motivated by these observations, this chapter explores a mode of "therapeutic editing" where one drug—the editor—suppresses or modulates the side effect promoted by the primary drug as the latter impacts off-target cells. Editor and primary drug may have overlapping or at least non-conflictive therapeutic impact, while the editor suppresses the downstream propagation of toxicity-related signaling triggered by the primary drug in off-target contexts.

In the realm of drug resistance, the power of combined mutually-editing drug therapies becomes apparent when an editing mechanism identified through systems

A. Fernández, *Physics at the Biomolecular Interface*, Soft and Biological Matter,
DOI 10.1007/978-3-319-30852-4_12

biology annotation enables the primary drug to edit out drug resistance mechanisms towards the second drug that is itself an inhibitor of drug resistance towards the first drug.

12.1 The Concept of Drug Editor for Side-Effect Removal

Side effects in molecular targeted anticancer therapy are common [1–5], hard to avoid [6], often unexpected [7] and potentially health-threatening [8], thereby introducing daunting difficulties to drug development. At least two major problems must be dealt with when addressing side effects [4–10]: (a) target proteins have paralogs with similar 3D-structure [9, 10] and (b) the target protein plays a role in off-target cells that is different from the role it plays in target (cancer) cells [5–7]. Accordingly, undesired cross-reactivities are expected due to (a), while (b) implies that target inhibition may be therapeutically efficacious in cancer cells, but toxic in off-target cells. This chapter will be limited to kinase inhibitors (KIs), the drugs that interfere with cell-signaling pathways, because of their promising prospects for anticancer therapy [9–12] and because such treatments require addressing the problems outlined above [5, 11–13].

Problem (a) motivated the exploitation of selectivity filters for drug design [9, 10], leading to an efficacious control of specificity through paralog discrimination [13]. This problem has been dealt with extensively in Chaps 7, 8 and 9. where we showed that dehydrons in proteins can been used to distinguish family members, and that such packing defects can be targeted using ligands engineered to "wrap" or shield them from interactions with water [10, 13], resulting in high specificity.

Dealing with problem (b) requires advanced network annotation [5, 14]. The problem can be dissected by comparing protein interaction networks across tissues and placing the target protein in different interacting contexts represented in various cell types [14]. In this chapter we focus on problem (b) and advocate a strategy to avert the resulting side effects.

The context-dependence problem has been effectively dealt with through formulations for tissue-specific delivery [15–17]. The strategies to enhance tissue-specificity include incorporating the drug molecules into nanosystems such as liposomes, coupling drugs to antibodies or ligands that target cell receptors, exploiting permeation enhancers or passive or active translocators, etc. These approaches often fail to be completely specific and introduce additional problems like limited delivery to the site of action.

In light of this situation, we explore an alternative systemic approach to deal with the context-dependence problem: a mode of "therapeutic editing" using two-component free-drug treatments. One drug—the editor—suppresses the side effect promoted by the primary drug, while both drugs overlap in their impact on therapeutically relevant targets. Thus, the editor selectively suppresses downstream propagation of toxicity-related signaling while circumventing the delivery issues that plague the conventional approaches to the context-dependence problem.

12.2 Editing Out Side Effects: Illustrations

Context-dependent functionalities introduce a risk of toxicity in molecular target therapy [4–7]. For example, the constitutively active chimera Bcr-ABL kinase is a clinical target for treating chronic myeloid leukemia (CML), as evidenced by the success of the KI imatinib (Gleevec) [18], but ABL-inhibition in cardiomyocytes (off-target cells in CML treatment) may introduce a cardiotoxicity risk [5–7], as shown in Fig. 12.1. This risk is more pronounced for promiscuous KIs like sunitinib [5, 8].

In general, prevention of side effects arising from context-dependence requires tissue-specific delivery [15–17] to selectively avoid inhibiting a target kinase in off-target cells. In practice, this level of selectivity is difficult to achieve. In view of this problem, we propose a two-component treatment in which one drug modulates a downstream component of the toxicity-signaling pathway recruited by the other.

Two-component treatments have been exploited to improve therapeutic impact [19–21]. In this chapter we advocate an alternative application with the aim of curbing the side effects of one drug by "editing" its impact in off-target cells with another drug. This requires that we introduce a dual-role editor, so that the two drugs overlap therapeutically in target tumor cells, but behave antagonistically in off-target cells (Fig. 12.2).

Two premises must hold for this editing therapy to apply: (I) the side effect of the primary drug is attributable to an alternative role of the target kinase in off-target cells; (II) target inhibition in the off-target cell activates a pathway that compromises cell survival. Consequently, the editor must fulfill the following conditions (Fig. 12.2): (i) in cancer cells, drug and editor have overlapping impact; (ii) inhibition of a primary target activates a detrimental pathway in off-target cells; (iii) in off-target cells, the editor has an antagonistic effect through downstream inhibition, impairing recruitment of the deleterious pathway triggered by the action of primary drug; and (iv) this antagonistic effect is not carried over to cancer cells, where the biochemical flow is restored through alternative (redundant) pathways that warrant robustness vis-à-vis drug-induced perturbations [14].

We focus on editing cardiotoxicities of KIs, which are significant side effects [5–8] and fulfill the applicability premises described above (cf. Fig. 12.1). In general, KI cardiotoxicity remains a controversial subject [5–8, 22, 23]. However, *in-vivo* and animal-model studies [13] and clinical studies [8] corroborate the original characterization of KI cardiotoxicities [5–7]. The discrepancy with alternative reports that focus on imatinib [22, 23] is mainly due to the fact that the latter are not prospective studies, as patients did not have baseline functional studies and clinical endpoints were not delineated.

The therapeutic concept is illustrated by considering possible editors for sorafenib, a major KI successfully exploited to treat Renal Cell Carcinoma (RCC) [24]. Sorafenib entails a cardiotoxicity risk traced to RAF1 inhibition in cardiomyocytes (Fig. 12.3 [5]). The drug binding to RAF1 hinders the MST2-inhibitory RAF1-MST2 association [5, 25], as suggested by the results described

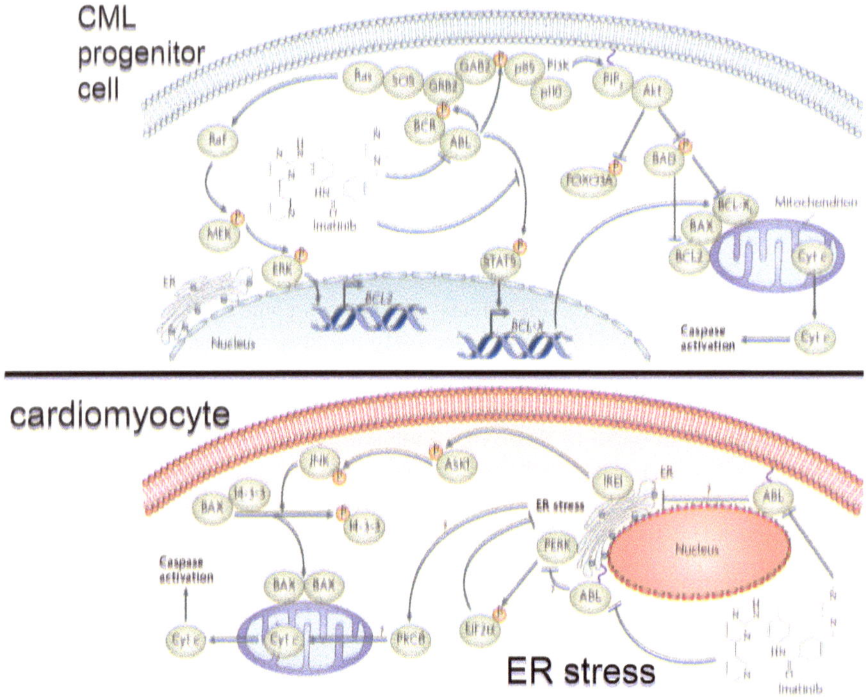

Fig. 12.1 Molecular dissection of imatinib side effects due to the context-dependent functionality of one of its primary targets. On display is the constitutive signaling in chronic myeloid leukaemia (CML) progenitor cell through the cytoplasmic BCR–ABL tyrosine kinase, which leads to activation of Ras–ERK (extracellular signal-regulated kinase), phosphatidylinositol 3-kinase (PI3K)–Akt and signal transducer and activator of transcription 5 (STAT5) pathways. Several of these pathways converge on anti-apoptotic mechanisms: the Ras–ERK pathway stimulates the expression of *BCL2*, STAT5 activates *BCL-X*, and Akt inhibits BCL2-antagonist of cell death (BAD) and forkhead box O3A (FOXO3A). Imatinib blocks all BCR–ABL-dependent phosphorylation and signalling events, leading to reversal of pro-survival effects and activation of apoptosis. On the other hand, ABL (localized to the endoplasmic reticulum, ER) seems to maintain ER homeostasis in cardiomyocytes. The ABL kinase inhibitor, imatinib, induces ER stress, leading to activation of the PKR-like ER kinase (PERK) and IRE1 pathways, and to overexpression of protein kinase Cδ (PKCδ). PERK phosphorylates the eukaryotic translation initiation factor 2α (EIF2α) as part of a protective response, and on sustained ER stress IRE1 activates Jun N-terminal kinases (JNKs), leading to phosphorylation of 14-3-3 and release of BAX followed by mitochondrial depolarization, ATP depletion, cytochrome *c* (Cyt *c*) release, and features of necrotic and apoptotic cell death. Notation: 14-3-3, 14-3-3 protein; ASK1, apoptosis signal-regulating kinase 1; BAX, BCL2 associated X protein; GAB2, GRB2-associated binding protein 1; GRB2, growth factor receptor-bound protein 2; MEK, mitogen-activated ERK kinase; PIP_3, phosphatidylinositol trisphosphate; SOS, son of sevenless. Reprinted from [5], copyright 2007 with permission from Macmillan Publishers Ltd

below. Thus, RAF1 inhibition activates pro-apoptotic pathways mediated by MST2 and further downstream by JNK, ultimately leading to cytochrome *c* release from mitochondria [5]. As shown in Fig. 12.3, two compounds are potentially capable of

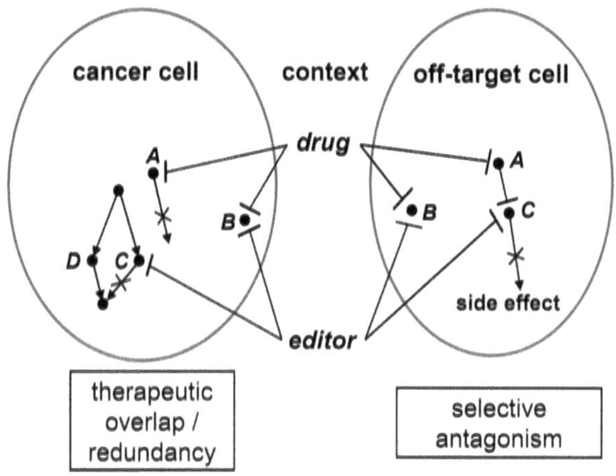

Fig. 12.2 Scheme of editing strategy. Drug and editor overlap therapeutically in on-target cancer cells, while the editor interferes selectively and antagonistically with the primary drug in off-target cells. Drug and editor have a common therapeutically relevant target B, while the drug inhibits additional target A. In addition, the editor inhibits target C which is redundant in cancer cells (biochemical flow may be restored through alternative signal transducer D), but becomes indispensable in the side-effect related pathway recruited by the action of the primary drug on target A in normal (off-target) cells. Thus, phosphorylation of C is promoted by the inhibitory action of the drug on target A (A normally inhibits C). In this way, the editor interferes to suppress a side effect resulting from the action of the primary drug in off-target cells, while this interference becomes innocuous in cancer cells due to a built-in redundancy of the latter (D acts as alternative transducer to C). Reprinted from [32], copyright 2009 with permission from Elsevier

interfering with this pathway, acting as sorafenib editors: the highly specific imatinib-derived compound WBZ_4, a JNK inhibitor with $K_D = 34$ nM [13], already introduced in Chap. 8, and SUDE, a sunitinib-derived editor described below that inhibits MST2 with $K_D = 11$ nM [9]. Both compounds overlap therapeutically with sorafenib while they can suppress its potential cardiotoxicity by blocking the sorafenib-recruited pro-apoptotic pathway (Fig. 12.3a, b).

As shown in Chap. 8, WBZ_4 is a re-engineered version of imatinib with an enhanced specificity needed to curb imatinib's potential cardiotoxicity. The inhibitory impact of WBZ_4 is restricted to KIT, PDGFR and JNK, all with nM affinity, as corroborated by screening against a T7-bacteriophage library of kinase displays (Ambit Biosciences, CA) [13].

JNK is an indispensable downstream transducer of pro-apoptotic signals in cardiomyocytes [5, 25, 26] (Fig. 12.3), thus, its inhibition by WBZ_4 should antagonize sorafenib's inhibition of RAF1. In cancer cells, both drugs overlap on therapeutic targets PDGFR and KIT [5, 13], while JNK-inhibition by WBZ_4 interferes with a redundant pro-apoptotic pathway that may be restored through p38 mediation [26]. The surrogate role of p38 may not be universal, but holds in the MAPK cancer-related contexts of proliferation, differentiation and apoptosis [26].

SUDE was designed using the wrapping-based selectivity filter [27] with two purposes in mind: (A) decrease cross-reactivity and (B) target MST2. As a sunitinib

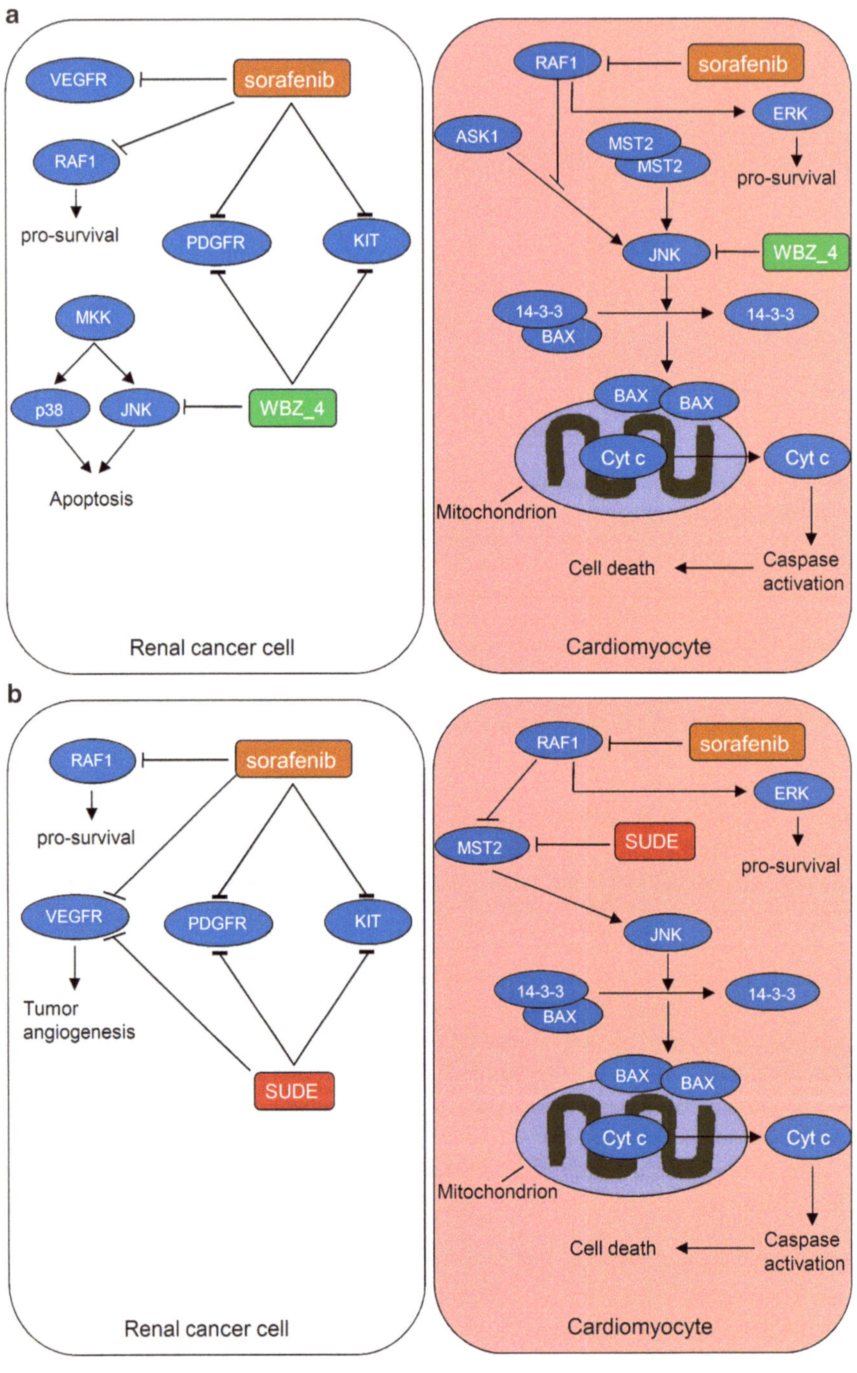

Fig. 12.3 Network level analysis of editing therapy, illustrated by sorafenib/WBZ_4 and sorafenib/SUDE combined activity. (**a**) Sorafenib edited by WBZ_4. In renal cancer cells sorafenib and WBZ_4 overlap therapeutically by inhibiting common clinically relevant targets (PDGFR and KIT). The inhibitory action of WBZ_4 on JNK confers editing power to this drug.

derivative [9], SUDE overlaps with sorafenib on therapeutic targets VEGFR, KIT, PDGFR, FLT1 and FLT3 [24, 28]. With MST2, these kinases constitute the complete target set for SUDE as corroborated by kinase phage-display screening, while SUDE becomes an MST2-inhibitor more potent than the parent compound ($K_D = 11$ nM vs 56 nM). Thus, SUDE should edit sorafenib, counteracting its deleterious pathways in cardiomyocytes through its anti-MST2 activity (Fig. 12.3b). This editing role was experimentally validated for RCC, as shown below, but may not hold for other cancers like ovarian carcinoma. In the latter, MST2 inhibition may actually promote cancer survival since the tumor suppressor LATS1 becomes activated by MST2 [29]. This fact introduces a limitation in the editing applicability of SUDE.

12.3 Epistructural Design of a Drug Editor

SUDE was primarily engineered to edit sorafenib by inhibiting the kinase MST2, a downstream transducer in the pro-apoptotic pathway recruited by sorafenib in cardiomyocytes (Fig. 12.3). Additionally, it was necessary for SUDE to overlap therapeutically with sorafenib in cancer cells, to avoid disrupting anticancer activity. Furthermore, SUDE needed to be specific enough to reduce the cardiotoxicity risks of its parent drug sunitinib, adopted as the scaffold (see Chap. 13 for details). Sunitinib overlaps with sorafenib on the clinical targets VEGFR, KIT, PDGFR, FLT1 and FLT3 [5, 24, 28]. Sunitinib was redesigned guided by the wrapping selectivity filter defined in Chap. 10. The first goal is to make it selective and less cardiotoxic, while the second is to turn it also into an MST2 inhibitor. To achieve the first goal, we exploit the wrapping differences across sunitinib's reported targets [30]. As further described in Chap. 13, alignment of wrapping patterns revealed a targetable dehydron in sunitinib's clinical targets that is absent in the cardiotoxicity-related targets AMPK and RSK (Fig. 12.4, [5, 8]). The distinct and targetable feature is the backbone hydrogen bond pairing residues N900 and E917 in VEGFR, a primary target of sunitinib (Fig. 12.4b), a bond significantly exposed to solvent in all primary targets, but well wrapped in AMPK and RSK (Fig. 12.4a). This targeted dehydron is also present in MST2 and involves residues K69 and E85 at the aligned position (Fig. 12.4a). Thus, a methyl group was incorporated to

Fig. 12.3 (continued) This inhibitory activity is likely inconsequential for cancer cells, as the biochemical flow is restored through p38, while in cardiomyocytes, the inhibitory action of WBZ_4 on JNK antagonizes a pro-apoptotic pathway activated by upstream sorafenib-promoted inhibition of RAF1. This pathway leads to a BAX-mediated cytochrome *c* release to the cytosol triggering the caspase cascade activation. (**b**) Sorafenib edited by SUDE. SUDE overlaps therapeutically with sorafenib on VEGFR, PDGFR and KIT in renal cancer cells, whereas in cardiomyocytes, the SUDE inhibition of MST2 impairs the recruitment of pro-apoptotic pathways activated by the upstream sorafenib-promoted inhibition of RAF1. Reprinted from [32], copyright 2009 with permission from Elsevier

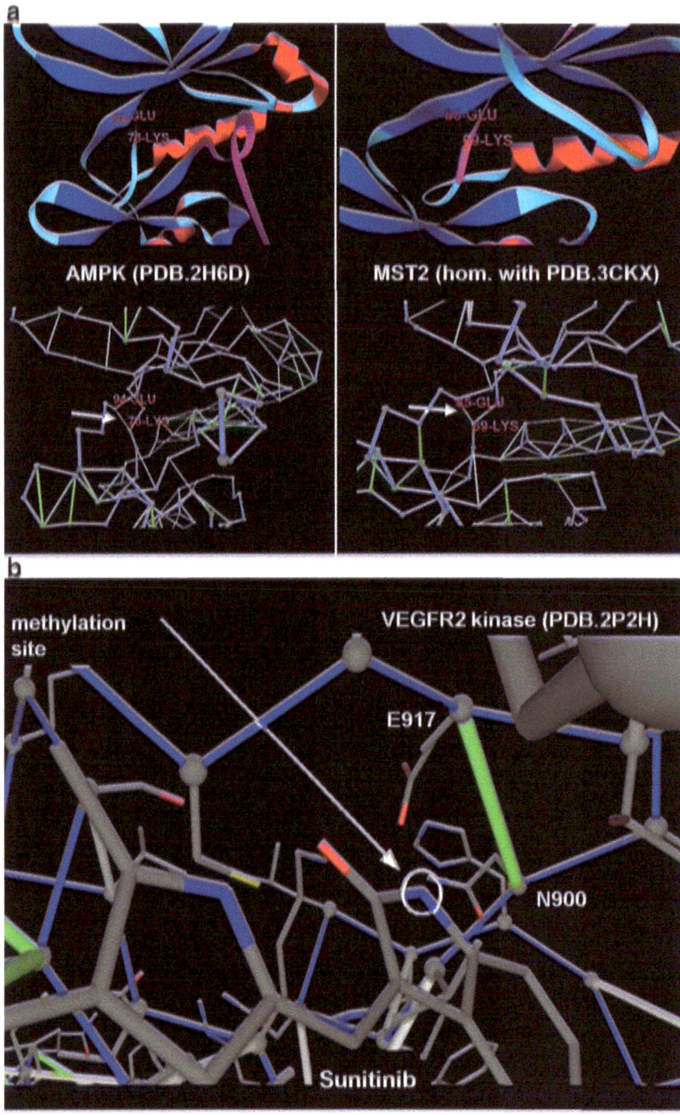

Fig. 12.4 Molecular design of selective sunitinib-derived editor (SUDE) guided by structural/wrapping comparison of sunitinib targets, by the need to remove cardiotoxicity-related anti-AMPK and anti-RSK activity, and by the need to incorporate the new target MST2 for editing purposes. Protein structures are also rendered in ribbon representation for visual aid. (**a**) Structures and dehydron patterns of sunitinib target AMPK (PDB.2H6D) and MST2 (homologous to PDB.3CKX) kinase in complex with sunitinib. These targets need to be discriminated by SUDE, which should have affinity for MST2 but not for AMPK (or RSK). The K69-E85 dehydron in MST2 is conserved in the high-affinity targets (KIT, PDGFR, VEGFR, FLT1 and FLT3) relevant to angiogenesis and cancer-cell proliferation. Crucially, it aligns with the well-wrapped hydrogen bond K78-E94 in AMPK2 (and with a well-wrapped hydrogen bond in RSK, not shown). *This important packing difference suggests the modification to sunitinib needed to selectively target*

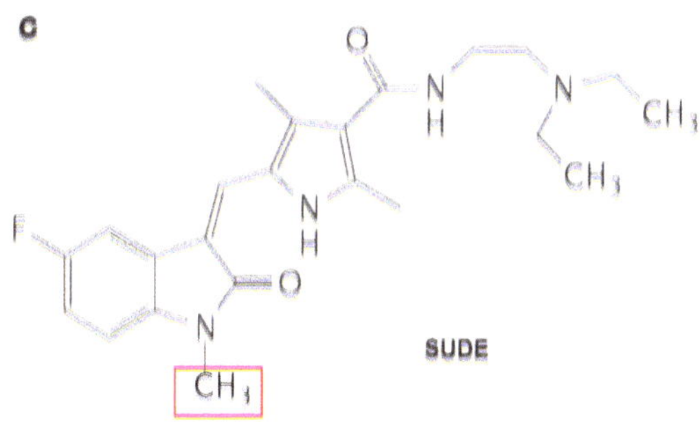

Fig. 12.4 (continued)

sunitinib to provide a "wrapping" protection to the MST2 dehydron, also present in all clinically relevant targets VEGFR, KIT, PDGFR, FLT1 and FLT3, but absent in AMPK and RSK. Thus, we redesigned sunitinib into SUDE (Fig. 12.4c), an MST2-inhibitor with controlled specificity.

The predicted SUDE target set {MST2, VEGFR, KIT, PDGFR, FLT1, FLT3} was confirmed using the AMBIT high-throughput assay that uses a phage-display kinase library [30]. With $K_D = 11$ nM, SUDE proved to be an MST2-inhibitor more potent than sunitinib ($K_D = 56$ nM, [30]). The higher affinity of SUDE is due to the additional wrapping protection on MST2 dehydron K69-E85.

12.4 Proof of Concept for Therapeutic Editing

At a cellular level, the purported sorafenib editors should be assayed using the renal cancer cell line RCC-786-O and NRVMs (Neonatal Rat Ventricular Monocytes). In RCCs, the sorafenib/WBZ_4 combination yields agonistic synergy, marked by an increased inhibition of cell proliferation when compared to sorafenib-alone levels at equivalent bulk concentrations (Fig. 12.5a): The dose-dependent inhibition is greater than the Loewe-additive values [20, 21] (Fig. 12.5b). By contrast, the sorafenib-induced recruitment of the pro-apoptotic pathway in cardiomyocytes

Fig. 12.4 (continued) MST2: methylation at the N1 indol position turns the drug into SUDE, a wrapper of the MST2 dehydron K69-E85. Thus, by inhibiting MST2, a signal transducer in the pro-apoptotic pathways triggered by sorafenib, SUDE overlaps antagonistically with sorafenib. (**b**) SUDE retains activity against clinically relevant sunitinib targets KIT, PDGFR, VEGFR, FLT1 and FLT3. Structure and packing pattern of VEGFR2 (PDB.2P2H) kinase in complex with sunitinib. The N900-E917 dehydron in VEGFR2 is conserved in other clinically relevant targets of sunitinib and hence the modification to sunitinib serves as a wrapper to this common dehydron. (**c**) Chemical structure of SUDE. The predicted SUDE cross-reactivities were confirmed using the AMBIT high-throughput assay that uses a phage-display kinase library. Reprinted from [32], copyright 2009 with permission from Elsevier

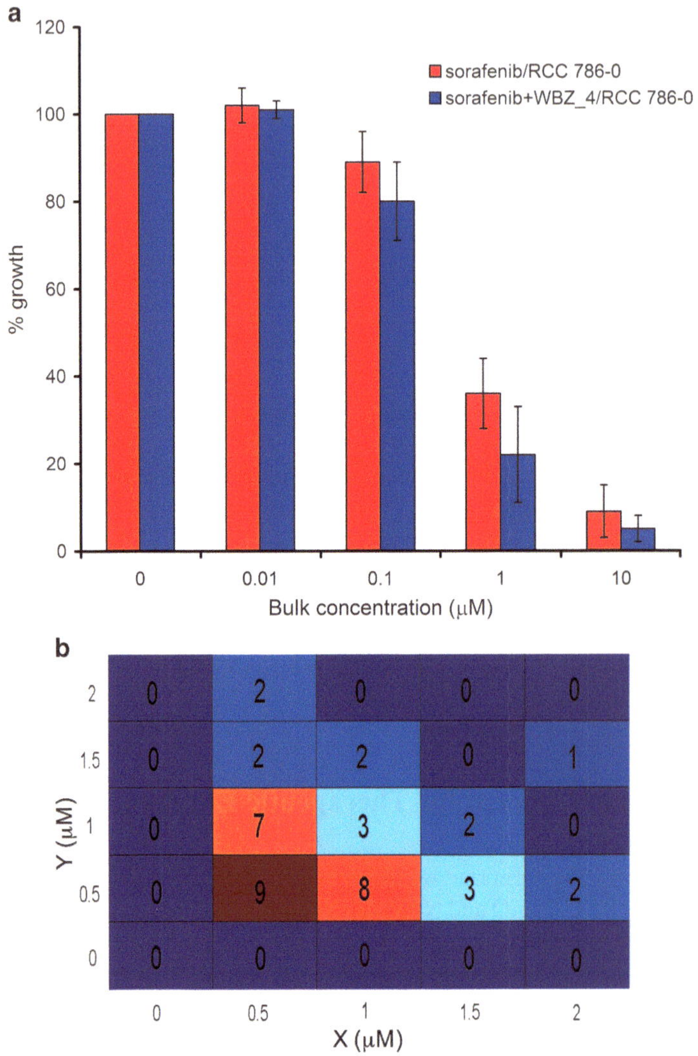

Fig. 12.5 (**a**) Cell proliferation assays for renal cancer cell line RCC 786-O for sorafenib/WBZ_4 combination treatments. The cells were seeded in 96 well plates (4×10^4 cells per well), cultured for 24 h and treated for 48 h with sorafenib alone and in 50–50 % combinations with WBZ_4 at over-all 0, 0.01, 0.1, 1 and 10 μM bulk concentrations. Cell proliferation was determined by Alamar Blue assay (Bio Source International, Camarillo, CA). Plates were read at dual wavelength (570–595 nm) in an Elisa plate reader (Molecular Devices Corporation, Sunnyvale, CA). Cell proliferation is expressed as the percentage of proliferating cells relative to untreated cells. The sorafenib/WBZ_4 combination treatment led to an increased inhibition of cell proliferation when compared to the sorafenib-alone levels at equivalent bulk concentrations. Also, the dose-dependent inhibition surpassed the Loewe-additive values indicating agonistic synergy. The sorafenib/SUDE combination treatment (not shown) suppressed RCC proliferation to sorafenib-alone levels at identical overall concentrations. In this case, the dose-dependent inhibition fits the Loewe additive scheme indicating therapeutic reinforcement without synergy. (**b**) Agonistic synergy matrix for sorafenib/WBZ_4 combinations assayed in cell-proliferation assays. Matrix elements are defined as $M(X,Y) = I(X,Y) - I'(X,Y)$, where X = sorafenib concentration in μM,

should be impaired by the downstream interference of WBZ_4 through JNK inhibition (Fig. 12.3a). Consequently, cytosolic release of cytochrome *c* should also be impaired. A dose-dependent study revealed antagonistic synergy: If Y_{50} denotes the WBZ_4 concentration yielding 50 % reduction in sorafenib-induced cytochrome *c* release, we consistently obtained $Y_{50} = Y_{50}(X)$ values (X = sorafenib concentration, $1\ \mu M \leq X \leq 10\ \mu M$) ~10 % below the additive (linear) (X, Y_{50}) -plot. Furthermore, the cardio-protective effect of the editor should become apparent as NRVM sarcolemmal integrity and ATP-levels are restored to normal values.

Alternatively, the sorafenib/SUDE combination suppressed RCC proliferation to sorafenib-alone levels at identical overall concentrations. This dose-dependent inhibition fits the Loewe additive scheme [20, 21] indicating therapeutic reinforcement without synergy. In NRVMs, SUDE inhibits MST2, hence it should block the pro-apoptotic pathway triggered by sorafenib (Fig. 12.3b). The detailed mechanism of the sorafenib-induced activation of MST2 is not known precisely, but it is likely that drug binding to RAF1 impairs its association with MST2 known to inhibit the latter [25]. Furthermore, in the combined sorafenib/SUDE treatment we expect reduction of cytosolic release of cytochrome *c*, restored sarcolemmal integrity and restored ATP-levels all confirming SUDE's role as sorafenib editor.

The clinical relevance of the editing therapy remains to be established. The potential sorafenib-induced cardiotoxicity [5] and the possibility of antagonist interference exploiting WBZ_4 or SUDE hints at the therapeutic value of this mode of side-effect removal. We already established the cardioprotective role of WBZ_4 [13], while that of SUDE is expected to be established on the same premises: antagonistic interference on a pro-apoptotic pathway recruited by an upstream kinase inhibition. A decisive test on the restoration of cardiac function through editing therapy requires determining whether the left ventricular ejection fraction (LVEF) decreases significantly upon sorafenib treatment and is restored to normal levels as sorafenib cardiotoxicity is antagonized downstream by WBZ_4 or SUDE.

The editors impact a set of therapeutic targets that is more limited than that of the parent compounds ([13]), probably precluding the editors from serving as therapeutic agents by themselves. This is evident in RCC786-O-proliferation assays. Thus, in contrast with sorafenib effects (Fig. 12.5a), cell growth decreases by a meager 9.5 % and 4.5 % respectively when these cells are treated solely with SUDE or WBZ_4 at 10 μM.

Fig. 12.5 (continued) Y = WBZ_4 concentration in μM, I(X,Y) = observed inhibition for the (X, Y) combination, $I'(X,Y)$ = predicted inhibition according to the Loewe additive scheme [20, 21] which requires the (X, Y) concentration pair to satisfy the linear equation $X/X_{I'} + Y/Y_{I'} = 1$, where $X_{I'}$ = sorafenib concentration yielding inhibition I' and $Y_{I'}$ = WBZ_4 concentration yielding inhibition I'. For clarification, a two-component treatment involving two drugs at respective concentrations X and Y is said to obey Loewe additivity if the concentration-dependent inhibition I(X, Y) is equal to $I'(X, Y)$, the predicted inhibition for the concentration pair (X, Y) that satisfies the linear equation $X/X_{I'} + Y/Y_{I'} = 1$, where $X_{I'}$ and $Y_{I'}$ are the single-agent effective concentrations that produce inhibition I'. Reprinted from [32], copyright 2009 with permission from Elsevier

12.5 Editing Therapy in the Epistructural Biology Era

To assess the applicability breadth of this editing strategy we surveyed signaling pathways presumably causative of side effects and recruited through the action of other anticancer KIs. In an effort to curb side effects through therapeutic editing, we explored the editing potential of WBZ_4, exploiting its selective activity against tyrosine kinases and its additional impact on a downstream kinase (JNK). We predict that WBZ_4 could also serve as editor of primary tyrosine KIs like imatinib (Fig. 12.6) and dasatinib [31]. WBZ_4 is a potential editor because it inhibits JNK, a downstream non-redundant transducer of pro-apoptotic pathways recruited by imatinib or dasatinib inhibition of ABL in cardiomyocytes (Figs. 12.1 and 12.6). Moreover, WBZ_4 overlaps therapeutically with imatinib and dasatinib in cancer cells on the clinical targets PDGFR and KIT [13]. The synergies of a possible imatinib/WBZ_4 treatment are described in Fig. 12.6.

To summarize, therapeutic editors may become a realistic engineering possibility as we harness the molecular factors controlling drug specificity (cf. Chap. 9).

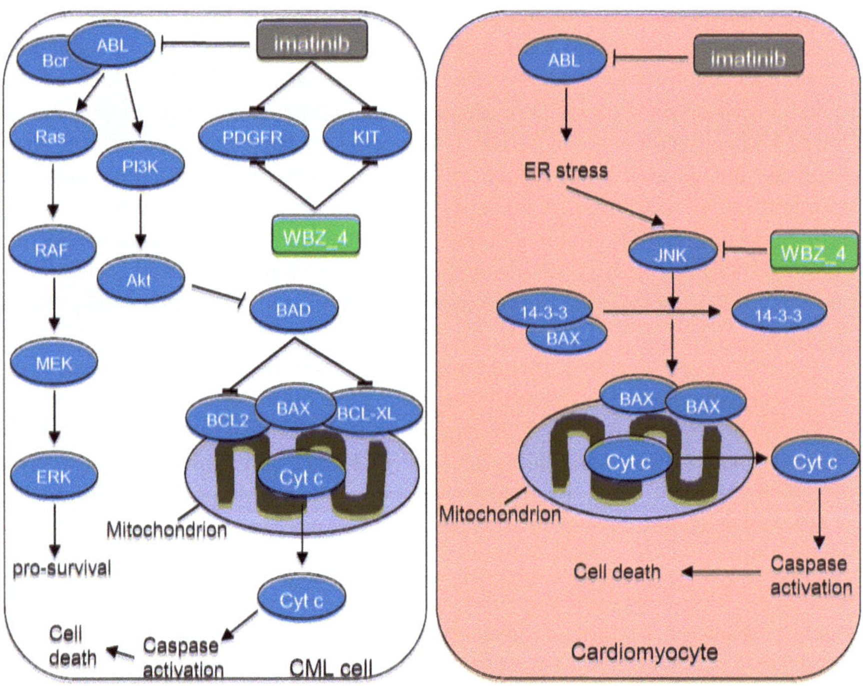

Fig. 12.6 Network level analysis of the proposed editing therapy involving imatinib/WBZ_4 combination treatment. In the CML cells imatinib and WBZ_4 overlap therapeutically by inhibiting common clinically relevant targets PDGFR and KIT, whereas in cardiomyocytes, the inhibitory action of WBZ_4 on JNK, antagonizes the pro-apoptotic pathways activated by upstream inhibition of ABL by imatinib. Reprinted from [32], copyright 2009 with permission from Elsevier

Thus, the success of the editing examples described depends pivotally on the fact that the high-throughput screening of the engineered kinase inhibitors fit the modeling predictions and the drug synergies found experimentally [32].

The therapeutic editor is a systemic concept whose clinical impact will ultimately be unraveled as the *in vitro* and *in vivo* assays described pave the way to clinical trials. This development may encounter difficulties arising, for instance, from differences in bioavailability, half-life, pharmacodynamics and pharmacokinetics between primary drug and editor. Thus, the engineering of adequate delivery vehicles and possibly even drug redesign to tune these parameters will become an essential component of the editing strategy in the advanced stages of development.

12.6 Editing Out Resistance to Inhibition of Drug Resistance

The focal adhesion kinase (FAK) is a 125-kDa non-receptor protein tyrosine kinase overexpressed in many tumors, including those derived from the head and neck, colon, breast, prostate, liver, and thyroid [33]. Furthermore, FAK is also overexpressed in most ovarian cancers and is clearly predictive of poor clinical outcome [33]. FAK has been shown to play a significant role in cancer cell survival, migration, and invasion and its overexpression is typically indicative of cancer metastasis and of the onset of the invasive phenotype. Decreased FAK phosphorylation by the dominant negative FAK-related non-kinase resulted in reduced migration and invasion of ovarian cancer cells *in vitro* [33]. FAK overexpression has also been suggested to protect cells from stressors such as chemotherapy by activation of the phosphoinositide-3-kinase–AKT survival pathway and induction of inhibitor-of-apoptosis proteins. In the specific context of ovarian cancer treatment, FAK is cleaved after treatment with docetaxel chemotherapy in a caspase-3-dependent manner, and that FAK down-regulation promoted docetaxel cytotoxicity in ovarian cancer cells [33], enhancing the therapeutic efficacy of chemotherapy. Moreover, *in vivo* FAK silencing using small interfering RNA (siRNA) resulted in antitumor and antiangiogenic effects in animal models of ovarian carcinoma. In both studies, FAK silencing/inhibition enhanced the effects of chemotherapy. Based on the known role of FAK in ovarian cancer pathogenesis and the encouraging results with FAK gene silencing, there is compelling interest in developing novel and powerful FAK inhibitors. As we shall show in detail in Chap. 17, epistructure-based drug design plays a central role in the development of such anti-FAK therapeutics. Since we shall be optimizing FAK inhibitor TAE226 (Pfizer) [33] exploited as a lead to improve safety and to enhance therapeutic impact, we should also be concerned with the other on-target kinases of therapeutic value for TAE226 (Fig. 12.7a): JNK [34] and IGF1R (Chap. 6, Fig. 6.5), both well recognized cancer targets, especially relevant to the ovarian context. We shall name the epistructure-based TAE226-optimization FAK/JNK/IGF1R-wrapper. The

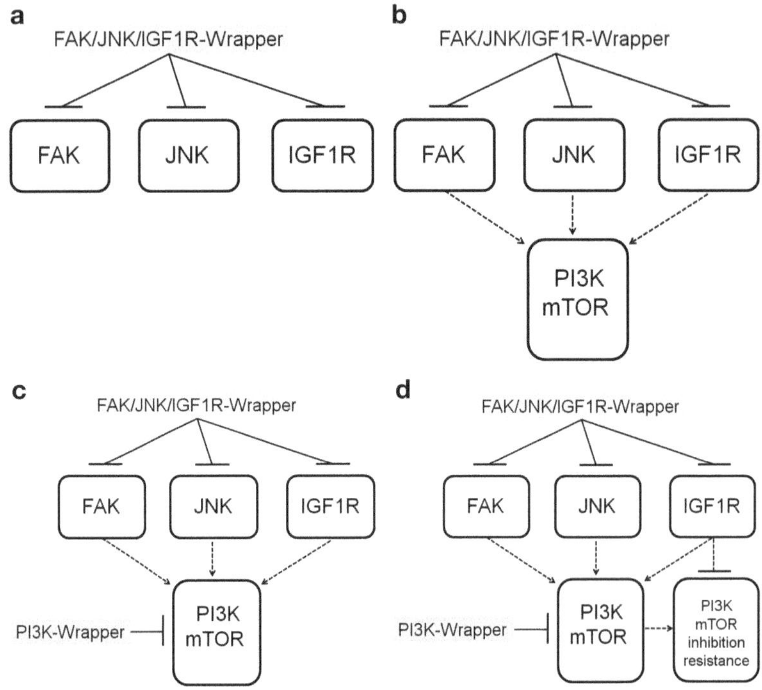

Fig. 12.7 (**a**) Primary inhibitory impact of FAK/JNK/IGFR-Wrapper, a safer and optimized version of TAE226 where off-target kinases (INR, LCK, etc.) have been dialed out through epistructural filtering (Chap. 17). (**b**) Drug Resistance to FAK/JNK/IGFR-Wrapper treatment is somatically articulated through recruitment of the PI3K/mTOR pathway. (**c**) A second drug, consisting of an optimized PI3K inhibitor, named PI3K-Wrapper (Chap. 17), exerts an inhibitory impact on the drug resistance mechanism. (**d**) Inhibition of IGF1R by the first drug exerts an editing effect upon the second drug since IGF1R inhibition abrogates resistance to PI3K-Wapper-based inhibition of the PI3K/mTOR drug-resistance mechanism

efficacy of a drug treatment based on FAK inhibition, like that of any other drug treatment, is ultimately greatly undermined by drug resistance, articulated through the recruitment of pathways like the PI3K/mTOR axis [35], as shown in Fig. 12.7b–d. As we shall see in Chap. 17, it is possible to develop an optimized safe inhibitor of the PI3K/mTOR drug-resistance pathway, hereby named PI3K-wrapper. However, in late stages of cancer progression, it is likely that even resistance to inhibition of drug resistance may develop. Yet, fortunately FAK/JNK/IGF1R-Wrapper may act as editor of PI3K-Wrapper because the former's inhibitory impact on IGF1R abrogates resistance to inhibition of PI3K/mTOR-based drug resistance [36], as shown in Fig. 12.7d. This case illustrates the power of combined drug therapies when an *editing mechanism identified by systems biology annotation enables the primary drug to edit out drug resistance mechanisms elicited by the second drug that is itself an inhibitor of drug resistance towards the first drug.*

12.7 Curing Cancer by Editing Out Mechanisms to Cope with Metabolic Stress

Since the days of biochemist Otto Warburg, it has been noted that cancer cells have much higher energetic demands than normal cells, and this makes them vulnerable to metabolic stress. Hence, the sensitivity of cancer cells to sustained decreases in the AMP/ATP ratio is now widely established [37] and recent therapeutic strategies are attempting to harness this fact. In this regard, the type-2 diabetes drug *metformin* is been carefully evaluated as a therapeutic anticancer agent due to its inhibitory impact on the mitochondrial respiratory chain, suppression of hepatic glucose production and decrease of the insulin-induced suppression of fatty acid oxidation. It is now well established that metformin inhibits the electron transport chain complex 1, thus becoming a major exogenous disruptor of energy homeostasis in the cell.

Yet, the efficacy of metformin as an anticancer agent *per se* is heavily compromised by its agonistic action on the kinase AMPK, a major regulator of energy homeostasis. The functionally competent AMPK complex is allosterically activated by AMP and inhibited by ATP, thus becomes a sensor of energetic imbalance. When activated, AMPK inhibits anabolic processes, such as fatty acid synthesis and protein synthesis, while promoting catabolic pathways for fatty acid oxidation and glucose transport through vesicle → plasma membrane translocation of GLUT4. Thus, increase in AMPK activation levels helps the cell to cope with metabolic stress. In this context, AMPK becomes clearly an oncogene, in sharp contrast with the tumor suppressor role often attributed to it [37]. On the other hand, metformin is an AMPK agonist simply because it alters the AMP/ATP ratio but has no direct effect on AMPK. *This agonistic role severely impairs the efficacy of metformin as an anticancer agent since it agonizes the metabolic switch that enables the cancer cell to cope with the metabolic stress that metformin induces.*

Ergo, to position metformin, or simply intense physical exercise, as anticancer agents acting by imposing metabolic stress on the cancer cell we need to edit out the agonistic effect exerted by these agents upon AMPK. In other words, we need to combine metformin or intense physical exercise with an AMPK inhibitor that would act as a therapeutic editor by suppressing the stress-coping mechanism triggered by the allosteric activation of AMPK. This combination strategy is likely to work, since it is well established that the loss of energy homeostasis leads to apoptosis in cells lacking AMPK [37].

There are several leads that may be potentially redesigned to turn them into selective AMPK inhibitors. The most interesting lead is TAE-684 [30], a powerful AMPK and FAK inhibitor whose value as therapeutic editor is severely compromised due to its cross reactivity against the insulin receptor (InsR) kinase. It would be necessary to exploit the selectivity filters provided by epistructural analysis to dial out the anti-InsR activity of TAE-684 while retaining or possibly enhancing its inhibitory impact against AMPK and FAK through wrapping-based redesign.

12.8 Problems

12.1. Evaluate the potential of compounds CEP-701 and R406 as therapeutic editors of *imatinib* in regards to removing its cardiotoxicity. The kinome-wide screening of the two compounds is presented in Table 12.1. The affinities are given in nM units.

Table 12.1 Kinome-wide high-throughput screening of compounds CEP-701 and R406

Kinase	CEP-701	R406
AAK1	3.1	410
ABL1(E255K)-phosphorylated	840	220
ABL1(F317I)-nonphosphorylated		
ABL1(F317I)-phosphorylated	4900	
ABL1(F317L)-nonphosphorylated	7600	
ABL1(F317L)-phosphorylated	2200	
ABL1(H396P)-nonphosphorylated	1000	170
ABL1(H396P)-phosphorylated	870	390
ABL1(M351T)-phosphorylated	780	390
ABL1(Q252H)-nonphosphorylated	2500	630
ABL1(Q252H)-phosphorylated	830	430
ABL1(T315I)-nonphosphorylated	950	670
ABL1(T315I)-phosphorylated	160	190
ABL1(Y253F)-phosphorylated	910	1800
ABL1-nonphosphorylated	3200	530
ABL1-phosphorylated	1000	340
ABL2	6500	720
ACVR1	75	4800
ACVR1B	86	1100
ACVR2A	2500	
ACVR2B	3800	
ACVRL1	2200	
ADCK3		1100
ADCK4		64
AKT1	4400	
AKT2	2100	
AKT3	6200	
ALK	71	240
AMPK-alpha1	25	300
AMPK-alpha2	56	750
ANKK1	32	330
ARK5	3.7	75
ASK1	70	

(continued)

Table 12.1 (continued)

Kinase	CEP-701	R406
ASK2	170	
AURKA	16	160
AURKB	16	110
AURKC	4.1	41
AXL	35	82
BIKE	16	1900
BLK	160	25
BMPR1A	950	
BMPR1B	57	720
BMPR2	1500	
BMX	200	180
BRAF		240
BRAF(V600E)		380
BRK	1700	360
BRSK1	25	
BRSK2	16	
BTK	66	190
CAMK1	130	230
CAMK1D	47	200
CAMK1G	470	460
CAMK2A	10	1100
CAMK2B	59	
CAMK2D	8.8	3000
CAMK2G	26	920
CAMK4	3000	
CAMKK1	23	
CAMKK2	15	
CASK	330	
CDC2L1		
CDC2L2		
CDC2L5	1000	
CDK11		
CDK2	20	
CDK3	420	
CDK4-cyclinD1	30	
CDK4-cyclinD3	35	
CDK5	410	
CDK7	87	7700
CDK8	8000	
CDK9	530	
CDKL1	490	2100
CDKL2	43	5900

(continued)

Table 12.1 (continued)

Kinase	CEP-701	R406
CDKL3	430	
CDKL5	10	
CHEK1	2.8	2500
CHEK2	210	
CIT	85	440
CLK1	9.6	1200
CLK2	44	1600
CLK3		1900
CLK4	3	360
CSF1R	110	31
CSK	260	160
CSNK1A1		
CSNK1A1L	7100	
CSNK1D		
CSNK1E		
CSNK1G1		
CSNK1G2		
CSNK1G3		
CSNK2A1	2600	64
CSNK2A2	970	14
CTK	970	
DAPK1	80	55
DAPK2	63	62
DAPK3	82	13
DCAMKL1		
DCAMKL2	4500	
DCAMKL3	1100	600
DDR1	510	9.3
DDR2	400	43
DLK	840	
DMPK	240	
DMPK2	2500	
DRAK1	3.4	32
DRAK2	5	9.2
DYRK1A	20	260
DYRK1B	29	
DYRK2	22	470
EGFR	1100	300
EGFR(E746-A750del)	400	1100
EGFR(G719C)	1200	
EGFR(G719S)	1100	
EGFR(L747-E749del, A750P)	320	260

(continued)

Table 12.1 (continued)

Kinase	CEP-701	R406
EGFR(L747-S752del, P753S)	1400	560
EGFR(L747-T751del,Sins)	560	340
EGFR(L858R)	860	250
EGFR(L858R,T790M)	7.8	61
EGFR(L861Q)	800	190
EGFR(S752-I759del)	1400	450
EGFR(T790M)	16	72
EIF2AK1		
EPHA1	380	67
EPHA2	790	140
EPHA3	160	360
EPHA4	1700	150
EPHA5	5900	860
EPHA6	1200	120
EPHA7	260	39
EPHA8		1300
EPHB1	5200	390
EPHB2	6900	880
EPHB3		
EPHB4	6900	600
EPHB6	98	310
ERBB2		
ERBB3	5500	
ERBB4	1400	600
ERK1	9800	
ERK2		
ERK3		
ERK4		
ERK5	1100	
ERK8	4.9	640
ERN1	5.7	1600
FAK	1100	250
FER	28	130
FES	370	200
FGFR1	310	44
FGFR2	660	73
FGFR3	580	56
FGFR3(G697C)	730	67
FGFR4	1800	350
FGR	52	33
FLT1	120	16
FLT3	8.5	0.71

(continued)

Table 12.1 (continued)

Kinase	CEP-701	R406
FLT3(D835H)	0.66	0.65
FLT3(D835Y)	0.57	0.37
FLT3(ITD)	1.5	0.54
FLT3(K663Q)	5.4	0.68
FLT3(N841I)	1.4	1.5
FLT3(R834Q)	7	4.4
FLT4	17	9.1
FRK	740	68
FYN	84	28
GAK	7.3	250
GCN2(Kin.Dom.2,S808G)	4.1	3.3
GRK1	24	870
GRK4	12	450
GRK7	7.2	2000
GSK3A	2300	2500
GSK3B	1300	120
HCK	270	150
HIPK1	590	820
HIPK2	370	4800
HIPK3	250	3200
HIPK4	150	2800
HPK1	38	1400
HUNK	570	
ICK	39	
IGF1R	4200	1300
IKK-alpha	9	120
IKK-beta	12	71
IKK-epsilon	11	13
INSR	1100	1100
INSRR	1800	4600
IRAK1	6.1	9.7
IRAK3	16	11
IRAK4	1.7	140
ITK	290	400
JAK1(JH1domain-catalytic)	8.8	21
JAK1(JH2domain-pseudokinase)		
JAK2(JH1domain-catalytic)	3.7	3.5
JAK3(JH1domain-catalytic)	2.3	36
JNK1	11	38
JNK2	320	47
JNK3	12	37
KIT	150	6.8

(continued)

Table 12.1 (continued)

Kinase	CEP-701	R406
KIT(A829P)	35	3.5
KIT(D816H)	11	6.5
KIT(D816V)	7.1	2.5
KIT(L576P)	40	5.6
KIT(V559D)	76	5.7
KIT(V559D,T670I)	45	11
KIT(V559D,V654A)	380	410
LATS1	30	
LATS2	1	4000
LCK	430	30
LIMK1	1600	620
LIMK2		1400
LKB1	57	3400
LOK	5.7	70
LRRK2	3.6	36
LRRK2(G2019S)	2.5	27
LTK	490	800
LYN	190	26
LZK	340	
MAK	1700	
MAP3K1	4400	
MAP3K15	3.4	
MAP3K2	4.8	94
MAP3K3	6	150
MAP3K4	2600	840
MAP4K2	4.7	670
MAP4K3	110	2400
MAP4K4	23	
MAP4K5	43	
MAPKAPK2	230	
MAPKAPK5	930	
MARK1	83	270
MARK2	14	280
MARK3	49	1900
MARK4	120	89
MAST1	19	
MEK1	3.5	320
MEK2	3.9	730
MEK3	5	100
MEK4	3.8	200
MEK5	7	41
MEK6	44	480

(continued)

Table 12.1 (continued)

Kinase	CEP-701	R406
MELK	26	
MERTK	32	170
MET	290	280
MET(M1250T)	480	230
MET(Y1235D)	41	760
MINK	3.6	
MKK7	80	
MKNK1	56	230
MKNK2	1.4	130
MLCK	12	
MLK1	3.5	4.3
MLK2	15	3.8
MLK3	18	11
MRCKA	9200	
MRCKB	1000	
MST1	3	
MST1R	8200	470
MST2	2.4	3000
MST3	760	
MST4	230	
MTOR		
MUSK	4100	100
MYLK	13	6500
MYLK2	56	
MYLK4	29	810
MYO3A	260	
MYO3B	770	
NDR1	220	9400
NDR2	370	
NEK1	2500	170
NEK11		170
NEK2	4900	260
NEK3		170
NEK4		1800
NEK5	5600	13
NEK6	4400	6.3
NEK7	4000	11
NEK9	5400	16
NIM1	1800	660
NLK	770	
OSR1	8000	3100
p38-alpha		

(continued)

Table 12.1 (continued)

Kinase	CEP-701	R406
p38-beta		
p38-delta	3500	690
p38-gamma	370	390
PAK1	30	260
PAK2	43	340
PAK3	35	690
PAK4	24	96
PAK6	9.9	310
PAK7	24	69
PCTK1	27	
PCTK2	720	330
PCTK3		
PDGFRA	350	60
PDGFRB	29	3.3
PDPK1	3.4	120
PFCDPK1(*P. falciparum*)	5.6	
PFPK5(*P. falciparum*)	5600	
PFTAIRE2	1300	
PFTK1	720	
PHKG1	0.39	
PHKG2	1.7	940
PIK3C2B	5200	
PIK3C2G	230	
PIK3CA	4900	
PIK3CA(C420R)	4900	
PIK3CA(E542K)	5900	
PIK3CA(E545A)	5600	
PIK3CA(E545K)	7400	
PIK3CA(H1047L)	49	
PIK3CA(H1047Y)	650	
PIK3CA(I800L)		
PIK3CA(M1043I)	1100	
PIK3CA(Q546K)	5400	
PIK3CB	6700	
PIK3CD	4000	4200
PIK3CG	350	
PIK4CB	290	930
PIM1	200	
PIM2	250	
PIM3	18	
PIP5K1A	36	1900
PIP5K1C		130

(continued)

Table 12.1 (continued)

Kinase	CEP-701	R406
PIP5K2B	18	1200
PIP5K2C		120
PKAC-alpha	220	600
PKAC-beta	160	510
PKMYT1		
PKN1	5.3	660
PKN2	1.8	2600
PKNB(*M. tuberculosis*)	55	34
PLK1	220	830
PLK2	330	34
PLK3	2300	5.1
PLK4	1.5	4.2
PRKCD	240	4100
PRKCE	160	
PRKCH	140	
PRKCI	1800	
PRKCQ	100	
PRKD1	14	
PRKD2	8.1	
PRKD3	8.9	3900
PRKG1	370	
PRKG2	25	3600
PRKR	120	180
PRKX	42	
PRP4	550	
PYK2	41	29
QSK	120	
RAF1		840
RET	20	4.1
RET(M918T)	14	4.1
RET(V804L)	6.7	2.2
RET(V804M)	4.7	3.1
RIOK1	9	340
RIOK2	6700	3400
RIOK3	7.7	210
RIPK1		
RIPK2	640	970
RIPK4	13	220
RIPK5	5.3	
ROCK1	56	590
ROCK2	71	560
ROS1	26	29

(continued)

Table 12.1 (continued)

Kinase	CEP-701	R406
RPS6KA4(Kin.Dom.1-N-terminal)	62	
RPS6KA4(Kin.Dom.2-C-terminal)	1300	44
RPS6KA5(Kin.Dom.1-N-terminal)	42	
RPS6KA5(Kin.Dom.2-C-terminal)	960	150
RSK1(Kin.Dom.1-N-terminal)	28	490
RSK1(Kin.Dom.2-C-terminal)	2400	3800
RSK2(Kin.Dom.1-N-terminal)	290	8000
RSK3(Kin.Dom.1-N-terminal)	88	
RSK3(Kin.Dom.2-C-terminal)	1200	
RSK4(Kin.Dom.1-N-terminal)	110	
RSK4(Kin.Dom.2-C-terminal)	9100	
S6K1	69	
SBK1	200	93
SgK110	120	190
SGK3	5.8	
SIK	82	53
SIK2	15	78
SLK	35	33
SNARK	1	65
SNRK		
SRC	640	8.4
SRMS	820	610
SRPK1	10	590
SRPK2	51	
SRPK3	13	400
STK16	19	1.7
STK33	26	970
STK35	810	1400
STK36	3800	230
STK39	1100	
SYK	9.5	19
TAK1	2.9	49
TAOK1	70	490
TAOK2	500	640
TAOK3	86	4600
TBK1	18	23
TEC	1200	1600
TESK1	290	360
TGFBR1	120	2600
TGFBR2	540	330
TIE1	680	76
TIE2	630	310

(continued)

Table 12.1 (continued)

Kinase	CEP-701	R406
TLK1	110	490
TLK2	64	380
TNIK	5.4	1300
TNK1	9.3	25
TNK2	6.6	70
TNNI3K	2100	310
TRKA	8.3	170
TRKB	14	900
TRKC	14	570
TRPM6		
TSSK1B	98	1000
TTK	6.5	1100
TXK	660	140
TYK2(JH1domain-catalytic)	15	190
TYK2(JH2domain-pseudokinase)	910	
TYRO3	650	1900
ULK1	1000	14
ULK2	60	19
ULK3	4.9	7.1
VEGFR2	220	40
VRK2		
WEE1	1200	890
WEE2	7900	9400
YANK1	2100	2300
YANK2	5900	
YANK3	8000	
YES	160	35
YSK1	97	9200
YSK4	0.52	67
ZAK	6700	
ZAP70	16	21

The profiling was performed and reported by AMBIT Biosciences (KINOME scan™, [30]). Affinities are given in nM units

References

1. Widakowich C, de Castro G, de Azambuja E, Dinh P, Awada A. Review: side effects of approved molecular targeted therapies in solid cancers. Oncologist. 2007;12:1443–55.
2. Schmidinger M, Zielinski CC, Vogl UM, et al. Cardiac toxicity of sunitinib and sorafenib in patients with metastatic renal cell carcinoma. J Clin Oncol. 2008;26:5204–12.
3. Lacouture ME. Mechanisms of cutaneous toxicities to EGFR inhibitors. Nat Rev Cancer. 2006;6:803–12.

4. Verheul HM, Pinedo HM. Possible molecular mechanisms involved in the toxicity of angiogenesis inhibition. Nat Rev Cancer. 2007;7:475–85.
5. Force T, Krause D, van Etten RA. Molecular mechanisms of cardiotoxicity of tyrosine kinase inhibition. Nat Rev Cancer. 2007;7:332–44.
6. Force T, Kerkela R. Cardiotoxicity of the new cancer therapeutics—mechanisms of, and approaches to, the problem. Drug Discov Today. 2008;13:778–84.
7. Kerkela R, Grazette L, Yacobi R, et al. Cardiotoxicity of the cancer therapeutic agent imatinib mesylate. Nat Med. 2006;12:908–16.
8. Chu TF, Rupnick MA, Kerkela R, et al. Cardiotoxicity associated with tyrosine kinase inhibitor sunitnib. Lancet. 2007;370:2011–9.
9. Fernández A, Crespo A, Tiwari A. Is there a case for selectively promiscuous anticancer drugs? Drug Discov Today. 2009;14:1–5.
10. Zhang X, Crespo A, Fernández A. Turning promiscuous kinase inhibitors into safer drugs. Trends Biotechnol. 2008;26:295–301.
11. Dancey J, Sausville EA. Issues and progress with protein kinase inhibitors for cancer treatment. Nat Rev Drug Discov. 2003;2:296–313.
12. Levitzki A, Gazit A. Tyrosine kinase inhibition: an approach to drug development. Science. 1995;267:1782–8.
13. Fernández A, Sanguino A, Peng Z, Ozturk E, Chen J, Crespo A, Wulf S, Shavrin A, Qin C, Ma J, Trent J, Lin Y, Han HD, Mangala LS, Bankson JA, Gelovani J, Samarel A, Bornmann W, Sood AK, Lopez-Berestein G. An anticancer c-kit kinase inhibitor is reengineered to make it more active and less cardiotoxic. J Clin Invest. 2007;117:4044–54.
14. Kitano H. A robustness-based approach to systems-oriented drug design. Nat Rev Drug Discov. 2007;6:202–9.
15. Torchilin VP. Multifunctional nanocarriers. Adv Drug Deliv Rev. 2006;58:1532–55.
16. Langer R. Drug delivery: drugs on target. Science. 2001;293:58–9.
17. Allen TM, Cullis PR. Drug delivery systems: entering the mainstream. Science. 2004;303:1818–22.
18. Deninger M, Buchdunger E, Druker BJ. The development of imatinib as a therapeutic agent for chronic myeloid leukemia. Blood. 2005;105:2640–53.
19. Dancey JE, Chen HX. Strategies for optimizing combinations of molecularly targeted anticancer drugs. Nat Rev Drug Discov. 2006;5:649–59.
20. Keith CT, Borisky AA, Stockwell BR. Multicomponent therapeutics for networked systems. Nat Rev Drug Discov. 2005;4:71–8.
21. Zimmermann GR, Lehar J, Keith CT. Multi-target therapeutics: when the whole is greater than the sum of the parts. Drug Discov Today. 2007;12:34–42.
22. Verweij J, Casali P, Kotasek D, et al. Imatinib does not induce cardiac left ventricular failure in gastrointestinal stromal tumours patients: analysis of EORTC-ISG-AGITG study 62005. Eur J Cancer. 2007;43:974–8.
23. Atallah E, Durand JB, Kantarjian H, Cortes J. Congestive heart failure is a rare event in patients receiving imatinib. Blood. 2007;110:1233–7.
24. Escudier B, Eisen T, Stadler WM, et al. Sorafenib in advanced clear-cell renal-cell carcinoma. N Engl J Med. 2007;356:125–34.
25. O'Neill E, Rushworth L, Baccarini M, Kolch W. Role of the kinase MST2 in suppression of apoptosis by the proto-oncogene product Raf-1. Science. 2004;306:2267–70.
26. Dhillon AS, Hagan S, Rath O, Kolch W. MAP kinase signalling pathways in cancer. Oncogene. 2007;26:3279–90.
27. Chen J, Zhang X, Fernández A. Molecular basis for specificity in the druggable kinome: sequence-based analysis. Bioinformatics. 2007;23:563–72.
28. Faivre S, Demetri G, Sargent W, Raymond E. Molecular basis for sunitinib efficacy and future clinical development. Nat Rev Drug Discov. 2007;6:734–45.
29. O'Neill EE, Matallanas D, Kolch W. Mammalian sterile 20-like kinases in tumor suppression: an emerging pathway. Cancer Res. 2005;65:5485–7.

30. Karaman MW, Herrgard S, Treiber DK, et al. A quantitative analysis of kinase inhibitor selectivity. Nat Biotechnol. 2008;26:127–32.
31. Kantarjian H, Jabbour E, Grimley J, Kirkpatrick P. Dasatinib. Nat Rev Drug Discov. 2006;5:717–8.
32. Fernández A, Sessel S. Selective antagonism of anticancer drugs for side effect removal. Trends Pharmacol Sci. 2009;30:403–10.
33. Halder J, Lin YG, Merritt WM, Spannuth WA, Nick AM, Honda T, Kamat AA, Han LY, Kim TJ, Lu C, Tari AM, Bornmann W, Fernandez A, Lopez-Berestein G, Sood AK. Therapeutic efficacy of a novel focal adhesion kinase inhibitor TAE226 in ovarian carcinoma. Cancer Res. 2007;67:10976–83.
34. Vivas-Mejia P, Benito JM, Fernández A, Han HD, Mangala L, Rodriguez-Aguayo C, Chavez-Reyes A, Lin YG, Carey MS, Nick AM, Stone RL, Kim HS, Claret FX, Bornmann W, Hennessy BT, Sanguino A, Peng Z, Sood AK, Lopez-Berestein G. c-Jun-NH2-kinase-1 inhibition leads to antitumor activity in ovarian cancer. Clin Cancer Res. 2010;16:184–94.
35. Martz CA, Ottina KA, Singleton KR, Jasper JS, Wardell SE, Peraza-Penton A, Anderson GR, Winter PS, Wang T, Alley HM, Kwong LN, Cooper ZA, Tetzlaff M, Chen PL, Rathmell JC, Flaherty KT, Wargo JA, McDonnell DP, Sabatini DM, Wood KC. Systematic identification of signaling pathways with potential to confer anticancer drug resistance. Sci Signal. 2014;7: ra121.
36. Muranen T, Selfors LM, Worster DT, Iwanicki MP, Song L, Morales FC, Gao S, Mills GB, Brugge JS. Inhibition of PI3K/mTOR leads to adaptive resistance in matrix-attached cancer cells. Cancer Cell. 2012;21:227–39.
37. Liang J, Mills GB. AMPK: a contextual oncogene or tumor suppressor? Cancer Res. 2013;73:2929–35.

Chapter 13
Epistructure-Based Design of Drugs with Controlled Promiscuity

The complex etiology of cancer has inspired drug developers to seek multi-pronged molecular targeted therapies. The one-disease-one-target paradigm, illustrated by chronic myeloid leukemia (CML) and the chimeric product Bcr-Abl, a resultant of aberrant chromosomal translocation, is representative of the exception rather than the rule in cancer therapeutics. On the other hand, multi-drug treatments, with their diverse and often incompatible pharmacodynamics and pharmacokinetics profiles, do not offer attractive therapeutic possibilities for a multi-pronged attack, except in the context of therapeutic editing and reciprocal therapeutic editing delineated in Chaps 12 and 17, respectively. Fueled by the output of systems biology, the burgeoning interest in multi-target drugs is motivating a re-assessment of the therapeutic value of controlled promiscuity. While drug efficacy may not correlate with specificity, it would be risky to take the opposite stance and welcome promiscuous compounds indiscriminately, without a rational strategy to funnel their therapeutic impact. In this chapter we survey epistructure-based approaches to control the therapeutic impact of multi-target kinase inhibitors to fulfill the therapeutic imperatives of human cancer. Thus, we advocate for the application of wrapping design to turn multi-target kinase inhibitors into clinical opportunities through judicious chemical modifications.

13.1 Is Cancer Systems Biology Calling for Controlled Promiscuity in Molecular Targeted Therapy?

The need to reduce toxic side effects obviously makes specificity of small-molecule drugs a desired attribute in molecularly targeted therapy [1–5]. The side effects, arising from drug cross-reactivity and from the diverse roles played by the target protein in off-target cellular contexts, prompted researchers to advocate for the "magic-bullet" paradigm [6], epitomized by compounds with high binding

A. Fernández, *Physics at the Biomolecular Interface*, Soft and Biological Matter,
DOI 10.1007/978-3-319-30852-4_13

specificity. However, this paradigm faces serious challenges since there is no obvious correlation between drug specificity and therapeutic index (lethal dose (LD50) over therapeutically effective dose (ED50)) [6–11]. Thus, the causal complexity of cancer, with its manifold and heterogeneous scenarios for the unfolding and progression of the disease, clearly advocates for a multi-pronged treatment often translated into a multi-target molecular therapy.

Be as it may, welcoming "dirty drugs" as a means to fulfill the therapeutic imperatives of human cancer entails a high level of risk, and is obviously not the paradigm shift that most researchers in the field have in mind. In principle, multi-target drugs are more prone to trigger side effects than their more specific counterparts. Thus, while there is a perception that a shift from the magic bullet paradigm is needed, there is no clue on what the right replacing concept should be. Here we advocate for controlled multi-targeting, a concept that may be implemented by assessing the degree of conservation of dehydrons across paralogs. At first sight, this last sentence may seem puzzling, as we have shown that dehydron patterns are not conserved while the overall folded structure across homologous proteins tends to be conserved. Yet, this lack of conservation of the overall dehydron pattern is relative, and there is a whole hierarchy of dehydron conservation within the wrapping pattern itself. Some dehydrons are common to a subset of paralogs and hence, targeting them may ensure a controlled multi-prong treatment. The whole issue is whether this dehydron-guided impact is the desired one vis-à-vis the therapeutic imperative that one is addressing.

Uncertain clinical outcome resulting from health-threatening side effects is a central issue for therapies based on multi-target kinase inhibitors since kinases play different signal-transduction roles in different cellular contexts and a favorable inhibition in one scenario may prove fatal in another [12]. For instance, a specific inhibitor of the Abelson (Abl) kinase may be assumed to be most efficacious in treating chronic myeloid leukemia (CML), since an aberrantly deregulated Abl kinase is a recognized primary target for treating this malignancy [5]. However, as extensively described in Chaps 8 and 9, a systems biology assessment [12] has recently revealed that Abl inhibition is also a likely culprit of cardiotoxicity: Abl inhibition recruits a signaling cascade that leads to mitochondrial depolarization, ATP-depletion and ultimately apoptosis in cardiomyocites [12]. Thus, a more cross reactive drug, i.e., one that *also* inhibits another kinase along the pro-apoptotic pathway triggered by Abl inhibition (hence suppressing it) is expected to have a higher therapeutic efficacy in treating CML [13]. In this regard, an inhibitor that also impacts JNK would be highly desirable, as described in Chaps 9 and 11. The rationale for this assertion is that higher therapeutic doses may be tolerable due to the removal of cardiotoxicity, *albeit* at the expense of sacrificing specificity [13–15]. One possible downturn of such an approach is that JNK inhibition may introduce uncertain clinical outcomes in off-target cells (i.e. cells other than cardiomyocytes). Given the downstream position of JNK in most signal transduction cascades, this is not likely to be the case, as JNK may be replaced by alternative and redundant signal transducers in other cellular contexts.

The causal complexity of cancer and other diseases has motivated much effort directed at reassessing the therapeutic value of promiscuity. This conceptual shift is in part motivated by telling cases. For instance, multi-target kinase inhibitors such as *Sutent* (sunitinib) or *Nexavar* (sorafenib) have recently received FDA approval, *albeit* with important caveats highlighting their side effects, especially their cardiotoxicity [7, 12]. Thus, the possibility of exploiting drug promiscuity as a therapeutic opportunity is being considered in novel approaches to treat complex disorders such as cancer, depression and cardiovascular disease [7]. The simultaneous modulation of multiple targets is often required to alter a clinical phenotype as biological redundancies and alternative pathways can often bypass the inhibition of a single target or of multiple targets along a single pathway [16]. Thus, in contrast with the magic bullet, a "magic-shotgun" targeting multiple proteins may in some instances possess a higher therapeutic index than a specific drug [6].

Another argument favoring promiscuity arises with the observation that highly cross-reactive drugs may be more resilient against drug-resistant mutations [17]. Cross reactivity arises because, upon association with the target protein, such drugs typically make interactions with evolutionarily conserved residues and with backbone groups, and less interactions with the residues that mutate to confer drug resistance [17, 18]. Targeting conserved residues or unspecific parts of the chain (backbone) begets promiscuity.

The safety of less selective drugs, and especially that of promiscuous kinase inhibitors, is often addressed by assessing the full extent of their cross reactivity [19]. This assessment has been facilitated by novel high-throughput screening assays such as: (a) the kinase assay using the T7-phage expression panels from Ambit Biosciences [20, 21]; (b) a thermal stability shift assay using a 60-Ser/Thr kinase panel [22]; and (c) the living-cell assays of pathway inhibition that assess the impact of the drug on the protein-recruiting capability of the target [23].

In Vitro assays of cross reactivity are affected by the complexities of tissue distribution and, generally, by pharmacodynamic issues [24]. Yet, target affinity, being governed by thermodynamics of ligand binding, represents a telling parameter independent of *in vivo* heterogeneities, except for allosteric antagonism. On the other hand, crowding, membrane adsorption and other effects can modulate *in vivo* drug concentrations, increasing local levels in different spatial locations. Thus, certain cross-reactivities undetectable *in vitro* may surface *in vivo*, causing unexpected side effects [24, 25]. Conversely, tissue and subcellular distribution may prevent a ligand from binding an *in-vitro*-established target, introducing another caveat in the interpretation of high-throughput screening results.

In spite of timely efforts to establish a paradigm shift, promiscuous drugs obviously carry a higher risk of inducing side effects than their more specific counterparts. Even the successful anticancer drug *Gleevec* (imatinib), with a reduced spectrum of primary targets, i.e. the kinases Bcr-Abl, C-Kit, Lck, PDGFR, and CSF1R [20, 21], has been recently shown to be potentially cardiotoxic [12, 26], and labeled as such by the FDA (see: http://www.fda.gov/medwatch/safety/2006/Gleevec_DHCP_10-19-2006.htm). Not surprisingly, the more promiscuous anticancer kinase inhibitors sunitinib and sorafenib are also cardiotoxic, and to a larger extent than imatinib [12, 26]. In Ambit screenings [20] sunitinib was shown to bind 79 kinases out of 119 assayed, while sorafenib binds to 41.

Hence, it is forbiddingly risky to welcome promiscuous compounds into the therapeutic arena without a rational strategy to control their specificity and therapeutic index. Such control may be achieved if we can identify features in the target structure that are unique to the target, i.e. "selectivity filters", and drug modifications that promote interactions with such nonconserved features. Thus, selectivity filters may serve as guidance to rational drug design [13–15, 27–29]. We advocate for this type of control of therapeutic impact to clean less selective drugs following an integral assessment of the diverse functional roles of targeted proteins.

13.2 Epistructure-Based Control of Drug Promiscuity

As described in Chaps 6 and 9, most protein drug targets have paralogs, that is, proteins that share a common ancestor with the target and have diverged away from it after speciation [30]. Thus, kinases belong to common-ancestry groups (families) which typically share the same fold and basic structural features that often yield unexpected cross reactivities [20, 21, 31].

In principle, much of this cross reactivity could be removed by drug redesign guided by the identification of the common structural features that promote promiscuity and the nonconserved features that enable paralog discrimination. As shown in Chap. 9, this approach has been successfully attempted on the cross-reactive drug imatinib with promising results [13–15]. Its further utilization to clean truly promiscuous drugs requires further analysis, as described in the next section, and will enable us to propose its extended use to clean kinase inhibitors.

The evolutionary latitudes to differentiate or group paralogs may be exploited in drug design and may be traced to the response to the selection pressure imposed on duplicate genes depending on the level of wrapping of the parental gene. Thus, a gene coding for a very under-wrapped product (large proportion of dehydrons) will have high dosage sensitivity because it is highly reliant on associations with other proteins coded by other genes to maintain its structural integrity, whereas a gene with well-wrapped products will have a low dosage sensitivity and the paralogs will be under low pressure to differentiate. As shown in Chap. 6, gene duplicates with high dosage sensitivity will tend to differentiate significantly, exploiting various evolutionary possibilities while conserving the protein fold: (a) Wrapping; (b) Transcriptional pattern; and (c) Post-transcriptional pattern (microRNA targeting). Thus, to control drug cross reactivity, one can of course exploit the wrapping pattern similarities and differences across paralogs. Furthermore, when targeting more than one paralog through common dehydrons, researchers may exploit the paralog differences at the transcriptional and post-transcriptional level to ensure that inhibition of an on-target protein in an off-target tissue does not occur (Fig. 13.1a).

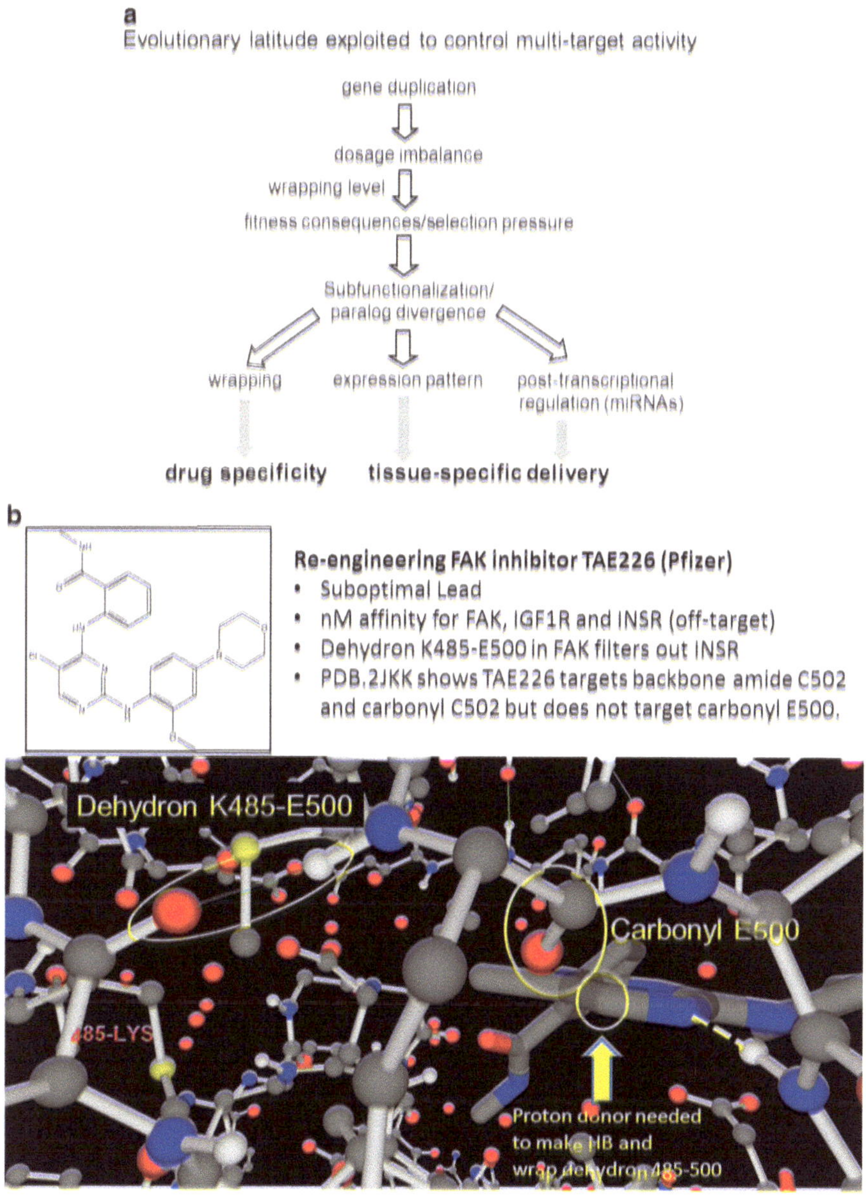

Fig. 13.1 (**a**) Evolutionary latitudes for paralog differentiation arising after the gene duplication event that may be exploited to control drug cross reactivity on a selected subset of paralog proteins. (**b**) Epistructural selectivity filter (FAK dehydron K485-E500, PDB.2JKK) exploited to clean suboptimal compound TAE226 in order to turn it into a therapeutic agent to treat cancer metastasis

A number of paralog-discriminating nonconserved features different from wrapping have been identified and, although not amenable of systematic application, some can be exploited as selectivity filters [27, 28]. One approach uses high-resolution crystal structures of kinases in complex with non-ATP ligands to identify unique structural motifs in the purported targets. For example, a unique helical insert has been found in the activation loop of the NEK2 and MPSK1 kinases adjacent to the highly conserved DFG catalytic triad [27].

Specific inhibitors that bind to the inactive kinase conformations have also been developed targeting the 'DFG out' conformation [32]. In this unique conformation, the position of the phenylalanine (F) residue which is located at the start of the activation loop, is flipped with respect to the active conformation, so that it points inwards within the ATP pocket. This mode of association has been confirmed by the binding of imatinib to the inactive Abl kinase [33]. In addition to the DFG-out motif, nonconserved structural features within the inactive ensemble should perhaps be exploited to achieve paralog specificity. The inactive conformations of kinases make them more discernible, while the active conformation reveals fewer discriminatory features since it is constrained to be catalytically functional and hence more conserved. The substrate-discriminatory amino acid variations that define the different kinase functions are mostly located in loopy regions framing the ATP-pocket, rather than in the pocket itself, making them less accessible targets. While targeting the inactive conformation may be a logical choice, there are also advantages in targeting the active conformation. The latter requires structure conservation for functional purposes, and hence it is less tolerant to drug-resistant mutations. Thus, a kinase that binds to a drug in the active conformation will seldom develop drug-resistant mutations since the latter can compromise the activity.

The selectivity filter advocated throughout this book is in principle of universal validity, as shown in Chap. 10, and may be used systematically. It arises from comparison of the wrapping patterns of protein targets [30]. The wrapping pattern may be turned into an operational selectivity filter for two reasons: (a) dehydrons may be targeted by drugs that further wrap them by bringing nonpolar groups to their proximity upon association; and (b) the overall dehydron pattern is **not** conserved across paralogs, as shown in Chap. 10. Thus, differences in dehydron distribution constitute promising selectivity filters to clean promiscuous inhibitors, whereas conservation of specific dehydrons across paralogs enables us to group them using wrapping drugs.

For example, the Pfizer compound TAE226 (Chap. 17) has shown promising activity in treating cancer metastasis though a blockade of the focal adhesion kinase (FAK) pathway, recognized in abundant system biology annotation as a primary target to block the onset of the invasive phenotype. The compound also shows nanomolar (nM) activity against primary anticancer target IGF1R (insulin-like growth factor 1 receptor) kinase. However, TAE226 has not been found suitable as a therapeutic agent due to its nanomolar activity against an IGF1R kinase paralog, the InsR (insulin receptor) kinase. The latter activity induces a diabetic coma in the patient, fatally compromising the therapeutic value of

TAE226. However, it is possible to apply the dehydron selectivity filter in order to dial out anti-InsR activity through an epistructure-based re-engineering of TAE226, as described in Chap. 17. As shown in Fig. 13.1b, structural alignment of the TAE226/FAK complex (PDB.2JKK) with a PDB-based epistructure for InsR kinase reveals that FAK dehydron K485-E500 constitutes an appropriate selectivity filter since it becomes a well-wrapped hydrogen bond at the aligned position in the InsR kinase. Thus, an expansion of the chemical structure of TAE226 to turn the ligand into a wrapper of the the K485-500 dehydron would clean TAE226, funneling its therapeutic impact solely on desirable targets such as FAK and IGF1R.

13.3 Cleaning Dirty Drugs with Dehydron Filters

In principle, most kinase inhibitors are susceptible of being turned into selective wrappers of packing defects through minor chemical modification and without altering their chemotype. Thus, clinically relevant compounds with considerable cross reactivities such as sunitinib (inhibiting 55 % of the 290 kinases screened [21]), dasatinib (30 %), EKB-569 (20 %), sorafenib (20 %), erlotinib (15 %), gefitinib (6 %) and imatinib (4 %) may be redesigned into drugs with enhanced specificity using the wrapping design concept. The generic strategy consists of modifying the parental compound to turn it into a wrapper of unique dehydrons while removing potential sources of cross reactivity. The latter arise since chemical groups in the ligand often interact with conserved structural features in the protein such as backbone moieties or side chains that remain invariant within the target family. The wrapping strategy may in principle be applied to enhance the specificity of promiscuous drugs like sunitinib, dasatinib, EKB-569, sorafenib and erlotinib towards their respective primary targets: KIT/VEGFR2, ABL/SRC, EGFR, VEGFR2, EGFR [21].

To illustrate the operational value of this approach, we focus on redesigning a truly promiscuous drug, EKB-569 (Wyeth-Ayerst, [34]) to significantly enhance its selectivity. This drug was launched as a major inhibitor of the epidermal growth factor receptor (EGFR) kinase (IC50 = 38.5 nM). It gained major therapeutic interest for the treatment of non-small cell lung cancer (NSCLC), colorectal neoplasia and other EGFR-dependent solid tumors. Phase I and II trials for such therapeutic applications are currently in progress and closed to new patients [34, 35]. Preliminary high-throughput screening using a battery of 119 T7-phage expressed kinases [20] revealed 25 sub-micromolar targets for EKB-569, making it a promiscuous drug with likely side effects. Other compounds such as *Iressa* (gefitinib) and *Tarceva* (erlotinib) share the same "4-anilinoquinoline" chemotype [36], yet they are more specific EGFR inhibitors [20]. The latter two gained FDA approval as anti-NSCLC agents.

EKB-569 promiscuity can be traced to its intermolecular interactions with highly conserved residues within the EGFR kinase family. As shown in

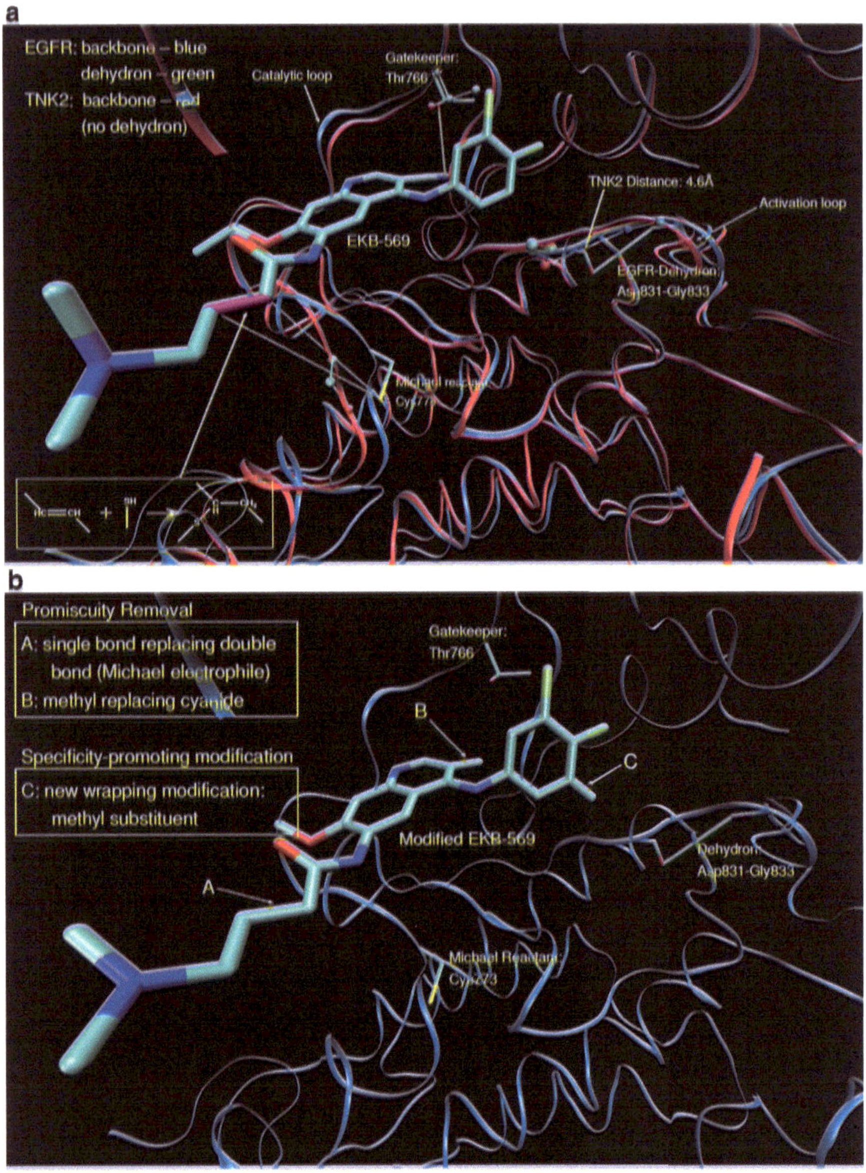

Fig. 13.2 Structural features promoting promiscuity and selectivity in EGFR kinase guiding EKB-569 cleaning redesign. (**a**) Structural alignment of EGFR kinase (*blue* ribbon representation, atoms in licorice) and the paralog TNK2 kinase (*red* ribbon representation, atoms in balls and sticks), complexed with EKB-569 (licorice). Atoms are depicted following standard color convention (chlorine in *green*, fluorine in *light green*). The wrapping pattern of EGFR includes the poorly conserved Asp831-Gly833 dehydron that may be targeted to achieve selectivity since TNK2 lacks this dehydron. TNK2 contains the same two promiscuity-fostering features, while lacking the dehydron at the locus where EGFR contains the specificity-promoting feature. Thus, targeting the latter will ensure a discriminatory binding of EGFR without hitting TNK2. (**b**) EGFR

Fig. 13.2a, the terminal acryl group in the ligand is a source of promiscuity because it reacts with the *conserved* residues Cys/Ser in the EGFR paralogs, acting as electrophile in an irreversible Michael adduct. Another source of EKB-569 promiscuity is the intermolecular electrostatic interaction between its cyanide group and the gatekeeper residue (Thr or Met), typically conserved within the family (Fig. 13.2a).

Thus, we may remove these sources of promiscuity by introducing the following modifications:

(a) Replace the double bond (the Michael acceptor) in the acryl group with a single bond.
(b) Replace the cyanide group with a methyl to retain the chemotype while removing the electrostatic interaction with the conserved gatekeeper.

To promote selectivity, we may introduce a wrapping modification to target a non-conserved dehydron in the primary target EGFR. A comparative structural analysis of the EGFR family reveals that the least conserved EGFR dehydron is Asp831-Gly833. To wrap this dehydron, we may append a methyl group at position 3 of the terminal benzene ring (Fig. 13.2b). This wrapping modification is likely to possess inhibitory activity *exclusively* against the EGFR paralogs that retain the dehydron in its original position: EGFR, Aurora2 (STK6), CLK3, ERBB2, PIM1, PIM2, PTK2 (FAK), SLK, STK10. The modified drug is expected to bind a subset of this target group since steric hindrances may arise as a result of the chemical medication. We thus provide a rational approach to clean the promiscuous inhibitor EKB-569 based on a selectivity filter, reducing its inhibitory spectrum from 25 confirmed targets [20] to only 9.

To validate this design strategy, we established a correlation between the affinities of EKB-569 for the 38 paralogs of EGFR reported in PDB and the extent of residue conservation at the Michael reaction site and at the gatekeeper position. To do so, we aligned each paralog structure with the EGFR structure and examined residues that align with Cys773 (Michael reactant) and Thr766 (gatekeeper). A statistical model was built to assess such correlation, and revealed that the EKB-569 affinity is indeed dictated by these two sources of promiscuity (P-value = 0.0003). Thus, the terminal acryl group and the cyanide group in EKB-569 (Fig. 13.2a) are indeed the "dirty" moieties responsible for promiscuity.

Fig. 13.2 (continued) kinase structure (same representation as above) complexed with the prototype EKB-569 re-designed inhibitor (licorice representation). To reduce EKB-569 promiscuity, the acrylic double bond (Michael electrophile) is replaced by a single bond and the gatekeeper-interacting cyanide is replaced by a methyl. To selectively target EGFR, a methyl group is added to the terminal benzene ring as a wrapper of the barely conserved Asp831-Gly833 dehydron. (**c**) Retro-synthetic pathway of the re-designed EKB-569 inhibitor. The synthesis recapitulates the EKB-569 pathway, albeit with different reactants [36]. Reprinted from [37], copyright 2008 with permission from Elsevier

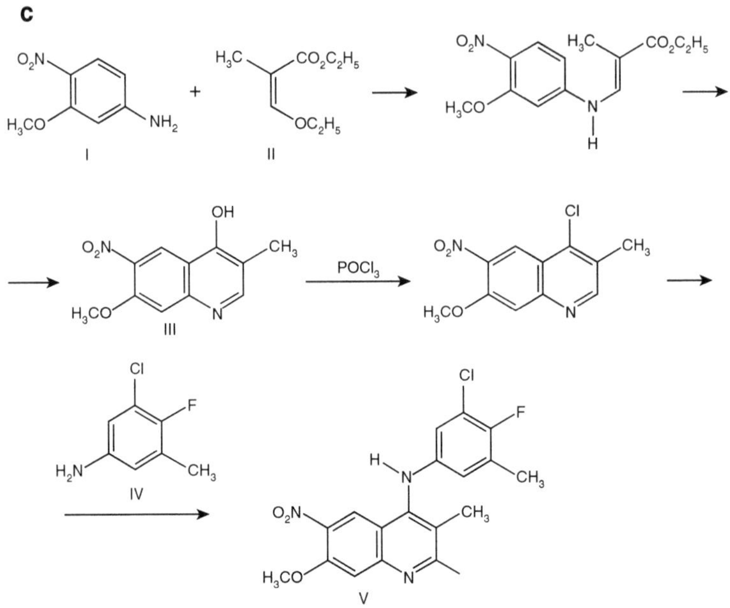

I: 4-nitro-3-methoxyaniline
II: ethyl(ethoxymethylene)methylacetate
III: 7-methoxy-4-hydroxy-6-nitroquinoline-3-methyl
IV: 3-methyl-4-fluoro-5-chloro-aniline
V: 4-(3-methyl-4-fluoro-5-chlorophenylamino-7-methoxy-6-nitroquinoline-3-methyl

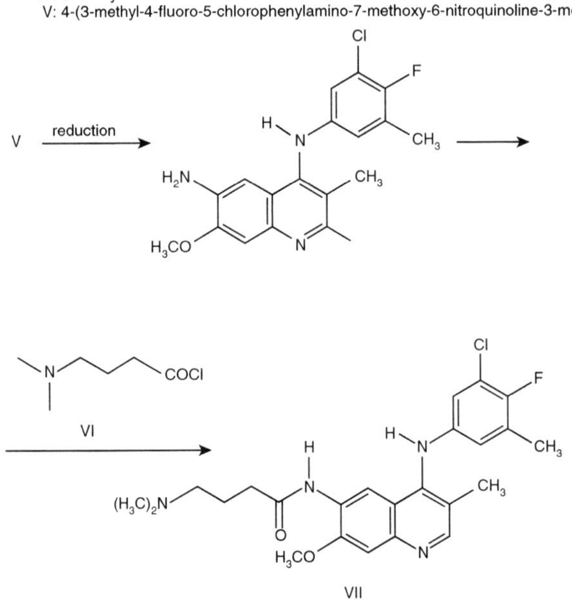

VI: 4-dimethylamino-butanoyl-chloride
VII: 4-dimethylamino-butanoic-acid-[4-(5-chloro-4-fluoro-3-methylphenylamino)
-3-methyl-7-methoxy-quinoline-6-yl]-amide

Fig. 13.2 (continued)

Of the four dehydrons within the binding pocket of EGFR: Asp831-Gly833, Gly697-Phe699, Gly697-Gly700, Gly772-Val821. By examining the conservation of these four dehydrons across the 38 EGFR-paralogs, we found the least conserved is dehydron Asp831-Gly833, retained only in the 9 paralogs mentioned above. Thus, we choose this dehydron as the nonconserved selectivity feature to be targeted. We then appended a methyl group at position 3 of the terminal benzene ring that acts as a wrapper or protector of such feature (Fig. 13.2b). This re-designed EKB-569 molecule was synthesized following a pathway that recapitulates the EKB-569 synthesis [37], as shown in Fig. 13.2c.

Based on the selectivity filter provided by the wrapping patterns, we predict the affinity profile of our prototype. The prediction is based on the conservation of the EGFR dehydron Asp831-Gly833 wrapped by the prototype (*but not by EKB-569*) and the existence of steric hindrances with nonpolar groups in the catalytic loop of the target structures. We predicted as "hits" only those targets with a conserved dehydron in such position and no steric hindrance. Only six hits are predicted (Table 13.1): CLK3, EGFR, ERBB2, PTK2 (FAK), SLK, STK10. In the cases where the residues aligning with EGFR's Asp831-Gly833 are not engaged in a dehydron or a well-wrapped hydrogen bond, we examined whether such dehydron can be *induced* upon ligand binding with minimal structural adaptation (this issue is further discussed in Chap. 11). Only in BTK and LCK kinases this dehydron can be induced upon drug binding with no steric hindrance with the catalytic loop, so they represent other "possible" hits (Table 13.1). The experimentally obtained affinity profile (Fig. 13.3) for the prototype agrees almost perfectly with our predicted profile, since the drug actually binds only to the six hits inferred with certainty and the two likely hits.

In this exercise, we have successfully cleaned the dirty inhibitor EKB-569 using an approach based on a selectivity filter. This was accomplished by first removing the chemical features that promote promiscuity. Subsequently, we introduced a wrapping modification to target a non conserved dehydron in the intended target and made the prototype more selective than the parental compound: out of the 211 kinases experimentally screened [17, 25], our compound binds at nM levels to 6 kinases only, whereas EKB-569 binds to 19.

13.4 Controlling *Staurosporine* Promiscuity Through the Dehydron Filter

In a more stringent test, staurosporine, the most promiscuous kinase inhibitor known to man [21] was re-engineered to elicit an inhibitory impact with controlled and enhanced specificity [38]. Staurosporine binds ($K_D < 3$ μM) to ~90 % of the 290 human kinases recently screened [21]. To pursue a structure-based "cleaning" of staurosporine through a wrapping modification of the ligand, we compared all 28 PDB structures of kinases in complex with the ligand. A comparative wrapping

Table 13.1 Predicted and experimental affinity profiles for the "clean version" of kinase inhibitorpEKB-569

Kinase	Dehydron conservation	Steric hindrance	Predicted affinity	Experimental affinity
ABL1	–		–	–
ACK1(TNK2)	–		–	–
Aurora2 (AURKA)	+	–	–	–
Aurora3 (AURKC)	–		–	–
BTK	Possibly induced	+	Possible hit	+
CDK2	–		–	–
CDK5	–		–	–
CLK1	–		–	–
CLK3	+	+	+	+
CSK	–		–	–
DAPK2	Possibly induced	–	–	–
DAPK3	–		–	–
EGFR	+	+	+	+
EPHA2	–		–	–
ERBB2	+	+	+	+
FGFR1	Possibly induced	–	–	–
FGFR2	–		–	–
HCK	–		–	–
INSR	–		–	–
JAK2	–		–	–
JNK1	–		–	–
JNK3	–		–	–
KIT	–		–	–
LCK	Possibly induced	+	Possible hit	+
MKNK2	–		–	–
NEK2	–		–	–
PAK1	–		–	–
PAK4	–		–	–
PAK6	–		–	–
PAK7(PAK5)	Possibly induced	–	–	–
PIM1	+	–	–	–
PIM2	+	–	–	–
PTK2(FAK)	+	+	+	+
RPS6KA5	–		–	–
SLK	+	+	+	+
SRC	–		–	–
STK10	+	+	+	+

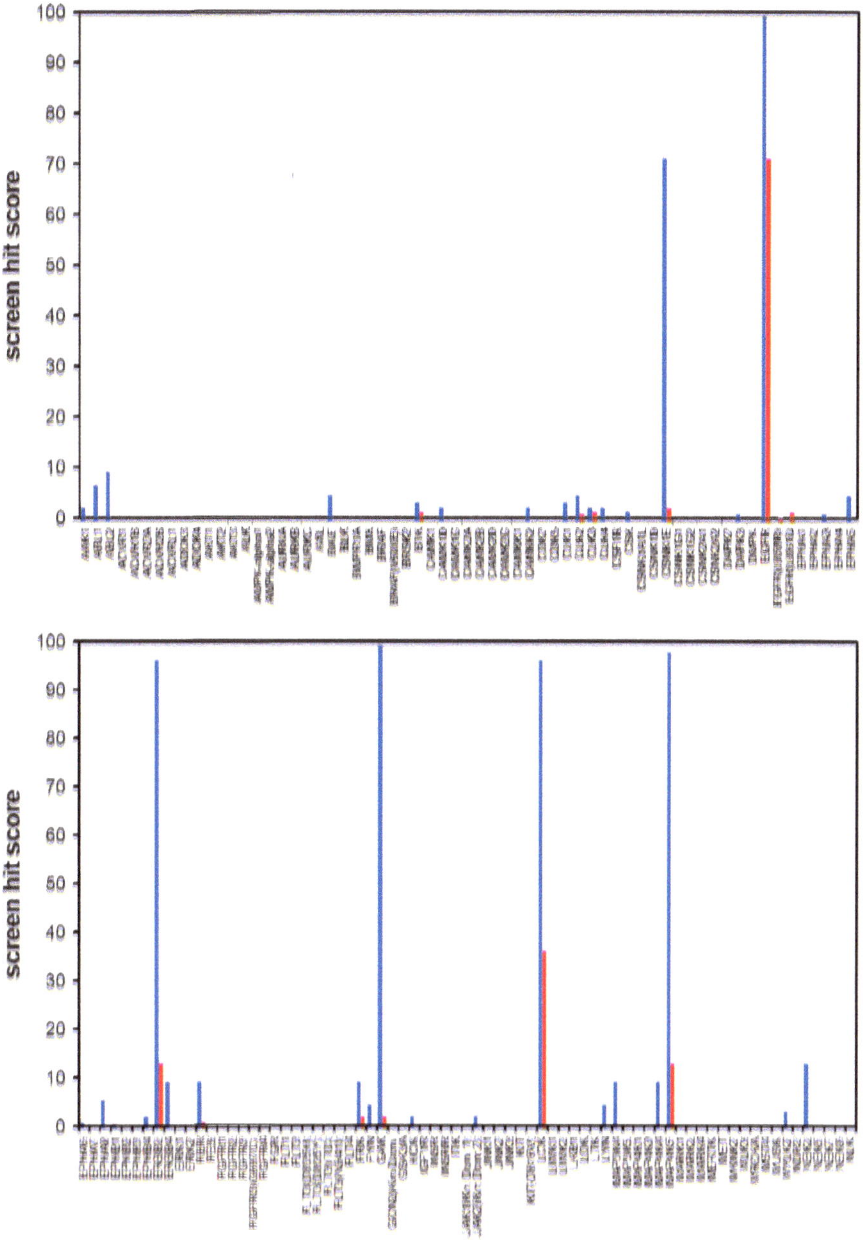

Fig. 13.3 Affinity profile of the re-designed EKB-569 prototype inhibitor. High-throughput screening at 10 mM EKB-569 (*blue*) and prototype (*red*) over a battery of 228 human kinases displayed in a T7-bacteriophage-expressing library. Hit values are reported as percentage bound kinase. Reprinted from [37], copyright 2008 with permission from Elsevier

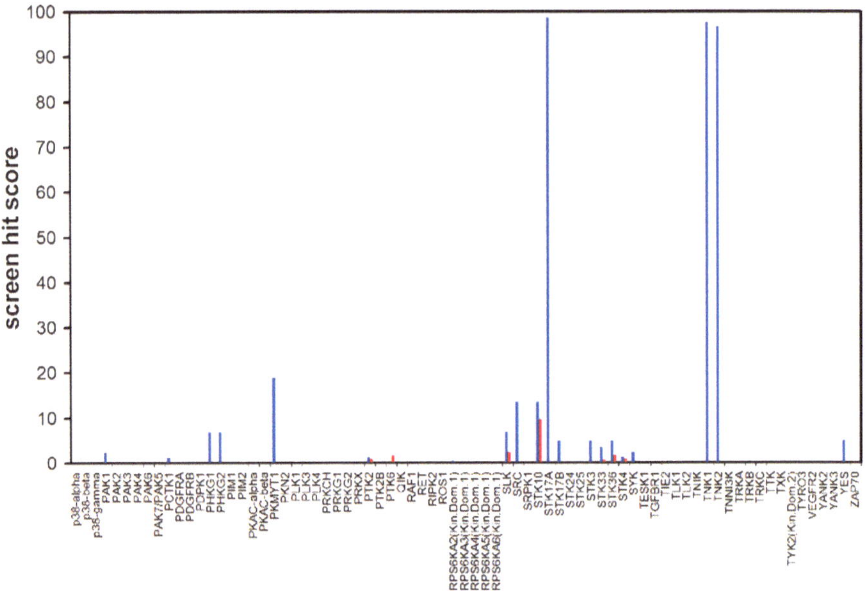

Fig. 13.3 (continued)

analysis of their respective aligned ATP-pockets prompted us to focus on the Src dehydron Q250-E267 which is the least conserved dehydron in this structural assortment. This dehydron is conserved in 7 out of 28 kinase structures, SRC, ABL, LCK, EGFR, ASK, TAO2 and GSK3, as indicated in Table 13.2 and Fig. 13.4a, b. The dehydron has been targeted by wrapping it through a specific methylation of staurosporine at the imide N6-position of the indole ring (Fig. 13.4a) [38]. Thus, by redesigning staurosporine to turn it into a wrapper of the Src Q250-E267 dehydron, we can significantly restrict its inhibitory impact to the 7 kinases that share the dehydron at the aligned position. The screening of the wrapping derivative of stuarosporine is shown in Fig. 13.5a–c: only ~11 %, or 26 out of the 223 kinases assayed represent a hit, indicating a massive enhancement in specificity. *Most significantly, our structure-based prediction of affinity based on the presence or absence of the dehydron at the aligned position has only one false positive and not a single false negative over 15 instances (Table 13.2) where packing prediction can be contrasted with experiment.* These results demonstrate that the packing differences across kinase homologs may guide molecular design to significantly enhance specificity in a highly controllable manner and even for the most promiscuous drugs.

Table 13.2 Wrapping comparison of PDB-reported kinases in complex with staurosporine at position aligned with Src dehydron Q250-E267

Group[a]	Protein	PDB	Wrapping classification	Hit in screening	Match prediction/ experiment
TK	SRC	1BYG	DH	YES	YES
	ABL	2HZ4	DH	YES	YES
	LCK	1QPD	DH	YES	YES
	EGFR	2ITW (wt)	DH	YES	YES
		2ITQ (G719S)	HB	Not screened	
		2ITU (L858R)	HB	Not screened	
	FYN	2DQF	HB	NO	YES
	JAK3	1YVJ	HB	Not screened	
	SYK	1XBC	HB	NO	YES
	ZAP70	1U59	HB	NO	YES
	ITK	1SNU	HB	NO	YES
		1SM2	HB		
TKL	IRAQ4	2OIC	HB	Not screened	
		2NRY	HB		
AGC	PDK1	1OKY	HB	NO	YES
	PKAα	1STC	HB	NO	YES
	PKCθ	1XJD	HB	Not screened	
STE	ASK (MAP3K5)	2CLQ	DH	NO	NO
	TAO2	2GCD	DH	Not screened	
CAMK	CHK1	1NRV	HB	NO	YES
	MAPKAPK2	1NXK	HB	Not screened	
		2PZY	HB		
	MNK2	2HW7	HB	NO	YES
	DAPK1	1WVY	HB	Not screened	
	PIM1	1YHS	HB	NO	YES
CMCG	CDK2	1AQ1	HB	NO	YES
	GSK3β	1Q3D	DH	Not screened	
Unclassified	MPSK1 (STK16)	2BUJ	No HB	Not screened	

DH dehydron, *HB* well-wrapped hydrogen bond. The wrapping variant is predicted to bind ONLY to the kinases that have a dehydron at the aligned position, and the prediction is contrasted with the experimental screening of the wrapping variant of staurosporine

[a]Group names (Manning et al. 2002): *AGC* containing PKA, PKG, PKC families; *CAMK* calcium/calmodulin-dependent protein kinase; *CK1* Casein kinase 1; *CMGC* containing CDK, MAPK, GSK3, CLK families; *STE* homologs of yeast Sterile 7, Sterile 11, Sterile 20 kinases; *TK* tyrosine kinase; *TKL* tyrosine kinase-like

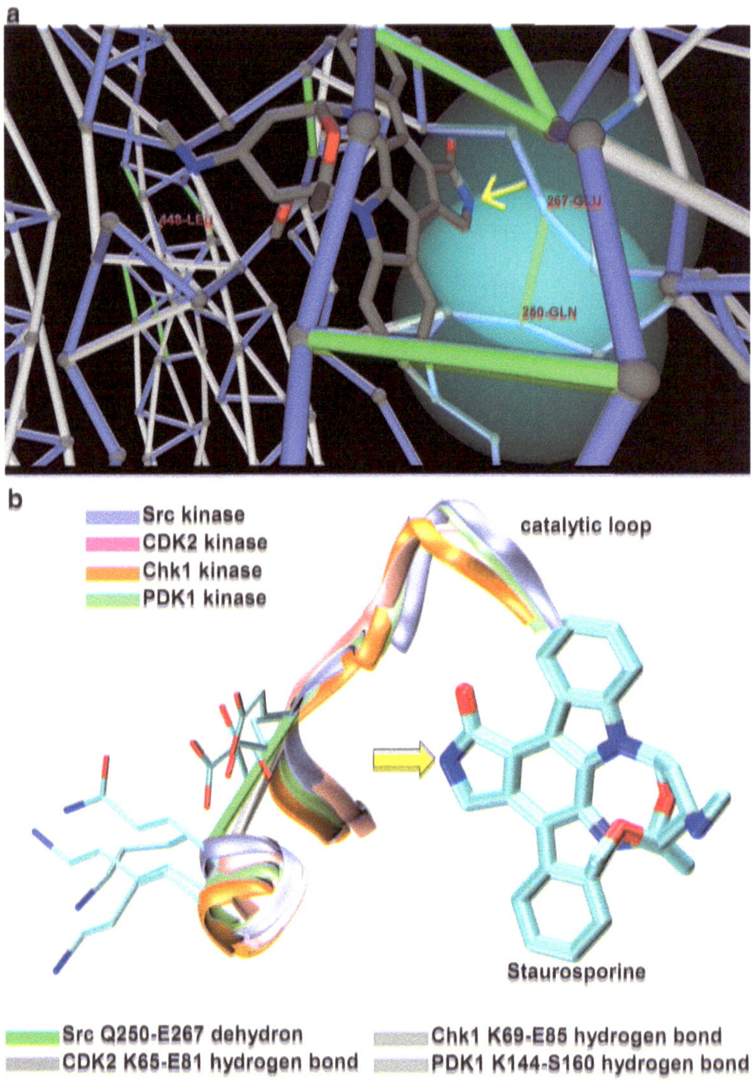

Fig. 13.4 (**a**) Relative position of Src kinase packing defects around the ligand indole region in a staurosporine-kinase complex (PDB.1BYG). Microenvironment of dehydron Q250-E267 in Src kinase framed by the desolvation spheres centered at the α-carbons of Q250 and E267. Methylation at the indole N5-position (indicated by the *yellow arrow*) would turn the ligand into a wrapper of the nonconserved packing defect. (**b**) Selected aligned backbones (ribbon representation) of Src (PDB.1BYG, *ice blue*), CDK2 (PDB.1AQ1, *pink*), Chk1 (PDB.1NVR, *orange*) and PDK1 (PDB.1OKY, *lime*) kinases complexed with staurosporine. The Src Q250-E267 dehydron (*green* virtual bond joining α-carbons) maps into well-wrapped backbone hydrogen bonds (*grey* virtual bond joining α-carbons): K65-E81 in CDK2, K69-E85 in Chk1 and K144-S160 in PDK1

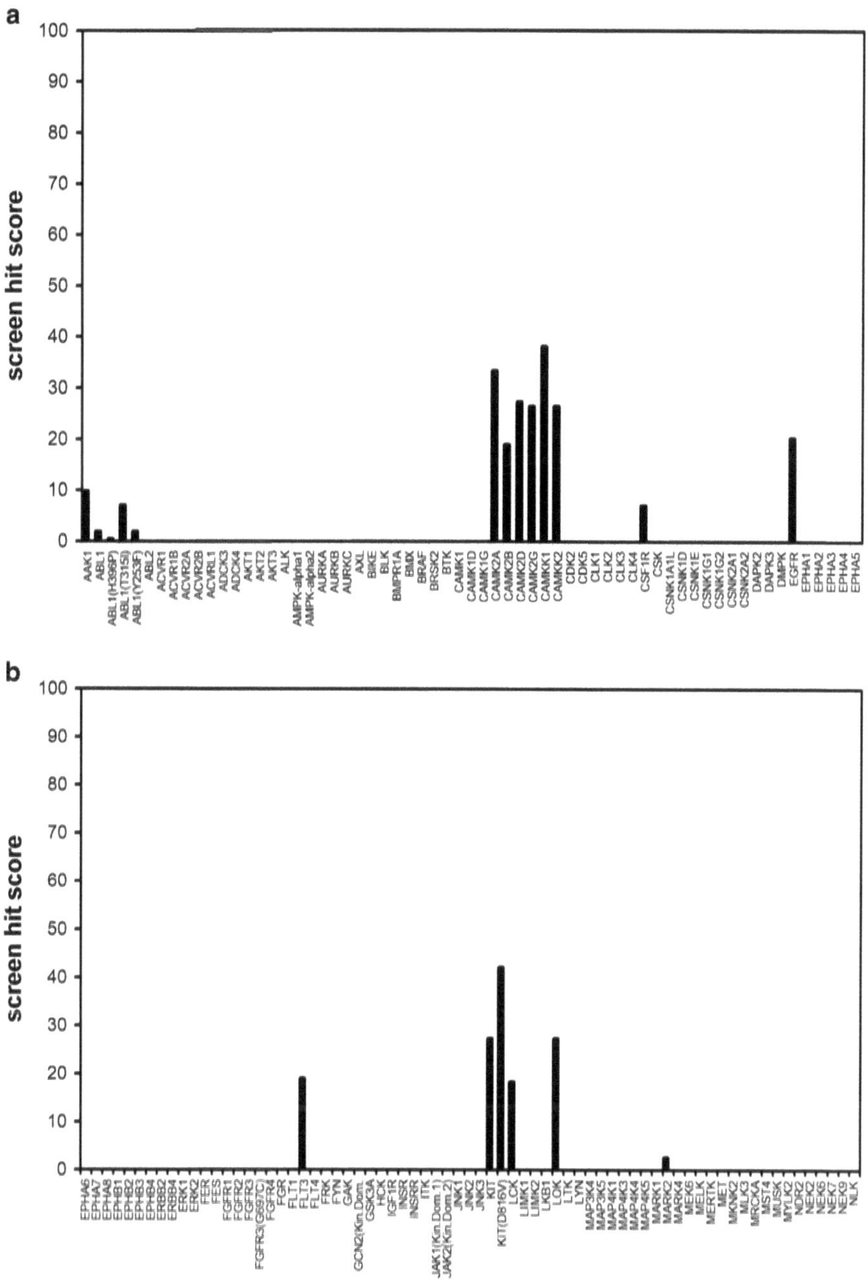

Fig. 13.5 (**a–c**) High-throughput screening at 10 mM for the wrapping variant of staurosporine over a battery of 223 human kinases displayed in a T7-bacteriophage library. Hit values are reported as percentage bound kinase

13.5 Systems Biology Broadens the Platform for Drug Discovery

Systems biology, that is, the multi-scale integration of the functional components of organismic complexity, is likely to broaden the bioinformatics platform of drug discovery [39, 40]. The modulation of clinical targets in living systems remains a central challenge and, for obvious reasons, systems biology is entrusted with providing meaningful approaches to face this challenge [41]. This emerging discipline is expected to bring meaningful insights to the issue of controlled promiscuity in anticancer therapy, yet its practical impact in this arena remains unclear and far from obvious. For all the interest in dirty or promiscuous therapeutic agents, drug discovery is still ingrained on the premise that successful therapeutic agents will result from targeting *single* disease-causing molecules, as the paradigmatic example of imatinib illustrated.

Tumors are perceived to be complex systems whose survival and propagation is a consequence of their resilience to external perturbations [41]. In the context of this discussion, these perturbations materialize as drug-based therapeutic interferences with the cell-signaling apparatus. Thus, when facing a drug-induced node failure arising from the *specific* inhibition of a signal transducer, the tumor is likely to resort to basic robustness-enhancing possibilities that will ensure survival. These options arise especially from the resilient architecture of the signaling network and the genetic instability of the tumor [42], a feature that promotes sufficient target heterogeneity to render most selective drugs useless.

From a systems biology perspective, the features conferring robustness include [42–44]:

- Network modularity to hinder the propagation of effects arising from drug-induced node failure.
- Network redundancy to maintain a biochemical flux in the face of a drug-induced node failure.
- Feedback adaptive controls that ensure the habilitation of alternative pathways as one pathway is blocked by drug action.
- Heterogeneity in the tumor cell population, promoted by genetic instability and ensuring the adaptive survival of a cell subpopulation which defines a drug-resistant trait.

Thus, since the tumor is a masterful survivor, capable of displaying such a gamut of drug-resistant adaptations and of introducing manifold obstacles to molecular therapeutic intervention, it is hard to fathom what would be the advantage of a single-target therapeutic agent. It surely appears that a single drug-induced node failure can be readily overcome by merely resorting to the adaptive resilience of the network itself, unless the node happens to coincide with a point of network fragility. On the other hand, if the latter happens to be the case, the tumor can ultimately resort to its biological heterogeneity as an adaptive property. Thus, the genetic instability of the tumor is likely to promote a heterogeneous cellular population whereby the specific subpopulation that expresses drug-resistant kinase mutants or genetic variants are likely to survive drug treatment [42].

These considerations lead to the perception that an effective drug should target biochemical fluxes embedded in networks rather than specific molecules [45]. This goal may well be operationally unfeasible from the rational perspective of structure-based drug design. Furthermore, abandoning the single-target paradigm plunges us into the daunting challenges of controlling side effects described in the previous sections of this chapter.

The multiplicity of context-dependent roles for a target protein poses a danger of side effects *even in highly selective drugs*, such as the monoclonal antibodies *trastuzumab* (Herceptin, Genentech) [46] and *bevacizumab* (Avastin, Genentech/Roche) [47]. These therapeutic agents targeting respectively ERBB2 and VEGFA kinases are cardiotoxic and exhibit additional side effects like neutropaenia, hemorrhage, hypertension and thromboembolism.

As suggested by the systems biology assessment of robustness, a simultaneous modulation of multiple targets is probably necessary to alter a clinical phenotype as biological redundancies and alternative pathways can often bypass the inhibition of a single target. Thus, in spite of the issues pertaining to the control of side effects, a "magic-shotgun" targeting multiple proteins may in some instances possess a higher therapeutic index than a more specific agent.

As noted previously in this chapter, multi-target drugs may be more resilient against drug-resistant mutations [17], providing another motivation to exploit promiscuity. This assertion readily follows from the fact that cross-reactive drugs typically make interactions with evolutionarily conserved residues and fewer interactions with residues that need to mutate to confer drug resistance.

Even if systems biology advocates the exploitation of multi-target kinase inhibitors, the latter probably entail a higher risk of promoting severe side effects than more specific compounds simply because they target more kinases. Yet, to counterbalance the argument, the highly specific anti-VEGFA bevacizumab introduces multiple and severe toxicities (hemorrhage, hypertension, cardiotoxicity) comparable to sunitinib [12]. In fact, nearly all drug-targeted therapies directed against a constitutively deregulated kinase or a target downstream of such a kinase present side effects. This is likely due to inhibition of "bystander" targets not essential for cancer cell killing or primary targets that may be essential to kill the cancer cell but play crucial roles associated with cell survival in other contexts.

13.6 Taming Sunitinib Promiscuity to Enhance Safety and Therapeutic Efficacy

From a systems biology perspective, the highly promiscuous but powerful anticancer agent sunitinib [48, 49] comes close to the ideal drug. Sunitinib action illustrates the power of multi-targeted kinase inhibitors with a dual pharmacological impact covering tumor-angiogenesis targets VEGFR1-3 and PDGFR and tumor-proliferation/survival targets KIT, CSF1R, FLT3 and RET. This level of promiscuity obviously entails a greater risk of cardiotoxicity due to the likelihood of inducing cardiomyocyte apoptosis through activation of pathways that compromise ATP replenishment or promote mitochondrial dysfunction [12, 50]. Intriguingly, treated patients with advanced GIST and metastatic renal carcinome often reveal significant reduction in left ventricular ejection fraction, a condition enhanced by other sunitinib-induced comorbidities, especially hypertension [50].

Thus, a wrapping-based redesign of sunitinib poses a far more serious challenge than that presented by imatinib (Chap. 9) because it is a far more promiscuous drug, because it targets major cancers (highly resilient solid tumors) and because of its confirmed severe side effects. We may regard the sunitinib redesign to substantially decrease its toxicity as the true proof of principle of the wrapping design concept.

A systems-biology assessment of bystander suntinib targets points to AMPK and RSK inhibition as culprit for cardiotoxicity [12]. Thus, RSK inhibition releases the pro-apoptotic factor BAD with activation of BAX and cytochrome c release, in turn promoting apoptosis and ATP depletion. The inhibition of AMPK promotes ATP depletion by preventing downstream inhibition of EEF2, mTOR and ACC which are normally inhibited by AMPK in energy-compromising settings. In addition, AMPK inhibition affects cardiomyocyte survival if hypoxia occurs, since AMPK inhibits the energy-consuming processes of protein translation and lipid biosynthesis. Finally, the exacerbated activity of mTOR and EEF2 leads to cardiac hypertrophy [12, 50].

The differences in dehydron patterns across targets guide the re-design sunitinib into a discriminating therapeutic agent capable of retaining the inhibitory impact against the clinically relevant sunitinib targets (VEGFR, PDGFR, KIT, CSF1R, FLT3 and RET), while significantly reducing the activity towards the cardiotoxicity-related targets AMPK and RSK. The challenge from a structural biology standpoint is to identify a molecular feature that enables us to tell apart these two sets of targets, and to harness on this difference to construct a therapeutically selective KI.

At first glance, a comparison of the wrapping patterns of primary sunitinib targets through structure alignment leads us to focus on a particular dehydron: the unburied backbone hydrogen bond N900-E917 in the VEGFR kinase, sunitinib's primary antiangiogenic target (Fig. 13.6a, b). This dehydron is conserved in all high-affinity targets relevant to angiogenesis, cancer-cell proliferation and other clinically relevant processes, including PDGFR, CAMK1G, CAMK1D, CSF1R, FLT3, KIT and RET. By contrast, the dehydron aligns with the well-packed hydrogen bond K78-E94 in AMPK2 (Fig. 13.7a, b) and with the well packed hydrogen bond T116-D132 in RSK. *That is, differences in the packing of the region that aligns with dehydron N900-E917 in VEGFR may enable discrimination between clinically relevant and cardiotoxicity-promoting targets.*

Thus, a promising wrapping difference between therapeutic and side-effect related sunitinib targets is apparent as we contrast Figs. 13.6 and 13.7. This wrapping difference suggests possible modifications to sunitinib needed to remove

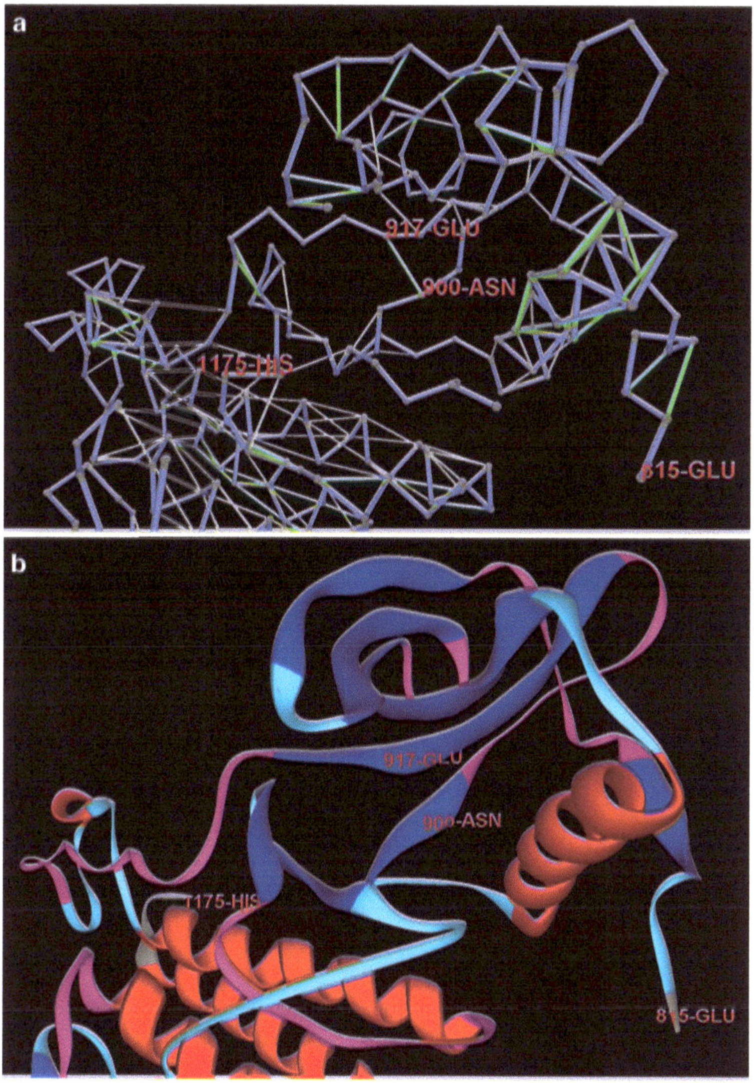

Fig. 13.6 (**a**) Wrapping of VEGFR kinase (PDB.2P2H), in the usual representation with backbone in *blue*, well-wrapped hydrogen bonds in *grey* and dehydrons in *green*. All bonds, including the backbone virtual bonds, are represented as lines joining the α-carbons of the paired residues. The N900-E917 dehydron is a *unique feature* conserved across all sunitinib's clinically relevant targets but aligned with well-wrapped hydrogen bonds in the cardiotoxicity-related targets AMPK and RSK. (**b**) Ribbon representation of VEGFR

its cardiotoxicity and at the same time, probe the potential of AMPK and RSK as cardiotoxicity biomarkers.

According to the premises of the wrapping technology, the methylation at the position indicated in Fig. 13.8 would turn the drug into a selective wrapper of

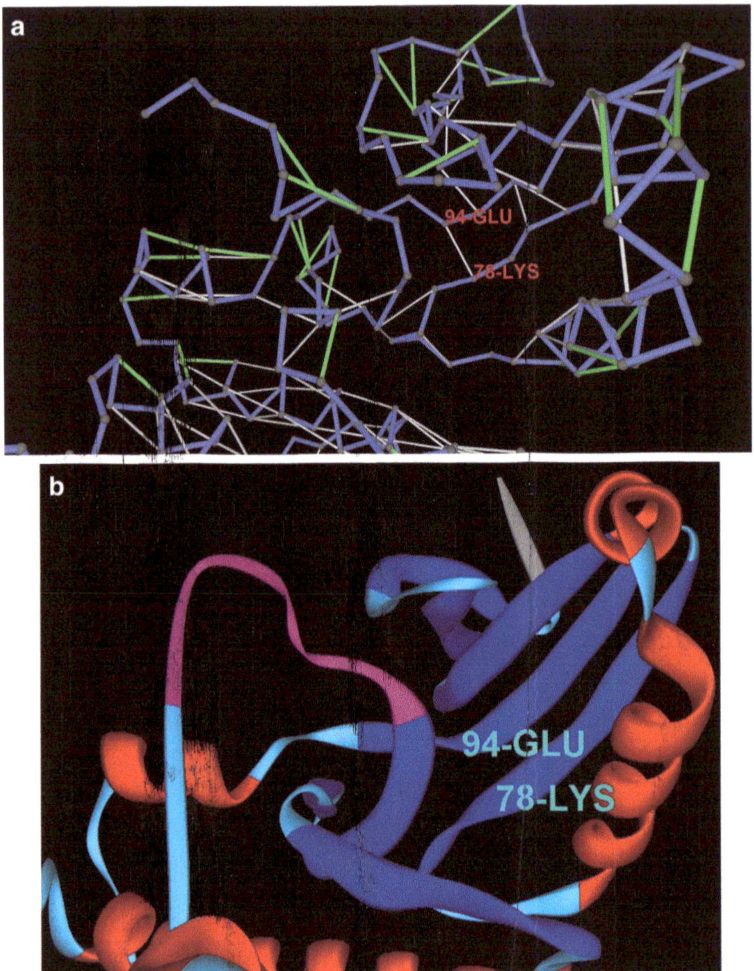

Fig. 13.7 (**a**) Wrapping of AMPK, highlighting the well-wrapped backbone hydrogen bond K78-E94. This bond aligns with a dehydron in all therapeutically relevant sunitinib targets. (**b**) Ribbon representation of AMPK structure

dehydron N900-E917 in VEGFR. This drug is likely to bind selectively to all clinically relevant targets since the latter contain a dehydron at the position that aligns with N900-E917 in VEGFR. By contrast, this wrapping modification will produce a steric clash upon association with kinases RSK and AMPK2. Following this reengineering, a series of experimental studies will be necessary to demonstrate the target specificity, efficacy (*in vitro* and *in vivo*), and safety (with primary emphasis on cardiac safety using *in vitro* and *in vivo* models).

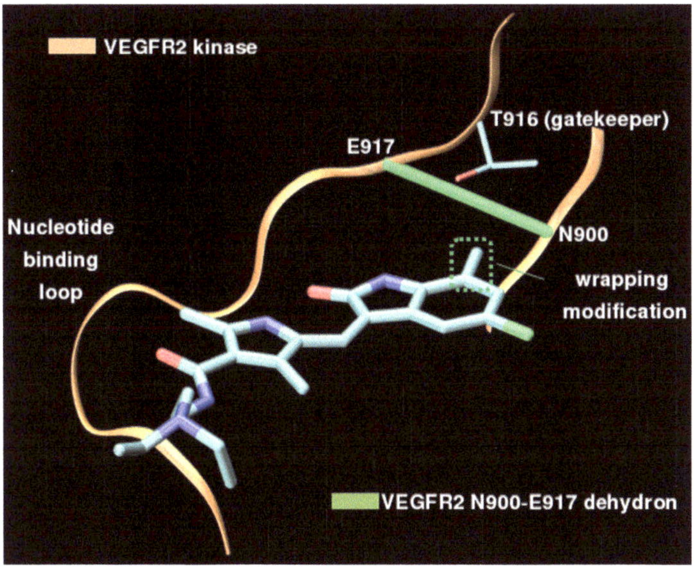

Fig. 13.8 Wrapping modification to sunitinib, making it selective for clinically relevant targets VEGFR1-3, PDGFR, KIT, CSF1R, FLT3 and RET, while avoiding AMPK and RSK

13.7 The Therapeutic Paradigm of Controlled Promiscuity

Mounting evidence from the treatment of complex disorders and malignancies reveals that therapeutic efficacy does not correlate with drug specificity. This observation has motivated a reassessment of the therapeutic value of promiscuity and may well trigger a paradigm shift, from magic bullets to multi-target therapies. These conceptual leaps are supported by novel high-throughput screening technologies to assess cross reactivity, hidden phenotypes and side effects.

While a paradigm shift may be perceived as a necessity, it is far from becoming a reality. The major impediment arises because we do not have a clear picture of the therapeutic advantages of the novel concepts that would replace drug specificity. Knowing the realm of activity of "magic shotguns" is insufficient to warrant broad therapeutic application. The clinical uncertainties and risks they introduce are likely broader than the more limited side effects associated with more specific drugs.

Novel systems biology insights are leading to the belief that multi-target or even promiscuous drugs would be more effective to deal with complex diseases since dynamical objects like biochemical fluxes and not the molecules themselves are now perceived as the appropriate targets. However, the greater danger of side effects arising in multi-target therapies clearly counterbalances the recent faith bestowed upon them. Thus, we propose the use of selectivity filters to guide the rational control of specificity towards targets of clinical relevance within the systems biology frame of inference that dictated the use of multi-target drugs in the first place. It is clear that tumors may be more efficiently tackled by opening several therapeutic fronts but only a *selectively promiscuous drug* would be operative in such scenario.

At this juncture, multi-target molecular therapies could only be welcomed if their target selectivity can be controlled to curb side effects and clinical uncertainties. Thus, there is a niche in emerging biotechnologies for novel approaches to clean promiscuous drugs in order to achieve tighter specificity control. We have shown that such approaches are in principle feasible by redesigning highly cross-reactive kinase inhibitors, guided by novel selectivity filters. The drug-redesign exercises and the proof of principle surveyed in this chapter suggest that cleaning a promiscuous drug guided by basic new concepts is in principle possible and will hopefully inspire further efforts in this regard. Promiscuity may become a welcomed feature in drug design, and may well propel the needed paradigm shift, but only if cross reactivity can be held under tight control through the effective implementation of selectivity filters.

13.8 Problems

13.1. Two compounds named CEP-701 and R406 have been identified as possible therapeutic editors to remove imatinib cardiotoxicity (Problem 12.1). The kinome-wide profiling of these compounds is given in Table 12.1. Both compounds appear to be promiscuous, having low selectivity. Discuss molecular engineering strategies to remove immunosuppressive effects in these compounds. Which of the two compounds deserves particular attention in this regard?

References

1. Dancey J, Sausville EA. Issues and progress with protein kinase inhibitors for cancer treatment. Nat Rev Drug Discov. 2003;2:296–313.
2. Levitski A, Gazit A. Tyrosine kinase inhibition: an approach to drug development. Science. 1995;267:1782–8.
3. Tibes R, Trent J, Kurzrock R. Tyrosine kinase inhibitors and the dawn of molecular cancer therapeutics. Annu Rev Pharmacol Toxicol. 2005;45:357–84.
4. Gibbs J, Oliff A. Pharmaceutical research in molecular oncology. Cell. 1994;79:193–8.
5. Donato NJ, Talpaz M. Clinical use of tyrosine kinase inhibitors: therapy for chronic myelogenous leukemia and other cancers. Clin Cancer Res. 2000;6:2965–6.
6. Roth BL, Sheffler DJ, Kroeze WK. Magic shotguns versus magic bullets: selectively non-selective drugs for mood disorders and schizophrenia. Nat Rev Drug Discov. 2004;3:353–9.
7. Frantz S. Drug discovery: playing dirty. Nature. 2005;437:942–3.
8. Keith CT, Borisy AA, Stockwell BR. Multicomponent therapeutics for networked systems. Nat Rev Drug Discov. 2005;4:71–8.

9. Mencher SK, Wang LG. Promiscuous drugs compared to selective drugs (promiscuity can be a virtue). BMC Clin Pharmacol. 2005;5:3–9.
10. McGovern SL, Helfand BT, Feng B, Shoichet BK. A specific mechanism of nonspecific inhibition. J Med Chem. 2003;46:4265–72.
11. Feng BY, Shelat A, Doman TN, et al. High-throughput assays for promiscuous inhibitors. Nat Chem Biol. 2005;1:146–8.
12. Force T, Krause D, van Etten RA. Molecular mechanisms of cardiotoxicity of tyrosine kinase inhibition. Nat Rev Cancer. 2007;7:332–44.
13. Fernández A, Sanguino A, Peng Z, Ozturk E, Chen J, Crespo A, Wulf S, Shavrin A, Qin C, Ma J, Trent J, Lin Y, Han HD, Mangala LS, Bankson JA, Gelovani J, Samarel A, Bornmann W, Sood AK, Lopez-Berestein G. An anticancer c-kit kinase inhibitor is reengineered to make it more active and less cardiotoxic. J Clin Invest. 2007;117:4044–54.
14. Demetri GD. Structural reengineering of imatinib to decrease cardiac risk in cancer therapy. J Clin Invest. 2007;117:3650–3.
15. Crunkhorn S. Anticancer drugs: redesigning kinase inhibitors. Nat Rev Drug Discov. 2008;7:120–1.
16. Hopkins AL, Mason J, Overington J. Can we rationally design promiscuous drugs? Curr Opin Struct Biol. 2006;16:127–36.
17. Hampton T. "Promiscuous" anticancer drugs that hit multiple targets may thwart resistance. JAMA. 2004;292:419–22.
18. Hopkins AL, Ren J, Milton J, et al. Design of non-nucleoside inhibitors of HIV-1 reverse transcriptase with improved drug resistance properties. J Med Chem. 2004;47:5912–22.
19. Owens J. Screening: dirty drugs' secrets uncovered. Nat Rev Drug Discov. 2006;5:542.
20. Fabian MA, Biggs WH, Treiber DK, et al. A small molecule kinase interaction map for clinical kinase inhibitors. Nat Biotechnol. 2005;23:329–36.
21. Karaman MW, Herrgard S, Treiber DK, et al. A quantitative analysis of kinase inhibitor selectivity. Nat Biotechnol. 2008;26:127–32.
22. Fedorov O, Marsden B, Pogacic V, et al. A systematic interaction map of validated kinase inhibitors with Ser/Thr kinases. Proc Natl Acad Sci U S A. 2007;104:20523–8.
23. MacDonald ML, Lamerdin J, Owens S, et al. Identifying off-targets effects and hidden phenotypes of drugs in human cells. Nat Chem Biol. 2006;2:329–37.
24. Brunton LL, Lazo JS, Parker KL. Goodman & Gilman's the pharmacological basis of therapeutics. 11th ed. New York: McGraw-Hill; 2006.
25. Rishton GM. Failure and success in modern drug discovery: guiding principles in the establishment of high probability of success drug discovery organizations. Med Chem. 2005;1:519–27.
26. Kerkela R, Grazette L, Yacobi R, et al. Cardiotoxicity of the cancer therapeutic agent imatinib mesylate. Nat Med. 2006;12:908–16.
27. Fedorov O, Sundstrom M, Marsden B, Knapp S. Insights for the development of specific kinase inhibitors by targeted structural genomics. Drug Discov Today. 2007;12:365–72.
28. Bogoyevitch MA, Fairlie DP. A new paradigm for protein kinase inhibition: blocking phosphorylation without directly targeting ATP binding. Drug Discov Today. 2007;12:622–33.
29. Crespo A, Fernández A. Kinase packing defects as drug targets. Drug Discov Today. 2007;12:917–23.
30. Chen J, Zhang X, Fernández A. Molecular basis for specificity in the druggable kinome: sequence-based analysis. Bioinformatics. 2007;23:563–72.
31. Griffin JD. Interaction maps for kinase inhibitors. Nat Biotechnol. 2005;23:308–9.
32. Liu Y, Gray NS. Rational design of inhibitors that bind to inactive kinase conformations. Nat Chem Biol. 2006;2:358–64.
33. Noble ME, Endicott JA, Johnson LN. Protein kinase inhibitors: insights into drug design from structure. Science. 2004;303:1800–5.
34. Torrance CJ, Jackson PE, Montgomery E, et al. Combinatorial chemoprevention of intestinal neoplasia. Nat Med. 2000;6:1024–8.

35. Erlichman C, Hidalgo M, Boni JB, et al. Phase I study of EKB-569, an irreversible inhibitor of the epidermal growth factor receptor, in patients with advanced solid tumors. J Clin Oncol. 2006;24:2252–60.
36. Wissner A, Overbeek E, Reich MF, et al. Synthesis and structure–activity relationships of 6,7-disubstituted 4-Anilinoquinoline-3-carbonitriles. The design of an orally active, irreversible inhibitor of the tyrosine kinase activity of the epidermal growth factor receptor (EGFR) and the human epidermal growth factor receptor-2 (HER-2). J Med Chem. 2003;46:49–63.
37. Zhang X, Crespo A, Fernández A. Turning promiscuous kinase inhibitors into safer drugs. Trends Biotechnol. 2008;26:295–301.
38. Fernández A, Maddipati S. The a-priori inference of cross reactivity for drug targeted kinases. J Med Chem. 2006;49:3092–100.
39. Butcher EC. Systems biology in drug discovery. Nat Biotechnol. 2004;22:1253–9.
40. Butcher EC. Can cell systems biology rescue drug discovery? Nat Rev Drug Discov. 2005;4:461–7.
41. Kitano H. A robustness-based approach to systems-oriented drug design. Nat Rev Drug Discov. 2007;6:202–9.
42. Volgestein B, Kinzler KW. Cancer genes and the pathway they control. Nat Med. 2004;10:789–99.
43. Kitano H. Biological robustness. Nat Rev Genet. 2004;5:826–37.
44. Carlson JM, Doyle J. Complexity and robustness. Proc Natl Acad Sci U S A. 2002;99:2538–45.
45. Hellerstein MK. Exploiting complexity and the robustness of network architecture for drug discovery. J Pharmacol Exp Ther. 2008;325:1–9.
46. Seidman A. Cardiac dysfunction in the trastuzumab clinical trials experience. J Clin Oncol. 2002;20:1215–21.
47. Martel CL. Bevacizumab-related toxicities: association of hypertension and proteinuria. Commun Oncol. 2006;3:90–3.
48. Motzer RJ, Michaelson MD, Redman BG, et al. Activity of SU11248, a multitargeted inhibitor of vascular endothelial growth factor receptor and platelet-derived growth factor receptor, in patients with metastatic renal cell carcinoma. J Clin Oncol. 2006;24:16–24.
49. Joensuu H. Cardiac toxicity of sunitinib. Lancet. 2007;370:1978–80.
50. Chu TF, Rupnick MA, Kerkela R, et al. Cardiotoxicity associated with tyrosine kinase inhibitor sunitinib. Lancet. 2007;370:2011–9.

Chapter 14
Synergizing Engineered Immunotherapy with Molecularly Targeted Cancer Treatment

In this chapter we develop molecular design strategies to empower drug-based anticancer therapy by aligning an engineered immune response with the drug-induced antigenic activity. We first focus on re-engineering immunosuppressive anticancer drugs where the impact on the immune system is an undesirable outcome, betraying their own *raison d'être*. Subsequently, we outline genetic engineering strategies to tailor an adoptive immune arm to the antigenic repertoire generated by the drug-based treatment. The first goal is to remove the immunosuppressive effects caused by drug treatment through molecular redesign, more specifically, to restore the adaptive T-cell-related immune response and natural killer (NK) cell cytotoxicity compromised by the drug treatment. The redesign strategy enables us to create synergies that may empower drug-based anticancer therapy, aligning the immune response with the molecularly targeted treatment. The second goal is to ensure therapeutic success of synergistic immunotherapies based on the genetic manipulation and infusion of autologous or allogeneic immune cells into patients subject to antigen-generating drug therapy. This success hinges on the ability to genetically tailor the modified T-cell receptor (TRC) and chimeric antigen receptor (CAR) to the antigenic repertoire of the drug-based treatment.

14.1 Molecularly Engineered Removal of Drug-Induced Immunosuppressive Effects in Anticancer Drug Therapy

As described in Chaps 8, 9, 10, 11, 12 and 13, kinase inhibitors (KIs) that interfere selectively with cell signaling pathways represent a therapeutic opportunity for cancer treatment [1–11]. Yet, most drug-based anticancer therapies are incomplete and do not provide a decisive cure [12]. A well-known culprit for the ultimate failure arises from the somatic evolution of patterns of drug resistance that often

A. Fernández, *Physics at the Biomolecular Interface*, Soft and Biological Matter,
DOI 10.1007/978-3-319-30852-4_14

materialize as site mutations or constitutively deregulated chimeras. Somatic mutations compromise the affinity of the drug for its target or, alternatively, increase the affinity for the kinase natural ligand ATP, in the case of ATP-competitive inhibitors [12]. A far less acknowledged culprit for drug failure arises because these targeted therapies are typically not tailored to operate synergistically with the immune system, and often antagonize it in spite of generating antigenic activity [13–16]. KIs are often of limited applicability, since drug treatments are marred by episodes of relapse and by the development of drug resistance and intolerance [12, 17]. One reason for this incomplete success is the lack of synergy between the drug and the immune system, with the latter often incapacitated at crucial junctures during treatment due to antagonistic drug-induced effects [13, 14, 18–21]. In fact, these immune-antagonistic effects are prevalent because of the multiple roles that a protein target may play in off-target cellular contexts, a systems biology issue described in Chaps 6, 8 and 9. Thus, a therapeutic imperative arises from the need to recruit cooperative immune responses concomitant with the molecular targeted treatment. The goal is to design anticancer drugs that inhibit targeted cellular functions and steer the immune system to harness the antigenic products of the drug-induced apoptosis of tumor cells. To fulfill this imperative of therapeutic integration, we propose re-designs of anticancer drugs that fulfill three constraints: (a) nanomolar activity against anticancer targets; (b) reversal of tumor-induced immunomodulation; and (c) removal of drug-induced immunosuppressive activity.

The drug-design strategies introduced to address the therapeutic imperative of immuno-synergy have the potential to revolutionize cancer treatment and the understanding of the adaptive immune response by steering it with molecular targeted therapy. We are counting on the premise that by restoring the adaptive immune response to drug-induced antigenic activity, we shall be able to create synergies that will reciprocally empower the immune system and drug-based anticancer treatment. Novel therapeutic possibilities to harness and manipulate the immune system will probably transpire from the evaluation of immuno-synergic drugs, as described in the subsequent sections.

14.2 Therapeutic Shortcomings of Immunosuppressive Anticancer Drugs

The level of molecular fine tuning required to redesign an anticancer drug into an immuno-synergic drug cannot be achieved within the current drug-discovery paradigm [17, 22]. Rather, a rational design approach is needed [23], an area where novel molecular filters can be exploited to better control drug specificity [24, 25]. Thus, an unprecedented control of specificity [17] is required to design therapeutic agents capable of discriminating between cancer-related targets and immunosuppressive targets. The reward in building immuno-synergic molecular therapies through rational design is potentially immense, since these targeted therapies will then engage a formidable ally—the immune response—in their anticancer activity.

Well established anticancer KIs like *imatinib* [1–3] or *dasatinib* [4–9] are also known to be directly immunosuppressive [13, 14, 18–21] through their powerful blockade of upstream signaling in the adaptive immune response, and yet, as shown in this study, they hold promise as chemical scaffolds that can be turned into immuno-synergic drugs. The choices are justified since these KIs are nanomolar inhibitors of the major anticancer target c-Kit kinase [2, 3, 8, 9], the kinase of the stem-cell factor (SCF) receptor, and hence the KI treatment is expected to have the additional effect of reversing tumor-induced immunosuppression, at least in certain tumor environments where tumor-secreted SCF is deployed to hijack the immune system [26]. The latter effect is promoted by the accumulation of myeloid derived suppressor cells (MDSC) and T-regulatory cells (Tregs), whose development/activation requires the SCF expressed by the tumor cells. However, both KIs are also direct immunosuppressants [13–15, 18–21], and this role is clearly antagonistic to their anticancer activity, thus requiring careful removal through a molecular remodeling of the parent compounds using stringent selectivity filters [23–25].

More precisely, these KIs are powerful (nanomolar) inhibitors of two crucial kinases implicated as upstream signal transducers controlling the immune response: the lymphocyte-associated kinase Lck and the colony-stimulating factor-1 receptor (CSF1-R) kinase [17, 22]. In fact, dasatinib is the most powerful inhibitor of Lck known to date ($K_D = 0.2$ nM) [22]. Specifically, Lck is a signal transducer for the signaling cascade that originates in the CD4/8 receptors (CD = cluster of differentiation) expressed on the surface of T-cells in antigen-triggered T-cell differentiation (Fig. 14.1a, from KEGG, the Kyoto Encyclopedia of Genes and Genomes) and in natural killer (NK) cell-mediated cytotoxicity (Fig. 14.1b, KEGG, release 8/2/2013) [27]. On the other hand, the CSF1-R kinase is implicated in the development of the monocyte/macrophage (M/M) lineage [27]. Hence, both KIs suppress antigen-specific T-cell effector functions and NK cytotoxicity and inhibit the hematopoietic development of crucial components of the immune response.

Both imatinib and dasatinib are likely to hamper the adaptive immune response triggered by the very antigenic activity they generate, and compromise the antigen presentation by precluding macrophage development. The profound inhibition of all antigen-specific T-cell effector functions at therapeutically relevant concentrations has been mechanistically tracked down to the early blockade of signal transduction events promoted by Lck inhibition [13, 14]. In fact, Lck is central in the transduction of T-cell receptor (TCR) signaling in response to MHC-I and MHC-II antigen presentation (MHC = major histocompatibility complex) that ultimately promotes T-cell differentiation and proliferation. Thus, the drug-induced immunosuppressive effects caused by Lck inhibition -purposely engineered in the case of dasatinib [28]- actually betray the purpose of these drugs as anticancer agents, possibly enabling the development of episodes of relapse and drug resistance and impacting the frequency of opportunistic infections. In the case of *dasatinib*, the original drug-discovery pursuits seemed to focus squarely on

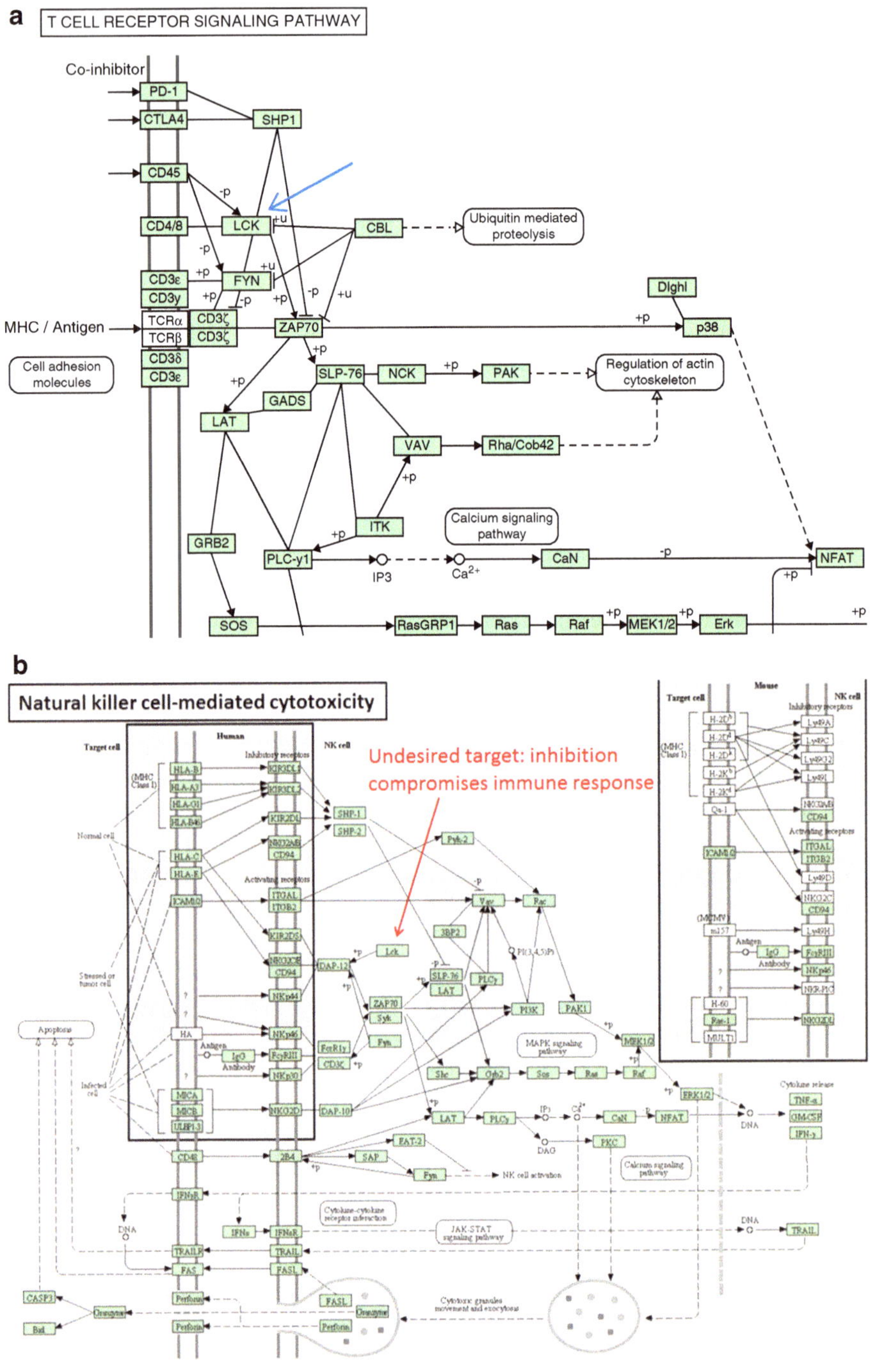

Fig. 14.1 (**a**) Receptor signaling pathways implicating Lck in the antigen-triggered T-cell response (adaptive immune response, *Source*: KEGG, Kyoto Encyclopedia of Genes and Genomes). (**b**) The role of Lck in signaling pathways recruited in natural killer (NK) cell-mediated cytotoxicity (*Source*: KEGG, Kyoto Encyclopedia of Genes and Genomes)

modulating the immune response [28]. The repositioning of *dasatinib* as an anticancer agent seemed to have arisen as an afterthought, as the precursor drug *imatinib* showed cross-reactivity against Lck as well as the cancer-associated targets Bcr-Abl, c-Kit and PDGFR kinases. Furthermore, the Lck-related immunosuppression by both KIs is reinforced by the blockade of macrophage development -and thereby of antigen presentation- caused by CSF1-R kinase inhibition (both compounds are nanomolar inhibitors of CSF1-R).

These immuno-antagonistic effects are ostensibly at odds with the need to maintain an uncompromised immune response to fight cancer, unless the kinases targeted for immunosuppression happen to be also relevant anticancer targets. Thus, the immunosuppressive effects are likely to doom much of the ongoing efforts to apply these KIs to combat cancers other than hematologic malignancies, where immunosuppression may become adjuvant. Even in those applications for which FDA approval has been obtained, i.e. the treatment of CML and Philadelphia chromosome-positive acute lymphoblastic leukemia (Ph+ ALL), opportunistic infections have been reported in patients treated at therapeutic doses [19].

A kinome-wide screening of KIs [17, 22] reveals that potential side effects and immunosuppressive complications due to nanomolar inhibition of Lck (dissociation constant $K_D < 50$ nM) are likely to be a concern for several compounds of therapeutic interest. Thus, a significant drug-induced impairment of adaptive immune responses, mostly affecting T-cell and NK-cell activation, is readily expected for the following KIs (K_D's from [22]): NVP-AST487 (Novartis, $K_D = 11$ nM); *nintedanib* (BIBF-1120, Vargatef®, Boehringer-Ingelheim, $K_D = 6.2$ nM); *crizotinib* (Xalkori®, Pfizer, $K_D = 30$ nM); *dasatinib* (Sprycel®, Bristol-Myers Squibb, $K_D = 0.2$ nM); *foretinib* (GSK1363089, GlaxoSmithKline ($K_D = 6$ nM), *imatinib* (Gleevec®, Novartis, $K_D = 40$ nM); *nilotinib* (Tasigna®, Novartis, $K_D = 47$ nM); PD-173955 (Parke-Davis, $K_D = 1.1$ nM); *bosutinib* (SKI-606, Bosulif®, Wyeth/Pfizer, $K_D = 0.59$ nM); NVP-TAE684 (Novartis, $K_D = 49$ nM); and *vandetanib* (Caprelsa®, AstraZeneca, $K_D = 17$ nM).

While anticancer activity is a desired clinical outcome for these KIs, the reported structural similarities between Lck and validated anticancer targets like c-KIT, PDGFR and Src-family kinases, introduce extremely undesirable cross-reactivities. In fact, while these KIs are being actively evaluated as anticancer agents, their nanomolar inhibition of Lck truly compromises the immune system, depriving the patient of a key endogenous resource to fight the disease. This adverse aspect of treatment seems to have been overlooked except in the cases of *imatinib* and *dasatinib*, as indicated previously. This chapter advocates then for the need to address the imperative of building a therapeutic alliance between immune response and molecularly targeted therapy. This goal may be achieved by exploiting a selectivity filter purposely created to guide the re-design of anticancer drugs to dial out targets related to immunosuppression.

14.3 Immuno-Synergic Cancer Drugs

To redesign an anticancer drug to make it immuno-synergic, an unprecedented level of control of drug specificity, well beyond the current drug-discovery standard, is required. This is so because anticancer activity must be retained while targets related to immunosuppression must be selectively removed from the drug affinity profile. In this regard, we may exploit the effective selectivity filter provided by the protein's unique dehydron pattern to guide the molecular re-design [12, 23–25, 29], as described in Chaps 8, 9 and 10. As we know, the dehydron pattern of proteins is not conserved even in cases of very high structural similarity, as between anticancer targets like c-Kit, PDGFR or c-Src kinases, and immunosuppression-related targets like Lck.

Thus, the problem of target discrimination tailored to our therapeutic challenge may well exploit this new selectivity filter that can transcend structural similarity (cf. Figs. 14.2 and 14.3). Thus, if a dehydron is present in an anti-cancer target protein (i.e. c-Kit) while the hydrogen bond at the same location in a homologous off-target protein (i.e. Lck) is well-wrapped, we may re-design the original drug with its undesired cross-reactivity into a new drug that can tell apart the two targets, binding only to the desired target. As described in Chap. 9, the "dehydron-wrapping" drug, whereby the drug improves the dehydron packing upon association, becomes an expansion of the original chemical scaffold and is intended to locally exclude water from the specific dehydron present in the desired target and absent in the undesired target at the same location [12]. Specifically, the dehydron- wrapping capability of a drug is built by expanding the original parental compound with the addition of a nonpolar group (usually methyl) that can penetrate the desolvation domain of the preformed targetable dehydron upon drug-target association. Since the incorporation of a methyl group usually compromises bioavailability, a compensatory change increasing drug solubility is often introduced at an innocuous site removed from the protein/ligand interface [12].

To illustrate this operational strategy, let us consider the problem of redesigning *imatinib* into a new drug that retains anticancer efficacy against gastrointestinal stromal tumor (GIST) [2] and other c-Kit-dependent cancers (c-Kit dependent melanoma) [24] while becoming immune-synergic by selective removal of *imatinib*'s anti-Lck activity. The dehydron patterns for the *imatinib*-binding region of the desired target c-Kit (PDB accession code 1T46) and the undesired target Lck (PDB accession code 2PL0) are displayed in Figs. 14.2 and 14.3, respectively. Of crucial importance as guidance to *imatinib* remodeling is the dehydron (green stick) pairing residues Cys673 and Gly676 in the nucleotide-binding loop of c-Kit (Fig. 14.2b). This dehydron becomes a well-wrapped hydrogen bond (grey segment) pairing residues Met319 and Gly322 in the *imatinib* target Lck (Fig. 14.3). Thus, the uniqueness of the c-Kit Cys673-Gly676 dehydron guides us to re-design *imatinib* into a more selective dehydron-wrapping drug by expanding the chemical scaffold at the position highlighted in the terminal ring (Figs. 14.2b and 14.3). Thus, the resulting compound WBZ_4 [30] arises from methylation of *imatinib* at the

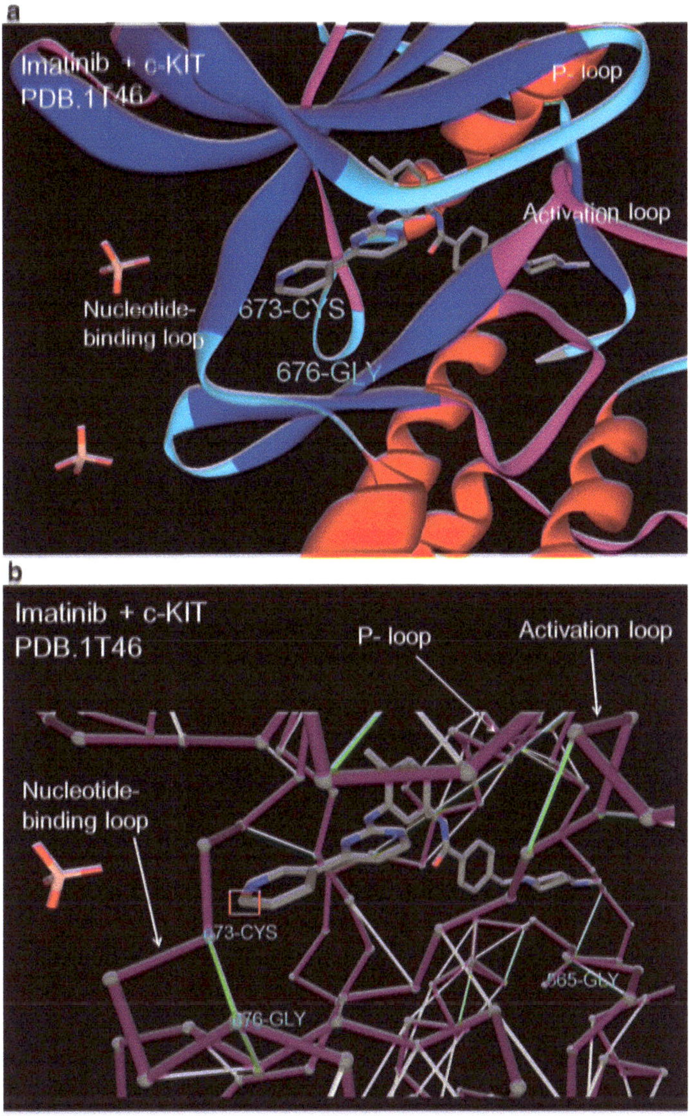

Fig. 14.2 Structure and dehydron pattern of anticancer target c-Kit kinase in the complex with KI *imatinib* (PDB entry 1T46). (**a**) The protein is displayed in ribbon representation colored conventionally according to the secondary structure motif (*red* = α-helix, *blue* = β-strand, *light blue* = turn, *pink* = loop) while the drug is displayed in bond-sticks colored conventionally according to chemical element (*grey* = carbon or hydrogen, *blue* = nitrogen, *red* = oxygen, *pink* = phosphorus). (**b**) Dehydron pattern for the protein in complex with the drug. The protein backbone is displayed in magenta, with virtual bonds joining adjacent α-carbons, well wrapped (buried) hydrogen bonds are represented as grey sticks joining the α-carbons of the paired residues, while dehydrons are shown as green sticks. Of interest is the dehydron Cys673-Gly676 present in the c-Kit kinase that becomes a well-wrapped hydrogen bond pairing residues Met319 and Gly322 in Lck. This selectivity feature unique to c-Kit directs the remodeling of the *imatinib* chemical scaffold by methylation at the location highlighted by the *red* rectangle. Reprinted from [36], copyright 2014 with permission from Elsevier

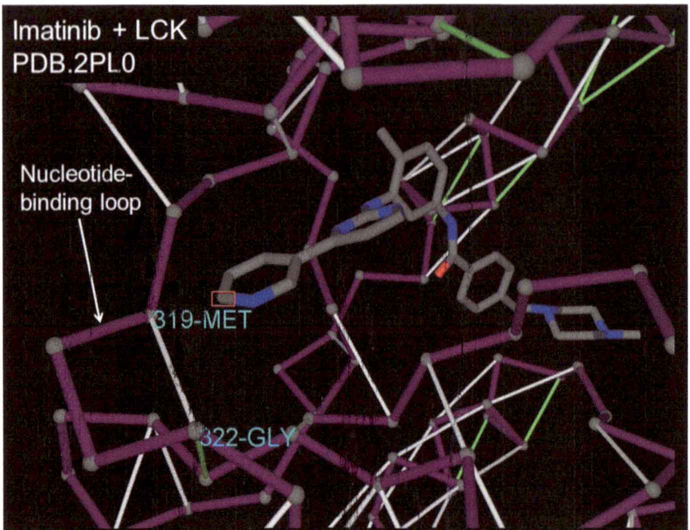

Fig. 14.3 Dehydron pattern for Lck within the complex with *imatinib* with structural coordinates obtained from PDB entry 2PL0. The representation convention of Fig. 14.2b is followed. Upon structural alignment, the dehydron Cys673-Gly676 in the c-Kit kinase becomes the well wrapped hydrogen bond Met319-Gly322 in Lck. Reprinted from [36], copyright 2014 with permission from Elsevier

position indicated in Figs. 14.2b and 14.3 and its chemical structure is given in Fig. 9.2d. As shown in Chap. 9, WBZ_4 is a dehydron-wrapping drug that stabilizes the dehydron 673–676 of the c-Kit kinase upon association with the target, whereas it unfavorably promotes water removal from the surroundings of the well-wrapped 319–322 hydrogen bond in off-target Lck. Therefore, WBZ_4 retains nanomolar activity against c-Kit while showing no detectable inhibitory activity against Lck [25]. The immuno-synergic properties of WBZ_4 await validation but its higher selectivity relative to the parent compound is indicative that there will be no impairment of the adaptive immune response triggered by the antigenic activity generated during drug treatment.

Similarly, a re-design of *imatinib* discriminating between desired target SCFR (c-Kit) and CSF1-R kinases is essential to secure development of the macrophage lineage during molecular therapy and may be also carried out by exploiting the dehydron filter. We can take advantage of local differences in the dehydron patterns and align this wrapping representation for PDB entries 1T46 (c-Kit) and 3LCO (CSF-1R kinase).

14.4 Therapeutic Evaluation of Immuno-Synergic Drugs

The immuno-synergic drugs should undergo testing at all levels, from *in silico* to clinical trials. Using the dehydron selectivity filter [12, 30], the parent drugs may be reworked into compound series with nanomolar affinity towards the original cancer targets with immuno-synergic properties arising from: (A) removal of activity against Lck, (B) removal of activity against the CSF1R kinase; (C) simultaneous removal of both cross-reactivities.

To test class-A drugs, we may assay for T-cell receptor (TCR)-mediated T-cell proliferation and activation in a dose-dependent manner using the parent compounds in the control assays. Besides testing the redesigns in proliferation assays using cancer cell lines, we may test for selective removal of the inhibition of antigen-specific T-cell effector functions. In accord with the predictions set forth in this study, the impact on antigen-triggered CD4+ T-cell proliferation and T-cell effector function [31] by the purportedly immune-synergic WBZ_4 should be preliminarily evaluated *ex vivo* using *imatinib* as control. The results are expected to reveal a minimal level of modulation of the immune response exerted by WBZ_4 when compared with the parental drug [32].

Class-B compounds may be tested by examining the development of the monocyte/macrophage (M/M) lineage from bone marrow progenitors at concentrations of class-B or anticancer drugs of the order of 0.5–1 micromolar [26]. Based on the KEGG annotation cited above, we conjecture that anticancer drug impairment of M/M development is the result of drug inhibition of kinase CSF1-R, so the impairment should vanish as the hematopoietic stem cells (HSCs) are treated with class-B drugs. Class-C compounds should be tested following all the protocols mentioned above and are expected to restore antigen-induced T-cell response, M/M development and macrophage-mediated antigen presentation, hence they should be the optimally suited to be combined with immunotherapy, as they are likely to restore cross-talk between different immune responses.

In all cases, we may corroborate retention or enhancement of anticancer activity as the parent drugs are molecularly re-engineered exploiting the dehydron filter, following the same *in vitro*, *ex vivo* and *in vivo* assays that have led to assert the therapeutic value of the parent drugs.

The novel targeted therapies will bring about significant antigenic activity by promoting apoptosis of tumor cells without compromising the immune response. This will make them amenable to be synergistically combined with immunotherapies. The choice of the latter will depend on the changes in the components of the innate and adaptive immune system caused by the targeted therapy and relative to the parent drug treatment. The immune response triggered by targeted therapy should be determined in animal models and the immunotherapies will likely include depletion of regulatory T-cells by treatment with CD25 antibody, targeting the cytotoxic T-lymphocyte-associated antigen 4 (CTLA4) with monoclonal antibodies like *ipilimumab* (suited for c-Kit-dependent melanomas to be treated by class-A variants of *imatinib*) [33], etc.

14.5 Anticancer Drugs to Overcome HIV-1 Induced Immunosuppression

Probably the need for immuno-synergic anticancer drugs is nowhere more pressing than in the context of viral cancers within AIDS, an immunosuppressive setting induced by the human immunodeficiency virus-1 (HIV-1) [34]. Drug-induced immunosuppression would be a highly undesirable effect especially in this therapeutic context where the immune system is already compromised by endogenous factors triggered by the HIV-1 infection. One of the primary cancers concurrent with HIV-1-induced immunosuppression is the Kaposi sarcoma, caused by a facilitated herpes viral infection [35]. As recently demonstrated, new lines of treatment to prevent the reactivation of Kaposi's sarcoma herpesvirus and invasiveness of infected endothelial cells involve targeting the inflammation-related kinase MAP4K4 [35]. Direct inspection of the screening data reported in [17, 22] enables us to identify two very powerful nanomolar inhibitors of the MAP4K4 kinase: NVP-AST-487 (Novartis, $K_D = 15$ nM) and *bosutinib* (SKI-606, Bosulif®, Wyeth/Pfizer, $K_D = 8.2$ nM). Yet, none of them should be used in this specific therapeutic context because they are also extremely powerful inhibitors of Lck ($K_D = 11$ nM and 0.59 nM, respectively), as already noted in this chapter. Thus, to treat Kaposi sarcoma in HIV-1-positive patients, one must first attempt a wrapping-based molecular remodeling of both inhibitors to retain nanomolar activity against MAPK4K but suppress as much as possible their anti-Lck activity. This molecular re-engineering would entail assessing the differences in the dehydron patterns for the on-target kinase MAPK4K versus the off-target kinase Lck, and redesign the parent chemical scaffolds to discriminate the two proteins according to the premises of the wrapping technology [12].

Most drug-based anticancer therapies are incomplete and do not provide a decisive cure. Molecularly targeted therapy aimed at blocking cancer-related signaling pathways seldom recruits the immune system but rather antagonizes it. This problem is magnified when the therapeutic context for cancer treatment involves HIV-1-induced immunosuppression, as in AIDS patients. Thus, we make it a therapeutic imperative to recruit cooperative immune responses triggered by molecularly targeted treatment. The main challenge posed in this chapter is to engineer drugs that inhibit targeted cellular functions and without an adverse interference, steer the immune system to harness the antigenic products that the drugs generate. To achieve this synergy we need to remove immuno-antagonistic effects generated by the drug treatment while retaining the anticancer potency of the drug. This goal demands a paradigm shift in drug design.

Through their inhibition of anticancer target c-Kit, drugs like *imatinib* or *dasatinib* are likely to elicit antigenic activity as they induce apoptosis of tumor cells, and evidence suggests that c-Kit inhibition may reverse tumor-induced immunosuppression. Thus, these inhibitors partially fulfill the therapeutic imperative. Yet, because they are also nanomolar inhibitors of Lck and CSF1-R kinase, they suppress antigen-specific T-cell effector functions and the development of the macrophage lineage, hampering antigen presentation. Ultimately, these drugs compromise the immune response, partially defeating their own purpose. Besides imatinib and dasatinib, this chapter reports on nine additional compounds of clinical oncological interest that are likely to also impair the adaptive immune response precisely at the juncture when it is most indispensable.

In consonance with theimperative of therapeutic integration, this chapter advocates for the design of immuno-synergic anticancer drugs. The synergic designs must fulfill three constraints: (a) nanomolar inhibition of anticancer targets; (b) reversal of tumor-induced immunosuppression; and (c) removal of drug-induced immunosuppressive activity [36]. To fulfill these goals, an unprecedented level of control of drug specificity is required and is provided by the exploitation of an effective selectivity filter to guide molecular design: the dehydron pattern. The therapeutic challenge of immuno-synergy requires this new generation of selectivity filters for target discrimination. The proposed strategy has the potential to revolutionize anticancer treatment and the understanding of the adaptive immune response by steering it along with molecularly targeted treatment. We are counting on the premise that restoration of the adaptive immune response triggered by drug-induced antigenic activity will empower drug-based anticancer treatment [36]. Novel possibilities for therapeutic manipulation of the immune system are likely to arise from the evaluation of the immuno-synergic drugs along the lines presented in this chapter.

14.6 Genetically Modified Adoptive Immunotherapy Synergized With Drug-Based Targeted Anticancer Treatment

Innovative immunotherapies are being developed to optimize the cancer-targeted response through genetic modification. By harnessing and steering the function of immune cells, the possibility has materialized to enhance the immune response to achieve persistent cancer regression. Synergies of such immunotherapies with drug-based targeted therapies required to generate cancer-related antigens through induced apoptosis of the tumor cell will surely generate powerful combination treatments. Immunotherapies based on the manipulation and infusion of autologous or allogeneic immune cells into patients, have been satisfactorily tested in clinical trials. These include genetically modified T-cell immunotherapies [37]. As highlighted in Fig. 14.1a, combinations of such T-cell based immunotherapies with antigen-generating anticancer drug therapies may prove fruitful if the latter has low or no inhibitory impact on the major signal transducer Lck of T-cell antigen response.

T-cells contain T-cell receptors (TCRs) on the cell surface. The TCR is a multi-subunit transmembrane complex that mediates the antigen-specific activation of T cells (cf. Fig. 14.1a). The TCR is composed of two different polypeptide chains, the

TCR α and β chains. Both chains have an N-terminal variable region and a constant region. The chains are linked by a disulfide bond, with each receptor providing a single antigen-binding site. The TCR recognizes an antigen comprised of a short contiguous amino acid sequence of a protein presented on the target cell by the major histocompatibility complex (MHC) molecule. Other adhesion molecules expressed by T-cells, such as CD4 for MHC class II and CD8 for MHC class I, are also involved. The TCR interacts with the ligand by making contacts with both the MHC molecule and the antigen peptide. Signal transduction occurs through the associated CD3 complex (cf. Fig. 14.1a), which is composed of four different CD3 proteins that form two heterodimers (CD3δε and CD3γε) and one homodimer (CD3ζζ), as shown in Fig. 14.1a. As previously noted in this chapter, signal transduction in the T-cell triggered by antigen recognition is transduced through Lck. Thus, for such synergistic combinations to become effective it becomes absolutely essential that the antigen-generating drug-based treatment avoids inhibiting Lck.

Following contact with the cognate peptides presented by the MHC, naive $CD8^+$ cytotoxic T cells proliferate and acquire the phenotype of effector T cells. Thus, the adaptive immune response is directed by the TCR recognition and for this event to materialize it is essential that the antigen-generating arm of the therapy does not block the signaling pathway recruited by TCR-antigen recognition (cf. Fig. 14.1a).

The concept of transferring T-cells to patients (adoptive T-cell transfer) to treat disease has been now been established. Thus, the antigen specificity of T-cells can be manipulated by genetic modification and redirected to successfully target antigens that are generated by the complementary antitumor drug-based therapy. In particular, T-cells can be engineered to express modified TCRs (the so-called TCR therapies) or protein-fusion-derived chimeric antigen receptors (CARs) that have enhanced selective affinity for the drug-related antigenic repertoire. These "tailored" genetically modified T-cell-based approaches are predicted to be highly efficient at targeting tumors in synergistic combinations the antigen-generating drug treatment that induces tumor cell apoptosis (Fig. 14.4).

Genetically modified TCR therapies are based on altering T-cell specificity through the expression of specific TCR α and β chains, which mediate the antigen-recognition process. On the other hand, chimeric antigen receptors (CARs) combine antibody-like recognition with T-cell-activating function [38]. They are composed of an antigen-binding region, typically derived from an antibody, and a transmembrane domain to anchor the CAR to the T-cell.

The therapeutic success of synergistic immunotherapies based on the genetic manipulation and infusion of autologous or allogeneic immune cells into patients combined with antigen-generating drug therapy (Fig. 14.4) will depend pivotally on the ability to genetically tailor the TRC and CAR to the antigenic repertoire of the drug-based treatment.

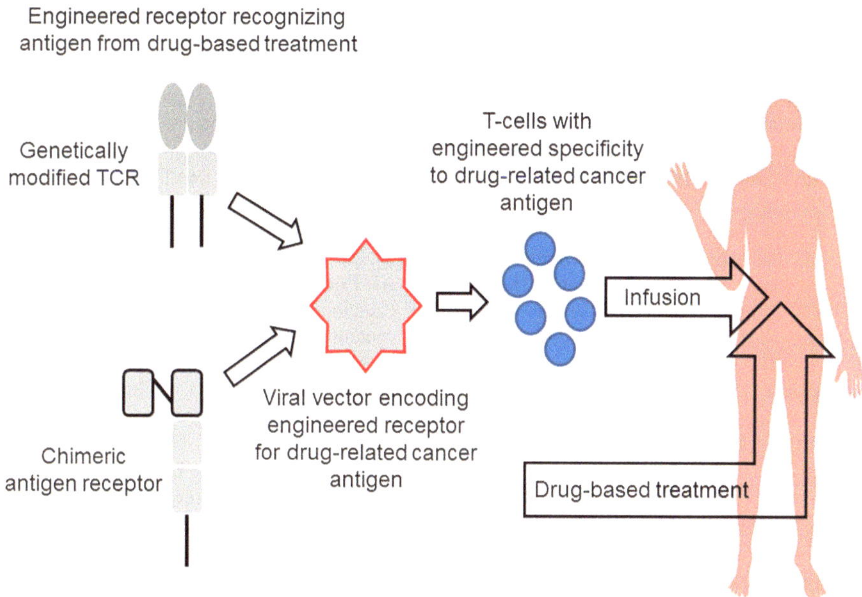

Fig. 14.4 Genetically modified adoptive immunotherapy synergized with drug-based targeted anticancer treatment

14.7 Problems

14.1. Design *in vitro*/*ex vivo* experimental tests for class-A immuno-synergic anticancer drugs, assaying for T-cell receptor (TCR)-mediated T-cell activation in a dose-dependent manner with adequate controls.

References

1. Druker BJ, Lydon NB. Lessons learned from the development of an Abl tyrosine kinase inhibitor for chronic myelogenous leukemia. J Clin Invest. 2000;105:3–7.
2. Demetri GD, von Mehren M, Blanke CD, Van den Abbeele AD, Eisenberg B, Roberts PJ, Heinrich MC, Tuveson DA, Singer S, Janicek M, Fletcher JA, Silverman SG, Silberman SL, Capdeville R, Kiese B, Peng B, Dimitrijevic S, Druker BJ, Corless C, Fletcher CD, Joensuu H. Efficacy and safety of imatinib mesylate in advanced gastrointestinal stromal tumors. N Engl J Med. 2002;347:472–80.
3. Hodi FS, Friedlander P, Corless CL, Heinrich MC, Mac Rae S, Kruse A, Jagannathan J, Van den Abbeele AD, Velazquez EF, Demetri GD, Fisher DE. Major response to imatinib mesylate in KIT mutated melanoma. J Clin Oncol. 2007;26:2046–51.
4. Gnoni A, Marech I, Silvestris N, Vacca A, Lorusso V. Dasatinib: an anti-tumor agent via Src inhibition. Curr Drug Targets. 2011;12:563–78.
5. Li J, Rix U, Fang B, Bai Y, Edwards A, Colinge J, Bennett KL, Gao J, Song L, Eschrich S, Superti-Furga G, Koomen J, Haura EB. A chemical and phosphoproteomic characterization of dasatinib action in lung cancer. Nat Chem Biol. 2010;6:291–9.

6. Pichot CS, Hartig SM, Xia L, Arvanitis C, Monisvais D, Lee FY, Frost JA, Corey SJ. Dasatinib synergizes with doxorubicin to block growth, migration, and invasion of breast cancer cells. Br J Cancer. 2009;101:38–47.
7. Koreckij T, Nguyen H, Brown LG, Yu EY, Vessella RL, Corey E. Dasatinib inhibits the growth of prostate cancer in bone and provides additional protection from osteolysis. Br J Cancer. 2009;101:263–8.
8. Schittenhelm MM, Shiraga S, Schroeder A, Corbin AS, Griffith D, Lee FY, Bokemeyer C, Deininger MW, Druker BJ, Heinrich MC. Dasatinib (BMS-354825), a dual SRC/ABL kinase inhibitor, inhibits the kinase activity of wild-type, juxtamembrane, and activation loop mutant KIT isoforms associated with human malignancies. Cancer Res. 2006;66:473–81.
9. Woodman SE, Trent JC, Stemke-Hale K, Lazar AJ, Pricl S, Pavan GM, Fermeglia M, Gopal YN, Yang D, Podoloff DA, Ivan D, Kim KB, Papadopoulos N, Hwu P, Mills GB, Davies MA. Activity of dasatinib against L576P KIT mutant melanoma: molecular, cellular, and clinical correlates. Mol Cancer Ther. 2009;8:2079–85.
10. Martin P, Oliver S, Kennedy SJ, Partridge E, Hutchison M, Clarke D, Giles P. Pharmacokinetics of Vandetanib: three phase I studies in healthy subjects. Clin Ther. 2012;34:221–37.
11. Cui JJ, Tran-Dubé M, Shen H, Nambu M, Kung PP, Pairish M, Jia L, Meng J, Funk L, Botrous I, McTigue M, Grodsky N, Ryan K, Padrique E, Alton G, Timofeevski S, Yamazaki S, Li Q, Zou H, Christensen J, Mroczkowski B, Bender S, Kania RS, Edwards MP. Structure based drug design of Crizotinib (PF-02341066), a potent and selective dual inhibitor of mesenchymal–epithelial transition factor (c-MET) kinase and anaplastic lymphoma kinase (ALK). J Med Chem. 2011;54:6342–63.
12. Fernández A. Transformative concepts for drug design: target wrapping. Heidelberg: Springer; 2010.
13. Seggewiss R, Loré K, Greiner E, Magnusson MK, Price DA, Douek DC, Dunbar CE, Wiestner A. Imatinib inhibits T-cell receptor-mediated T-cell proliferation and activation in a dose-dependent manner. Blood. 2005;105:2473–80.
14. Weichsel R, Dix C, Wooldridge L, Clement M, Fenton-May A, Sewell AK, Zezula J, Greiner E, Gostick E, Price DA, Einsele H, Seggewiss R. Profound inhibition of antigen-specific T-cell effector functions by dasatinib. Clin Cancer Res. 2008;14:2484–91.
15. Wölfl M, Langhammer F, Wiegering V, Eyrich M, Schlegel PG. Dasatinib medication causing profound immunosuppression in a patient after haploidentical SCT: functional assays from whole blood as diagnostic clues. Bone Marrow Transplant. 2013;48:875–7.
16. Heine A, Held SA, Bringmann A, Holderried TA, Brossart P. Immunomodulatory effects of anti-angiogenic drugs. Leukemia. 2011;25:899–905.
17. Davis MI, Hunt JP, Herrgard S, Ciceri P, Wodicka LM, Pallares G, Hocker M, Treiber DK, Zarrinkar PP. Comprehensive analysis of kinase inhibitor selectivity. Nat Biotechnol. 2011;29:1046–51.
18. Wolf D, Tilg H, Rumpold H, Gastl G, Wolf AM. The kinase inhibitor imatinib—an immunosuppressive drug? Curr Cancer Drug Targets. 2007;7:251–8.
19. Sillaber C, Herrmann H, Bennett K, Rix U, Baumgartner C, Böhm A, Herndlhofer S, Tschachler E, Superti-Furga G, Jäger U, Valent P. Immunosuppression and atypical infections in CML patients treated with dasatinib at 140 mg daily. Eur J Clin Invest. 2009;39:1098–109.
20. Kreutzman A, Juvonen V, Kairisto V, Ekblom M, Stenke L, Seggewiss R, Porkka K, Mustjoki S. Mono/oligoclonal T and NK cells are common in chronic myeloid leukemia patients at diagnosis and expand during dasatinib therapy. Blood. 2010;116:772–82.
21. Ravandi F. Dasatinib, an immunomodulator? Blood. 2010;116:673–4.
22. Karaman MW, Herrgard S, Treiber DK, Gallant P, Atteridge CE, Campbell BT, Chan KW, Ciceri P, Davis MI, Edeen PT, Faraoni R, Floyd M, Hunt JP, Lockhart DJ, Milanov ZV, Morrison MJ, Pallares G, Patel HK, Pritchard S, Wodicka LM, Zarrinkar PP. A quantitative analysis of kinase inhibitor selectivity. Nat Biotechnol. 2008;26:127–32.

23. Crunkhorn S. Anticancer drugs: redesigning kinase inhibitors. Nat Rev Drug Discov. 2008;7:120–1.
24. Demetri G. Structural reengineering of imatinib to decrease cardiac risk in cancer therapy. J Clin Invest. 2007;117:3650–3.
25. Fernández A, Sanguino A, Peng Z, Ozturk E, Chen J, Crespo A, Wulf S, Shavrin A, Qin C, Ma J, Trent J, Lin Y, Han HD, Mangala LS, Bankson JA, Gelovani J, Samarel A, Bornmann W, Sood AK, Lopez-Berestein G. An anticancer c-kit kinase inhibitor is reengineered to make it more active and less cardiotoxic. J Clin Invest. 2007;117:4044–54.
26. Kao J, Ko EC, Eisenstein S, Sikora AG, Fu S, Chen SH. Targeting immune suppressing myeloid-derived suppressor cells in oncology. Crit Rev Oncol Hematol. 2011;77:12–9.
27. Kanehisa M, Goto S, Sato Y, Kawashima M, Furumichi M, Tanabe M. Data, information, knowledge and principle: back to metabolism in KEGG. Nucleic Acids Res. 2014;42: D199–205.
28. Das J, Brystol-Myers Squibb (2003) Cyclic protein tyrosine kinase inhibitors. U.S. Patent 6,596,746
29. Crespo A, Fernández A. Kinase packing defects as drug targets. Drug Discov Today. 2007;12:917–23.
30. Fernández A, Bornmann W, Lopez-Berestein G, Sanguino A, Peng Z, Sood AK, The Board of Regents of the University of Texas System (2013) Methods and composition of matter related to wrapping of dehydrons. U.S. Patent 8,466,154
31. Fei F, Yu Y, Schmitt A, Rojewski MT, Chen B, Greiner J, Götz M, Guillaume P, Döhner H, Bunjes D, Schmitt M. Dasatinib exerts an immunosuppressive effect on CD8+ T cells specific for viral and leukemia antigens. Exp Hematol. 2008;36:1297–308.
32. Sondel P (2012) Unpublished results (personal communication)
33. Weber J. Overcoming immunologic tolerance to melanoma: targeting CTLA-4 with ipilimumab (MDX-010). Oncologist. 2008;13:16–25.
34. Gallo RC, Sarin PS, Gelmann EP, Robert-Guroff M, Richardson E, Kalyanaraman VS, Mann D, Sidhu GD, Stahl RE, Zolla-Pazner S, Leibowitch J, Popovic M. Isolation of human T-cell leukemia virus in acquired immune deficiency syndrome (AIDS). Science. 1983;220:865–7.
35. Haas DA, Bala K, Büsche G, Weidner-Glunde M, Santag S, Kati S, Gramolelli S, Damas M, Dittrich-Breiholz O, Kracht M, Rückert J, Varga Z, Keri G, Schulz TF. The inflammatory kinase MAP4K4 promotes reactivation of Kaposi's sarcoma herpesvirus and enhances the invasiveness of infected endothelial cells. PLoS Pathog. 2013;9:e1003737.
36. Fernández A. Synergizing immunotherapy with molecularly targeted anticancer treatment. Drug Discov Today. 2014;19:1427–32.
37. Maus MV, Grupp SA, Porter DL, June CH. Antibody-modified T cells: CARs take the front seat for hematologic malignancies. Blood. 2014;123:2625–35.
38. Maher J. Clinical immunotherapy of B-cell malignancy using CD19-targeted CAR T-cells. Curr Gene Ther. 2014;14:35–43.

Chapter 15
Quantum Mechanical Concepts for Epistructural Drug Design

In this chapter we show that a quantum mechanics (QM) approach to the epistructural physics of ligand-target association significantly broadens the technological base of the wrapping technology for drug design. We specifically incorporate electron correlation effects to empower the paradigmatic concept of "dehydron-wrapping drug". The QM approach provides the required scientific grounds for the incorporation of halogens as wrapping groups in the chemical scaffold of the drug exploited for molecular targeted therapy. The chapter explores the possibility that the group that wraps exogenously a dehydron upon drug binding may also effectively interact through quantum mechanical forces with the carbonyl oxygen paired by the preformed dehydron in the target protein. This type of interaction involves dispersion forces that induce an anisotropic electron distribution on the halogen sigma orbital named "sigma hole". This electron anisotropy promotes the formation of a halogen bond with the carbonyl oxygen in the target protein. Thus, the intermolecular halogen bond is coupled to the dehydron-wrapping interaction, and the wrapping group is a halogen with significant electron polarizability (Cl < Br < I), capable of inducing a significant dispersion force through QM displacement of electron density within the sigma orbital. This novel modality of ligand association involves two coupled drug-target interactions branching from the same drug substituent. The inclusion of polarizable halogens as wrappers is likely to significantly empower drug design as it reinforces the dehydronic drag exerted on the drug through favorable electron-correlation effects. The QM-based epistructural design is illustrated by modifying the chemotype of anticancer drug *imatinib* to incorporate a wrapping halogen in order to overcome drug resistance due to mutation in the imatinib target c-Kit kinase.

A. Fernández, *Physics at the Biomolecular Interface*, Soft and Biological Matter,
DOI 10.1007/978-3-319-30852-4_15

15.1 Quantum Mechanical Approach to Epistructural Drug Design

The standard physico-chemical analysis of drug-target associations tends to ignore high-level quantum mechanical effects arising from dispersion forces that realize electron correlation effects. Thus, docking algorithms are typically based on force field approaches that are inadequate to handle quantum mechanical effects [1]. To illustrate the magnitude of the problem, we shall focus on one of the major shortcomings of force fields: the assignment of one partial charge to each atom [2]. This picture fails to describe the anisotropic nature of electron density induced on polarizable halogens like X = Cl, Br, I, through dispersion forces. These effects arise from induced deformations of the electron density distribution along molecular orbitals. Such deformations are best illustrated by the so-called "sigma-hole", a positively charged region at the tip of the sigma orbital along the C–X bond, representing electron-withdrawal induced by the spatial proximity of a nucleophilic moiety such as a carbonyl or carboxyl oxygen. These dispersion forces are only operative under stringent orientation and spatial demands: the X---O distance must be within 0.2–0.3 Å of the sum of van der Waals radii of the two atoms (~3 Å for the I-bond) and the C–X---O approach must be nearly collinear, with a C–X---O angle within the range $180° \mp 25°$ to contribute energetically within the thermodynamically meaningful range in the interval [RT, 3.5 kcal/mol]. The upper bounds in enthalpies of halogen-bond formation correspond to halogens involved in aryl halogenates, as the electron delocalization -and thereby polarizability- introduced by the π-cloud enhances significantly the deformability of the electron distribution on the halogen sigma bond, enlarging the sigma hole [1].

These quantum mechanical effects have led scientists to postulate the existence of the so called C–X---O "halogen bonds" or "X-bonds". Such interactions cannot be captured by density functional theory (DFT) approaches, which have long been recognized as inadequate to account for dispersion forces [3]. They would require high-level (and costly) computational approaches such as those provided by coupled cluster methods (MP2) or third-generation dispersion-corrected DFT [1]. However, such methods would be again inadequate for docking as required by high-throughput molecular design. A provisional compromise that appears promising arises from the incorporation of empirical scoring functions for the recognition of halogen bonds in drug-target associations [4]. Yet no scoring method is able to incorporate the *cooperativity arising from X-bond and wrapping* that would materialize if the halogen is X-bonded to a carbonyl oxygen which is simultaneously engaged in a pre-formed hydrogen bond with an amide group (Fig. 15.1). This novel wrapping motif involves highly orientational quantum mechanical effects and would enable the use of halogens as wrapping groups, as described in the next section.

15.2 Halogen Bond Synergizing with Dehydron Wrapping: A Novel Motif in Drug Design

This section explores the possibility that a halogen in the drug may be used as a wrapper of a preformed dehydron on the surface of the target protein, as depicted in Fig. 15.1. The geometry of this motif should be specified with considerable particularity attending to the stringent orientation demands of the X-bond. Thus, a halogen with high polarizability (Cl, Br, I) may be used as a wrapping group in a drug if the structure of the drug-target complex places it within the geometric constraints described in Sect. 15.1, and approximately orthogonal to the dehydron. Thus, the hereby termed "X-wrapper" has a dual interactive role as it expels labile interfacial water (Chap. 4) while interacting at the quantum level with the dehydron-paired carbonyl oxygen. This implies that the two quasi-orthogonal interactions, hydrogen bond and X-bond, are not really uncoupled from each other, as certain sources would claim, but rather engage in a cooperative interplay. This cooperativity only becomes apparent after acknowledging the wrapping role of the halogen as remover of interfacial water at the dehydron location.

As it turns out, the X-wrapping motif is abundant in the structural repositories of halogen-containing drugs: It is found in every X-bond-H-bond near-orthogonality that arises upon drug-target association. For instance, drugs designed to target protein helices by X-bonding backbone carbonyls are clear illustrations of X-wrapping. In Fig. 15.2 we recognize an X-wrapping interaction in the complex (PDB.2BXL) between 2-Hydroxyl-3,5-diiodobenzoic acid and human serum

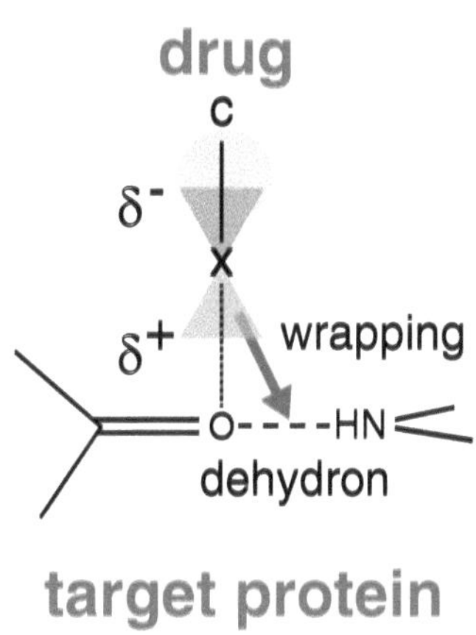

Fig. 15.1 The X-wrapping motif in drug-target associations. The polarizable halogen (X) is engaged in a dual role: (**a**) X-bonding with the dehydron-paired carbonyl oxygen; (**b**) Wrapping the dehydron that engages the X-bonded carbonyl oxygen. Thus, the X-bond and wrapping interaction become cooperative and materialize synergistically within a near-orthogonal X-bond-H-bond arrangement. The sigma hole with its dimpled electron distribution and induced charge separation (δ^+, δ^-) is represented as a truncated orbital with rotational symmetry along the C–X-axis

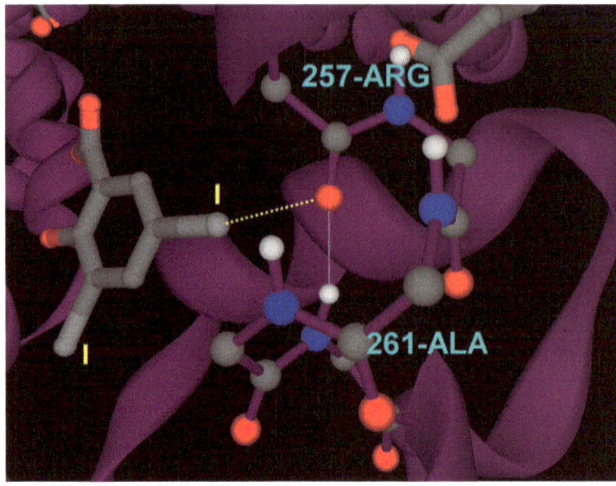

Fig. 15.2 Realization of the X-wrapping motif within the association of 2-Hydroxyl-3,5-diiodobenzoic acid with transport protein human serum albumin (HSA) in the PDB complex with entry 2BXL. The X-bond is represented by a dotted line and the drug iodine wraps the preformed Arg257-Ala261 dehydron

albumin (HSA), a crucial drug transporter. Thus, upon binding, the iodine in the drug is engaged in an X-bond with the carbonyl oxygen of residue Asp257, while it also wraps the Asp257-Ala261 dehydron, contributing to the helix stabilization.

The complexity associated with correcting scoring functions or force fields to incorporate X-wrapping is truly daunting. This is because this motif subsumes a cooperative interplay between quantum-mechanical and interfacial-tension effects which are obviously of a very different nature and operative at different length scales. Only empirical scoring and, provisionally, physico-chemical intuition, can guide lead optimization at this juncture in our understanding of biomolecular interfaces. The intent of this book has been precisely to invite and strengthen this kind of intuition.

15.3 QM-Based Redesign of *Imatinib* to Overcome Drug Resistance

In this section, the QM-based epistructural design is illustrated by modifying the chemical structure of the anticancer drug *imatinib* to incorporate a wrapping halogen in order to overcome a pattern of drug resistance arising in an imatinib-targeted kinase. Specifically, we focus on the drug resistance pattern created by somatic mutation D816V in the imatinib target c-Kit kinase [5].

As it is established [5], the mutation D816V confers drug resistance to the c-Kit-targeted molecular therapy that utilizes imatinib to treat GIST (Gastrointestinal Stromal Tumor). By raising the number of carbonaceous HB-wrapping groups in the protein chain, this mutation induces dehydron F811-A814 in the activation loop of c-Kit kinase (Fig. 15.3), which is significantly floppier in the wild type. Calculated with the methods given in Chap. 1, the interfacial free energy of the mutant $\Delta G_{if}(m)$ is significantly larger than that of the wild type $\Delta G_{if}(0)$, with $\Delta G_{if}(m) - \Delta$

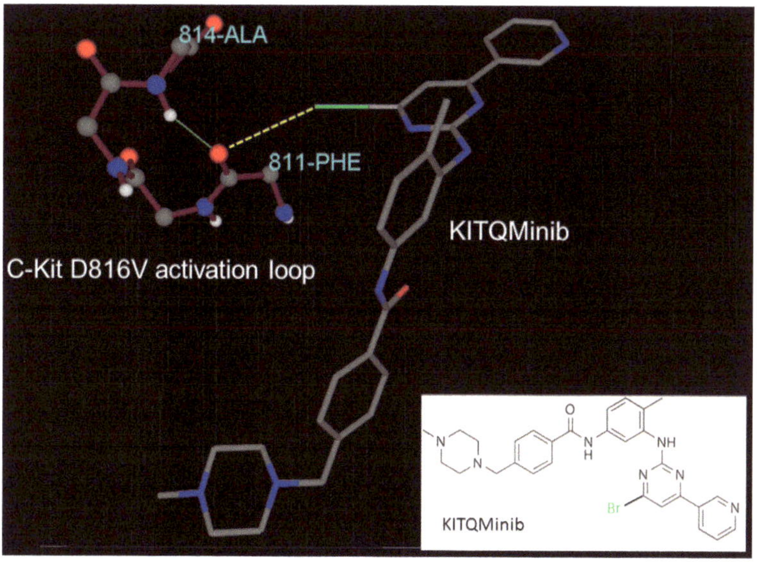

Fig. 15.3 Engineered halogen (Br) wrapper built into the chemical scaffold of imatinib to overcome drug resistance in imatinib target c-Kit. Imatinib redesign prototype KITQMinib becomes a wrapper of the preformed dehydron F811-A814 induced by drug-resistance mutation D816V in c-Kit. The wrapping effect is reinforced by the halogen bond (*dashed line*) involving the bromine in redesigned imatinib and the oxygen from the backbone carbonyl from residue F811. The inset shows the chemical structure of ligand KITQMinib

$G_{if}(0) \approx 2.3$ kcal/mol. Thus, an imatinib methylated variant named WBZ_7 [5] has been previously designed to wrap the interfacial tension generator F811-A814 dehydron in order to overcome drug resistance. The prototype WBZ_7 has been experimentally shown to be a nanomolar inhibitor ($K_D \approx 39$ nM) of the imatinib-resistant D816V c-Kit mutant while also retaining nanomolar activity against the wild-type c-Kit kinase ($K_D \approx 21$ nM) [5].

An even higher affinity for the wild type and drug-resistant c-Kit kinases may be achieved by engineering a halogen wrapper built into the chemical scaffold of imatinib, as shown in Fig. 15.3. Thus, the brominated version of imatinib, named KITQMinib, is engineered based on the QM concept of halogen wrapper previously introduced in this chapter. The prototype KITQMinib becomes a wrapper of the mutation-induced preformed dehydron F811-A814 and the wrapping effect is reinforced or buttressed by the halogen bond (dashed line, Fig. 15.3) pairing the bromine imatinib substituent and the oxygen from the backbone carbonyl from residue F811. The affinity of KITQMinib for the c-Kit D816V mutant is experimentally determined (see protocol in [5]) at $K_D \approx 32$ nM, thus it is somewhat enhanced beyond the "pure wrapping" level furnished by WBZ_7. The affinity of KITQMinib for the wild-type c-Kit kinase remains at the levels comparable to WBZ_7, with $K_D \approx 20$ nM.

15.4 Problems

15.1. Identify the X-wrapping bond between gefitinib and its target EGFR kinase in PDB entry 2ITY. The chemical structure of gefitinib emerges from the 3D rendering of the PDB structural coordinates and may be alternatively found in the answer display (Fig. P15.1).

15.2. Lead Optimization: Redesign the FAK (focal adhesion kinase) inhibitor TAE226 with chemical structure shown in Fig 15.4 to fulfill the following affinity-related demands (Fig 15.5):

(I) An X-bond with the backbone carbonyl of Leu567
(II) H-bond with the carbonyl oxygen in Ile428
(III) An H-bond with the backbone carbonyl of Glu500
(IV) Wrapping of dehydron Cys502-Gly505 by engineering an X-wrapper drug substituent

The PDB file for the FAK/TAE226 complex is 2JKK. In redesigning the TAE226 compound, take into account the torsional degrees of freedom that may be engineered in the drug to enable the putative interactions.

15.3. As demonstrated in Chap. 12, an AMPK inhibitor could be used in combined molecular cancer therapy. In the therapeutic context of interest, an AMPK inhibitor may be exploited to suppress the mechanism that cancer cells deploy to cope with drug-induced metabolic stress. An available sub-micromolar AMPK inhibitor is dorsomorphin, also known as "compound C", whose chemical structure is displayed in Fig. 15.6. Crystallographic evidence from

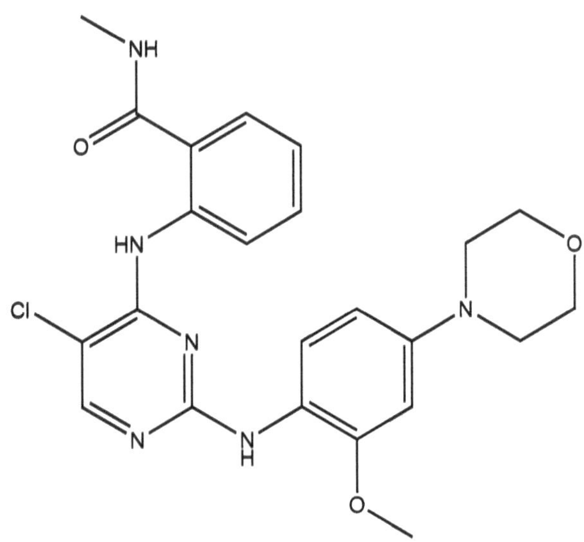

Fig. 15.4 Chemical structure of FAK inhibitor TAE226

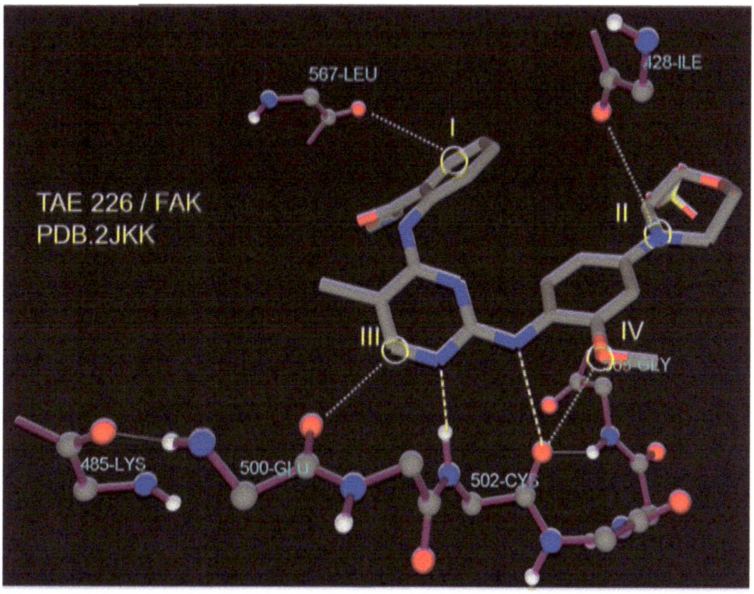

Fig. 15.5 Structural details of the complex PDB.2JKK between focal adhesion kinase (FAK) and its inhibitor TAE226 guiding a lead optimization

6-[4-[2-(1-Piperidinyl)ethoxy]phenyl]-3-(4-pyridinyl)pyrazolo[1,5-a]pyrimidine

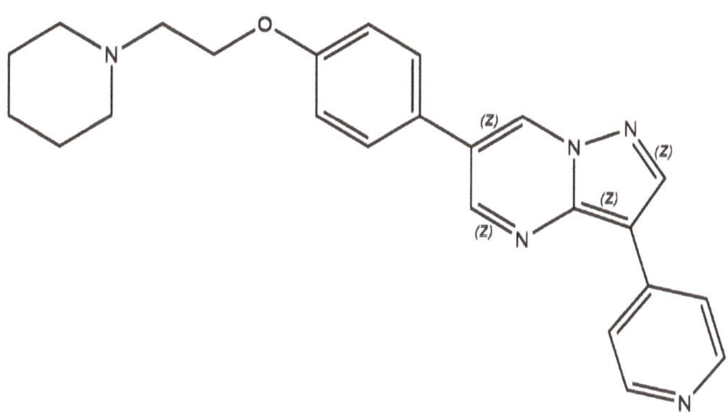

Fig. 15.6 Chemical structure of AMPK inhibitor dorsomorphin

the dorsomorphin/AMPK complex (PDB.3AQV) reveals extensive induced fit leading to the formation of helix 101–109 in the nucleotide-binding loop of AMPK (Fig. 15.7). Specifically, the terminal piperidine ring induces the formation of dehydrons D103-K107 and Y104-H108 by contributing nonpolar carbonaceous groups to their desolvation domains (Fig. 15.8).

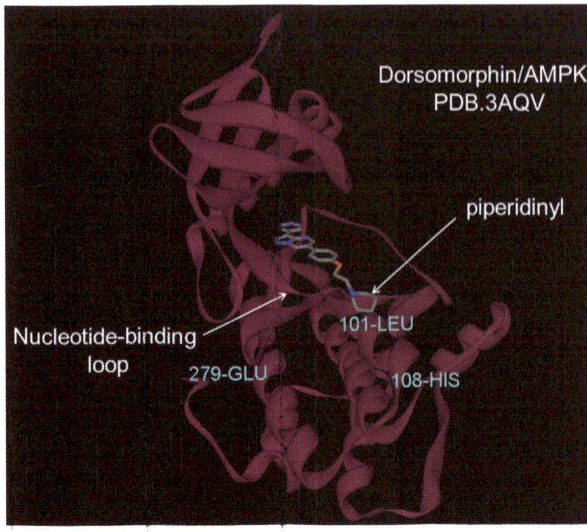

Fig. 15.7 Structural rendering of dorsomorphin in complex with target AMPK, from PDB entry 3AQV

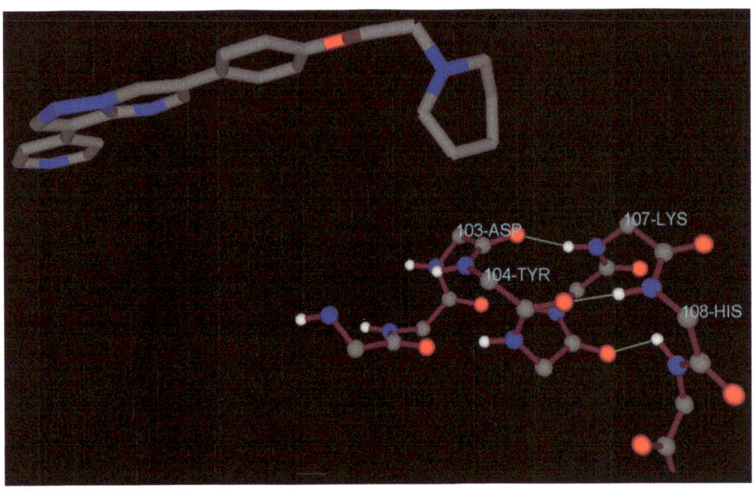

Fig. 15.8 Induced dehydrons in AMPK region 101–109 upon association with dorsomorphin

Given this structural context, optimize dorsomorphin to improve its AMPK-affinity through rational redesign based on the QM concept put forth in this chapter. To that effect, (a) exploit the concept of X-wrapping to construct an X-bond between the drug and D103 that simultaneously wraps the D103-K107 dehydron; and (b) design a cation-π interaction between K107 and the drug by judiciously replacing the terminal piperidinyl for a suitable aryl group (Fig. 15.9).

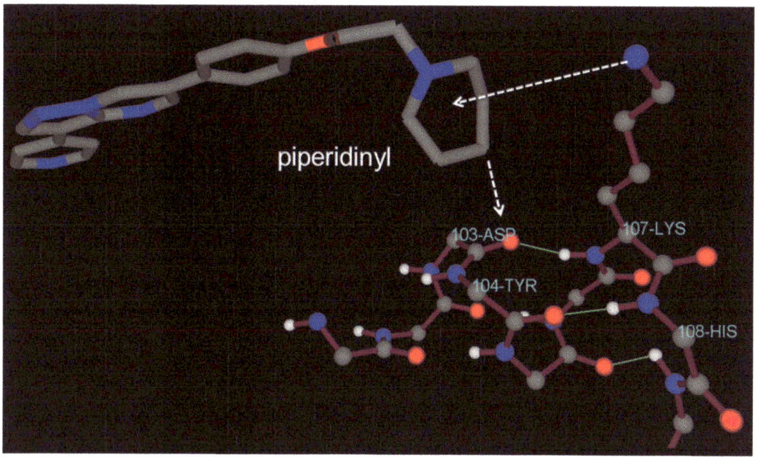

Fig. 15.9 Proposal to redesign dorsomorphin. The dashed arrows signify planned cation-π interaction and X-bond/X-wrapping

The chemical structure of optimized re-engineered dorsomorphin is displayed in Fig. P15.3 (Appendix B). The re-engineered compound displayed in Fig. P15.3 is an ATP-competitive inhibitor of AMPK with K_D=19.0 nM. Targeted AMPK dehydrons 103–107 and 104–108 are selectivity filters and consequently K_D>>100 nM for all other kinases in the human kinome according to a recent high throughput screening using a T7-bacteriophage library (protocols in Chaps. 9, 13).

References

1. Wilcken R, Zimmermann MO, Lange A, Joerger AC, Boeckler FM. Principles and applications of halogen bonding in medicinal chemistry and chemical biology. J Med Chem. 2013;56:1363–88.
2. Wang J, Wang W, Kollman PA, Case DA. Automatic atom type and bond type perception in molecular mechanical calculations. J Mol Graph Model. 2006;25:247–60.
3. Lu YX, Zou JW, Fan JC, Zhao WN, Jiang YJ, Yu QS. Ab initio calculations on halogen-bonded complexes and comparison with density functional methods. J Comput Chem. 2009;30:725–32.
4. Kuhn B, Fuchs JE, Reutlinger M, Stahl M, Taylor NR. Rationalizing tight ligand binding through cooperative interaction networks. J Chem Inf Model. 2011;51:3180–98.
5. Fernández A, Sanguino A, Peng Z, Crespo A, Ozturk E, Zhang X, Wang S, Bornmann W, Lopez-Berestein G. Rational drug redesign to overcome drug resistance in cancer therapy: imatinib moving target. Cancer Res. 2007;67:4028–33.

Chapter 16
Structure-Based Drug Discovery Without Structure: Working Around the Paradox to Disrupt Protein-Protein Associations

Drug discovery has been focusing for some time on protein-protein (PP) associations, the basic molecular events in biology. The recruitment of protein complexes is required to initiate and propagate signaling cascades, regulate enzyme activity, articulate and control mechanistic processes involving molecular motors, etc. When such associations engage altered binding partners, complex formation may lead to deregulation of biological functions and the drug-based disruption of the aberrant associations may represent new therapeutic opportunities.

However, the disruption of protein-protein interfaces (PPIs) remains a major challenge in drug discovery. The problem becomes daunting when the structure of the target protein is unknown and is further complicated when the interface is susceptible to disruptive phosphorylation. Based solely on protein sequence and on information on phosphorylation-susceptible sites within the PPI, a new technology is described to identify drug leads to inhibit protein associations in the absence of target structure. The novel technology is illustrated by a patented invention to treat heart failure (US Patent 9,051,387). The success of this technology as implemented by Drs. Richard L. Moss and Ariel Fernández shows that it is possible to generate drug leads in the absence of target structure.

In cardiomyocytes, the myosin-myosin-binding protein C (MyBP-C) association creates a molecular brake on contractility imposed by the repressive activity of the unphosphorylated form of MyBP-C on myosin. In heart failure, MyBP-C is phosphorylated minimally or not at all due to down-regulation of β-adrenergic receptors. Since phosphorylation of MyBP-C improves contraction, US patent 9,051,387 proposed to target the phosphorylation site on MyBP-C with a pharmaceutical designed to disrupt its interaction with myosin and thereby improve cardiac function. This problem involves designing an optimal MyBP-C-derived peptide. The peptide contains the motif responsible for interaction with myosin and hence disrupts the MyBP-C-myosin association as required for therapeutic action. The myosin-binding peptide-based therapeutic agent may be identified based on the premise that disruption of the myosin-MyBP-C interface would release the

A. Fernández, *Physics at the Biomolecular Interface*, Soft and Biological Matter,
DOI 10.1007/978-3-319-30852-4_16

molecular brakes on cardiomyocyte contractility imposed by the repressive activity of the unphosphorylated form of MyBP-C on myosin.

The 3D-structure of the 11-domain MyBP-C is unknown, and therefore it was decided to design the peptide based on the output from a sequence-based predictor of native disorder. The peptide was developed based on a prediction of the dehydron-rich region that constitutes the putative myosin-binding site. To identify the peptide sequence, a region in the twilight zone between order and disorder was identified based on the methodology described in Chap. 5. In this way, it became possible to identify the region containing the operational phosphorylation sites in cMyBP-C and determine the sequence of the patented peptide that may be used as a lead to the therapeutic agent to treat heart failure. The invention has been shown to significantly increase cardiac contractile force and frequency in the failing heart.

16.1 Therapeutic Disruption of Protein-Protein Interfaces

Drug discovery endeavors have been focusing for some time on protein-protein (PP) associations, the basic molecular events in biology [1]. The recruitment of protein complexes is required to initiate and propagate signaling cascades, regulate enzyme activity, articulate and control mechanistic processes involving molecular motors, etc. When such associations engage altered binding partners, complex formation may lead to deregulation of biological functions and the drug-based disruption of the aberrant associations may represent new therapeutic opportunities [2–4].

Major problems arise in the identification of drug leads and optimization strategies for small compounds involved in the disruption of protein-protein interfaces (PPIs) [4]. The latter tend to have low surface curvature and often extend over more than 1000 $Å^2$ on the protein surface, in contrast with the smaller cavities where natural ligands typically bind [5]. The absence of obvious leads and the sheer size of the binding surface make it difficult to identify candidate compounds that would disrupt PPIs. In spite of these obstacles, it becomes possible in some cases to identify suitable leads and even implement optimization strategies. For example, in the MDM2/p53 complex, MDM2 (murine double minute 2) has been identified as the E3-ligase responsible for the ubiquitin-related degradation of the master tumor suppressor p53 [6], and the disruption of the MDM2/p53 complex promotes the onset of many cell-fate processes that may halt cancer development and progression [6, 7]. In tumors, MDM2 is overexpressed and altered so that cell processes associated with senescence, cell-cycle arrest and apoptosis triggered by p53 activity are suppressed through untiring modulation of p53 via an aberrantly persistent MDM2/p53 complex. In this case, crystallographic structure-based analysis coupled with high-throughput screening has generated useful leads, the so-called Nutlins [7]. These leads ultimately steered the discovery of compounds of low molecular weight that hold great promise as anticancer agents through the disruption of the MDM2/p53 PPI [6, 8].

To address many of the problems related to the epitope size in PP associations, methodologies have been implemented for the identification of "hot spots" or sites that make a significant contribution to binding [9]. Such approaches are typically based on alanine scanning, assessing the impact of single-residue alanine substitutions (beta-carbon truncation except for glycines) on the binding free energy. Thus, an effective epitope significantly smaller than the PPI and consisting of the residues with most significant contribution to binding is determined [4, 10]. Once the size of the epitope has been significantly reduced, fragment-based lead discovery may be utilized to generate promising candidates for competitive binding [4, 11, 12]. Once in the optimization phase, the discovery process is aided by biophysical methods for structure analysis, including crystallography, disulfide tethering, surface plasmon resonance and nuclear magnetic resonance [12].

16.2 Therapeutic Inhibition of Protein-Protein Associations with Unknown Complex Structure

The problem of therapeutic disruption of a PP association becomes especially difficult when the structure of the targeted binding partner is unknown, since the biophysical methods mentioned cannot render useful information. The discovery process becomes even further complicated when the interface can be naturally disrupted by phosphorylation at a specific location on the epitope of one of the binding partners, as is often the case in the regulation of activities that recruit complexes. Even in such difficult cases, we uphold the opinion that it is possible to implement a drug discovery platform based solely on sequence-based predictors of binding epitopes endowed with chemical functionality.

The novel approach hinges on three conceptual tenets that will be described in the following sections and contrasted vis-à-vis current methodologies: (a) structural defects in proteins, known as *dehydrons*, promote water exclusion at the interface, and thus residues paired by dehydrons constitute hot spots promoting protein associations [13]; (b) dehydrons are identified as order-disorder twilight regions along the protein sequence [14] and therefore may be inferred utilizing a sequence-based predictor of intrinsic disorder [15], and (c) dehydrons functionalize PPI-disruptive phosphorylation sites in their proximity [16, 17]. The efficacy of the technology will be subsequently illustrated by a recently patented invention to treat heart failure through disruption of the myosin-MyBP-C interface [18]. MyBP-C refers to a multi-domain myosin-binding protein with unreported structure that is a central regulator of cardiac contractility [19, 20]. MyBP-C molecules constitute molecular brakes modulating the displacement of myosin motors, with a brake-release mechanism hinging on site-specific phosphorylation. By sequence-based inference of dehydron-rich regions in MyBP-C, we identified drug leads to cure heart failure [18]. Here we propose to describe the technological advances utilized in the invention to enable therapeutic disruption of protein-protein associations in a generic context.

16.3 Binding Hot Spots and Dehydron-Rich Epitopes

As described in Chap. 5, the integrity of a soluble protein is contingent on the capability of its structure to exclude water from backbone amide-carbonyl hydrogen bonds. Water-exposed intramolecular hydrogen bonds, the so-called *dehydrons*, constitute structural defects taking the particular form of wrapping deficiencies [21], as previously described in the Chaps. 1–7 that deal with science fundamentals. By wrapping we mean the extent to which the backbone hydrogen bond is shielded from hydration as it is surrounded by side-chain carbonaceous groups. These defects favor removal of surrounding water as a means to strengthen and stabilize the underlying electrostatic interaction, and thus are predictably implicated in protein associations. By exogenously contributing to the wrapping of pre-formed hydrogen bonds, PP associations in effect remove the wrapping defects, thereby stabilizing the structure [21].

As described in Chap. 8, intramolecular hydrogen bonds that are not "wrapped" by a sufficient number of nonpolar groups in the protein itself may become stabilized and strengthened by the attachment of a ligand, i.e. a potential drug, or a binding partner that further contributes to their dehydration. Ample bioinformatics evidence on the distribution of dehydrons at the interface of protein complexes support this physical picture [22]. Thus, dehydrons have been identified as decisive factors driving formation of protein complexes (Chaps. 1 and 5).

Dehydrons may be identified from structural coordinates using available software [23], and a code for dehydron identification written as a PyMol open source is presented in Appendix A [24]. To describe the extent of backbone shielding from hydration we introduce a quantifier of hydrogen-bond wrapping, ρ, indicating the number of nonpolar groups contained within a "desolvation domain" around the bond. Insufficiently wrapped bonds become deshielded and constitute dehydrons. As said, this approach requires structure and we need a way to identify dehydron locations from sequence alone. One alternative is to predict structural coordinates from sequence [25] but unfortunately too many decoys with significant wrapping variability are often generated.

Dehydron-rich regions in soluble proteins are typical hot spots for protein associations because of their propensity towards further dehydration. A functional perspective reinforces this view, since dehydrons constitute vulnerabilities that need to be "corrected" to maintain the integrity of the protein structure and its functional competence. Thus, specific residues of the binding partner contribute to the stabilizing dehydration of preformed dehydrons, as they penetrate their desolvation domain upon association.

A solvent-centric perspective is even more informative on the role of dehydrons as hot spots driving PP associations. The water molecules partially occluded in the dehydron nano-environment are frustrated in their hydrogen-bond coordination, and hence they generate interfacial tension. This tension is in turn released upon PP association, as the frustrated water molecules are removed from the epitope [13]. Thus, the residues pairing or significantly wrapping preformed dehydrons in a binding partner are in fact expected and verified to be hot spots driving the PP association [13, 21].

16.4 Identifying Binding Epitopes Using a Sequence-Based Predictor

The structural integrity of a soluble protein is contingent on its capacity to exclude water from backbone amide-carbonyl hydrogen bonds. This implies that proteins with dehydron-rich regions must rely on binding partnerships to maintain their structural integrity [26].

As described in Chap. 5, dehydron-rich regions identified on the protein sequence may be characterized as belonging to a "twilight zone" between order and native disorder [14, 22]. This characterization is suggested by a strong correlation between wrapping of intramolecular hydrogen bonds (ρ) and propensity for structural disorder (f_d). The correlation reflects the fact that a local incapacity to exclude water intramolecularly from pre-formed hydrogen bonds is causative of a local loss of structural integrity, whereby full backbone hydration becomes structurally disruptive.

The local disorder propensity may be accurately quantified by a sequence-based score generated by a predictor of native disorder propensity such as PONDR-VLXT [15] or other software [27] that takes into account residue attributes such as hydrophilicity, aromaticity, and their distribution within the window of the protein sequence interrogated [15]. The disorder score ($0 \leq f_d \leq 1$) is assigned to each residue within a sliding window, representing the predicted propensity of the residue to be in a disordered region ($f_d = 1$, certainty of disorder; $f_d = 0$, certainty of order). Only 6 % of 1100 nonhomologous PDB proteins gave false positive predictions of disorder in sequence windows of 40 amino acids [22]. The strong correlation (over 2806 nonredundant nonhomologous PDB domains) between disorder score of a residue and extent of wrapping of the hydrogen bond engaging the residue (if any) implies that dehydrons correspond to structurally vulnerable regions [22]. Hence the characterization of dehydrons as belonging to the order-disorder twilight range $0.35 < f_D < 0.8$ and flanked by ordered and disordered regions is warranted.

A major caveat applies to the use of disorder predictors to infer dehydrons [28]. Dehydrons require detailed information resolved at the residue level, whereas disorder predictors such as PONDR provide a smeared out signal as a window of fixed length slides along the protein sequence. In practice, learning technologies for resolution enhancement (LREs) such as *Twilighter* [28] are required to deconvolute the PONDR signal. The LREs use a training set constructed from protein sequence windows combined with reliable structure-based dehydron identification within the windows and PONDR plots for PDB-reported proteins.

Both order and native disorder are well-characterized structural attributes. However, the vulnerable regions in a soluble structure that promote protein associations [13, 14] belong to the novel category of "tamed disorder". One may say that neither order nor disorder is an adequate category to describe dehydron-rich regions.

To illustrate the efficacy of sequence-based dehydron inference [15, 22] (Fig. 16.1), we contrast it against structure-based determination, considering the HIV-1 protease in its catalytically competent dimer (PDB.1A30). The free monomers have a large structural twilight content (54 %, Fig. 16.1), as expected since the protein is natively dimeric. Two dehydron-rich regions are detectable with the software described in [23]: a catalytic region and a flap region (Fig. 16.2). Dehydrons Asp25-Ala28 ($\rho = 22$) and Ala28-Arg87 ($\rho = 15$) are spatially adjacent to the catalytically competent Asp25, and water exclusion promoted by these dehydrons serves the dual purpose of fostering the association of the peptide substrate and enabling the nucleophilic attack by Asp25 by lowering the local dielectric permittivity and preventing the hydrolytic reversal of the enzyme-substrate adduct back to reactants. The flap dehydron Gly49-Gly52 ($\rho = 11$) is required to confer flexibility to the flap region, as required for the mechanistic

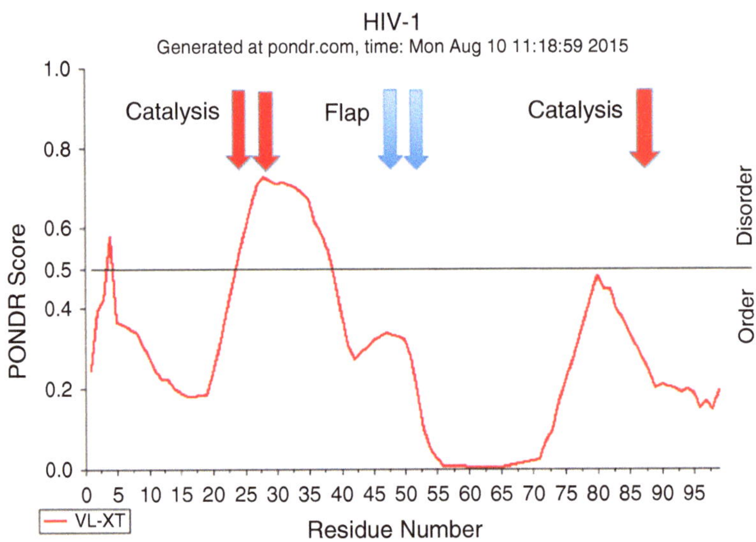

Fig. 16.1 Disorder propensity plot for a subunit of the HIV-1 protease functional homodimer generated by PONDR-VLXT [15]. The residues in the order-disorder twilight region representing dehydrons are marked by *arrows* and correspond to functionally relevant locations on the structure of the homodimeric complex, as shown in Fig. 16.2. Reprinted from [Fernandez A, Scott R. Drug leads for interactive protein targets with unknown structure. Drug Discovery Today 2016; 21:531–5] copyright 2016 with permission from Elsevier

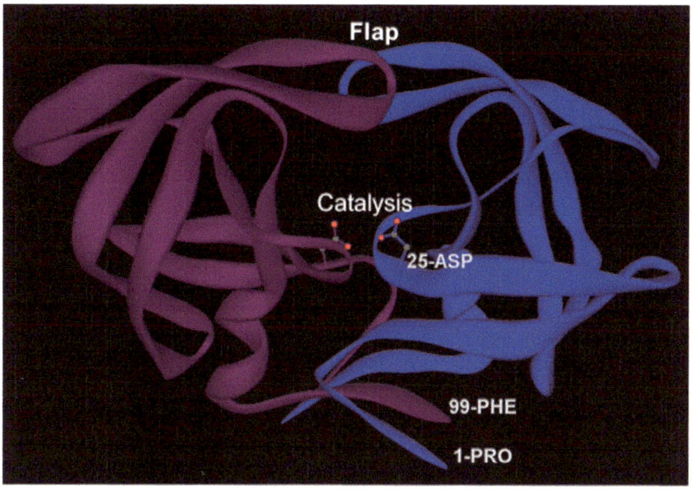

Fig. 16.2 Ribbon representation of the functionally competent homodimeric HIV-1 protease (PDB.1A30). The individual protein chains are displayed in *magenta* and *blue* with indication of catalytic and flap regions. The catalytic residue Asp25 is displayed with full side chain. Reprinted from [Fernandez A, Scott R. Drug leads for interactive protein targets with unknown structure. Drug Discovery Today 2016; 21:531–5], copyright 2016 with permission from Elsevier

processivity of the enzyme. In a sequence-based analysis, the dehydrons that shape the catalytic and flap regions may be directly inferred from the order-disorder twilight regions in the PONDR plot for the protease primary sequence (Fig. 16.1). Notice that the flap and catalytic dehydrons lie precisely at the lower and upper boundaries of the twilight region, a common occurrence since PONDR only gives a smeared out signal [28].

Although the sequence-based determination of dehydron-rich regions is less precise than a structure-based approach, it is precise enough to identify lead candidates of interest (cf US Patent 9,051,387). As said, a significant improvement of the dehydron signal in sequence-based prediction may be achieved using the Twilighter LRE, as described in [28].

16.5 Learning Technologies to Discover Epitopes for Protein Associations in the Absence of Structure

As indicated in Chap. 5, the program PONDR or other predictors of native disorder generate a smeared-out plot resulting from inference of disorder propensity upon sliding a window of fixed length along the protein sequence. The detailed analysis required in drug design demands a higher resolution of the smeared-out signal to obtain individual residue values of disorder propensity. This fine structure is missed in the PONDR output.

A protocol to resolve PONDR signals to the level of fine graining needed for drug design was provided in Chap. 5. The raw data uses PONDR to infer dehydrons as twilight regions in accord with Fig. 5.1, but uses also structural information that serve as templates in a training set to generate the desired level of resolution of the dehydron signal. Thus, the output from a structure-based dehydron calculator (Appendix A) is needed to supplement the twilight inference provided by PONDR. In this way, the PONDR output is read in such a way that the machine "learns to infer dehydrons", becoming a learned resolution enhancer (LRE). In this way, a supervised learning resource named *Twilighter*, is trained with [sequence/PONDR-score/(# of dehydrons)] windows extracted from the PDB and from the PONDR plots generated for PDB-reported proteins [28]. The Twilighter maps the number of dehydrons onto a sequence/PONDR-score window to generate the wrapping parameter $\nu = D/H$ (D = number of dehydrons, H = number of backbone hydrogen bonds). In this way, the dehydron microstructure embedded in the PONDR signal emerges when PONDR output is analyzed by the LRE.

Thus, Twilighter adopts as training set the completely reliable structure-based dehydron identification obtained using Dehydron Calculator (Appendix A) for a large proportion (typically 66 %) of the proteins reported in the PDB combined with the PONDR plot output on the protein sequences for the same PDB-entries. In this way, the learning machine is trained to interpret each PONDR signal from a PDB reported protein as containing a well determined number of dehydrons that would not be identifiable from direct reading of the PONDR plot. This training is then used to make sequence-based dehydron inferences based on PONDR scores for sequence windows of optimized length. The testing set for this learning machine is the rest (approximately 33 %) of PDB reported proteins, where we can directly contrast a learned dehydron prediction with the actual identification of dehydrons obtained using Dehydron Calculator.

16.6 Dehydron Epitopes Activate Phosphorylation-Susceptible Residues

As extensively described in Chap. 7, where the foundations of frustration-based chemistry are laid, dehydrons in the proximity of residues susceptible of getting phosphorylated have been found to chemically functionalize such residues, enabling their nucleophilic attack on the terminal phosphoester linkage of ATP [16, 17]. This chemical role is mediated by an interfacial water molecule around a dehydron that becomes frustrated due to impaired hydrogen bond coordination, thus behaving as a chemical base. Thus, the dehydron-associated interfacial water molecule promotes proton acceptance [17], thereby enhancing the nucleophilicity of a phosphorylation-prone residue in the dehydron proximity. These findings suggest that we can target a PPI susceptible of disruptive phosphorylation by exogenously wrapping the pre-formed dehydrons that are part of the binding epitope and enable phosphorylation in one of the binding partners.

16.7 Structure-Based Discovery of Drug Leads in the Absence of Target Structure

As shown, dehydrons are main determinants of epitopes (Chap. 1), can be inferred from sequence (Chap. 5) and may activate residues susceptible of phosphorylation (Chap. 7). These properties make it intuitively appealing to identify dehydron-rich regions with peptide-based leads to disrupt PPIs when the structure of the binding partner is unknown and the PPI may be naturally disrupted through site-specific phosphorylation. To illustrate the power of the discovery technology, we focus on the lead identification for disruption of a specific PPI, a patented invention *en route* to cure heart failure [18].

Myosin binding protein C (MyBP-C) is a central regulator of cardiac contraction [19, 20]. In murine models, genetic ablation or phosphorylation of MyBP-C by kinases PKA or CAMKII accelerates contraction and increases its force in cardiac muscle. Most recently, it has been shown that CAMKII phosphorylation of MyBP-C at Ser282 and Ser302 in the mouse and Ser284 and Ser304 in the human underlies the increase in myocardial contractile force as heart rate is increased, according to the so-called "staircase phenomenon". Proof for this statement was obtained by observing that (1) phosphorylation at these two residues is increased when stimulus frequency is increased, but other potential phosphorylation sites are not, and (2) the staircase phenomenon is absent in hearts in which these residues are replaced by residues not susceptible of phosphorylation. MyBP-C normally reduces the speed and strength of contraction by means of its interaction with the contractile protein motor myosin and thereby reduces the probability of myosin sliding along actin filaments [19] (Fig. 16.3). Ablation of MyBP-C or phosphorylation of MyBP-C by PKA or CAMKII disrupts this interaction and relieves MyBP-C repression of myosin [20]. Once phosphorylated, MyBP-C no longer binds to myosin, myosin slides along actin, the probability of myosin binding to actin increases, and the contraction speed and strength increase.

In heart failure, MyBP-C is phosphorylated minimally or not at all due to down-regulation of β-adrenergic receptors. Since phosphorylation of MyBP-C improves contraction, US patent 9,051,387 [18] proposed to target the CAMKII site on MyBP-C with a pharmaceutical designed to disrupt its interaction with myosin and thereby improve cardiac function. This problem involves designing an optimal MyBP-C-derived peptide. The peptide contains the motif responsible for interaction with myosin and hence disrupts the MyBP-C-myosin association as required for therapeutic action (Fig. 16.3). The myosin-binding peptide-based therapeutic agent may be identified based on the premise that disruption of the myosin-MyBP-C interface would release a molecular brake on cardiomyocyte contractility imposed by the repressive activity of the unphosphorylated form of MyBP-C on myosin.

The 3D-structure of the enormous 11-domain protein MyBP-C is unknown, and therefore it was decided to design the peptide based on the output from PONDR® or another predictor of native disorder [15] (cf. Fig. 16.1). More specifically, the peptide

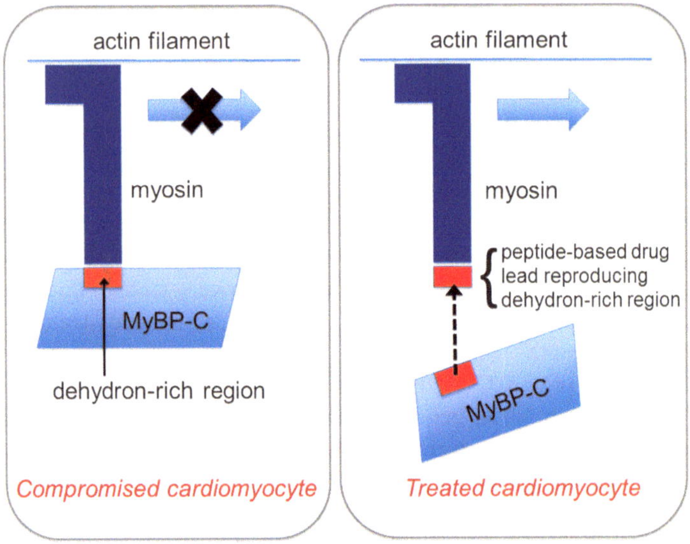

Fig. 16.3 Schematic view of the therapeutic disruption of the MyBP-C/myosin interface to treat heart failure. The drug lead (*red*) consists of a small peptide mimicking the dehydron-rich region in MyBP-C that binds to myosin. To prevent detachment, the peptide is modified through Ser302Ala and Ser307Ala substitutions to become non-susceptible to phosphorylation. As the peptide competitively binds to myosin, it displaces the myosin modulator MyBP-C, enabling the myosin motor to slide along the actin filament as needed for myocyte contractility. At the molecular level, the therapeutic effect of the drug lead represents an induced "brake release" for the compromised heart

was developed based on a sequence-based prediction of the dehydron-rich region that constitutes the putative myosin-binding site. To predict the peptide sequence, the inventors examined a region between the C1 and C2 domains of MyBP-C in the twilight zone between order and disorder. In this way, it became possible to identify the twilight region containing phosphorylation sites Ser302, Ser307 in the motif region intercalated between domains C1 and C2 of cMyBP-C and determine the sequence of the patented peptide ($_{293}$FSSLLKKRDSFRRDSKLF$_{310}$) that may be used as a lead to the therapeutic agent to treat heart failure. The invention has been shown to significantly increase cardiac contractile force and frequency in the failing heart [18]. To turn the peptide into a therapeutic agent still entails an arduous optimization process in order to improve bioavailability while avoiding peptidase digestion. Optimization strategies would need to be implemented to replace peptide bonds and create unnatural side chains that mimic the key binding components that may emerge from a structural characterization of the peptide-myosin association.

16.8 Problems

A single dehydron contributes $\Delta G_{if} \sim 2k_BT$ to the interfacial tension (Chap. 1). To verify the importance of dehydron-rich regions in promoting protein associations, a key fact exploited in US patent 9,051,387, examine the free subunits in 28 protein complexes (Table 16.1) with defined contact topologies represented by graphs. The total area of the P-P (protein-protein) interfaces for each complex is computed after identification of the subunit residues engaged in intermolecular contacts (Fig. P16.1, Appendix B). To explore the solvent-accessible envelope of the protein surfaces for the free subunits, cover each protein surface with a minimal set W of water-confining osculating spheres D_js, $j \in W$. These spheres must make first-order contact with the envelope obtained by sliding a water molecule along the surface, thus smoothening out its singularities (Chaps. 1–4). As a minimal covering, W must contain all interfacial water molecules and this property cannot hold if any osculating sphere is excluded from the set. Interfacial tension arises in D_j when the

Table 16.1 Representative topologies of inter-subunit contacts for 28 protein complexes reported in PDB

<table>
<tr><td>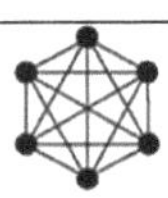 2af7(63)</td><td>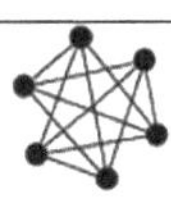 1fr3(61)</td><td>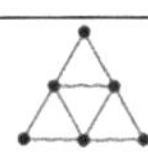 2siv(59)</td><td> 1ub4(53)</td><td> 2a1b(51)</td><td> 2c5r(40)</td><td> 4otc(57)</td></tr>
<tr><td> 1qoh(54)</td><td>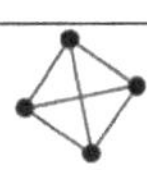 1dvz(57)
1jtk(55)</td><td> 1hho(64)</td><td> 1nys(39)</td><td>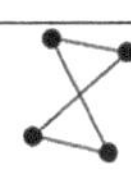 1uo5(54)
1vwe(53)</td><td> 1fvu(50)</td><td>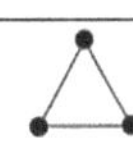 1zij(52)
1ufy(51)</td></tr>
<tr><td colspan="2"> 1wud(42)</td><td colspan="3"> 1u4j(48) 1hpo(56)
1htg(50) 1lve(47)
2oqq(60) 1v2z(51)
1btn(62) 2ien(59)
1ihb(57) 1p6z(72)</td><td colspan="2"> 3orc 1b8x 2gfh
1goc 2b65 1h5m
1may 1uga 1quz</td></tr>
</table>

The dataset is used to unravel the role of P-W interfacial tension patches as promoters of protein associations. In the topological representation of a complex, a ball represents a protein subunit and a stick between two balls represents a protein-protein (P-P) interface. Numbers in brackets denote % decrease in overall interfacial tension upon complexation starting with free subunits interfacial free energy associated with spanning the j-th contact region is $\Delta G_j = \Delta G_{if}(D_j) > 0$. Thus, for each D_j compute

$$\Delta G_{if}(\phi) = \int \left[\frac{1}{2}\omega||\vec{\nabla}\phi||^2 - k_B T(\rho_0 - \rho)\ln\left(1 - \frac{\phi}{4}\right)\right] d\vec{r}$$

(Eq. 1.7, Chap. 1) for each free protein subunit with charge and atomic radii assigned using the program PDB2PQR. Define filtered sets $W_n = \{j \in W: \Delta G_j \geq n k_B T\}$ (n = 1,2,...) of contributors to the protein-water (P-W) interfacial tension in the free subunits, with S_n, the total P-W interface associated with W_n, and A_n, the surface area of S_n.

Prove that interfacial solvent cavities spanned at the significant thermodynamic cost $\Delta G \geq 2k_B T$ are indeed the promoters of protein associations. To support this conclusion, establish a correlation between A_2 and the total P-P interfacial area of each complex. Prove that the correlation becomes weaker for S_1 with area A_1 and becomes negligible when we consider the total solvent-exposed surface area (SESA) of free protein subunits. Establish the fact that the correlation between the areas of surface patches that significantly destabilize the protein-water interface (A_2) and those of binding sites implies that proteins associate to reduce the P-W interfacial tension caused by dehydrons.

References

1. Pelay-Gimeno M, Glas A, Koch O, Grossmann TN. Structure-based design of inhibitors of protein-protein interactions: mimicking peptide binding epitopes. Angew Chem Int Ed Engl. 2015;54:8896–927.
2. Wells JA, McClendon CL. Reaching for high-hanging fruit in drug discovery at protein–protein interfaces. Nature. 2007;450:1001–9.
3. Nero TL, Morton CJ, Holien JK, Wielens J, Parker MW. Oncogenic protein interfaces: small molecules, big challenges. Nat Rev Cancer. 2014;14:248–62.
4. Arkin MR, Tang Y, Wells JA. Small-molecule inhibitors of protein-protein interactions: progressing toward the reality. Chem Biol. 2014;21:1102–14.
5. Hwang H, Vreven T, Janin J, Weng Z. Protein-protein docking benchmark version 4.0. Proteins. 2010;78:3111–4.
6. Khoo KH, Verma CS, Lane DP. Drugging the p53 pathway: understanding the route to clinical efficacy. Nat Rev Drug Discov. 2014;13:217–36.
7. Vassilev LT, Vu BT, Graves B, Carvajal D, Podlaski F, Filipovic Z, Kong N, Kammlott U, Lukacs C, Klein C, Fotouhi N, Liu EA. In vivo activation of the p53 pathway by small molecule antagonists of MDM2. Science. 2004;303:844–8.
8. Fry DC. Small-molecule inhibition of protein-protein interactions: how to mimic a protein partner. Curr Pharm Des. 2012;18:4679–84.
9. Kouadio JL, Horn JR, Pal G, Kossiakoff AA. Shotgun alanine scanning shows that growth hormone can bind productively to its receptor through a drastically minimized interface. J Biol Chem. 2005;280:25524–32.
10. King C, Garza EN, Mazor R, Linehan JL, Pastan I, Pepper M, Baker D. Removing T-cell epitopes with computational protein design. Proc Natl Acad Sci U S A. 2014;111:8577–82.

11. Erlanson DA, Wells JA, Braisted AC. Tethering fragment-based drug discovery. Annu Rev Biophys Biomol Struct. 2004;33:199–223.
12. Winter A, Higueruelo AP, Marsh M, Sigurdardottir A, Pitt WR, Blundell TL. Biophysical and computational fragment-based approaches to targeting protein-protein interactions: applications in structure-guided drug discovery. Q Rev Biophys. 2012;45:383–426.
13. Fernández A. Epistructural tension promotes protein associations. Phys Rev Lett. 2012;108:188102.
14. Fernández Stigliano A. Biomolecular interfaces: interactions, functions and drug design, Chap. 5. Heidelberg: Springer; 2015.
15. Xue B, Dunbrack RL, Williams RW, Dunker AK, Uversky VN. PONDR-FIT: a meta-predictor of intrinsically disordered amino acids. Biochim Biophys Acta. 2010;1804:996–1010.
16. Fernández A. Communication: chemical functionality of interfacial water enveloping nanoscale structural defects in proteins. J Chem Phys. 2014;140:221102.
17. Fernández A. Packing defects functionalize soluble proteins. FEBS Lett. 2015;589:967–73.
18. Moss R, Fernández A. Inhibition of MyBP-C binding to myosin as a treatment for heart failure, US patent 9,051,387; 2015.
19. Previs MJ, Beck Previs S, Gulick J, Robbins J, Warshaw DM. Molecular mechanics of cardiac myosin-binding protein c in native thick filaments. Science. 2012;337:1215–8.
20. Colson BA, Bekyarova T, Fitzsimons DP, Irving TC, Moss RL. Radial displacement of myosin cross-bridges in mouse myocardium due to ablation of myosin binding protein-C. J Mol Biol. 2007;367:36–41.
21. Fernández Stigliano A. Biomolecular interfaces: interactions, functions and drug design, Chap. 1. Heidelberg: Springer; 2015.
22. Pietrosemoli N, Crespo A, Fernández A. Dehydration propensity of order–disorder intermediate regions in soluble proteins. J Proteome Res. 2007;6:3519–26.
23. Fraser CM, Fernández A, Scott LR. WRAPPA: a screening tool for candidate dehydron identification. University of Chicago, Department of Computer Science Technical Report TR-2011-05. 2011. http://www.cs.uchicago.edu/research/publications/techreports/TR-2011-05.
24. Fernández Stigliano A. Biomolecular interfaces: interactions, functions and drug design, Appendix 1: Code for dehydron identification. Heidelberg: Springer; 2015.
25. Chen J, Zhang X, Fernández A. Molecular basis for specificity in the druggable kinome: sequence-based analysis. Bioinformatics. 2007;23:563–72.
26. Chen J, Liang H, Fernández A. Protein structure protection commits gene expression patterns. Genome Biol. 2008;9:R107.
27. Zeng Z, Shi H, Wu Y, Hong Z. Survey of natural language processing techniques in bioinformatics. Comp Math Methods Med. 2015;2015:674296.
28. Fernández Stigliano A. Biomolecular interfaces: interactions, functions and drug design, Chap. 5. Heidelberg: Springer; 2015.

Chapter 17
Epistructural Drug Design to Treat Cancer Metastasis and the Associated Drug Resistance

This chapter exploits translational aspects of epistructural physics to develop a multi-target molecular therapy against cancer metastasis. The goal is to optimize a therapeutic agent capable to realizing a dual blockade of the signaling pathway that uses the focal adhesion kinase (FAK) as signal transducer to prevent the onset of the invasive phenotype. Thus, the epistructure-based optimization of lead compound TAE226 (Pfizer) enables a careful control of the multi-target impact towards kinases of therapeutic interest to block cancer progression, including FAK, JNK and the IGF1R kinase. The molecular therapy requires synergy with a second drug that is optimized to block the recruitment of the PI3K/mTOR pathway conferring drug resistance. The uniqueness of the drug combination arises from the reciprocal editing (cf. Chap. 12), whereby the first drug abrogates resistance to the second drug which is itself an inhibitor of drug resistance towards the first drug.

17.1 Optimized FAK Inhibitor to Treat Cancer Metastasis

This section illustrates the power of epistructural physics as the foundational guidance to the rational development of kinase inhibitors with controlled specificity. Here we focus on a two-prong upstream/downstream targeting of the FAK (focal adhesion kinase) signaling axis to create novel cancer therapeutics. The main goals are the impairment of the phosphorylation-recruiting capability of FAK and the blockade of the downstream propagation of its signals in cancer cells to suppress tumor progression, metastasis and the onset of invasive phenotypes (see Chap. 12). The effort is motivated by the imperative of a molecular therapy that exploits drugs with controlled impact for effective target suppression through simultaneous upstream and downstream inhibition. The focus is justified because FAK overexpression is observed in most invasive ovarian cancers and is a predictor of poor outcome and because interference with FAK pathways is known to modulate tumor invasion and migration (Chap. 12 and references therein). The second

A. Fernández, *Physics at the Biomolecular Interface*, Soft and Biological Matter,
DOI 10.1007/978-3-319-30852-4_17

on-target kinase that should be inhibited by the FAK inhibitor is JNK, a downstream modulator of FAK-triggered signaling, to promote anoikis and apoptosis of ovarian cancer cells. The compound of therapeutic interest fulfilling the imperative of simultaneous upstream and downstream inhibition will be designed based on the tenets of epistructural design, combining in one molecule a selective anti-FAK and anti-JNK activity.

As described in Chap. 12, the functional roles of FAK motivate us to propose the exploitation of FAK-targeting inhibitors as chemosensitizers and anti-neoplastic agents suitable to treat cancer metastasis [1, 2]. Thus, based on the known role of FAK in ovarian cancer pathogenesis and the encouraging results of FAK gene silencing, there is compelling interest in developing novel and powerful FAK inhibitors. As we shall now demonstrate, epistructure-based drug design enables us to develop such FAK inhibitors through optimization of lead compounds such as TAE226 (Pfizer) [1, 2], known to have nanomolar activity against on-target kinases FAK, JNK [3] and IGF1R. The goal of there-engineering optimization process is to improve the safety and to control the therapeutic impact to avoid off-target reactivity. Both JNK [3, 4] and IGF1R [5] are targets of great therapeutic value in the ovarian cancer context and their inhibition makes ovarian cancer cells particularly susceptible to chemotherapy [3, 5]. We hereby name FAK/JNK/IGF1R-wrapper the epistructure-based optimized compound that exploits TAE226 as a lead compound. As noted in Chap. 12, there are two off-target kinases inhibited in the nM range by TAE226, that must be filtered out in the epistructure-based redesign: the insulin receptor kinase (INR), which is highly homologous to the on-target IGF1R (insulin-like growth factor 1 receptor) kinase, and the lymphocyte-associated kinase, Lck, whose inhibition yields immunosuppression, as documented in Chap. 14. Because it would induce a diabetic coma in the patient, the nanomolar INR inhibition by TAE226 renders this compound readily unsuitable for drug therapy. As said, Lck inhibition by TAE226 blocks the alignment of the immune system in its capacity to synergize with the drug treatment in the fight against cancer (Chap. 14).

There is another powerful reason to retain and enhance the dual FAK/JNK activity of TAE226 in the structural re-engineering of this molecule: Both kinases are targets in the blockade of the Wnt pathway (Fig. 17.1) essential for the maintenance and transition to pluripotency of cancer stem cells [4].

In order to filter out Lck and the INR (insulin receptor, InsR, IR) kinase, telling them apart from the on-target FAK-kinase through epistructural recognition, we may re-engineer the TAE226 to turn it into a wrapper of the dehydrons that constitute the appropriate selectivity filters (Fig. 17.2). The positions requiring wrapping substituents are marked by circles in Fig. 17.2. The structural alignment of the FAK/TAE226 complex (PDB.2JKK) with a PDB-reported structure for the INR kinase (Chap. 6, Fig. 6.5), reveals that dehydron K485-E500 in FAK, well wrapped in the INR kinase, constitutes an adequate selectivity filter to dial out the INR kinase by redesigning TAE226 into a wrapper of the dehydron (cf. Fig. 17.2). Similarly, structural alignment of PDB.2JKK with a PDB-reported structure for Lck (PDB.2PL0), reveals that dehydron C502-G505 in FAK becomes well-wrapped hydrogen bond M319-G322 in Lck (cf. Chap. 9, Fig. 9.2c). Hence, redesigning

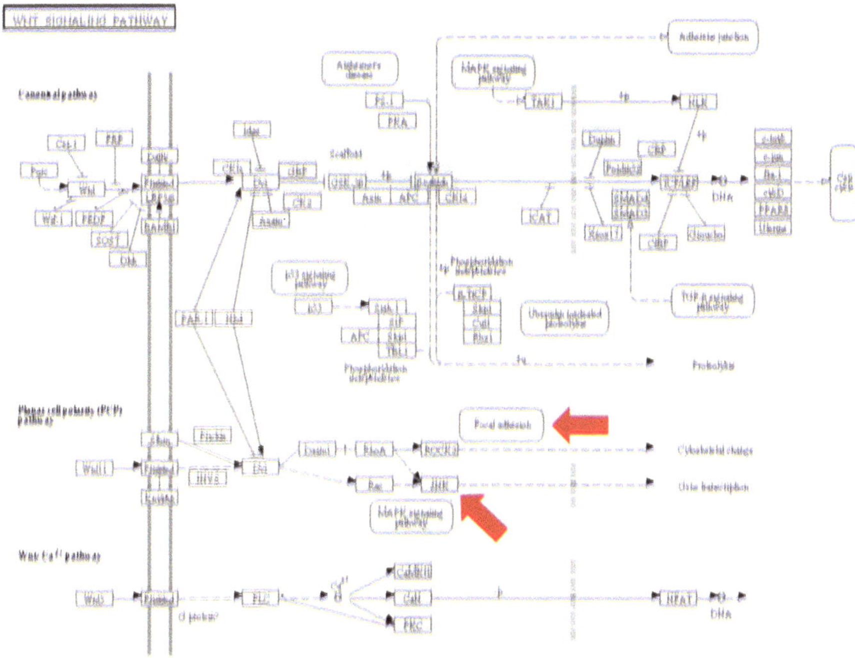

Fig. 17.1 Wnt signaling network in cancer stem cell

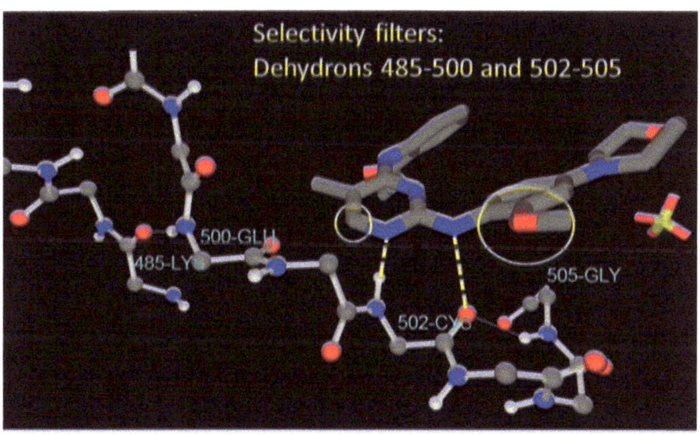

Fig. 17.2 Selectivity filters for FAK inhibition through optimization of parental lead TAE226. The epistructural display utilizes the structural coordinates from PDB file 2JKK

TAE226 into a wrapper of the C502-G505 dehydron dials out the undesired target Lck (Fig. 17.2). On the other hand, such a wrapping modification would also enhance the affinity towards on-target JNK through dehydron induction, as demonstrated in Chap. 11. In addition to targeting the dehydronic selectivity filters,

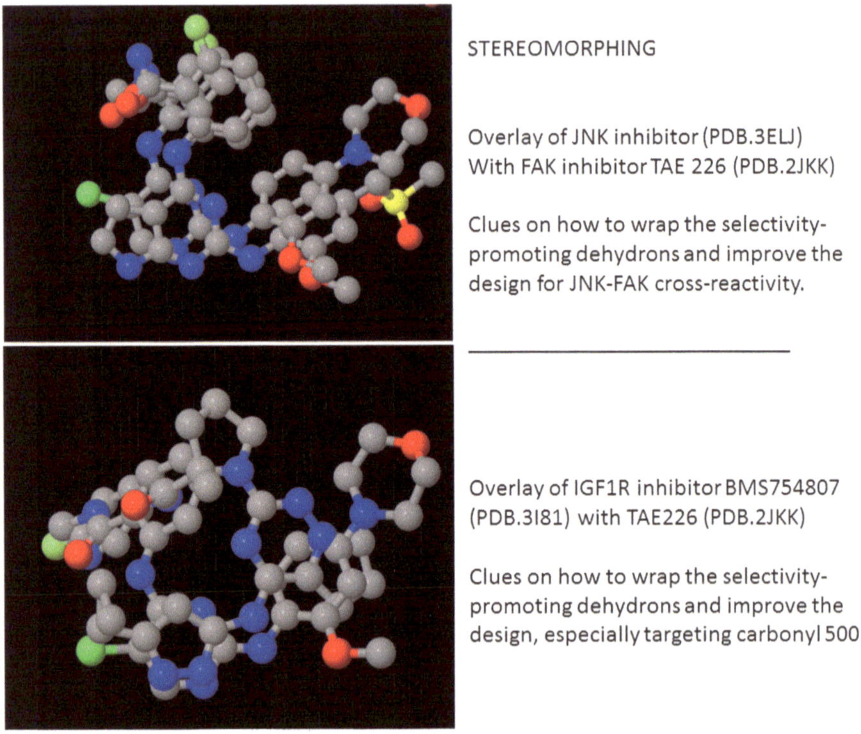

Fig. 17.3 Ligand stereomorphing corresponding to alignment of PDB-reported structures in entries 2JKK/3ELJ (*upper panel*) and 2JKK/3I81 (*lower panel*)

enhancement of FAK affinity requires engineering an intermolecular hydrogen bond that engages the unused backbone carbonyl from E500 in FAK (PDB.2JKK), as shown in Fig. 17.2.

To guide the re-engineering of lead compound TAE226 in order to fulfill the requirements described in the preceding paragraph, we resort to a novel technique to expand the drug chemical structure hereby termed "stereomorphing". Stereomorphing requires the aligned overlay of two chemical structures in order to guide the chemotype expansion of one ligand as it morphs into the other ligand. Thus, stereomorphing TAE226 from PDB.2JKK into the JNK inhibitor from PDB.3ELJ and into the IGF1R kinase inhibitor BMS754807 from PDB.3I81 (Fig. 17.3) guides the TAE226 redesign that filters out the undesired reactivity against LcK and INR kinase. This is accomplished by providing adequate wrapping of selectivity-promoting dehydrons K485-E500 and C502-G505 in FAK as TAE226 morphs into the JNK and IGF1R inhibitors that already avoid Lck and INR. At the same time, stereomorphing generates the hydrogen bond that pairs the re-engineered TAE226 with the backbone carbonyl of E500 (cf. Figs. 17.2 and 17.3). Thus, the epistructure-based re-engineering of TAE226 into the more powerful, safer and specificity-controlled inhibitor FAK/JNK/IGF1R-wrapper is presented in Fig. 17.4, overlaid onto the parental compound TAE226 for comparative analysis.

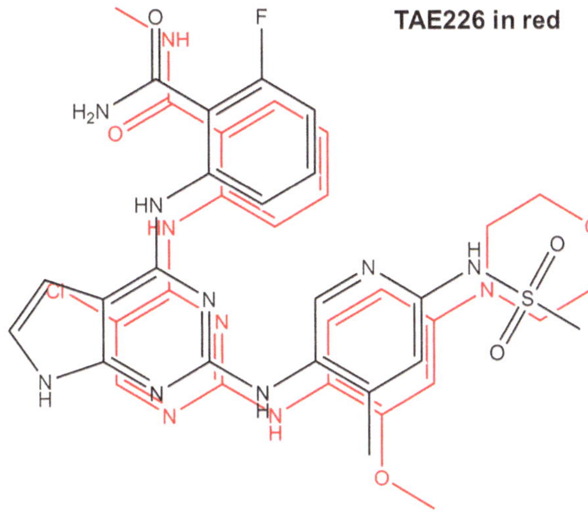

Fig. 17.4 Solution to TAE226 optimization problem (chemical structure of optimized compound in *black*, overlaid parental compound in *red*)

17.2 Overcoming Drug Resistance in the Anti-FAK Treatment of Cancer Metastasis

The efficacy of a drug treatment based on FAK inhibition is ultimately undermined by drug resistance, articulated through the recruitment of pathways like the PI3K/mTOR axis [6]. Based on epistructural design to target PI3K, it is possible to develop an optimized safe inhibitor of the PI3K/mTOR drug-resistance pathway, hereby named PI3K-wrapper. Illustrations of extant PI3K inhibitors are *Verastem* (Pfizer) and *CH5132799*, all targeting the ATP-binding region of PI3K and exploiting the morpholine chemotype (Fig. 17.5). These designs are somewhat deficient in that they do not hydrogen-bond to the carbonyl of E880 which would be wrapped by Y867, significantly enhancing affinity for PI3K without compromising specificity (Fig. 17.6). An optimized compound fulfilling such requirements is presented in Fig. 17.7. The considerable improvement becomes apparent in the salt form, whereby a π-cation interaction (see Chap. 7) between the ligand amino group and the aromatic side chain of Y867 reinforces the wrapping by Y867 of the intermolecular hydrogen bond sustained between the backbone carbonyl of E880 and the ligand amino group (compare Figs. 17.6 and 17.7). An alternative optimization of the PI3K inhibitor is presented in Fig. 17.8. In this case, an intramolecular dehydron in the ligand (Fig. 17.9) contributes to the PI3K affinity through intermolecular wrapping.

In late stages of FAK-targeted treatment of cancer progression, it is likely that even resistance to PI3K-targeted inhibition of drug resistance could develop. However, the combined FAK/JNK/IGF1R-Wrapper-PI3K-Wrapper proves to be particularly resilient in this regard. In fact, FAK/JNK/IGF1R-Wrapper may act as editor of PI3K-Wrapper because the former's inhibitory impact on IGF1R abrogates resistance to inhibition of PI3K/mTOR-based drug resistance [7], as shown in Fig. 17.10. This case illustrates the power of combined drug therapies when an

Fig. 17.5 Morpholine-based extant PI3K inhibitors

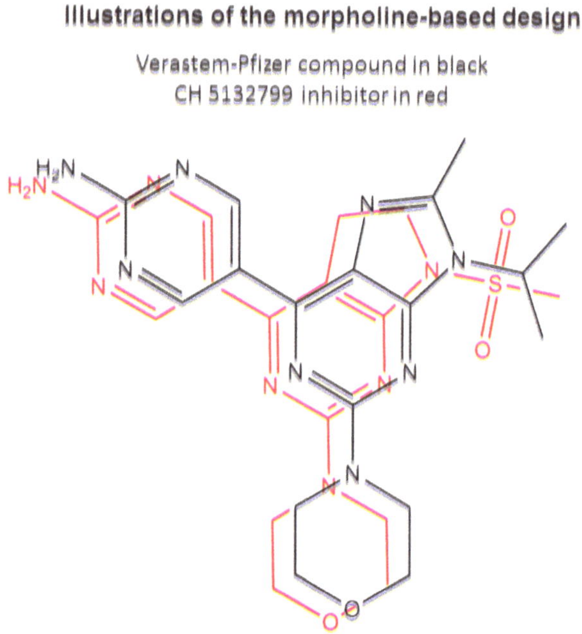

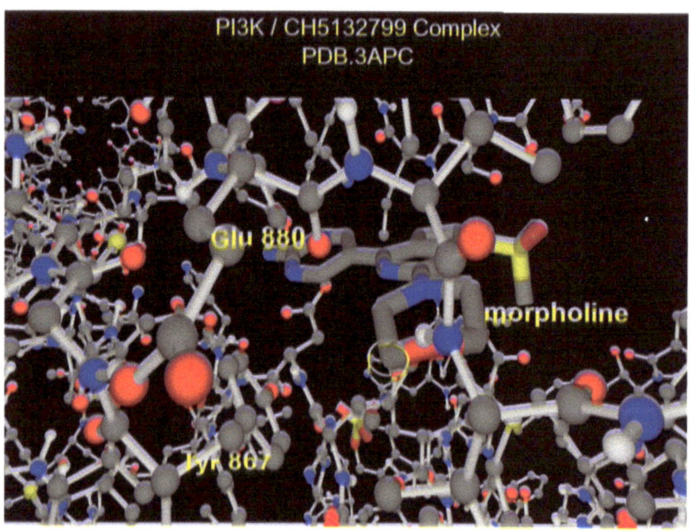

Fig. 17.6 Epistructural analysis of PI3K in complex with CH5132799 guiding ligand optimization

editing mechanism, in this case anti-IGF1R activity, enables the primary drug (FAK/JNK/IGF1R-Wrapper) *to abrogate drug resistance elicited by the PI3K-targeted inhibitor of drug resistance towards the primary drug*. This is an example of combined anticancer molecular therapies exerting reciprocal inhibition of drug resistance.

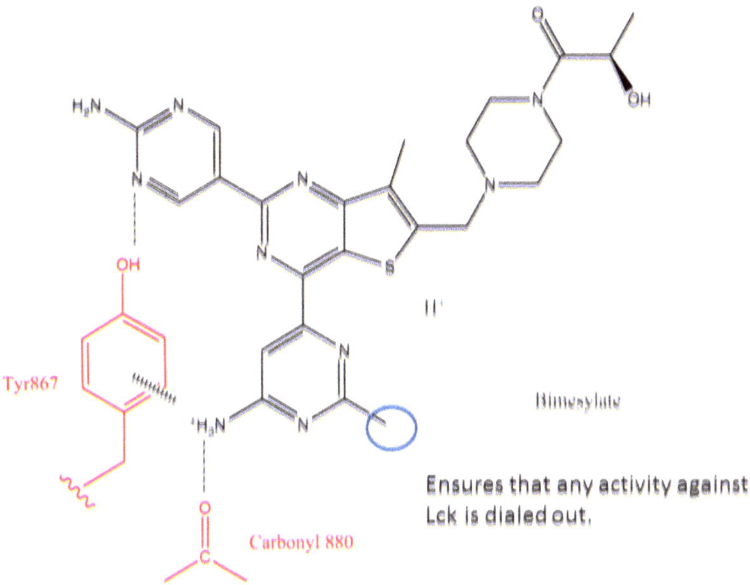

Fig. 17.7 Optimized PI3K inhibitor obtained by morpholine replacement exploiting epistructural design

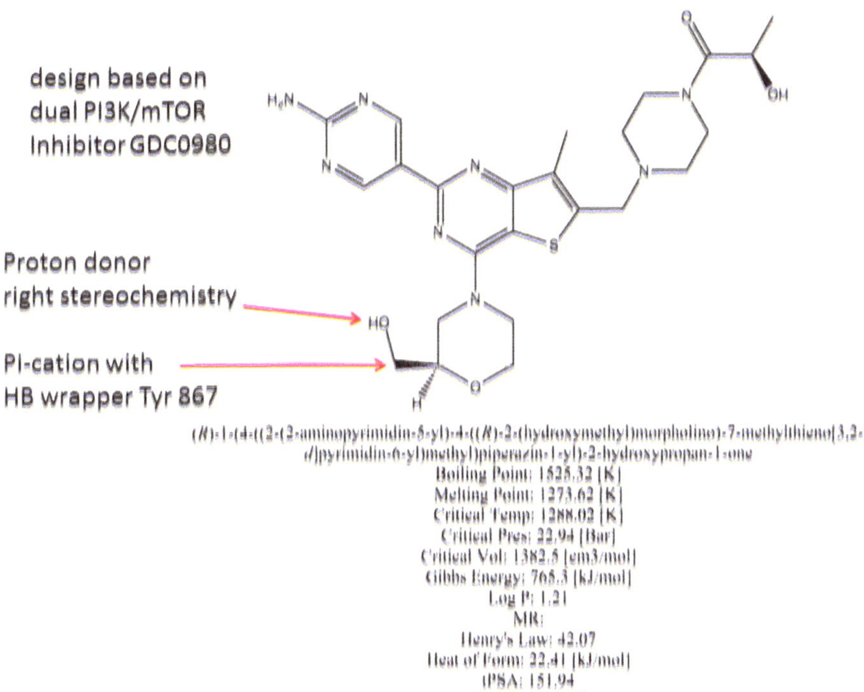

Fig. 17.8 Alternative solution to the PI3K-inhibitor optimization problem based on substituted morpholine

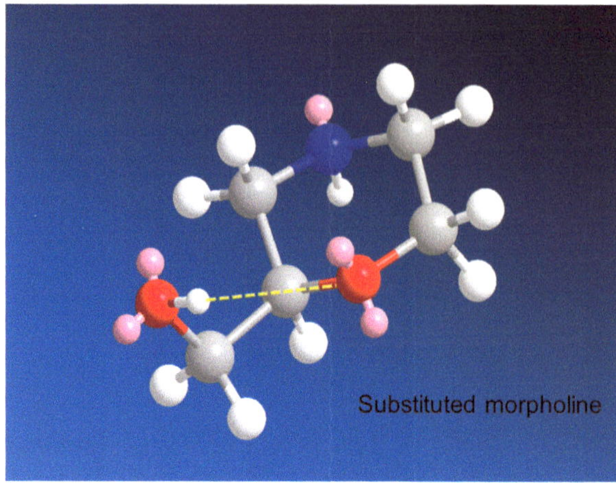

Fig. 17.9 Stereoview of intramolecular dehydron (*dashed line*) in modified morpholine ring for compound specified in Fig. 17.8

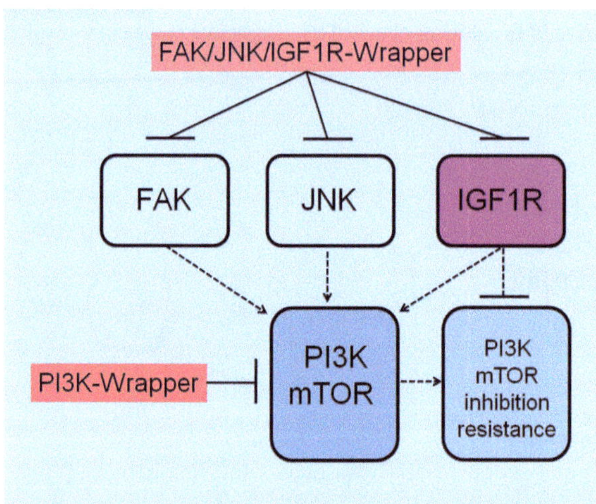

Fig. 17.10 Scheme of two synergistic anticancer molecular therapies exerting reciprocal inhibition of drug resistance. The inhibitory impact of primary drug FAK/JNK/IGFR-Wrapper is shown together with the pattern of drug resistance to FAK/JNK/IGFR-Wrapper treatment somatically articulated through recruitment of the PI3K/mTOR pathway. An optimized PI3K inhibitor, named PI3K-Wrapper, exerts an inhibitory impact on the drug resistance mechanism, while inhibition of primary target IGF1R by the first drug exerts an editing effect upon the second drug abrogating resistance to PI3K-Wapper-based inhibition of the PI3K/mTOR drug-resistance mechanism

17.3 Problems

17.1. Provide at least two alternative optimizations of FAK inhibitor TAE226 fulfilling the constraints given in Sect. 17.1.

17.2. Provide an alternative optimized PI3K inhibitor by replacing the morpholine ring of extant inhibitors in accord with the premises given in Sect. 17.2.

References

1. Lietha D, Eck MJ. Crystal structures of the FAK kinase in complex with TAE226 and related bis-anilino pyrimidine inhibitors reveal a helical DFG conformation. PLoS One. 2008;3, e3800.
2. Halder J, Lin YG, Merritt WM, Spannuth WA, Nick AM, Honda T, Kamat AA, Han LY, Kim TJ, Lu C, Tari AM, Bornmann W, Fernandez A, Lopez-Berestein G, Sood AK. Therapeutic efficacy of a novel focal adhesion kinase inhibitor TAE226 in ovarian carcinoma. Cancer Res. 2007;67:10976–83.
3. Vivas-Mejia P, Benito JM, Fernández A, Han HD, Mangala L, Rodriguez-Aguayo C, Chavez-Reyes A, Lin YG, Carey MS, Nick AM, Stone RL, Kim HS, Claret FX, Bornmann W, Hennessy BT, Sanguino A, Peng Z, Sood AK, Lopez-Berestein G. c-Jun-NH2-kinase-1 inhibition leads to antitumor activity in ovarian cancer. Clin Cancer Res. 2010;16:184–94.
4. Holland JD, Klaus A, Garratt AN, Birchmeier W. Wnt signaling in stem and cancer stem cells. Curr Opin Cell Biol. 2013;25:254–64.
5. Bast RC, Hennessy B, Mills GB. The biology of ovarian cancer: new opportunities for translation. Nat Rev Cancer. 2009;9:415–28.
6. Martz CA, Ottina KA, Singleton KR, Jasper JS, Wardell SE, Peraza-Penton A, Anderson GR, Winter PS, Wang T, Alley HM, Kwong LN, Cooper ZA, Tetzlaff M, Chen PL, Rathmell JC, Flaherty KT, Wargo JA, McDonnell DP, Sabatini DM, Wood KC (2014) Systematic identification of signaling pathways with potential to confer anticancer drug resistance. Science Signal. 7:ra121.
7. Muranen T, Selfors LM, Worster DT, Iwanicki MP, Song L, Morales FC, Gao S, Mills GB, Brugge JS. Inhibition of PI3K/mTOR leads to adaptive resistance in matrix-attached cancer cells. Cancer Cell. 2012;21:227–39.

ERRATUM TO

Solution to the Protein Folding Problem

Ariel Fernández

A. Fernández, *Physics at the Biomolecular Interface*, Soft and Biological Matter,
DOI 10.1007/978-3-319-30852-4_3

DOI 10.1007/978-3-319-30852-4_3

Below is the equation given in chapter 3 of page 90.

$$K\left(\vec{r},\vec{r}',,\{\vec{r}_j\}\right)=\left[\int \exp,\left\{i\vec{k}.\left(\vec{r}-\vec{r}'\right)\right\},L,\left(\vec{k}\rightarrow\right),d\vec{k}\right].\left[1+\sum_{j=1}^{n}\Gamma_j\left(\vec{r},\vec{r}'\right)\right]$$

The Equation 3.3 should be changed as follows:

$$K\left(\vec{r},\vec{r}',\{\vec{r}_j\}\right)=\left[\int \exp\left\{i\vec{k}.\left(\vec{r}-\vec{r}'\right)\right\}L\left(\vec{k}\right)d\vec{k}\right].\left[1+\sum_{j=1}^{n}\Gamma_j\left(\vec{r},\vec{r}'\right)\right]$$

The online version of the updated chapter can be found at
http://dx.doi.org/10.1007/978-3-319-30852-4_3

A. Fernández, *Physics at the Biomolecular Interface*, Soft and Biological Matter,
DOI 10.1007/978-3-319-30852-4_18

Epilogue

If all you have is a hammer everything looks like a nail.

Mark Twain

E1. Epistructural Physics May Revitalize Marcus' Rate Theory of Electron Transfer

As this book reveals, biological function understood in a broad sense is not completely enshrined in protein structure: the epistructural ("around-the-structure") aqueous interface needs to be brought into the analysis. Furthermore, the subtle dynamic entanglement of proteins and their partially structured aqueous environment attests to the overdue recognition that biomolecular phenomena cannot be fully grasped without dealing with interfacial behavior. Concepts pertaining to the realm of interfacial physics, such as hydrogen-bond frustration, interfacial tension, non-Debye water dielectrics and variational thermodynamics must be carefully waved into the discourse and background of the molecular biophysicist and the biomedical engineer. This has been essentially the scope of this book.

As the book reveals, the difficulties in providing a suitable statistical-mechanical apparatus to handle the complexities of biological interfacial phenomena may prove daunting. Yet, after covering the vast and interwoven intellectual territory that separates "interfacial tension" (Chap. 1) from "the design of selective drugs to treat cancer metastasis" (Chap. 17), we may at least state that the book is well positioned to address some of the core problems in biophysics and molecular pharmacology. These problems range from providing a solution to the protein folding problem, to unravelling the physico-chemical basis of enzyme catalysis and protein associations, to delineating the molecular aetiology of aberrant protein aggregation, to rationally designing molecular targeted therapies. This is indeed a

A. Fernández, *Physics at the Biomolecular Interface*, Soft and Biological Matter,
DOI 10.1007/978-3-319-30852-4

vast range of problems, yet they all share a central concept pertaining to the realm of interfacial physics: *epistructural tension*.

In what pertains to provide the physical underpinnings of biomolecular interfacial phenomena, the task set forth in this book is not complete, of course. A rigorous operational dielectric theory of water under nanoscale confinement is still missing. In this sense, the physical apparatus provided in the book should be regarded as provisional, much in the same sense as the Bohr model of the atom represented in its day a transitional yet valuable precursor to quantum mechanics.

This discussion brings up the question: *Where to start in the quest to develop a rigorous dielectric theory for water under nanoscale confinement?* To address the question, we first note that it is possible to apply the methods set forth in this book to "relatively simple problems" in relatively well understood areas of molecular kinetics/dynamics, such as the Marcus theory of electron transfer [1]. Substantive progress in this area would demand detailed information of the solvent structural dynamics and a rigorous characterization of the dielectric response applicable at nanoscales.

To be precise, let us focus on the simplest case of outer-shell electron transference between two groups with quantum mechanically decoupled electronic structure, where the dominant term in the transition-state free energy $\Delta G^{\neq}$ is the reversible work needed to create the solvent configuration conducive to electron relocation (Fig. E1). As it is well known, $\Delta G^{\neq}$ is related to the polarization energy of the solvent that tends to align with the field sustained by the charges generated by

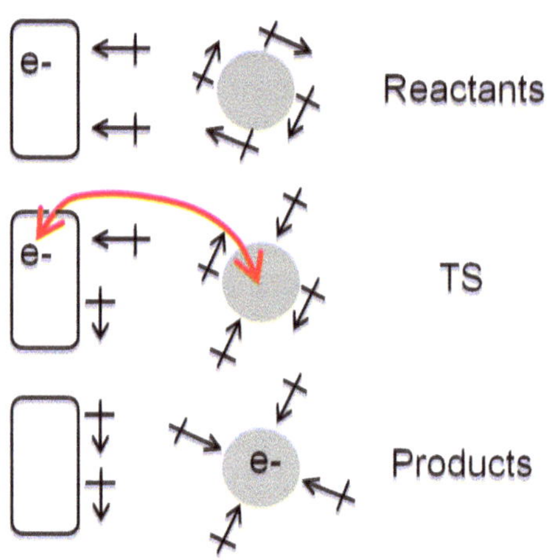

Fig. E1 Fluctuations in the solvation structure of reactants yield a transition state (TS) solvent configuration conducive to an outer-shell electron transference between two quantum mechanically decoupled groups. Water dipoles are represented as *arrows* with the net positive charge located on the *arrow head*. The electron acceptor (*grey ball*) is considered nonpolar in the reactants basin of the potential energy surface, and obviously negatively charged after the transfer event. The solvation structure of reactants, TS and products is schematically represented by dipole orientations vis-à-vis net solute charges as they develop along the reaction. The solvation structure of the TS is conducive to electron delocalization marked by the *red double-headed arrow*

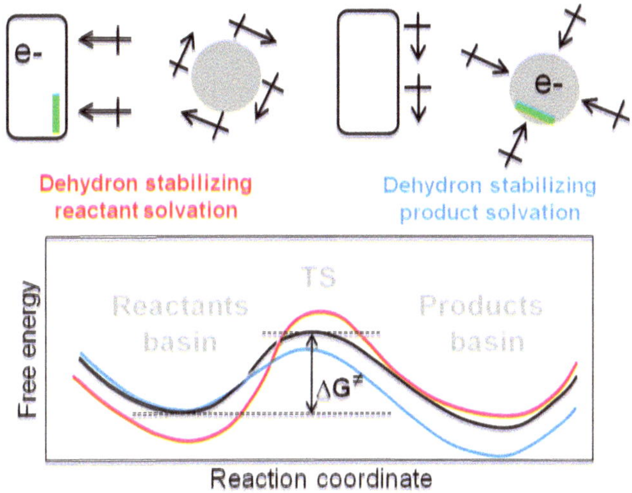

Fig. E2 The location of a dehydron (represented by a *green* segment) within the vicinity of the pathway for electron transference may significantly alter the kinetics of the reaction, as it perturbs the solvation pattern along the reaction coordinate. The non-Debye dielectric response, marked by local water polarization that is not aligned with the electric field, is schematically enshrined in the orientation of the dehydron-hydrating dipoles. Depending on spatial location, the dehydron may stabilize the reactant solvation pattern, slowing down electron transference (*red line plot* in the potential energy surface), or may accelerate the event (*blue line plot*) by stabilizing the solvation pattern of the products

electron relocation. However, what would happen if an enzymatic dehydron is located in the vicinity of the pathway for electron transference? In that case, the computation of $\Delta G^{\neq}$ is unlikely to be as straightforward as in the elegant Marcus theory, since solvent reorganization is unlikely to be controlled solely by Debye-type polarization. In fact, it is likely that non-Debye polarization effects (as described in Chaps. 1–4), statistically and causally independent of the alignment with the electric field, may significantly raise $\Delta G^{\neq}$ (Fig. E2). This would be so except in the rare case where the dehydron is strategically positioned to organize the solvent in a way that may propitiate electron relocation, for example if it stabilizes the hydration of the products, as shown in Fig. E2.

In any case, and in contrast with Marcus' theory, no simple expression where the solvent dielectric response is subsumed into a permittivity coefficient is likely to hold for $\Delta G^{\neq}$. In fact, in enzymatic reactions, the rate of electron transference is probably significantly tuned by the presence/absence of nearby dehydrons, as schematically shown in Fig. E2. This is so because of the influence of the latter on the solvent structural fluctuations. Furthermore, there is symptomatic evidence that water frustration, typically caused by dehydrons (Chaps. 1–4), may enhance electron transfer rates [2]. Precisely because of the highly detailed information on interfacial solvent structure and dynamics that is required, we envision this setting as ideal to develop the definite rigorous theory of epistructural interfacial physics suited for biological systems.

E2. Closing Thoughts

At this point, some closing thoughts are in order. This book was motivated by the need to tackle major challenges in molecular biophysics, a field that has not achieved—in the author's opinion—the level of intellectual maturity required to become operationally efficacious for the molecular engineer, drug designer or medical doctor. Thus, all or most of the major problems in the field remain virtually as daunting as when they were first formulated nearly five decades ago. The protein folding problem, the prediction of binding sites, the rational design of drugs and ligands and the engineering and optimization of enzymatic catalysts, all stand—or I now dare say, stood—in defiance to research efforts. Such efforts comprised a colossal amount of work enshrined in tens of thousands of papers published in the course of five decades. Whoever believes this formidable output has resulted in decisive progress is being delusional.

Perhaps even more worrisome is the conceptual stagnation of the field, enslaved by concepts like "folding intermediates", "folding pathway", "folding funnel in free energy surface" or "effective permittivity in soluble proteins", that are either irrelevant, fatally flawed or in contradiction with the basic laws of physics.

Creative evolution often entails destruction, and nowhere is this principle better represented as in Hinduism [3]. Figure E3 represents lord Siva, the destructive power whose participation is inherent to the creation cycle, according to the Indian

Fig. E3 Hindu deity Siva here represented as Lord of the Dance. Siva embodies the destructive impulse inherent to creation [3]

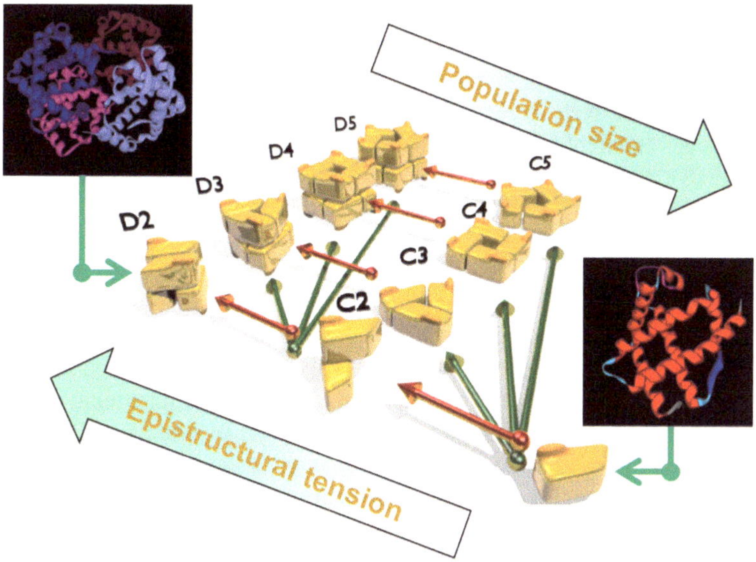

Fig. E4 Origins of biological complexity as revealed by the interfacial physics introduced in this book. A structural degradation of proteins through dehydron enrichment in complex species increases interfacial tension between protein and water for the protein subunit, which promotes higher levels of complexation as we compare proteins in species with lower population with those in unicellular species. For example, "haemoglobin" (depicted in ribbon representation) in a unicellular species tends to be monomeric, while in species with significantly lower population haemoglobin becomes a tetramer endowed with D_2-symmetry. The biological self-assembling yielding higher levels of symmetry $C_n \rightarrow D_m$, with $m \geq n$, in homomeric proteins is associated with a correlated increase in epistructural tension for the protein subunit that in turn results from the structural degradation that arises from deleterious mutations more readily fixed in species with lower population

cosmogony dating back to Vedic times [3]. If molecular biophysics ever strives to achieve the level of success reached by its sister discipline, structural biology, it would have to subject its conceptual background to a much more rigorous level of scrutiny.

This book portends to have achieved precisely that. By introducing the concept of frustrated interfacial water and exploring its impact in the realms of nanoscale dielectrics, epistructural tension and enzyme catalysis, the book has re-written the physics of biomolecular interfaces. This task entailed leaving aside or "destroying"—in a Vedic sense—some of the ill-conceived concepts that pervaded much of the thinking in biophysics in order to attempt a solution of the major standing problems in the field.

Strikingly, the biggest impact of the conceptual transformation put forth in this book, both for molecular engineering and for medicine, is achieved when the novel interfacial physics is endowed with the evolutionary axis inherent to biology (Chap. 6). This intellectual enrichment places the dehydron at the center of molecular evolution when examined from a structural perspective, while epistructural tension becomes the dominant factor driving biological self-organization. Both these aspects are synthetically depicted in Fig. E4.

The dwindling in species population arising at higher organismic complexity increases the influence of random genetic drift as a sculptor of the molecular phenotype [4]. Thus, when examining a protein across its established orthology in species of progressively higher complexity, the increasing importance of random drift becomes apparent and results in increasing structural degradation arising from a progressive enrichment in dehydrons (Chap. 6, [4]). This is so because random evolutionary change is usually deleterious to the protein function. The structural degradation in turn increases the interfacial tension between protein and water which promotes higher levels of complexation (Chaps. 1, 2, 4–6). For example, "haemoglobin" in a unicellular species tends to be monomeric while in species with significantly lower population, such as humans, haemoglobin is tetrameric and endowed with D_2-symmetry (Fig. E4). This level of complexation endows haemoglobin with allostery and regulation, the hallmarks of functional complexity. The biological self-assembling yielding higher levels of symmetry $C_n \rightarrow D_m$, with $m \geq n$, in homomeric proteins (Fig. E4) is invariably associated with a correlated increase in epistructural tension for the protein subunit. This increase in interfacial tension between the protein and water in turn results from the structural degradation that arises from deleterious mutations that are more easily fixed in the lower population of more complex species.

This is an illustration of the way biological complexity arises and evolves through enabled self-assembling as revealed by the interfacial physics introduced by this book.

References

1. Marcus RA, Sutin N. Electron transfers in chemistry and biology. Biochim Biophys Acta. 1985;811:265–322.
2. Remsing RC, McKendry IG, Strongin DR, Klein ML, Zdilla MJ. Frustrated solvation structures can enhance electron transfer rates. J Phys Chem Lett. 2015;6:4804–8.
3. Flood G. An introduction to Hinduism. Cambridge: Cambridge University Press; 1996.
4. Fernández A. Non-adaptive origins of interactome complexity. Nature. 2011;474:502–6.

Appendix A

A1.1 Code for Dehydron Identification

The software "YapView" (yet another protein view), currently incarnated as "Dehydron Calculator", is used to identify the dehydrons from structural coordinates of soluble proteins. This program and two equivalent programs, "WRAPPA" and "dehydron calculator in PyMOL plugin version", are currently freely downloadable from the site:

http://people.cs.uchicago.edu/~ridg/softwaredigbio.html

The code for the Dehydron Calculator in Python provided as plugin for PyMol is provided this Appendix. The open source is found at

https://raw.github.com/Pymol-Scripts/Pymol-script-repo/master/plugins/dehydron.py

YAP View/Dehydron Calculator and the "desolv" plugin within it used to calculate hydrogen bond burial and identify dehydrons are currently hosted at:

http://sourceforge.net/projects/protlib/files/yapview/0.6.8/

The main installer for the windows version (YAPView-0.6.8-1-Installer.exe) comes with a pre-compiled version of the desolv plugin that should be accessible from the UI (user interface).

Code for Dehydron Calculator as PyMOL Plugin (in Python)

```
Dehydron: A dehydron calculator plugin for PyMOL
Version: 1.7
Described at PyMOL wiki:
http://www.pymolwiki.org/index.php/dehydron
```


A. Fernández, *Physics at the Biomolecular Interface*, Soft and Biological Matter,
DOI 10.1007/978-3-319-30852-4

```
Author : Osvaldo Martin
email: aloctavodia@gmail.com
Date : March 2013
License: GNU General Public License
Acknowledgement: The H-bond detection code is based on the
list_mc_hbonds.py
script from Robert L. Campbell
http://pldserver1.biochem.queensu.ca/~rlc/work/pymol/
'''

import Tkinter
from Tkinter import *
import Pmw
from pymol import cmd
from pymol import stored

def __init__(self):
 """Add this Plugin to the PyMOL menu"""
 self.menuBar.addmenuitem('Plugin', 'command',
  'dehydron',
  label = 'Dehydron',
  command = lambda : mainDialog())

def mainDialog():
 """ Creates the GUI"""

 def get_dehydrons():
 angle_range = float(angle_value.get())
 max_distance = float(dist_cutoff_value.get())
 desolv = float(desolv_sphere.get())
 min_wrappers = float(min_value.get())
 selection = sel_value.get()
    dehydron(selection, angle_range, max_distance, desolv,
min_wrappers)

 master = Tkinter.Tk()
 master.title(' dehydron ')
   w = Tkinter.Label(master, text = 'dehydron calculator\nOsvaldo Mar-
tin - aloctavodia@gmail.com',
  background = '#000000',
  foreground = '#cecece',
  #pady = 20,
  )
 w.pack(expand=1, fill = 'both', padx = 4, pady = 4)
 Pmw.initialise(master)
 nb = Pmw.NoteBook(master, hull_width = 420, hull_height=280)
 p1 = nb.add('Main')
```

```
p2 = nb.add('About')
nb.pack(padx=5, pady=5, fill=BOTH, expand=1)
############################ Main TAB ##################################
### hydrogen bond settings
group = Pmw.Group(p1,tag_text='Hydrogen bond Settings')
group.pack(fill='x', expand=1, padx=20, pady=1)
Label(group.interior(), text='angle range').grid(row=2, column=0)
angle_value = StringVar(master=group.interior())
angle_value.set(40)
 entry_angle = Entry(group.interior(),textvariable=angle_value,
width=10)
entry_angle.grid(row=2, column=1)
entry_angle.configure(state='normal')
entry_angle.update()
 Label(group.interior(), text='max distance').grid(row=3, column=0)
dist_cutoff_value = StringVar(master=group.interior())
dist_cutoff_value.set(3.5)
entry_dist = Entry(group.interior(),textvariable=dist_cutoff_value,
width=10)
entry_dist.grid(row=3, column=1)
entry_dist.configure(state='normal')
entry_dist.update()
### dehydron settings
group = Pmw.Group(p1,tag_text='Dehydron Settings')
group.pack(fill='x', expand=1, padx=20, pady=5)
Label(group.interior(), text='desolvatation sphere radius').grid
(row=2, column=2)
desolv_sphere = StringVar(master=group.interior())
desolv_sphere.set(6.5)
 entry_desolv=Entry(group.interior(),textvariable=desolv_sphere,
width=10)
entry_desolv.grid(row=2, column=3)
entry_desolv.configure(state='normal')
entry_desolv.update()
 Label(group.interior(), text='minimum wrappers').grid(row=3,
column=2)
min_value = StringVar(master=group.interior())
min_value.set(19)
 entry_min_value=Entry(group.interior(),textvariable=min_value,
width=10)
entry_min_value.grid(row=3, column=3)
entry_min_value.configure(state='normal')
entry_min_value.update()
### selection settings
group = Pmw.Group(p1,tag_text='Selection')
```

```
group.pack(fill='x', expand=1, padx=20, pady=5)
Label(group.interior(), text='selection').grid(row=4, column=2)
sel_value = StringVar(master=group.interior())
sel_value.set('all')
  entry_sel_value=Entry(group.interior(),textvariable=sel_value,
width=10)
entry_sel_value.grid(row=4, column=3)
entry_sel_value.configure(state='normal')
entry_sel_value.update()
### submit
  Button(p1, text="Calculate", command=get_dehydrons).pack(side=BOTTOM)
############################ About TAB #################################
group = Pmw.Group(p2, tag_text='About dehydron plug-in')
group.pack(fill = 'both', expand=1, padx = 5, pady = 5)
text =u"""For a brief introduction to the dehydron concept, you could
read http://en.wikipedia.org/wiki/dehydron

Citation for this plugin:
Martin O.A.; dehydron calculator plugin for PyMOL,
2012. IMASL-CONICET.

Citation for PyMOL may be found here:
http://pymol.sourceforge.net/faq.html#CITE

Citation for dehydrons (I think these could be used):
Fern\u00E1ndez A. and Scott R.; "Adherence of packing
defects in soluble proteins", Phys. Rev. Lett. 91, 018102
(2003).

Fern\u00E1ndez A., Rogale K., Scott R. and Scheraga H.A.;
"Inhibitor design by wrapping packing defects in HIV-1
proteins", PNAS, 101, 11640-45 (2004).

Fern\u00E1ndez A. "Transformative Concepts for Drug Design:
Target Wrapping" (ISBN 978-3642117916),
Springer-Verlag, Berlin, Heidelberg (2010).
"""
 #
 # Add this as text in a scrollable panel.
 # Code based on Caver plugin
 # http://loschmidt.chemi.muni.cz/caver/index.php
 #
 interior_frame = Frame(group.interior())
 bar = Scrollbar(interior_frame)
  text_holder = Text(interior_frame, yscrollcommand=bar.set,
foreground="#cecece",background="#000000",font="Times 12")
 bar.config(command=text_holder.yview)
```

```
text_holder.insert(END,text)
text_holder.pack(side=LEFT,expand="yes",fill="both")
bar.pack(side=LEFT,expand="yes",fill="y")
interior_frame.pack(expand="yes",fill="both")

master.mainloop()

def dehydron(selection='all', angle_range=40, max_distance=3.5,
desolv=6.5, min_wrappers=19, quiet=0):
    '''
DESCRIPTION

dehydron calculator

USAGE
 dehydron [ selection [, angle_range [, max_distance [, desolv [,
min_wrappers ]]]]]
    '''

angle, max_distance = float(angle_range), float(max_distance)
desolv, min_wrappers = float(desolv), int(min_wrappers)
quiet = int(quiet)

name = cmd.get_legal_name('DH_%s' % selection)
cmd.delete(name)
selection_hb = '((%s) and polymer)' % (selection)
hb = cmd.find_pairs("((byres "+selection_hb+") and n. n)","((byres
"+selection_hb+")    and    n.    o)",mode=1,cutoff=max_distance,
angle=angle_range)

if not quiet:
hb.sort(lambda x,y:(cmp(x[0][1],y[0][1])))
    print "-----------------------------------------
------"
print "--------------------Dehydron Results-----------
-----"
print "-----------------------------------------
----"
 print " Donor | Aceptor |"
print " Object Chain Residue | Object Chain Residue | # wrappers"
 cmd.select('_nonpolar', '(elem C) and not (solvent or (elem N+O)
extend 1)', 0)
try:
cmd.select('_selection', '%s' % selection, 0)
except:
pass
sel = []
for pairs in hb:
```

```
 wrappers = cmd.count_atoms('((%s and _nonpolar and _selection)
within %f of byca (%s`%d %s`%d))' %
 ((pairs[0][0], desolv) + pairs[0] + pairs[1]))
 if wrappers < min_wrappers:
 cmd.distance(name, pairs[0], pairs[1])
 if not quiet:
 cmd.iterate(pairs[0], 'stored.nitro = chain, resi, resn')
 cmd.iterate(pairs[1], 'stored.oxy = chain, resi, resn')
 print ' %12s%4s%6s%5d | %12s%4s%6s%5d |%7s' % (pairs[0][0], stored.
nitro[0], stored.nitro[2], int(stored.nitro[1]), pairs[1][0],
stored.oxy[0], stored.oxy[2], int(stored.oxy[1]), wrappers)
 sel.append(pairs[0])
 sel.append(pairs[1])
 cmd.delete('_nonpolar')
 cmd.delete('_selection')
 if len(sel) > 0:
 cmd.show_as('dashes', name)
 elif not quiet and len(hb) != 0:
 print ' - no dehydrons were found - '
 else:
 print ' - no hydrogen bonds were found - '
cmd.extend('dehydron', dehydron)
# vi:expandtab:smarttab
```

NOTE: The code also runs directly from the line of command. It is basically the same as the plugin but adapted to be called from a terminal. When executed as python dehydron_ter.py > log.out it will compute dehydrons for all PDB files contained in the same folder as "dehydron_ter.py" and it would download the results in the log.out file. The code can of course be modified to print the results on a single file (trivial in Python).

```
import __main__
__main__.pymol_argv = ['pymol','-qck']
import glob
import pymol
from pymol import cmd, stored
pymol.finish_launching()
'''
Dehydron: A dehydron calculator plugin for PyMOL
Version: 1.7
Described at PyMOL wiki:
http://www.pymolwiki.org/index.php/dehydron

'''
```

```
def dehydron(selection='all', angle_range=40, max_distance=3.5,
desolv=6.5, min_wrappers=19, quiet=0):
'''
DESCRIPTION

 dehydron calculator

USAGE

 dehydron [ selection [, angle_range [, max_distance [, desolv [,
min_wrappers ]]]]]
  '''

 angle, max_distance = float(angle_range), float(max_distance)
 desolv, min_wrappers = float(desolv), int(min_wrappers)
 quiet = int(quiet)

 name = cmd.get_legal_name('DH_%s' % selection)
 cmd.delete(name)

 selection_hb = '((%s) and polymer)' % (selection)
 hb = cmd.find_pairs("((byres "+selection_hb+") and n. n)","((byres
"+selection_hb+") and n.
  o)",mode=1,cutoff=max_distance,angle=angle_range)

 if not quiet:
 hb.sort(lambda x,y:(cmp(x[0][1],y[0][1])))
 print "-----------------------------------------
----"
 print "---------------Dehydron Results------------------
---"
 print "-----------------------------------------
----"
 print " Donor | Aceptor |"
 print " Object Chain Residue | Object Chain Residue | # wrappers"

 cmd.select('_nonpolar', '(elem C) and not (solvent or (elem N+O)
extend 1)', 0)
 try:
 cmd.select('_selection', '%s' % selection, 0)
 except:
 pass

 sel = []
 for pairs in hb:
 wrappers = cmd.count_atoms('((%s and _nonpolar and _selection)
within %f of byca (%s'%d %s'%d))' %
 ((pairs[0][0], desolv) + pairs[0] + pairs[1]))
 if wrappers < min_wrappers:
```

```
cmd.distance(name, pairs[0], pairs[1])
if not quiet:
cmd.iterate(pairs[0], 'stored.nitro = chain, resi, resn')
cmd.iterate(pairs[1], 'stored.oxy = chain, resi, resn')
print ' %12s%4s%6s%5d | %12s%4s%6s%5d |%7s' % (pairs[0][0], stored.
nitro[0], stored.nitro[2], int(stored.nitro[1]), pairs[1][0],
stored.oxy[0], stored.oxy[2], int(stored.oxy[1]), wrappers)
sel.append(pairs[0])
sel.append(pairs[1])
cmd.delete('_nonpolar')
cmd.delete('_selection')
if len(sel) > 0:
cmd.show_as('dashes', name)
elif not quiet and len(hb) != 0:
print ' - no dehydrons were found - '
else:
print ' - no hydrogen bonds were found - '
############ main routine ###########
proteins = glob.glob('*.pdb')
proteins.sort()
for protein in proteins:
cmd.load(protein)
  dehydron(selection='all', angle_range=40, max_distance=3.5,
desolv=6.5, min_wrappers=19, quiet=0)
cmd.delete("all")
```

A1.2 Quantum Mechanics Computations for Dehydron Catalysis (Chap. 7)

In Chap. 7 we determine the free energy change associated with the chemical event of dehydron-induced proton transference through multiple steering molecular dynamics computations as delineated by Jarzynski [1]. The nucleophilic group and nearby proton-receptive water molecules at the dehydron interface are treated within a quantum mechanical (QM) scheme while the rest of the molecule and explicit solvent are treated through classical molecular mechanics (MM) in accord with a QM-MM hybrid approach. The QM/MM computation is performed with SIESTA [2–4], a package for efficient electronic structure calculations and ab initio molecular dynamics simulations with the code accessible from url:

http://departments.icmab.es/leem/siesta/CodeAccess/Code/downloads.html

The efficiency of SIESTA arises from the fact that it exploits strictly localized basis sets and from the implementation of linear-scaling algorithms to treat large molecular systems within a first-principle approach to electronic-structure. The SIESTA fundamentals and implementation details have been covered in previous work [2--4]. SIESTA uses flexible basis sets consisting of linear combinations of finite atomic orbitals defined in a real space grid and is optimized to yield an order N scaling. The algorithms perform numerical computations, so that basis functions are not constrained to predetermined analytical functional form while providing a suitable matching of the radial wave function to the core region, with the latter described by pseudopotentials. The pseudo atomic orbitals (PAOs) are the eigenfunctions of the atomic pseudopotential problem confined within an infinite potential with spherical symmetry [5]. This confinement generates finite-range PAOs that yield higher kinetic energies than the unconstrained eigenfunctions. The basis set is constructed in a balanced way by spreading the kinetic energy increment up to the same level over all PAOs. Split-valence bases are generated by combining suitable numerical orbitals on top of the minimal basis described above. The nuclei and inner electrons are represented by norm-conserving pseudopotentials to avoid the laborious computation of core states. Additionally, this procedure smoothens the valence charge density thereby relaxing the grid constraints. The Kohn-Sham Hamiltonian $\mathcal{H}_{KS}$ adopted in our SIESTA computations is

$$\mathcal{H}_{KS} = K + \sum_{\alpha} \left[V_{\alpha}^{loc} + V_{\alpha}^{nl}\right] + V^{H}(\mathbf{r}) + V^{xc}(\mathbf{r})$$

Here K is the kinetic energy operator, V^{H} and V^{xc} denote respectively the Hartree-Fock and exchange-correlation potentials, the dummy index α labels atoms, and the pseudopotential V results from two contributions: $V = V^{loc} + V^{nl}$, where V^{loc} indicates a local term independent of the angular momentum of the core electrons and determines the long-range interaction, and V^{nl} constitutes the nonlocal term that operates exclusively on the valence electrons. Details on the SIESTA-based numerical computation of the Kohn-Sham energy are provided elsewhere [2, 4]. The computation of Kohn-Sham energies were performed on contracted Gaussian basis sets of double-zeta valence polarized (DZVP) quality (pseudoatomic orbital energy shift = 30 meV, grid cutoff = 135 Ry).

Within the SIESTA scheme, the dehydron-associated water molecule is identified by a distance between backbone carbonyl and water oxygen atoms < 3.3 Å and is treated jointly with the purported dehydron-functionalized nucleophilic group as QM entities. The solvent and biomolecular environment are represented with an appropriate classical potential adopting the force field parametrization by Wang et al. [6]. This classical parametrization determines the intramolecular energy by setting force constants for bonds, angles and torsions, and for electrostatic and van der Waals interactions involving atom pairs separated by at least three covalent consecutive bonds. Save for the specified dehydron-associated water molecule that may eventually trigger the chemical event of proton transference, water is treated with the classical TIP3P model [7].

To cover long timescales relevant to folding events such as the mutation-induced dehydron formation in the D816V variant of the KIT kinase treated in the paper, the

MM region is treated through all-atom realizations of underlying coarse "protodynamics" as detailed in [8]. Within this scheme for accelerated MM simulation, torsional degrees of freedom of the backbone and side chains are coarse grained modulo basins of attraction (energetically allowed regions) in the potential energy surface, known as Ramachandran regions in local conformation space. Thus, the specification of the Ramachandran basin for each residue and its transition to another basin provide a coarse dynamic description of the protein chain that evolves according to a Markovian stochastic process (see [8] and references therein). This coarse realization of the chain is fleshed out to a full all-atom description subject to the parametrization given in [6] to determine the next coarse step. Thus, to cover relevant timescales for dehydron induction and dehydron-related chemical events ($\sim$100 ns to $\sim$ 1 μs), the dynamics are entrained by the coarser "protodynamics", where the backbone (Φ, Ψ) dihedral torsions are specified "modulo basins of attraction" in the potential energy surface. Coarse moves are defined as transitions between basins of attraction (R-basins) in the Ramachandran torsional (Φ, Ψ)-map for each residue. Thus, each residue is assigned an R-basin after a coarse move, and the coarse state of the chain becomes a conformational ensemble, with each conformation generated by selecting individual (Φ, Ψ)-coordinates within the assigned R-basins.

The total energy E_{TOT} is $E_{TOT} = E_{KS} + E_{MM} + E_{QM-MM}$, where E_{KS} is the Kohn-Sham energy [4], E_{MM} is the molecular mechanics energy, and the hybrid coupling term E_{QM-MM} is decomposed as

$$E_{QM-MM} = \sum_{i=1}^{C} q_i \int \frac{\rho(\mathbf{r})}{|\mathbf{r} - \tau_i|} \mathbf{dr} + \sum_{i=1}^{C} \sum_{\alpha=1}^{A} \frac{q_i Z_\alpha}{|\mathbf{R}_\alpha - \tau_i|} + E^{LJ}$$

Here $\rho(\boldsymbol{r})$ is the self-consistent quantum electronic density [2–4], C is the number of atoms in the classical region, with charges q_i determined by the classical force field adopted to model the biomolecular environment, while α indexes the A nuclei within the quantum subsystem with core charges Z_α. The position vectors $\boldsymbol{\tau}_i$, $\mathbf{R}_\alpha$ correspond to classical atoms and nuclei, respectively. The first term on the r. h. s. represents the electrostatic interaction between electrons and classical charges and results from numerical integration over the grid. The next term accounts for the electrostatic interactions between the nuclei in the quantum subsystem and the classical point charges, while E^{LJ} models the van der Waals interactions between the atoms in the quantum and classical regions modeled by a 6–12 Lennard-Jones potential with quantum atom LJ parametrization taken from the force field parametrization given in [6].

The forces acting on the QM nuclei are obtained by differentiation of the Kohn-Sham energy plus the coupling energy E_{QM-MM} with respect to the nuclei positions:

$$\boldsymbol{F}_\alpha = -\frac{\partial E_{KS}[\mathbf{R}]}{\partial \mathbf{R}_\alpha} - \frac{\partial E_{QM-MM}[\mathbf{R}, \tau]}{\partial \mathbf{R}_\alpha}$$

On the other hand, the forces acting on the classical atoms are computed as the respective derivative of E_{MM} plus E_{QM-MM} with respect to the classical atomic positions:

$$\mathbf{F}_i = -\frac{\partial E_{\mathrm{MM}}[\tau]}{\partial \tau_i} - \frac{\partial E_{\mathrm{QM-MM}}[\mathbf{R}, \tau]}{\partial \tau_i}$$

QM-MM restrained energy minimizations have been performed to determine the free energy change associated with the dehydron-induced proton transfer that functionalizes the QM nucleophilic groups. The proton-pulling coordinate is described in the main text as representing the intrinsic reaction coordinate within the multiple steering scheme delineated in [1]. In accord with this QM/MM thermodynamic computation, an additional elastic term has been added to the potential energy in order to perform the restrained minimization. This procedure enables the computation of the free energy change associated with dehydron-induced deprotonation, a computation susceptible of experimental validation by contrasting the computed thermodynamic value against experimental pKa measurements.

References

1. Jarzynski C. Nonequilibrium equality for free energy differences. Phys Rev Lett. 1997;78:2690–3. doi:10.1103/PhysRevLett.78.2690.
2. Soler JM, Emilio Artacho E, Gale JD, García A, Junquera J, Ordejón P, Sánchez-Portal D. The SIESTA method for ab initio order-N materials simulation. J Phys Condens Matter. 2002;14:2745. doi:10.1088/0953-8984/14/11/302.
3. Sanchez-Portal D, Ordejon P, Artacho E, Soler JM. Density-functional method for very large systems with LCAO basis sets. Int J Quantum Chem. 1997;65:453–61. doi:10.1002/(SICI)1097-461X(1997)65:5<453::AID-QUA9>3.0.CO;2-V.
4. Sanz-Navarro CF, Grima R, García A, Bea EA, Soba A, Cela JM, Ordejón P. An efficient implementation of a QM–MM method in SIESTA. Theor Chem Accounts. 2011;128:825–33. doi:10.1007/s00214-010-0816-5.
5. Sankey OF, Niklewski DJ. Ab initio multicenter tight-binding model for molecular-dynamics simulations and other applications in covalent systems. Phys Rev B. 1989;40:3979–95. doi:10.1103/PhysRevB.40.3979.
6. Wang J, Cieplak P, Kollman PA. How well does a restrained electrostatic potential (RESP) model perform in calculating conformational energies of organic and biological molecules? J Comput Chem. 2000;21:1049–74. doi:10.1002/1096-987X(200009)21:12<1049::AID-JCC3>3.0.CO;2-F.
7. Jorgensen WL. Quantum and statistical mechanical studies of liquids. 10. Transferable intermolecular potential functions for water, alcohols, and ethers. Application to liquid water. J Am Chem Soc. 1981;103:335–40. doi:10.1021/ja00392a016.
8. Fernández A. Fast track communication: water promotes the sealing of nanoscale packing defects in folding proteins. J Phys Condens Matter. 2014;26:202101. doi:10.1088/0953-8984/26/20/202101.

Appendix B

Answers to Problems

Chapter 1

1. (a) $\mu = k_B T \ln f$, where k_B = Boltzmann constant;
   (b) See Fig. P1.1.

1.2. Several expressions may be used to assess the departure from the Debye picture. Among them:

$$\frac{1}{2}\varepsilon_0{}^{-1}\int \|\vec{P^{\#}}\|^2 d\vec{r} = \int \frac{1}{2}\lambda \|\vec{\nabla} g\|^2 d\vec{r}$$

$$\left\{\int \left[\vec{\nabla}.\left(\vec{P^{\#}}\right)\right]^2 d\vec{r}\right\}^{1/2} = \left\{\int \left[\rho - \vec{\nabla}.\left(\varepsilon_0 \vec{E} + \vec{P^{\|}}\right)\right]^2 d\vec{r}\right\}^{1/2}$$

1.3. See Figs. P1.2 and P1.3. The motif is called PP2 and it leads to a high exposure and extensive hydration of inter-strand dehydrons. Thus, such over-exposed dehydrons become stabilized through hydration and do not cause interfacial tension, at least to the same degree as dehydrons in standard structural motifs. We therefore anticipate that the golden rule of protein architecture is not going to hold in this case since the over-exposed dehydrons do not constitute structural defects (i.e. producers of interfacial tension) that need to be compensated with disulfide bridges to maintain structural integrity.

1.4. At 67.65 % discrepancy between the estimated (Y = 32.35) and the real (Y = 100) Y-value, the antifreeze protein from the snow flea (PDB.2PNE, N = 81, two disulfide bonds or X = 2.47) is the single major outlier from the balance equation. Its pattern of structural deficiencies is shown in Figs. P1.2 and P1.3. The actual Y-value lies outside a loose confidence region containing

A. Fernández, *Physics at the Biomolecular Interface*, Soft and Biological Matter,
DOI 10.1007/978-3-319-30852-4

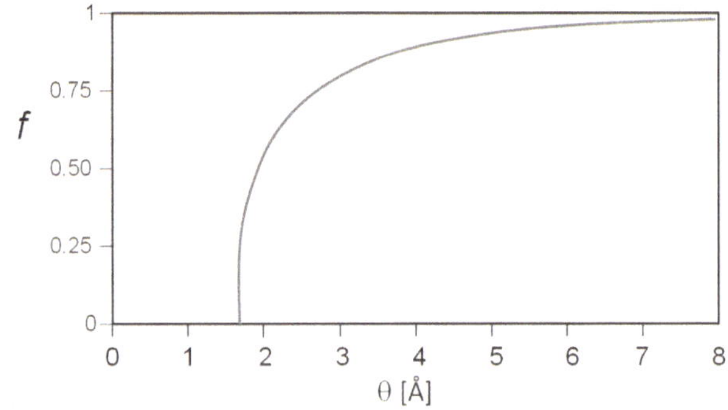

Fig. P1.1 The θ-dependence of interfacial fugacity (f)

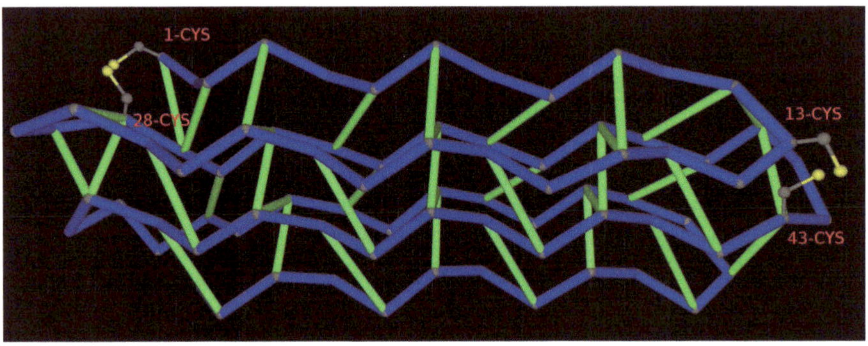

Fig. P1.2 Dehydron pattern for PDB.2PNE

all other data points and is defined by three standard deviations from the linear-regression fit $Y = 5.00X + 20.00$.

This anomalously large Y-value may be explained as we incorporate the polyproline II (PPII)-conformation content to the structural analysis. The hydrogen-bonded amide and carbonyls in the backbone of residues in PPII conformation maximize their solvent exposure and hence hydrogen bond burial is not required as a provider of thermodynamic compensation. At 80 %, the PPII content of the antifreeze protein PDB.2PNE (65 residues out of 81) is the highest in the PDB-derived database. The average PPII-content of a soluble protein is 11 %, with dispersion 6 %, and all soluble proteins have PPII-content $\leq$45 %, with the antifreeze protein PDB.2PNE being the sole exception. Hence, this protein maximizes its backbone exposure to solvent to a level where backbone hydrogen bonds no longer hinder the hydration of amides and carbonyls. At this anomalously high level of backbone hydration, the structure-destabilizing contribution of structural deficiencies represented by unburied backbone hydrogen bonds (dehydrons) is reduced significantly, and therefore, so is the need for disulfide-bond compensation.

The PPII content explains the anomalous Y-value for PDB.2PNE, but its extremely high PPII content places this protein in a class of its own, and does

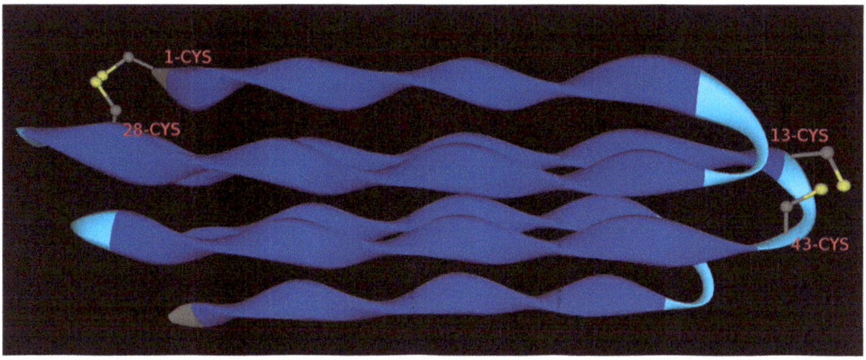

Fig. P1.3 Ribbon rendering of the structure in PDB.2PNE

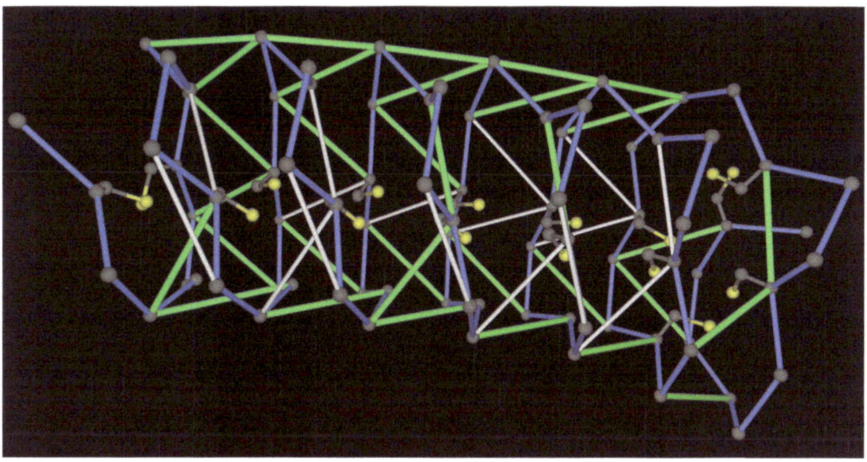

Fig. P1.4 Dehydron pattern for PDB.1EZG

not enable us to trace the deviations from the balance equation as correlated with PPII content. In fact, no such correlation exists.

1.5. See Figs. P1.4 and P1.5.
To establish adequate controls, all current 60 entries for PDB-reported antifreeze proteins excluding 2PNE were examined. Their average PPII-content was found to be 11 % and their respective (X, Y)-datapoints fit the balance equation to within 99 % accuracy. As an illustration, the beta-helix antifreeze protein from beetle tenebrio molitor (PDB.1EZG, $N = 81$, eight disulfide bonds, or $X = 9.87$) has seven residues in PPII conformation (<10 %) and 33 structural deficiencies out of 48 backbone hydrogen bonds. Thus, its actual Y-value is 68.75 %. The Y-value obtained from the balance equation is $Y = 5 \times 9.87 + 20 = 69.35$, a prediction within 99 % accuracy.

1.6. The expected number of carbonaceous HB-wrapping groups per residue is 3.

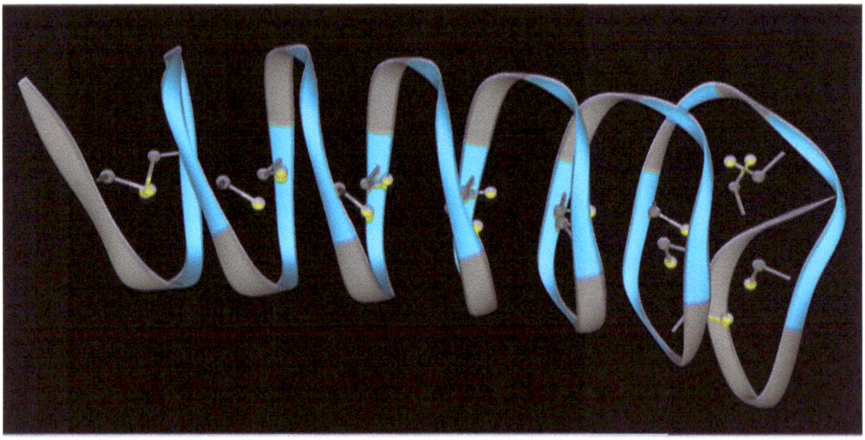

Fig. P1.5 Ribbon rendering of the structure in PDB.1EZG

Chapter 2

2.1. We first compute the elastic integral $\int \frac{1}{2}||\vec{\nabla}\phi||^2 d\vec{r}$ by integration by parts:

$$\begin{aligned}\int ||\vec{\nabla}\phi||^2 d\vec{r} &= -\int \phi\nabla^2\phi d\vec{r} + \oiint \phi\vec{\nabla}\phi . d\vec{\sigma} \\ &= -\int \phi\nabla^2\phi d\vec{r} + \int \vec{\nabla}.\left[\phi\vec{\nabla}\phi\right] d\vec{r} = -\int \phi\nabla^2\phi d\vec{r}\end{aligned} \tag{P2.1}$$

where the relation $\oiint \phi\vec{\nabla}\phi . d\vec{\sigma} = \int \vec{\nabla}.\left[\phi\vec{\nabla}\phi\right] d\vec{r} = 0$ follows from Gauss' divergence theorem with the surface integral evaluated on the closed surface indicated in Fig. 2.5. Multiplying (2.8) by ϕ and integrating yields

$$\int \phi\nabla^2\phi d\vec{r} = k\int \Gamma\phi^2 d\vec{r} \tag{P2.2}$$

Combining (P2.1) and (P2.2) we obtain the interfacial energy $\Delta U_\phi = U_{if}$ as follows:

$$\Delta U_\phi = \int \frac{1}{2}\lambda||\vec{\nabla}\phi||^2 d\vec{r} = -\frac{1}{2}\lambda k\int \Gamma\phi^2 d\vec{r} = -\frac{1}{2}\lambda\int \phi\nabla^2\phi d\vec{r} \tag{P2.3}$$

2.2. Anomalous polarization $P^{\#}$ creates a torque imposed by the electrostatic field $\vec{E} = \vec{E}(\vec{r})$ of the protein

$$\vec{E}(\vec{r}) \times \vec{\mu}_{P^{\#}}(\vec{r}) = \vec{E}(\vec{r}) \times \int \left(\vec{r'} - \vec{r}\right)\vec{\nabla}.\vec{P^{\#}}\left(\vec{r'}\right) d\vec{r'} \tag{P2.4}$$

where the anomalous polarization-associated dipole moment is

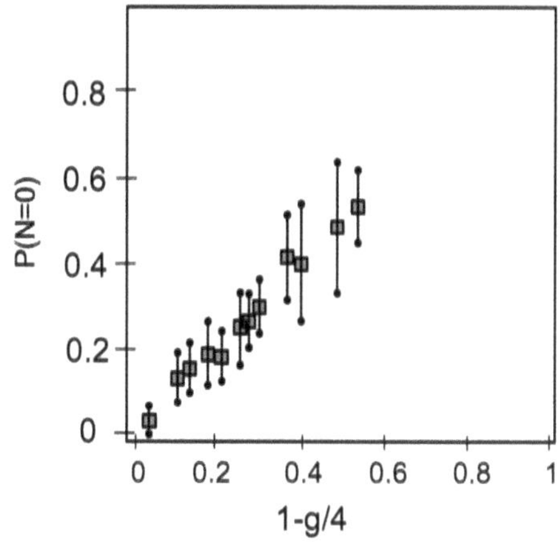

Fig. P2.1 Vacating probability, P(N = 0), plotted against deviation from bulk structure (1 – g/4) for the monomeric soluble proteins specified by PDB entries 1SRL, 1ESR, 1A8O, 1PIT, 1QGB, 1ATA, 1Q7I, 1PI2, 2PNE

$$\vec{\mu}_{P^{\#}}(\vec{r}) = \int \left(\vec{r'} - \vec{r}\right) \vec{\nabla} . \vec{P}^{\#}\left(\vec{r'}\right) d\vec{r'} \tag{P2.5}$$

Thus, as the protein folds, a minimization in this torque becomes operative, and this can only be achieved by minimizing the number of dehydrons (Fernandez A (2014) Fast Track Communication: Water promotes the sealing of nanoscale packing defects in folding proteins. J Phys Condens Matter 26:202101).

2.3. The vacating probability P(N = 0) for the solvation domain of each protein atom visited by water oxygens (Chap. 1) was obtained for the PDB-reported monomeric soluble proteins identified by entries 1SRL, 1ESR, 1A8O, 1PIT, 1QGB, 1ATA, 1Q7I, 1PI2, 2PNE. Independently, the deviation from bulk water structure quantified by 1 – g/4 was obtained for the same solvation domains. A linear correlation with slope 1 and $R^2 = 0.88$ was obtained upon plotting P(N = 0) vs. 1 – g/4, as shown in Fig. P2.1.

2.4 Since spatial location is taken to be fixed, we drop the spatial coordinate vector: $\phi(\vec{r}, t) = \phi(t); \varepsilon^{\#}(\vec{r}, f) = \varepsilon^{\#}(f)$. Defining an effective volume V by the average separation between interfacial hydrogen atoms, and denoting with "< >" the average over trajectories in phase space, we get:

$$\varepsilon^{\#}(f)^{-1} = 1 + \beta(\varepsilon_0 V)^{-1} \int_0^{\infty} e^{-2\pi i f t} \xi^2 \partial/\partial t\left[< \nabla\phi(0) \cdot \nabla\phi(t) > - < \nabla\phi>^2\right] dt$$

where $\xi^2[< \nabla\phi(0) \cdot \nabla\phi(t) > - < \nabla\phi>^2]$ is the time-correlation function for non-Debye polarization.

The expression for $\varepsilon^{\#}(f)$ given above has been validated by reproducing to nearly 96 % accuracy the perpendicular component of the experimental dielectric spectrum of water interfacing with a dehydron monolayer in the frequency range 1–100 GHz (AF Innovation, Gesellschaft mit beschränkter Haftung). The monolayer has composition 50 % pentadecan-2-ol and 50 % tridecane, and generates a dehydron layer at the interface through the hydrogen bonding of alcohol hydroxyl groups. The real part of $\varepsilon^{\#}(f)$ suffers significant attenuation at f → 0 from the known value 72 for bulk water to $\varepsilon^{\#}(f) = 38$ for interfacial water. The imaginary part of $\varepsilon^{\#}(f)$ for interfacial water is also attenuated relative to bulk water by 54 % at the absorption peak at f = 15 GHz. Furthermore, the imaginary part reveals an absorption peak for interfacial water at f = 53 GHz, dramatically blueshifted relative to the bulk. The blueshift is in accord with the acceleration of water molecules due to their frustration in the vicinity of dehydrons (Sierra MB, Accordino SR, Rodriguez-Fris A, Morini MA, Appignanesi GA, Fernández Stigliano A (2013) Eur Phys J E 36:62). Other portions of the dielectric spectrum, in particular around the hydrogen-bond bending mode at 1.8 THz, are reproduced to similar levels of accuracy (97 %) and reveal a similar blueshift in the absorption peak for the imaginary part. The anisotropy of the absorption spectrum has been probed by microwave dielectric relaxation spectroscopy using network analyzer HP8757A with non-Debye (orthogonal) permittivity measured by applying an electric field that oscillates in a plane perpendicular to the monolayer surface.

Chapter 3

3.1. The dehydronic field is most intense in the transition state (TS), since this ensemble is the richest in dehydrons. The fully hydrated backbone in the unfolded state has almost no dehydrons, while the folded chain minimizes the number of dehydrons because of their structure-destabilizing contribution. The dehydronic field commits the protein to fold since it exerts an attractive drag on nonpolar groups towards the TS dehydrons. Thus, the dehydronic field promotes folding cooperativity realized by three-body correlations involving the dehydron-paired residues and the wrapping residue. Thus, the local picture where backbone hydrogen bonds are most stable when fully exposed or fully wrapped validates the two-state folder model.

3.2. See Fig. P3.1.

3.3. We directly examined the coarse-grained (modulo Ramachandran basin) history of the conformational search of the villin headpiece subdomain simulated at the all-atom explicit solvent level for 1 μs using the AMBER simulation package [37, 38]. The topological history of the trajectory (Fig. 3.15) revealed that AMBER has no prospects of generating meaningful folding pathways, at least for this small autonomous folder (chain length = 36), for the following reasons:

- There is no indication of a cross-over behavior that would signal the transition from an exploratory phase to a commitment of the chain to fold.
- There is no indication that the simulation captures folding cooperativity. The AMBER force field seems so biased towards the helical turn that helices form immediately and spontaneously without any need for tertiary structure scaffolding.
- The trajectory does not converge to any structure or reveal any point of commitment to fold into a structure that is subsequently retained, rather the trajectory keeps switching between heavily biased conformations.

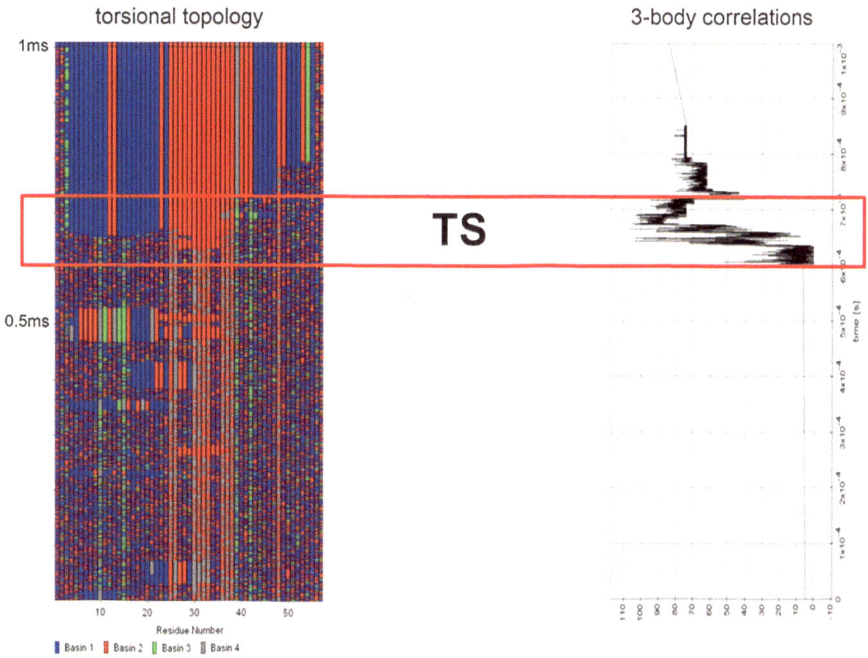

Fig. P3.1 Torsional topology, three body correlations and description of the transition state (TS) ensemble for the folding of the thermophilic variant of protein G. The boost in cooperativity at the TS results from the maximal intensity of the dehydronic field

Chapter 4

4.1. The total reversible re-reorientation work, $W^{\#}$ needed to align the confinement-induced dipole $\vec{\mu}_{P^{\#}}(\vec{r}) = \int \left(\vec{r}' - \vec{r}\right) \vec{\nabla}.\vec{P}^{\#}\left(\vec{r}'\right) d\vec{r}'$ along the direction of the electrostatic field, i.e. the work needed to locally restore the Debye alignment is:

$$W^{\#} = -\iint_{\theta(\vec{r})=\theta^{\#}(\vec{r})}^{\theta(\vec{r})=0} ||\vec{E}(\vec{r})||\; ||\vec{\mu}_{P^{\#}}(\vec{r})|| sin\theta(\vec{r}) d\theta(\vec{r}) d\vec{r}$$

$$= \int ||\vec{E}(\vec{r})||\; ||\vec{\mu}_{P^{\#}}(\vec{r})|| \left[1 - cos\,\theta^{\#}(\vec{r})\right] d\vec{r}$$

$$= \int \left[||\vec{E}(\vec{r})||\; ||\vec{\mu}_{P^{\#}}(\vec{r})|| - \vec{E}(\vec{r}).\vec{\mu}_{P^{\#}}(\vec{r})\right] d\vec{r}.$$

where $\theta(\vec{r})$ is the angle between anomalous polarization dipole and the electrostatic field and

$$cos\,\theta^{\#}(\vec{r}) = \vec{E}(\vec{r}).\vec{\mu}_{P^{\#}}(\vec{r})/[||\vec{E}(\vec{r})||||\vec{\mu}_{P^{\#}}(\vec{r})||]$$

Since this reversible work is in effect the potential energy change, we may assert that $K = W^{\#}$.

4.2. See Fig. P4.1.

4.3. See Fig. P4.2.

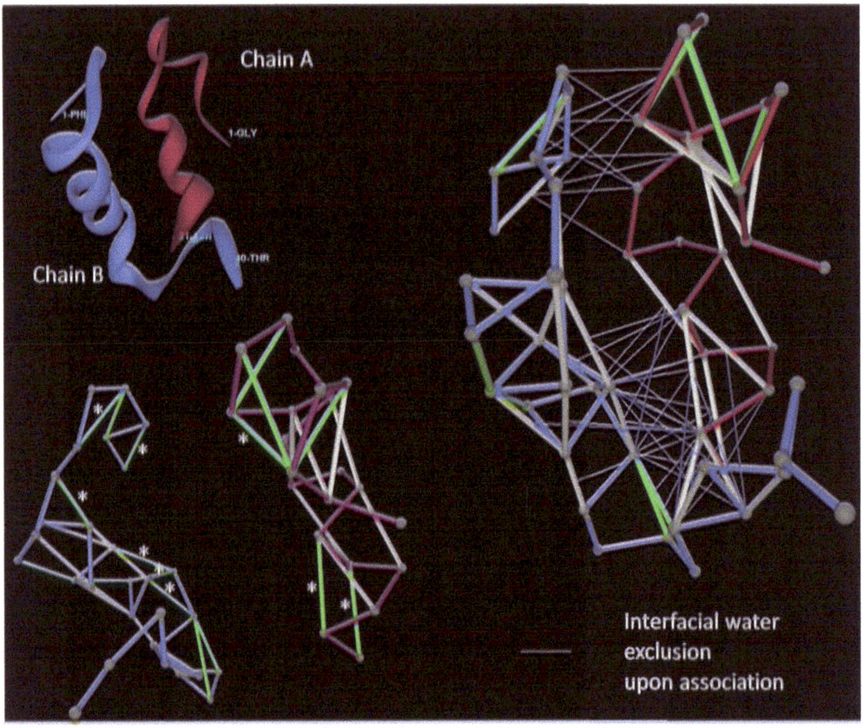

Fig. P4.1 Interfacial water exclusion upon complex formation in human insulin (PDB.3E7Y). The ribbon rendering of the chain conformations (*upper left*) are an aid to the eye. The free (uncomplexed) chains (A in *magenta*, B in *blue*) are shown on the *lower left*. The backbone is represented as virtual bonds joining the alpha-carbons of residues along the chain, with well-wrapped backbone hydrogen bonds and dehydrons shown as segments between the paired residues in grey and green, respectively. The location of hot interfacial water molecules ($g < 3$) is marked by asterisks and relative to the hydration vicinity of dehydrons in the free chains as specified in problem 4.2. The hot interfacial water molecules are excluded upon complex formation, as shown on the right, and the dehydrons vicinal to hot water molecules in the free chains turn into well-wrapped hydrogen bonds (*green* to *grey*) within the complex. The intermolecular water exclusion patterns are marked by *thin blue lines* joining the residue that excludes water upon binding and the center of the hydrogen bond whose vicinity is deprived of interfacial water upon association. All hot interfacial water molecules with $g < 3$ are excluded upon complexation human growth hormone / receptor (3HHR)

R43	E44	N72	T73	Q74	E75	W76	W80	S98	S102	I103	W104	I105	P106	C108	E120
S	X					#				#	#	#	X		

K121	C122	S124	D126	E127	D164	I165	Q166	K167	W169	V171	T194	T195	Q216	R217	N218
			X	S	X	#			#						

CD4/GP120 (1GC1)

S23	Q25	H27	K29	N32	Q33	K35	Q40	S42	L44	T45	K46	S49	N52	R59	S60	W62	D63	Q64	E85
			S						#		#			S		#			X

barnase/**barstar** (1BRS)

Y29	D35	W38	D39	T42	E76	E80
#	X	#	X		S	

P53/MDM2 (1YCR)

F19	W23	L26
#	#	#

barnase/barstar (1BRS)

K27	R59	R83	R87	H102
#	#	S	S	S

(2PTC)

G12	K15	I18	G36
X	#	#	X

(1A4Y)

Y434	D435	Y437
#	S	#

(1DFJ)

Y430	D432	Y433
#	X	#

trypsin inhibitor/ beta trypsin (2PTC)
ribonuclease inhibitor/ angiogenin (1A4Y)
ribonuclease inhibitor/ ribonuclease A (1DFJ)

colicin E9/DNase domain (1BXI)

E30	L33	V34	E41	S50	D51	Y54	Y55
X	#	#	S	X	X	#	#

Fig. P4.2 Comparison between alanine scanning and variational analysis of protein-protein interfaces combining the $\Delta\Delta G_a$ classifier (*upper entries*) and the g-classifier (*lower entries*) of interfacial residues. Hot-spot residues marked in *dark grey* (*bold font*), *white* and *light grey* correspond to the ranges $\Delta\Delta G_a \geq 3$ kcal/mol, 1 kcal/mol $\leq \Delta\Delta G_a < 3$ kcal/mol and $\Delta\Delta G_a < 1$ kcal/mol, respectively (*upper rows* for each PP interface). According to the second classifier (*lower rows*), residues are marked in *dark grey* (*bold font*), *white* and *light grey* if their vicinal water lies in the ranges $g < 3$, $3 \leq g < 4$, $g = 4$, respectively. The interfacial effect of the association upon the hydration vicinity of a residue is represented by "X" if water is displaced, "#" if the residue displaces water intermolecularly as a result of its side-chain penetration into the vicinity of another residue across the PP interface, "S" if an intermolecular salt bridge is formed across the PP-interface, and "blank space" if the hydration vicinity is unaltered

Chapter 5

5.1. The significant departure from the (X,Y) golden rule implies that the proteins are unable to compensate the dehydron-related destabilization of their native structure with disulfide bonds. Thus, if such proteins have a significant number of dehydrons in their soluble native structure, they would tend to relinquish that structure in search for a supramolecular organization that ensures the complete wrapping of dehydrons. All 100 % aggregation-prone proteins in ref. [42] of Chap. 5 satisfy the postulate previously given, while those that retain their soluble fold do not.

5.2. The 18-unit region 293-FSSLLKKRDSFRRDSKLE-310 was predicted as disordered in the 302,307-phosphorylated state but capable of a transition to an ordered state upon binding to myosin in the unphosphorylated state. This finding guided the molecular design of the peptide "302A": FSSLLKKRDAFRRDAKLE, with S→A substitutions at positions 302 and 307. Peptide 302A becomes a surrogate for a myosin-binding region of MyBP-C, susceptible to acquiring order upon binding to myosin, while incapable of being phosphorylated and thus of reversing back to the unbound state. These properties provide the rationale for its therapeutic impact to treat heart failure, based on its competitive binding to myosin and concurrent release of the "molecular brake" on contractility by precluding binding of MyBP-C.

Chapter 6

6.1. The implementation of automated dilution/concentration cycles is a possible way of modulating selection efficiency. As the population dwindles by pipetting a small sample of the evolving population into a large vessel, selection inefficiency sets in and mildly deleterious mutations that enrich the structure in dehydrons get a statistically significant chance to get fix in the population. As the population grows, the protein-protein interfaces that would materialize the dehydron-promoting associations get a chance to evolve and secure complex formation, as required for allosteric regulation.

Chapter 7

7.1.

(a) Figure P7.1
(b) Figure P7.2
(c) Figure P7.3

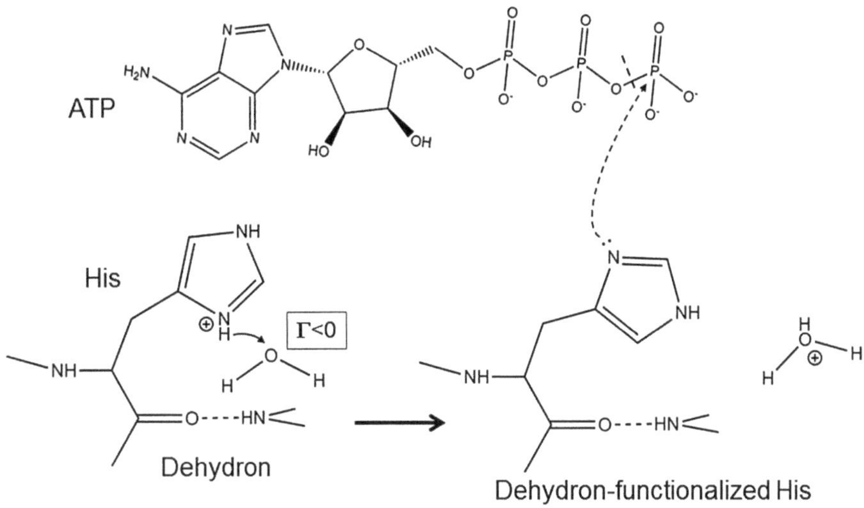

Fig. P7.1 Chemical mechanism for dehydron-based enablement of histidine first phosphorylation

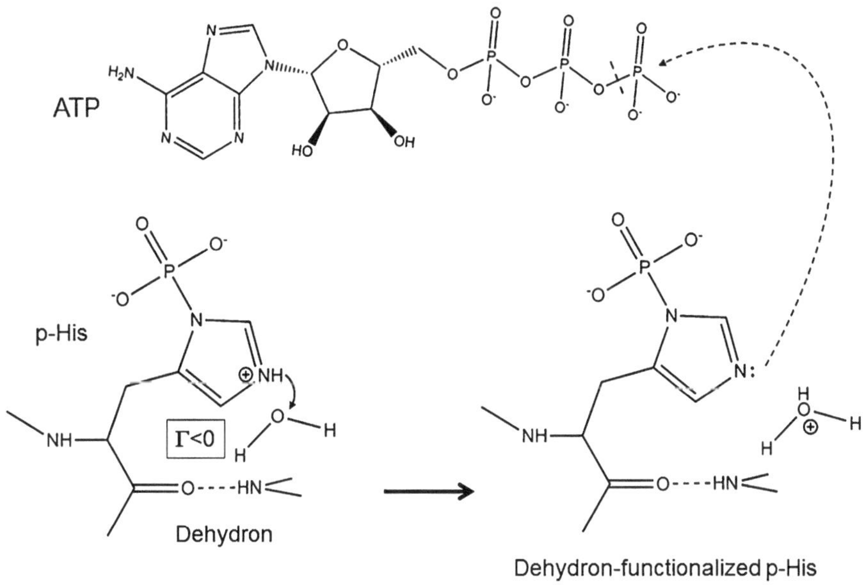

Fig. P7.2 Idem for second phosphorylation

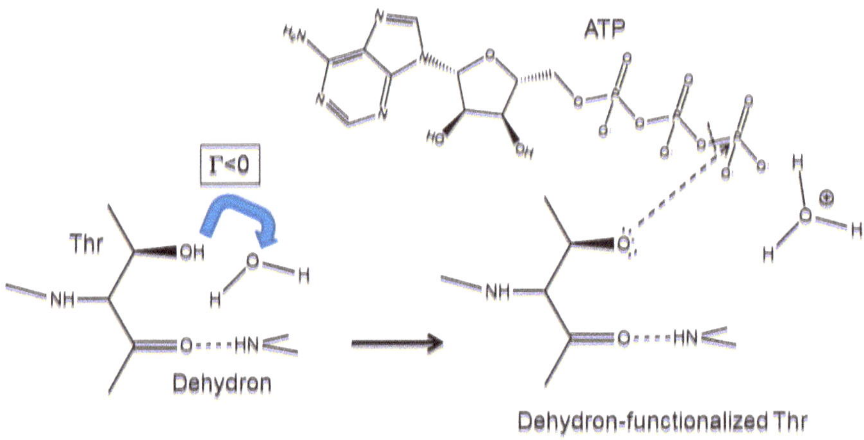

Fig. P7.3 Chemical mechanism for dehydron-based enablement of threonine phosphorylation

Chapter 8

8.1. The homologous kinases Lck and Lyn present important differences in the wrapping of the nucleotide-binding loop that may be exploited to remove anti-Lck activity. This may be achieved by following a molecular design strategy similar to the one adopted to rework imatinib into WBZ_4 (Chap. 9). The main wrapping difference lies in the Lyn-Lck alignment of the well-wrapped backbone hydrogen bond Met319-Gly322 in Lck. The hydrogen-bonded Lck residues align respectively with Met322 and Gly325 in the Lyn kinase (Figs. P8.1 and P8.2). The putative hydrogen bond pairing the latter residues in Lyn is so poorly wrapped that the bond does not form, but may be induced upon ligand binding provided the ligand is engineered to improve Lyn wrapping around this region. So, the dehydron 322-325 becomes "inducible" in Lyn. Ligand expansions designed to wrap the "ghost dehydron" 322-325 in Lyn are surely going to tell this kinase apart from its close homolog Lck.

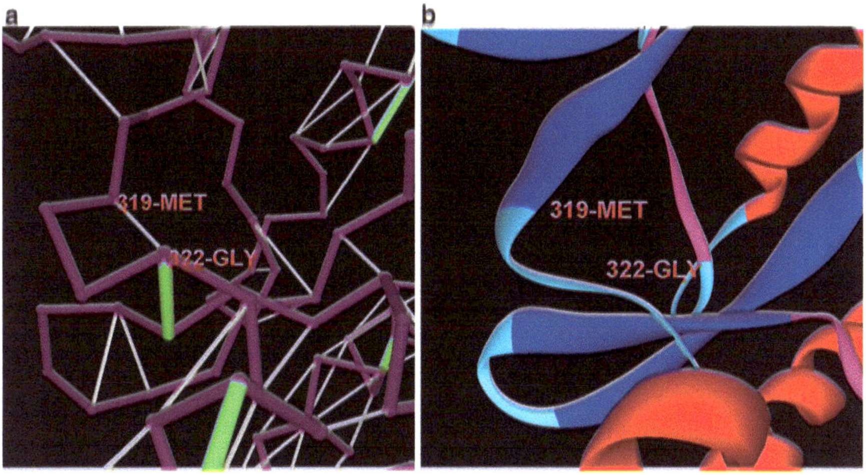

Fig. P8.1 (**a**) Dehydron pattern and (**b**) ribbon rendering of the nucleotide-binding loop region in Lck

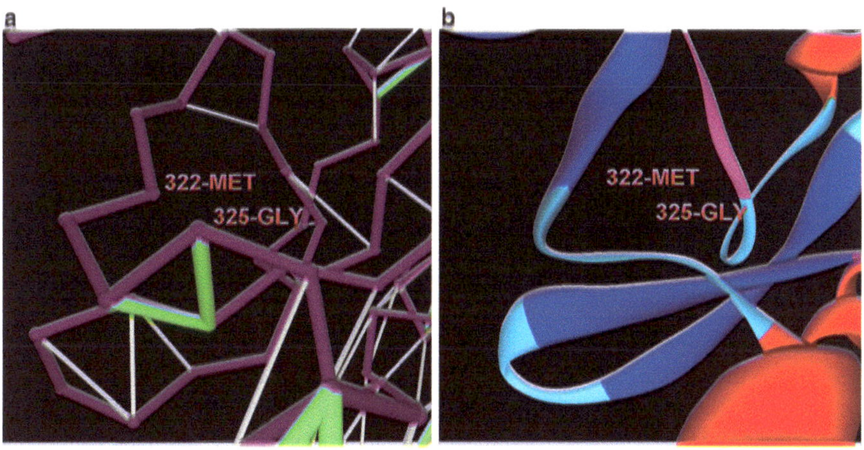

Fig. P8.2 (**a**) Dehydron pattern and (**b**) ribbon rendering of the nucleotide-binding loop region in Lyn kinase

Chapter 9

9.1. The compound WBZ_4, a reworking of imatinib designed using the wrapping technology, may be an ideal precursor or lead in the development of a chemosensitizer to treat PDA because WBZ_4 is more specific, safer, not cardiotoxic, and has a higher affinity towards c-Kit kinase than the parental compound. Furthermore, WBZ_4 has already been proven to synergize favorably with chemotherapy in the context of ovarian cancers with highly angiogenic phenotype (ref [17], Chap. 9). The chemosensitizers built based on WBZ_4 re-engineering should retain the exact affinity profile of WBZ_4 (c-Kit kinase, PDGFR kinase and JNK).

In designing the precursor WBZ_4, we exploited the wrapping selectivity filter that enabled us to radically reengineer imatinib to remove most of the immune-antagonistic effects and cardiotoxicity (a requirement given the massive dose of chemosensitizer needed to treat PDA), while enhancing its antimastocyte (anti-c-Kit) activity. This required that we specifically remove its inhibitory power against Lck and Bcr-Abl, while retaining or enhancing its activity against therapeutic targets c-Kit, JNK and PDGFR kinases. Because of the wrapping modification on the terminal pyridine ring, the anti-PDA drugs based on WBZ_4 should all induce upon association with JNK the formation of intramolecular backbone hydrogen bond Met111-Asn114, absent in other JNK complexes reported in PDB. This locally induced folding is achieved by molecularly expanding the parental compound imatinib to protect the JNK microenvironment from backbone hydration. Since the under-protected region in JNK aligns with a well-protected (wrapped) region in the Bcr-Abl and Lck kinases, WBZ_4 is not reactive against these imatinib targets, and hence it is neither cardiotoxic nor immunosuppressive. This side-effect removal may not be nearly as important in the therapeutic contexts where imatinib is effective as it is in the context of pancreatic cancer. Thus, the higher safety of WBZ_4 at least vis-à-vis these side effects enables a significant scale up of the WBZ_4 dose, as needed to treat the massive desmoplastic reaction in PDA. On the other hand, since the inducible bond in JNK aligns with a poorly wrapped hydrogen bond (Cys673-Gly676) in c-Kit and PDGFR, WBZ_4 retains full activity regarding these anticancer targets.

Molecular targeted monotherapies have all been unsuccessful against pancreatic cancer, especially against the ductal adenocarcinoma. On the other hand, a dual modality synergizing WBZ_4 with chemotherapy may only be effective if it achieves a significant sensitization of the tumor mass, a daunting task given the inflammatory nature of the pro-invasive phase in this type of cancer. By targeting mast-cell infiltration via the anti-c-Kit activity of WBZ_4, we expect to knock down the culprit for the invasive phenotype. In this way, WBZ_4 is expected to enhance susceptibility to chemotherapy to an unprecedented level. This goal demands a highly specific antimastocyte (anti-c-Kit) molecular agent that would not introduce further compromise to the immune system, and WBZ_4 fits this requirement. Furthermore, the therapeutic agent should meet a very high safety bar that enables massive dose scale-ups without the risks of side effects encountered by other c-Kit inhibitors of lower specificity such as imatinib. WBZ_4 may well serve as precursor for such a KI, positioned as a sensitizer of chemotherapy in the treatment of pancreatic cancer.

9.2. c-Jun-NH2-kinases (JNK) are serine/threonine protein kinase members of the mitogen-activated kinase family. Of the three isoforms described in the literature, JNK1 and JNK2 are ubiquitously expressed. Controversial effects have been reported for JNKs. JNKs has been shown to serve as mediators of apoptosis in response to cellular stress, have been show to sustain cell proliferation, and have been shown to promote survival in response to extracellular stimuli such as cytokines. JNKs can be activated by a variety of stimuli including environmental stress (UV and ionizing radiation, heat shock, osmotic or redox shock), inflammatory cytokines, and growth factors. Induction by their own phosphorylation leads to the activation of different transcription factors including c-Jun and JunD. Depending on the cellular context, this induction can have effects at the level of apoptosis, differentiation, survival, and carcinogenesis. In fibroblasts, JNK1 has a prominent role in c-Jun activation leading to cell proliferation, whereas JNK2 inhibits cell proliferation by promoting c-Jun degradation in unstimulated cells.

Recent findings suggest that JNK mediates oncogenic functions in several cancer types including head and neck, gastric, and breast cancers, and melanoma and thus it may be an attractive therapeutic target. JNK inhibition, mediated by the competitive inhibitor SP600125 or by JNK-specific siRNA, hampered growth of head and neck squamous cell carcinoma. The antitumor effects were thought to be mediated through effects on both tumor and endothelial cells. JNK1 has been shown to be important in promoting cell survival by controlling cell cycle arrest and apoptosis.

9.3. Using the variational computation described in Chap. 1 (Eq. 1.7), we obtain: $\Delta\Delta G_{if} = \frac{-1.48\ \mathrm{kcal}}{\mathrm{mole}}, \frac{0\ \mathrm{kcal}}{\mathrm{mole}}, \frac{-1.59\ \mathrm{kcal}}{\mathrm{mole}}$ for G676V in C-Kit (PDB.1T46), G322V in Lck (PDB.3LCK) and N114I in JNK (PDB.2G01), respectively. Since dehydrons raise interfacial free energy, this trend fits the one determined by the dehydron filter. This is so since G676V is a dehydron-suppressive mutation, while G322V is not dehydron-suppressive in Lck because the backbone hydrogen bond M319-G322 is already well-wrapped in wild-type Lck, and finally, N114I is dehydron-suppressive in JNK with an even more dramatic effect than mutation G676V in C-Kit because the wrapping of putative dehydron M111-N114 in wild-type JNK is already lower than that of the aligned dehydron C673-G676 in wild-type C-Kit. This problem illustrates the fact that a more quantitative selectivity filter for drug discovery could be introduced by exploiting the thermodynamic marker $\Delta\Delta G_{if}$ as a proxy of the actual level of wrapping of specific backbone hydrogen bonds.

Chapter 10

10.1. The chemical functionality of dehydrons established in Chap. 7 places an extra spin on the design paradigm of "drug as dehydron wrapper". Since dehydrons behave as catalytic stimulators, a novel strategy to deactivate a constitutively active signal transduction arises from the possibility of preventing the functionalization of the biomolecular interface by removal of dehydron-induced basic water. The particular therapeutic context of interest is one where the oncogenic mutation brings extra wrapping, generating a dehydron in an otherwise disordered region, such as the activation loop of a kinase (Fig. 10.10). If the mutation-generated dehydron is placed near a phosphorylation-susceptible residue such as Ser, Thr, Tyr or His, the oncogenic mutation may result in a constitutively active kinase responsible for transforming activity. This is so because the dehydron induces basicity at the aqueous interface, which in turn is responsible for promoting proton transference from the phosphorylation-susceptible group. This event surely promotes phosphorylation as it triggers the nucleophilic attack on the terminal phosphoester linkage of ATP. *Thus, the dehydron-based activation of the phosphorylation site is a novel transforming mode resulting from functionalization of the biomolecular interface.* To prevent such aberrant activity, the drug must be designed to remove the labile water molecule interfacing with the mutation-induced dehydron (Fig. 10.10). Such a design is enabled by the fact that dehydron-interfacing water is also labile and thus provides a blueprint for drug design, as shown in Chap. 4.

Chapter 11

11.1. The answers are contained in the Tables P11.1 and P11.2, respectively, where the wrapping patterns for the apo form, complex and complex with *in-silico*-removed drug are presented. The shaded rows in the tables correspond to hydrogen bonds fulfilling condition d.

Table P11.1 Wrapping pattern for the apo form, complex and complex with *in-silico*-removed drug for the imatinib/c-Kit association (*asterisk* indicates absence of bond)

	Apo enzyme	Inhibited enzyme		
HB	ρ	ρ without ligand	ρ with ligand	Wrappers in ligand
LEU595-VAL603	20	19	22	C1 C6 C20
GLY596-VAL603	18	21	22	C20
VAL603-GLY596	18	21	22	C20
VAL604-VAL622	30	26	29	C19 C18 C20
ALA621-THR670	32	27	31	C5 C19 C18 C20
VAL622-VAL604	30	26	29	C19 C18 C20
LYS623-VAL668	24	25	29	C19 C18 C17 C20
GLU640-ALA636	19	17	19	C28 C29
VAL643-SER639	24	22	23	C28
LEU644-GLU640	23	19	22	C27 C28 C29
LEU647-VAL643	27	27	28	C28
GLY648-LEU644	21	16	19	C27 C28 C29
VAL654-ILE808	26	23	25	C15 C25
LEU657-ILE669	28	29	32	C19 C18 C20
GLY658-ILE669	*	31	34	C19 C18 C20
VAL668-LYS623	24	25	29	C19 C18 C17 C20
ILE669-GLY658	*	31	34	C19 C18 C20
THR670-ALA621	32	27	31	C5 C19 C18 C20
GLY676-CYS673	*	16	19	C1 C6 C5
PHE681-ASP677	21	23	25	C1 C6
LEU800-GLY676	*	20	23	C1 C6 C5
ILE808-ASN652	*	23	24	C25
CYS809-ASN797	23	19	25	C15 C23 C25 C26 C27 C29
	pdb code 1PKG	pdb code 1T46		

Table P11.2 Wrapping pattern for the apo form, complex and complex with *in-silico*-removed drug for the sunitinib/VEGFR2 association (*asterisk* indicates absence of bond)

	Apo enzyme	Inhibited enzyme		
HB	ρ	ρ without ligand	ρ with ligand	Wrappers in ligand
LEU840-VAL848	17	18	22	C13 C3 C12 C19
GLY841-VAL848	15	19	21	C13 C3
VAL848-GLY841	15	19	21	C13 C3
ALA866-VAL916	31	30	31	C6
VAL899-ILE1044	29	23	24	C6
VAL916-ALA866	31	30	31	C6
GLU917-ASN900	21	20	21	C6
PHE921-LEU1036	22	24	25	C4
GLY922-CYS919	*	16	22	C4 C13 C20 C3 C12 C19
LEU1036-GLY922	16	16	22	C4 C13 C20 C3 C12 C19
CYS1045-ASN1033	24	24	26	C5 C6
ALA1050-PHE1047	*	14	16	C5 C7
	pdb code 1VR2	pdb code 4AGD		

11.2. Proteins are dynamic objects, capable of undergoing significant structural adaptation (*induced folding*) upon association with a ligand. The induced folded state is extremely hard to predict from basic physical principles, making the static conventional concept of structure-based drug design of limited applicability.

The majority of signaling proteins that constitute important targets for molecular cancer therapy contain floppy regions, capable of undergoing induced folding—or induced unfolding—in unpredictable manners. However unpredictable, such floppy regions pertain to a twilight zone between order and disorder, and hence are subject to a higher degree of genetic [1] and idiosyncratic [2] variability that make them desirable targets to achieve drug specificity and therapeutic personalization.

From a biophysical point of view, the induced folding problem, that is, the prediction of the structural adaptation of a protein upon binding to a ligand, is at least as hard as the protein-folding problem. This may suggest that there is little hope for rational drug design geared at targeting order-disorder twilight regions.

Yet, recent advances in molecular biophysics [1–6] suggest new and enticing possibilities for drug designs aimed at controlling or even harnessing the induced state of a target protein. The goal is to severely restrict the region where induced folding may take place, so as to be able to manipulate/control the structural adaptation of the target protein. Thus, dehydrons have been shown to be inducible within the floppy region of a protein by a drug purposely engineered to achieve high specificity and affinity [7]. In other words, the difficulty of controlling the induced folding in drug design could be at least locally surmounted.

Most importantly, because of its vulnerability, a dehydron becomes a dynamic object that may also be disrupted by purposely engineered drug-target mismatches, and this constitutes precisely the scope of this problem. This feature is expected to be the hallmark of the next generation of entropy-optimized drugs with high solubility and high bio-deliverability. The kinome-wide chemotype-wide exploration of this design possibility using a vast array of theoretical biophysical concepts [1–5], computational tools [7], thermodynamic calculations [8] and machine-learning technologies [2] fit into the scope of this problem. An overarching goal is to redesign every kinase inhibitor of clinical interest in accord with the paradigmatic strategy of induced dynamic state in order to further control the drug specificity, personalization, safety and bio-deliverability.

The possibility of enhancing affinity by increasing conformational entropy through engineered drug-target mismatches is intriguing from the point of view of drug design. It has never been rationally exploited or even attempted, although, several compounds of clinical interest share this characteristic. The idea is to create hydrophobic-polar mismatches between target and drug, so that targeted dehydrons are selectively disrupted by promoting their deshielding. This deprotection, and ultimate disruption, induced by the approaching polar group of the drug becomes necessary to enable hydration of the otherwise occluded polar group. The un-wrapping of the hydrogen bond upon association with the drug will result in an increase in conformational entropy with a consequent net increase in binding affinity. We therefore name this type of backbone hydrogen bonds *disruptons*.

As previously noted, standard structure-based design is of limited value, as target proteins may undergo an induced folding or boost conformational entropy when they associate with ligands, significantly impacting affinity. Even proteins with defined folds experience structural adaptation or conversely, may reduce their binding entropy penalty by enhancing conformational exploration. This binding mode allows for unconventional drugs that are not optimized vis-à-vis the binding enthalpy. The resulting induced states are poorly understood, often unpredictable. Yet, in the specific case where the drug is mismatched against dehydron-protective nonpolar groups, the induced unfolded or dynamic state can be controlled in order to boost the entropic content of the target protein. Thus, the dehydron is disrupted upon association by what literally amounts to an unwrapping process promoted by the approaching polar group of the drug.

There are clear illustrations of this peculiar "binding mode" (Fig. 11.7). As said, the limitations of structure-based drug design are particularly apparent in the field of therapeutic interference with signaling pathways, where the relevant targets, the kinases, present floppy regions prone to undergo induced conformational change. Many common kinase inhibitors of enormous clinical value, such as *sunitinib, gefitinib,* or *sorafenib* actually contain glaring hydrophobic-polar mismatches at the drug-target interface (see Fig. 11.7), a fact that seems at odds with the prevailing approaches that emphasize the maximization of favorable interactions with the target.

As discussed subsequently, this impression is reinforced by a telling example of compound PD173955 (Fig. 11.8) with a higher affinity than the well-known anticancer drug *imatinib* for the latter's primary target, which at the same time does not incorporate improvements in terms of enthalpy optimization relative to *imatinib* [9, 10]. In fact, this drug as the other three kinase inhibitors mentioned above are *not* enthalpically optimized (enthalpy content is not favored by mismatches), hence they stand in defiance to the prevailing approaches in drug design. At the same time, these drugs prompt us to incorporate new design principles based on dynamic considerations, as shown and advocated in this discussion.

The dynamic information on the target's induced state is often encoded in the protein-ligand complex and this information is typically overlooked in the analysis of the crystal structure. The history of this problem is long but poorly documented, with few references devoted to the subject [7]. The observation raises the enticing possibility that many compounds with pharmacological potential may have been optimized *vis a vis* the conformational entropy of the target (Fig. 11.7), rather than the binding enthalpy, although they were never reported in that light.

Thus, two molecular indicators—generally overlooked in standard structural analysis—should guide the conformational entropy optimization and therefore, should be intensively investigated through kinome-wide-and-PDB-wide data-mining efforts combined with machine-learning techniques: (a) mismatches across the protein/ligand interface and (b) induced folding or induced unfolding inferred by molecular dynamic analysis of drug-target complexes [2, 7].

The goal is to transcend standard enthalpic optimization by exploiting information accessible from NMR-relaxation studies on the induced conformational dynamics of protein-ligand associations. The intent is to design drugs that induce dynamic states and are thereby endowed with higher affinity and specificity and even capable of realizing the possibility of personalization.

Particular emphasis should be placed on the fact that this type of design may lead to an entropy optimization strategy, whereby the binding entropic cost is minimized by judicious control of localized structural changes upon binding. This paradigm will be contrasted against more conventional cases of compounds with low solubility fitting into rigid targets, where binding entropy changes are mostly associated with favorable solvent displacements and enthalpic optimization prevails as the dominant design approach.

The paradigmatic concept is rooted in specific molecular aspects of drug-target complexes, namely, hydrophobic-polar mismatches across the drug-target interface. These molecular indicators are hardly ever explicitly dealt with in the optimization of a lead compound. Rather, since they seem defiant of standard approaches to drug design, they are typically "swept under the rug". Such mismatches are at odds with the prevailing premise of enthalpy optimization, achieved by maximizing the number of favorable contacts between drug and target. Yet, basic thermodynamic considerations on drug-target complexes enable us to resolve what appear to be "anomalously high affinities".

To illustrate this phenomenon, researchers should focus on the perplexing higher affinity of PD173955 for the Bcr-ABL kinase when compared with that of *imatinib*, purposely designed to be the primary Bcr-ABL kinase inhibitor [9, 10]. A preliminary analysis based on sheer physical intuition leads us to the conclusion that drug-target mismatches can enhance the conformational entropy of the target as a means to enable the hydration of the drug polar moieties and that this reduction in the binding entropy cost enhances the drug affinity. Thus, PD173955 binds to its target with higher affinity in spite of the fact that it makes only two intermolecular hydrogen bonds with Bcr-ABL kinase, while imatinib makes six [9, 10]! The paradox may be resolved by taking into account that PD173955 is actually an entropically optimized compound capable of boosting the entropy content of its target as a means to promote its own favorable hydration (Fig. 11.8), while imatinib is not entropically optimized [10]. This conjecture should be ultimately tested through molecular dynamic analysis of the PD173955-Bcr-ABL complex deposited in PDB.1M52. Thus, the "PD173955 vs. imatinib paradox" leads to the novel paradigm of drugs inducing dynamic states in the target.

As hinted in Fig. 11.7, we are probably capable of providing a design rationale for the striking hydrophobic-polar mismatches between important kinase inhibitors *gefitinib, sunitinib* and *sorafenib* and their respective primary targets, the EGFR, KIT and P38MAP kinases. Chemical intuition suggests that these mismatches promote exploration of conformation space by the target protein as it associates with the drug. This crucial observation should be reinforced by thermodynamic computations [8] based on Molecular Dynamics [7]. Our rationale is that expansion of the accessible conformation space in the bound state of the protein is in turn necessary to enable proper hydration of the otherwise occluded polar moieties of the ligand.

These dynamic aspects do not surface in the standard structural underpinnings of the drug affinities. Clearly, none of these important therapeutic agents has been enthalpically optimized. Rather, they appear to fit a saddle-point type optimization (entropy is maximized while enthalpy is reduced), where a boost in target conformational entropy is balanced against the competing tendency to make as many favorable intermolecular contacts as possible.

Aided by a cross-validated combination of experimental affinity profiles, molecular dynamics of drug-protein complexes and computational thermodynamic analysis [2, 7, 8], a new agenda should be put in place to show how local unfolding of the target can be controllably induced by molecular redesign of therapeutically important drugs and exploited to increase the entropy of the complex and hence improve the drug affinity. Guided by the biophysical understanding gained from the analysis of the binding mode of the mentioned kinase inhibitors, we advocate for strategies for the redesign of all major anticancer kinase inhibitors aimed at controlling their selectivity, increase their therapeutic efficacy and safety and even personalize them as idiosyncratic dehydrons are selectively targeted for disruption. These translational discovery strategies should be benchmarked against *in-vitro, in-vivo* and preclinical data to be generated within suitable partnerships with the pharmaceutical industry and private sector in general. The intent is to confirm the importance of incorporating dynamic information on the induced state of the target protein to develop novel translational approaches to drug design.

We may conclude by highlighting the need for a paradigmatic shift in drug design strategies in order to harness the dynamic nature of induced states arising in therapeutically relevant targets. Accordingly, a massive computational effort should be launched to implement novel design concepts based on conformational entropy optimization. At this point we emphasize that many drugs of high therapeutic value fulfill these design premise, although they have not been characterized in that fashion. In terms of rational design, these drugs depart considerably from the usual dictates of enthalpy optimization that require a maximization of favorable contacts with the target.

We must also emphasize that dynamic information can be encoded in a useful and simple way to serve as a guidance to the drug designer and that such information is becoming increasingly accessible via NMR-relaxation analysis of protein-ligand complexes [11, 12]. This dynamic information has been often overlooked because it is signaled by mismatches across the target-drug interface. These mismatches, in turn, remain largely ignored, because they do not fit in the standard optimization strategies that guide the redesign of lead compounds. Nonetheless, as emphasized in this discussion, these singularities are actually the hallmarks of dynamic optimization, and as such, they should herald the advent of a new and powerful drug-design strategy that is expected to emerge from this research.

1. Fernández A, Lynch M. Nonadaptive origins of interactome complexity. Nature. 2011;474:502–5.
2. Fernández A. Transformative concepts for drug design: target wrapping. Heidelberg: Springer; 2010. ISBN: 978-3642117916.
3. Fernández A. Keeping dry and crossing membranes. Nat Biotechnol. 2004;22:1081–4.
4. Fernández A, Scott R. Adherence of packing defects in soluble proteins. Phys Rev Lett. 2003;91:018102.
5. Fernández A, Scheraga HA. Insufficiently dehydrated hydrogen bonds as determinants for protein interactions. Proc Nat Acad Sci U S A. 2003;100:113–8.

6. Fernández A, et al. An anticancer C-kit inhibitor is re-engineered to make it more active and less cardiotoxic. J Clin Investig. 2007;117:4044–54.
7. Fernández A, Bazan S, Chen J. Taming the induced folding of drug-targeted kinases. Trends Pharmacol Sci. 2009;30:66–71.
8. Andricioaei I, Karplus M. On the calculation of entropy from covariance matrices of the atomic fluctuations. J Chem Phys. 2001;115:6289–92.
9. Nagar B, Bornmann WG, Pellicena P, Schinlder T, Veach DR, Miller WT, Clarkson B, Kuriyan J. Crystal structures of the kinase domain of c-Abl in complex with the small molecule inhibitors PD173955 and Imatinib (STI-571). Cancer Res. 2002;62:4236–43.
10. Schindler T, Bornmann WG, Pellicena P, Miller WT, Clarkson B, Kuriyan J (2000) Structural Mechanism for STI-571 Inhibition of Abelson Tyrosine Kinase. Science 289:1938–1942
11. Diehl C, et al. Protein flexibility and conformational entropy in ligand design targeting the carbohydrate recognition domain of galectin-3. J Am Chem Soc. 2010;132:14577–89.
12. Stone MJ. NMR relaxation studies of the role of conformational entropy in protein stability and ligand binding. Acc Chem Res. 2001;34:379–88.

Chapter 12

12.1. According to information from the public domain, CEP-701 is the code name of the tyrosine kinase inhibitor *lestaurtinib*, a derivative of *staurosporine*, that inherits much of the promiscuity of the latter. On the other hand, R406 is the active metabolite of *fostamatinib* and is a nonspecific ATP-competitive inhibitor of spleen tyrosine kinase (Syk), a key signal transducer of activating Fc receptors and the B-cell receptor. Both inhibitors have potential as therapeutic editors of imatinib's cardiotoxicity. This is because of their overlapping impact on therapeutically relevant imatinib targets, their low affinity against the ABL and ABL variants, and because of their nano-molar affinity for JNK1/2 isoforms (Table 12.1), which enables them to selectively block pro-apoptotic pathways recruited by imatinib in the cardiomyocyte (Chap. 9). The compound CEP-701 exhibits very high affinity for isoform JNK1 (11 nM) but is a weaker inhibitor (sub-micromolar) of the isoform JNK2, which is also highly expressed in heart tissue (cf. Chap. 9). In this regard, it is a poorer therapeutic editor than R406, which exhibits significant nanomolar affinity against both JNK isoforms (Table 12.1). In regards to overlapping therapeutic activity relative to imatinib, CEP-701 is also a weaker therapeutic editor than R406, as its affinity for KIT is sub-micromolar, whereas R406 has a $K_D(\text{KIT}) = 6.8$ nM (Table 12.1). In addition, R406 is far more potent against both PDGFR isoforms (A/B). A serious potential shortcoming of R406 is its nanomolar potency against undesired target Lck, predicted to generate immunosuppressive activity (Chap. 9).

Chapter 13

13.1. Both compounds CEP-701 and R406 have low selectivity (Table 12.1) and yet they qualify as potential therapeutic editors of imatinib's cardiotoxicity. The affinity of CEP-701 for Lck is sub-micromolar (Table 12.1) and hence its potential immunosuppressive effects arising from a blockade of antigen-triggered adaptive response (Chap. 9) are not a serious source of concern. On these grounds, CEP-701 is not predicted to be worthy of structural remodeling to filter out the undesired cross-reactivity. On the other hand, the immunosuppressive effects arising from Lck inhibition by R406 are predicted to be far more significant, given the nanomolar affinity of this compound for Lck. In this case, the removal of the immunosuppressive side effect requires a remodeling of R406 guided by the wrapping selectivity filter. The re-design strategy requires targeting the dehydron or "inducible dehydron" (Chap. 11) that structurally aligns with the well wrapped hydrogen bond Met319-Gly322 in Lck displayed in Fig. P8.1a. To accomplish this goal, the chemical scaffold of R406 would need to be expanded through suitable substitution to increase the wrapping of the dehydron or inducible dehydron that structurally aligns with Met319-Gly322 in Lck.

Chapter 14

14.1. To test class-A drugs, we may assay for T-cell receptor (TCR)-mediated T-cell proliferation and activation in a dose-dependent manner using the parent compounds in the control assays. Besides testing the redesigns in proliferation assays using cancer cell lines, we may test for selective removal of the inhibition of antigen-specific T-cell effector functions. Purified CD3+ cells and virus-specific CD8+ T cells (naïve and memory T-cells) should be studied *ex vivo*, after antigen-specific effects are confirmed in defined T-cell clones. Functional outcomes including cytokine production, degranulation, activation, proliferation, apoptosis/necrosis induction and signal transduction mediated by Lck should all be assayed *ex vivo* on CD4+ and CD8+ T-cells treated with class-A variants of parent drugs and the results contrasted against those using parent drugs as controls (Fei F, Yu Y, Schmitt A, Rojewski MT, Chen B, Greiner J, Götz M, Guillaume P, Döhner H, Bunjes D, Schmitt M (2008) Dasatinib exerts an immunosuppressive effect on CD8+ T cells specific for viral and leukemia antigens. Experim Hematol 36:1297-1308). Immunoblot analysis of antigen-specific response should confirm that the massive inhibition by anticancer drugs of T-cell activation, proliferation, cytokine production and degranulation is the result of blockade of early TCR-induced signaling transduced by Lck. As the anti-Lck activity is specifically removed in class-A (and class-C compounds), we expect to have restored all antigen-induced T-cell functions. Thus, the class-A compounds would be readily and ideally suited to be combined with immunotherapies which should harness the initial antigenic activity of the molecular targeted therapy.

The impact on antigen-triggered CD4+ T-cell proliferation and T-cell effector function by the purported immune-synergic WBZ_4 is expected to reveal a minimal level of modulation exerted by WBZ_4 when compared with the parental drug.

Chapter 15

15.1. See Fig. P15.1.

15.2. Chemical Structure of optimized re-engineered TAE226 in Fig. P15.2.

15.3. Chemical structure of optimized re-engineered dorsomorphin in Fig. P15.3. The re-engineered compound displayed in Fig. P15.3 is an ATP-competitive inhibitor of AMPK with $K_D = 19.0$ nM. Targeted AMPK dehydrons 103–107 and 104–108 are selectivity filters and consequently $K_D \gg 100$ nM for all other kinases in the human kinome according to a recent high throughput screening.

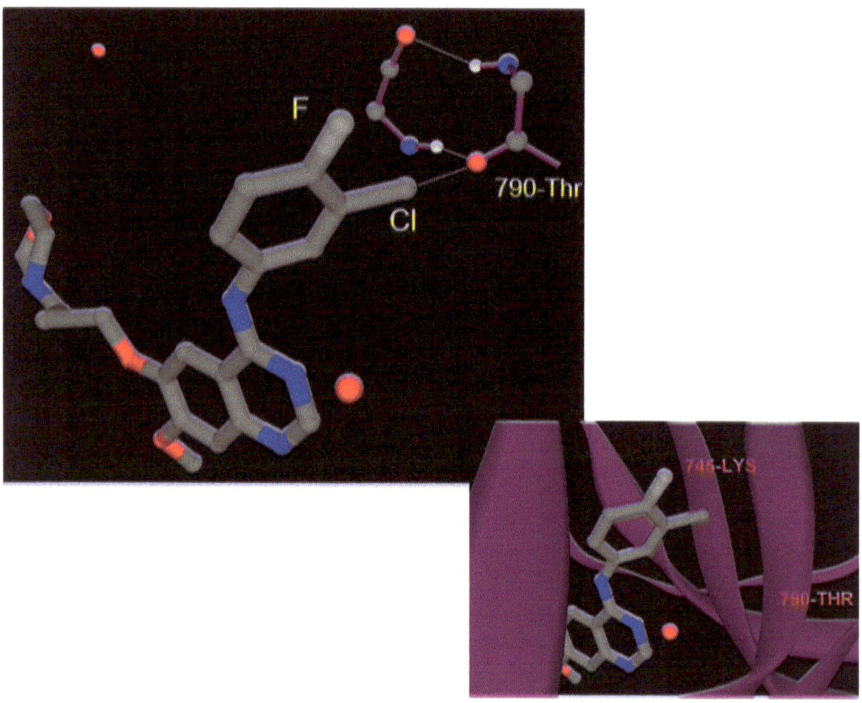

Fig. P15.1 X-wrapping bond between *gefitinib* and its target EGFR kinase, comprised of X-bond with carbonyl oxygen from Thr790 and wrapping interaction of dehydron Lys745-Thr790

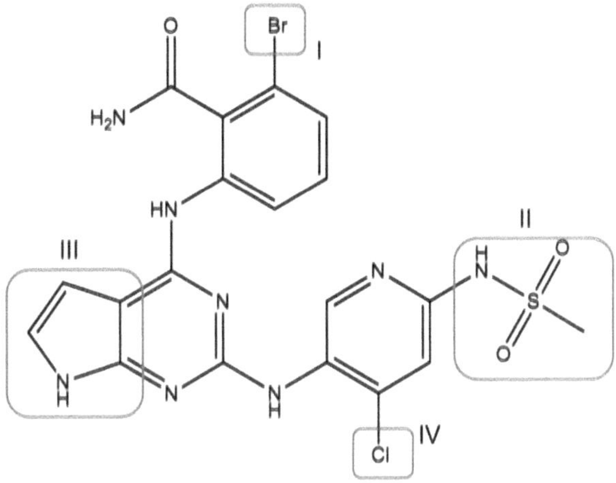

Fig. P15.2 Result of optimized reengineering of TAE226 in fulfillment of the constraints imposed in problem 15.2

6-(4-(2-(5-bromopyridin-2-yl)ethoxy)phenyl)-3-(pyridin-4-yl)pyrazolo[1,5-*a*]pyrimidine

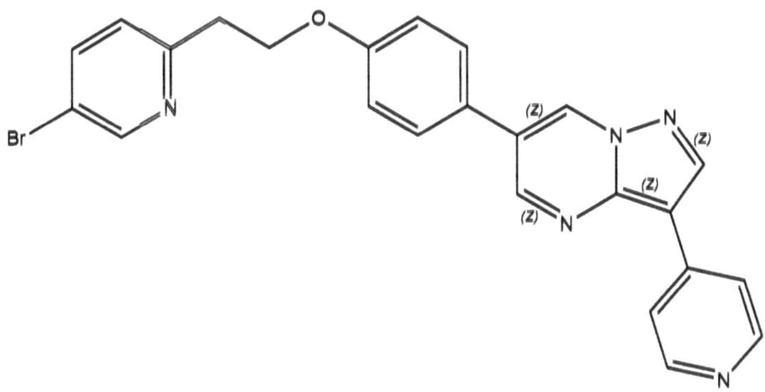

Fig. P15.3 Chemical structure of optimized reengineering of dorsomorphin in fulfillment of the constraints imposed in problem 15.3

Chapter 16

16.1. Results reported in Fig. P16.1.

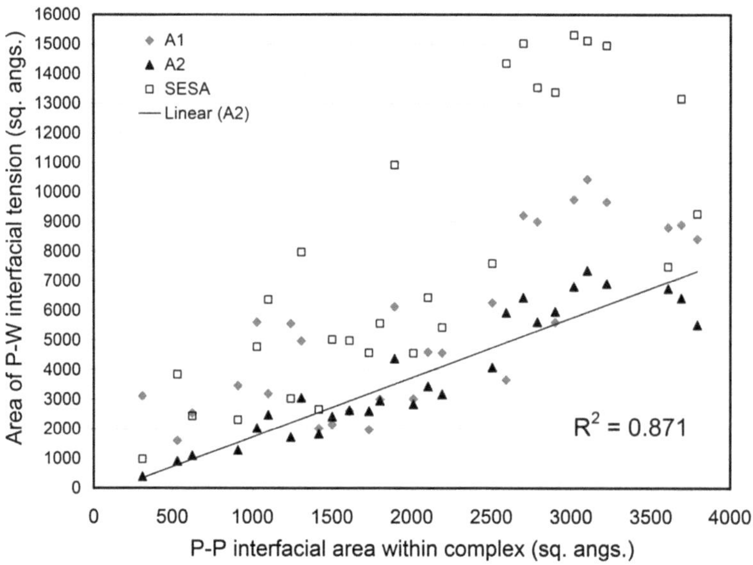

Fig. P16.1 Protein-water interfacial tension generated by dehydrons promotes biomolecular associations. Total area of surfaces generating protein-water (P-W) interfacial tension in free subunits (uncomplexed state) plotted against the total protein-protein (P-P) interfacial area for 28 protein complexes (Table 16.1). Datapoints with ordinates given by areas A_1, A_2 and solvent-exposed surface area (SESA) are represented with *diamonds*, *triangles* and *squares*, respectively. The linear fit with correlation coefficient $R^2 = 0.87$ for A_2-(P-P interface) points was obtained by linear regression

Chapter 17

17.1. Alternative optimizations of FAK inhibitor TAE226 are shown in Fig. P17.1.
17.2. An alternative optimal PI3K inhibitor is shown in Fig. P17.2.

a

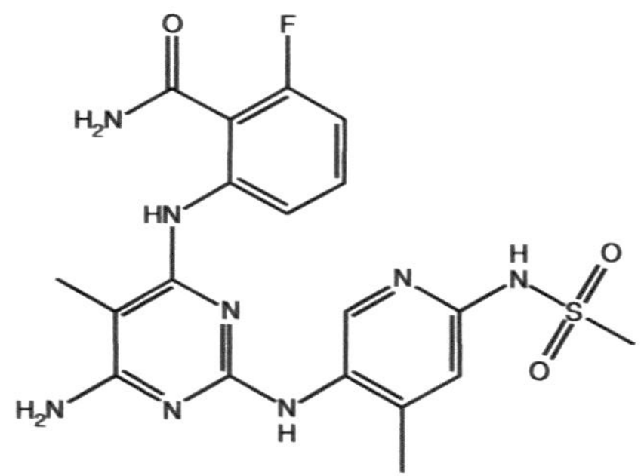

2-((6-amino-5-methyl-2-((4-methyl-6-(methylsulfonamido)pyridin-3-yl)amino)pyrimidin-4-yl)amino)-6-fluorobenzamide
Boiling Point: 1446.33 [K]
Melting Point:
Critical Temp:
Critical Pres:
Critical Vol:
Gibbs Energy:
Log P: 1.81
MR: 121.54 [cm3/mol]
Henry's Law: 25.45
Heat of Form:
tPSA: 176.42
CLogP: 2.71717
CMR: 11.7001
LogS: −5.865
pKa: none

b

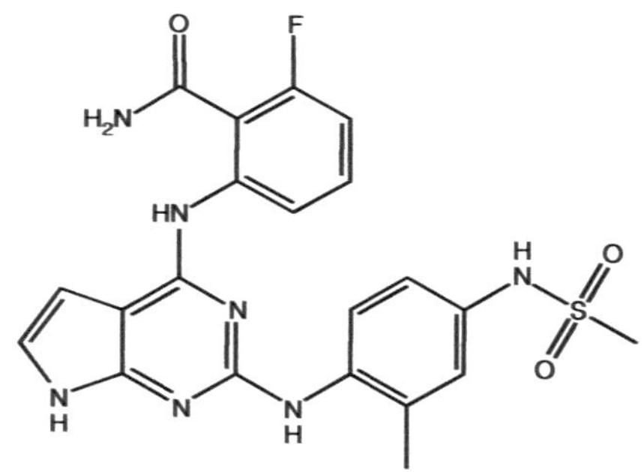

2-fluoro-6-((2-((2-methyl-4-(methylsulfonamido)phenyl)amino)-7*H*-pyrrolo[2,3-*d*]pyrimidin-4-yl)amino)benzamide
Boiling Point: 1410.37 [K]
Melting Point:
Critical Temp:
Critical Pres:
Critical Vol:
Gibbs Energy:
Log P: 2.31
MR: 123.8 [cm3/mol]
Henry's Law: 22.67
Heat of Form:
tPSA: 150.07
CLogP: 3.7471
CMR: 12.1962
LogS: −7.297
pKa: none

Fig. P17.1 Chemical structures of alternative FAK inhibitors resulting from optimization of parental compound TAE226

Alternative solution with amino-pyrimidine in lieu of morpholine: ATP-inspired base

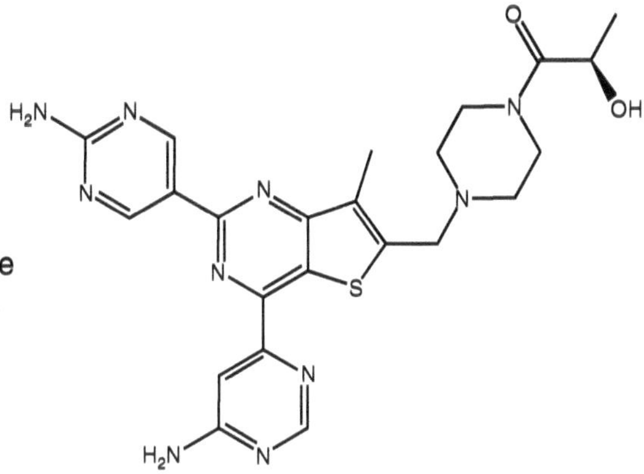

(*R*)-1-(4-((4-(6-aminopyrimidin-4-yl)-2-(2-aminopyrimidin-5-yl)-7-methylthieno[3,2-*d*]pyrimidin-6-yl)methyl)piperazin-1-yl)-2-hydroxypropan-1-one
Boiling Point: 1581.79 [K]
Melting Point: 1396.87 [K]
Critical Temp: 1271.69 [K]
Critical Pres: 28.35 [Bar]
Critical Vol: 1346.5 [cm^3/mol]
Gibbs Energy: 1155.82 [kJ/mol]
Log P: 1.65
MR:
Henry's Law: 39.76
Heat of Form: 523.7 [kJ/mol]
tPSA: 169.98
CLogP: 1.05783
CMR: 13.6019
LogS: −2.747
pKa: 6.038, 12.709

Fig. P17.2 Chemical structure of alternative PI3K inhibitor based on optimized morpholine replacement

Index

A. Fernández, *Physics at the Biomolecular Interface*, Soft and Biological Matter,
DOI 10.1007/978-3-319-30852-4

E

MIX
Papier aus verantwortungsvollen Quellen
Paper from responsible sources
FSC® C105338

If you have any concerns about our products,
you can contact us on
ProductSafety@springernature.com

In case Publisher is established outside the EU,
the EU authorized representative is:
Springer Nature Customer Service Center GmbH
Europaplatz 3, 69115 Heidelberg, Germany

Printed by Libri Plureos GmbH
in Hamburg, Germany